TRIPLE PLAY

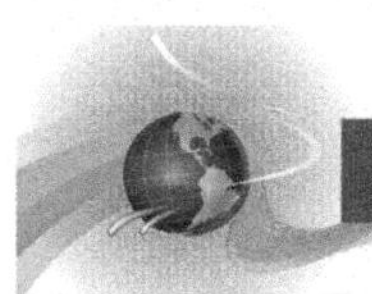

TELECOMS EXPLAINED - Unravel Emerging Technologies

Written by people in the know, the **Telecoms Explained Series** for Telecoms Professionals will

- Demystify the jargon of wireless and communication technologies
- Provide insight into new and emerging technologies
- Explore associated business and management applications
- Enable you to get ahead of the game in this fast-moving industry

Written in a concise and easy-to-follow format, titles in the series include the following:

Convergence: User Expectations, Communications Enablers and Business Opportunities Saxtoft
ISBN: 978-0-470-72708-9

Forthcoming titles include:

Why IPTV? Interactivity, Technologies, Services
Hjelm
ISBN: 978-0-470-99805-2

TRIPLE PLAY

Building the Converged Network for IP, VoIP and IPTV

Francisco J. Hens and José M. Caballero

Trend Communications Ltd., UK

John Wiley & Sons, Ltd

Telephone (+44) 1243 779777

Email (for orders and customer service enquiries): cs-books@wiley.co.uk
Visit our Home Page on www.wiley.com

Other Wiley Editorial Offices

John Wiley & Sons Inc., 111 River Street, Hoboken, NJ 07030, USA

Jossey-Bass, 989 Market Street, San Francisco, CA 94103-1741, USA

Wiley-VCH Verlag GmbH, Boschstr. 12, D-69469 Weinheim, Germany

John Wiley & Sons Australia Ltd, 42 McDougall Street, Milton, Queensland 4064, Australia

John Wiley & Sons (Asia) Pte Ltd, 2 Clementi Loop #02-01, Jin Xing Distripark, Singapore 129809

John Wiley & Sons Ltd, 6045 Freemont Blvd, Mississauga, Ontario L5R 4J3, Canada

Wiley also publishes its books in a variety of electronic formats. Some content that appears in print may not be available in electronic books.

Library of Congress Cataloging-in-Publication Data

Hens, Francisco J.
Triple play : building the converged network for IP, VoIP, and IPTV /
Francisco J. Hens, José M. Caballero.
p. cm.
ISBN 978-0-470-75367-5 (PB)
1. Computer networks–Standards. 2. Convergence (Telecommunication)
3. Internet telephony. 4. Internet television. 5. Computer network protocols.
6. Internet industry. I. Caballero, José Manuel. II. Title.
TK5105.55.H46 2008
384–dc22

2008000683

British Library Cataloguing in Publication Data

A catalogue record for this book is available from the British Library

ISBN 978-0-470-75367-5 (PB)

Typeset by 10/12 pt ZapfHumanist by Thomson Digital, Noida, India

Table of Contents

Preface .... **xi**

Chapter 1: Business Strategies **1**

1.1 Expanding Telco Businesses **2**

1.2 Triple Play Applications **5**

1.2.1 Television and Video Services 6

1.2.2 Video on Demand 8

1.2.3 New TV Receivers 8

1.2.4 Voice over Internet Protocol 9

1.2.5 VoIP Rollout 10

1.3 Driving Factors of Triple Play **11**

1.3.1 Business Redefinition 11

1.3.2 Competitive Pressure 11

1.4 Telcos Strategies **14**

1.4.1 Service Bundling and Network Convergence 15

1.4.2 Quadruple Play 16

1.4.3 VoD: the Key Difference 17

1.4.4 Making a Success Story 18

1.5 Infrastructures **18**

1.5.1 CPE Equipment 19

1.5.2 The First Mile 19

1.5.3 Network Convergence 22

1.6 Triple Play Market **26**

1.6.1 Warning: No Immediate Profit is Expected 27

1.7 Conclusions **28**

Chapter 2 IP Telephony **31**

2.1 Coding of Voice Signals **33**

2.1.1 Pulse Code Modulation 34

2.1.2 Adaptive Differential PCM 35

2.1.3 Code-excited Linear Predictive 36

2.1.4 Other Codecs 37

2.2 Network Performance Parameters . . . 37
2.2.1 Packet Loss and VoIP . . . 38
2.2.2 Delay and VoIP . . . 38
2.3 Opinion Quality Rating . . . 40
2.4 Objective Quality Assessment . . . 42
2.4.1 The E-model . . . 43
2.4.2 Speech-layer Models . . . 44
2.5 Market Segments . . . 45
2.5.1 Single User Solutions . . . 45
2.5.2 IP Telephony in Enterprise Networks . . . 46
2.5.3 IP Telephony in Service Provider Networks . . . 47

Chapter 3 Audiovisual Services . . . 51
3.1 Digital Television . . . 52
3.1.1 The Internet and Television . . . 52
3.2 Questioning the IPTV Business Models . . . 53
3.2.1 Strengths . . . 53
3.2.2 Opportunities . . . 54
3.2.3 Weaknesses . . . 55
3.2.4 Threats . . . 56
3.3 Regulatory Framework . . . 57
3.4 Architectural Design . . . 59
3.4.1 Television Services Rollout . . . 59
3.4.2 Business Model Definition . . . 60
3.4.3 Head-end . . . 62
3.4.4 Distribution Network . . . 62
3.4.5 Subscriber Site . . . 63
3.5 Television and Video Services and Applications . . . 65
3.5.1 IPTV Protocols . . . 65
3.5.2 Video-on-demand Services . . . 65
3.5.3 Personal Video Recording Services . . . 66
3.5.4 Converged Telephony . . . 67
3.6 Formats and Protocols . . . 68
3.6.1 Analogue TV . . . 68
3.6.2 Digital TV . . . 69
3.6.3 Audio and Video Codecs . . . 69
3.7 How a Codec Works . . . 71
3.7.1 MPEG-2 Levels and Profiles . . . 71
3.7.2 MPEG Compression . . . 73
3.7.3 MPEG Stream Generation Scheme . . . 79
3.7.4 The Transport Stream . . . 81
3.7.5 Packet Distribution and Delivery . . . 87
3.8 Windows Media and VC-1 . . . 88
3.8.1 VC-1 Profiles and Levels . . . 88

3.9 Service Provision **89**
3.9.1 Quality of Experience 89
3.9.2 Network Impairments 90
3.10 Service Assurance **92**
3.10.1 Content Faults 93
3.10.2 Network Impairments 93
3.10.3 Transaction Impairments 94
3.10.4 Transport Impairments 96
3.10.5 Media Delivery Index 98

Chapter 4 Signalling **101**
4.1 The Real-time Transport Protocol **102**
4.1.1 Synchronization Sources and Contributing Sources 103
4.1.2 Translators and Mixers 104
4.1.3 The RTP Packet 104
4.1.4 Stream Multiplexing 106
4.1.5 Security 108
4.2 The Real-time Control Protocol **108**
4.2.1 RTCP Packet Types and Formats 109
4.2.2 Quality of Service Monitoring 109
4.2.3 Source Description 113
4.2.4 Session Management: The BYE Packet 115
4.3 The Session Initiation Protocol **115**
4.3.1 Standardization 117
4.3.2 Architectural Entities 117
4.3.3 SIP Basic Signalling Mechanisms 118
4.3.4 The Session Description Protocol 121
4.3.5 Security Issues 123
4.3.6 Service Architecture and Protocol Extensions 125
4.3.7 Firewall Traversal 127
4.3.8 Interworking with the PSTN 133

Chapter 5 IP Multicasting **141**
5.1 IP Multicast Groups and their Management **142**
5.1.1 Multicasting in Ethernet Networks 143
5.1.2 Multicasting and the Internet Group Management Protocol 144
5.2 Multicast Routing **146**
5.2.1 Multicast Routing Algorithms 148

Chapter 6 QoS in Packet Networks **151**
6.1 QoS Basics **152**
6.1.1 Traffic Differentiation 152
6.1.2 Congestion Management 153

6.2 End-to-end Performance Parameters **154**
6.2.1 One-way Delay 155
6.2.2 One-way Delay Variation 157
6.2.3 Packet Loss 159
6.2.4 Bandwidth 160
6.3 Marking **161**
6.3.1 Traffic Flows 162
6.3.2 Traffic Classes 162
6.4 Scheduling **163**
6.4.1 First In, First Out Scheduler 163
6.4.2 Round Robin Scheduler 164
6.4.3 Weighted Fair Queuing 165
6.4.4 Priority Scheduler 165
6.5 Congestion Avoidance **165**
6.5.1 Admission Control 167
6.5.2 Resource Management 170
6.6 Congestion Control and Recovery **172**
6.6.1 Drop Tail 173
6.6.2 Partial Packet Discard 173
6.6.3 Early Packet Discard 174
6.6.4 Random Early Detection 174

Chapter 7 QoS Architectures **177**
7.1 QoS in ATM Networks **177**
7.1.1 Bandwidth Profile Characterization 178
7.1.2 Negotiated QoS Parameters 178
7.1.3 ATM Service Categories 179
7.1.4 SLA in ATM Networks 180
7.1.5 Resource Management 181
7.1.6 The Failure of ATM 182
7.2 QoS in IP Networks **182**
7.2.1 The Integrated Services Architecture 183
7.2.2 The Reservation Protocol 186
7.2.3 The Differentiated Services Architecture 194

Chapter 8 Broadband Access **203**
8.1 Broadband Services Over Copper **205**
8.1.1 The Limits of Copper Transmission 207
8.1.2 ADSL2 209
8.1.3 ADSL2+ 212
8.1.4 Bonded DSL 212
8.1.5 VDSL 214
8.1.6 VDSL2 215
8.2 The Passive Optical Network **221**
8.2.1 Basic Operation 222

8.2.2 Advantages 224
8.2.3 Broadband PON 225
8.2.4 Gigabit PON 227
8.2.5 Ethernet PON 230
8.3 Ethernet in the First Mile 233
8.3.1 Ethernet Over Telephone Copper Pairs 235
8.3.2 Ethernet in Optical Access Networks 237
8.4 Service Provisioning 238

Chapter 9 Quadruple Play 243
9.1 Cellular Communications Overview 244
9.1.1 The Global System for Mobile Communications 247
9.1.2 The Universal Mobile Telephone System 256
9.1.3 Long-term Evolution of 3GPP Networks 262
9.2 Wireless Communications Overview 264
9.2.1 Wireless Local Area Networks 265
9.2.2 Wireless Metropolitan Area Networks 272
9.3 The IP Multimedia Subsystem 280
9.3.1 Main Architectural Entities and Interfaces 282
9.3.2 Services 285
9.3.3 User Identity 287
9.3.4 AAA with Diameter 289
9.3.5 Policy and Charging Control 293
9.3.6 Basic Procedures 294
9.3.7 The Next-generation Network 298

Chapter 10 Carrier-class Ethernet 305
10.1 Ethernet as a MAN/WAN Service 306
10.1.1 Network Architecture 306
10.1.2 Ethernet Virtual Connections 308
10.1.3 Multiplexing and Bundling 308
10.1.4 Generic Service Types 309
10.1.5 Connectivity Services 310
10.2 End-to-End Ethernet 312
10.2.1 Optical Ethernet 315
10.2.2 Ethernet Over WDM 316
10.2.3 Ethernet Over SDH 316
10.3 Limitations of Bridged Networks 320
10.3.1 Scalability 322
10.3.2 Protection 322
10.3.3 Topologies 322
10.3.4 Quality of Service 324
10.4 Multiprotocol Label Switching 324
10.4.1 Labels 326
10.4.2 MPLS Forwarding Plane 328

10.4.3 Label Distribution 330
10.4.4 Martini Encapsulation 331
10.4.5 Pseudowires 335
10.4.6 Ethernet Pseudowires 338
10.4.7 Pseudowires and NG-SDH 344
10.4.8 Advantages of the MPLS 345
10.5 Migration 345
10.5.1 Migrating the Architecture 346
10.5.2 Legacy Services 348
10.5.3 Introduction to NG-SDH + Ethernet 348
10.5.4 NG-SDH + Ethernet Virtual Services 348
10.5.5 NG-SDH + MPLS + Ethernet 348
10.5.6 Service Interworking 349
10.5.7 Ethernet + MPLS – *urbi et orbe?* 349

Chapter 11 Next-generation SDH/SONET 351
11.1 Streaming Forces 351
11.2 Legacy and Next-generation SDH 352
11.2.1 Evolution of the Transmission Network 352
11.3 The Next-generation Challenge 353
11.3.1 The New Network Elements 354
11.4 Core Transport Services 355
11.4.1 Next-generation SDH 355
11.5 Generic Framing Procedure 356
11.5.1 Frame-mapped GFP 357
11.5.2 Transparent GFP 359
11.6 Concatenation 359
11.6.1 Contiguous Concatenation of VC-4 360
11.6.2 Virtual Concatenation 361
11.6.3 VCAT Setup 366
11.7 Link Capacity Adjustment Scheme 366
11.7.1 LCAS Protocol 366
11.7.2 Light Over LCAS 368
11.7.3 LCAS Applications 369
11.7.4 NG SDH Event Tables 373
11.8 Conclusions 373

Index 377

Preface

The telecommunications industry is undergoing important changes as service providers and users have already moved away from the model of the past century where voice, between fixed telephones, was the essential part of the business. Technological shifts, social changes and competition have persuaded operators and manufacturers to redefine the business model around new multiservice networks capable of delivering voice, TV and Internet access by means of unified infrastructures. Indeed, the whole world is changing faster than ever thanks to the new transport and telecom technologies that can move people, goods and information faster than the old galleons did. Harbours and routers are nothing but open windows to the interchange between human communities. The result has always been the same, globalization, which can be simplified as an increment of trading that is followed by a certain degree of cross cultures. Globalization is a controversial process, which is often perceived as a combination of feelings that swing from hope and interest, to fear and concern that new technologies may eventually simplify our cultural diversity. Despite the opinion everyone has, this will not be the first time that globalization has occurred.

History shows examples of how a communications network facilitates the establishment of permanent links first between towns and then between cultures that interact, enriching each other. For instance our phonetic alphabet was initially designed by the Phoenicians, then it was improved in Greece, adding the vowels that allowed the use of Greek as the first international language; finally these scripts were adopted and reshaped by Rome and spread its use across Europe and the Mediterranean countries. Since then only the lowercases have been invented before writing this book. The Mediterranean sea is indeed a good example of how a network generates synergies. Phoenicians, Egyptians Greeks, Romans, Byzantines, Arabs, Iberians, Ottomans and the French emerged as great civilizations at the rims of this sea, to extend the influence of their culture far from the borders of their home land. From Algeciras to Istanbul, from Nice to Alexandria, traders, soldiers, monks and families were sent to the other side of the coast to interchange goods, blood and culture, many times, under the threat of the sword. The maritime routes facilitated such a level of interchange that many cultural and racial borders were blurred, but it is interesting to realize that this process never produced a homogeneous nation because, despite the continuous changes and transformation, people's identity was kept because there were also forces that generated diversity.

Among all the aspects of the culture, cooking is one of the most interesting since it is less fastened to the impositions of the secular and religious authorities. Cooking is all about art and techniques, with the epicurean intention of feeding friends and relatives while maximizing the pleasure of eating and drinking. Cooking has been particularly important in the Mediterranean, where the kitchen occupies a central place at home. Time slows down for lunch and dinner, and an invitation to have food at home is the most common expression of hospitality. Cooking recipes are fascinating. No cuisine is established in isolation, which means that cooking in Egypt is not a carbon copy of that found in Lebanon, or in Greece, although all clearly show the Mediterranean influence. In a recipe you can trace not only climate and taste preferences, but also trading routes, migrations, invasions and many cultural aspects related to the calendar and believes. Mediterranean cooking styles change across regions. You may find many variations of the same recipe, even seamless transitions between two recipes apparently different. In any case there is always something that allows you to identify them as Mediterranean, which is the result of a crowded history of lends and borrows that the maritime network made possible.

The first civilizations, Phoenicians, Greeks and Romans, established a trinity of products, that is, olive trees, vineyards and wheat. For generations olive oil, wine and white bread have been signatures of Mediterranean cooking, totally embedded into lives and traditions, and leaving permanent footprints in all the civilizations developed on its shores.

Just two centuries after the Roman Empire collapsed, Arabs began to travel over the North of Africa and the South of Europe. They were great traders who introduced rice, sugar cane, oranges, pomegranates, spinach and aubergines from remote places like India and China. An extraordinary example of the Arabic heritage is couscous, which in Sicily is combined with fish. Sweets like marzipan, guirlache, panellets or turroni are no more but local versions of the outstanding variety of almond-based sweets that you can still enjoy in the arab medinas of Fez or Marrakech.

The Spanish expansion into America added new ingredients for cooks, including beans, peppers, potatoes, maize and tomatoes. Those territories integrated into the Aragonese crown quickly adopted the new vegetables, which could grow easily in their fields. Many recipes were transformed, like the paella in Valencia that would not exist without green beans, or the modest polenta in Italy, which was definitely improved by substituting the barley flour with corn flour. Could anyone imagine Venice today without grilled Polenta with calamari al nero?

We cannot forget the Ottomans that dominated most of East Europe and North Africa for several hundred years after the Byzantines were defeated in the fifteenth century. This is also an interesting case of melting pot, since this empire granted all type of exchanges. You have to be an expert to distinguish the differences between the small filo pastry cakes that you can have in Essahuira, further west in Morocco, in Athens in front of the orthodox Cathedral, or in the narrow streets of the old town of Jerusalem.

There are many examples, but the Sephardim were the protagonists of one of the most fascinating stories from the Mediterranean. It is believed that Sephardim Jews arrived and lived in relative tolerance in Iberia from the era of King Solomon (930 B.C.E.). However in 1492 the Edict of Expulsion was signed in Spain, for those who rejected conversion to Christianity. Large Sephardic communities were founded in Amsterdam, Geneva, London, Bordeaux and Hamburg as a result. Nevertheless, the majority preferred to look for a new home in the Ottoman territories, whose sultan Beyazit welcomed them. For centuries Sephardims preserved their culture, including the Jewish religion, the Ladino language, and of course their cooking style, which is a combination of the Iberian heritage and the Turkish, Greek or Arab flavours. The Sephardic kitchen is very Mediterranean and relies on appealing combinations of meat, vegetables and fish. Still today you may have mostachudos or huevos haminados, in the juderias of Tangier, Sofia, Thessalonika, Istanbul and Sarajevo. The deep-fried fish technique was introduced by them in Amsterdam after their expulsion, and then to England where it was initially known as the Jewish way of cooking fish. Except for the olive oil and the squeezed lemon, the recipe is today exactly the same as it was five hundred years ago. Curiously deep-frying was also adopted in Japan when Portuguese traders and the Jesuit Francisco Xavier mission settled in Nagasaki in the sixteenth century. Tempura, the popular Japanese dish, is a well-documented legacy of that time, in which seafood and vegetables are coated with batter and deep-fried in high *temperatura* oil. Surprisingly, two of the most specifically British and Japanese dishes have a common root anchored in the Mediterranean rim.

We have used the Mediterranean to highlight how a network does a matter, whether based on ships or IP packets, and facilitates all kind of exchanges between peoples. Multiculturalism is indeed intrinsic to the human being. Nothing lasts forever, we are just a picture of continuous transformation, but this fact does not necessarily mean uniformity across the globe. We want to be optimistic, we want to say that communication, in the widest sense of this word, means cultural enrichment and more opportunities to understand and to enjoy the differences of the world. It should be just a matter combining properly the new ingredients with those that are part of our backgrounds to maintain our peculiarities.

Regarding this book (which is about communications, not about cooking), we must remark how innovations in telecommunications have produced significant changes in our lives in a very short period of time. Now we talk about personal and mobile communications, and broadband Internet access, which is more flexible than ever, thanks to the technological convergence that has permitted the bundling of television with existing data and voice services. The convergence of IP-centric applications over unified infrastructures can generate significant benefits, but it is important to maintain performance and quality parameters by testing these networks, checking the availability of the services, and detecting early any issue that may affect the customers' experience. The audiovisual market is very mature, it has been served by broadcasters for years, therefore it is of key importance to match customers' expectations from the first day. Experience demonstrates that a trial and error strategy does not work in a consumer-led market.

José M. Caballero
Barcelona, Spain

Chapter 1: Business Strategies

Telecoms operators are now employing new strategies to deliver thrilling new services using next generation networks. The full package of services includes line rental and fixed line telephony with a combination of Internet access, IP television, video on demand (VoD), entertainment applications and, eventually, cellular phones. Using the terms adopted by the industry, we are talking about *Triple Play* and *network convergence*. In other words this means: multiple services, multiple devices, but one network, one vendor and one bill (see Figure 1-1).

This manoeuvre is much more than just a new commercial product. It is a consequence of the important changes the industry is undergoing, such as technological innovations, social changes and new regulations. These changes have persuaded operators to redefine their businesses based around a new unified network that should be able to support any type of telecoms service.

Beyond each particular strategy we can identify some of the common drivers, such as a declining voice business (see Figure 1-2), the flat profit perspectives on data access, the new regulations encouraging competition, and the technological achievements that have made network convergence possible.

Triple Play: Building the Converged Network for IP, VoIP and IPTV Francisco J. Hens and José M. Caballero

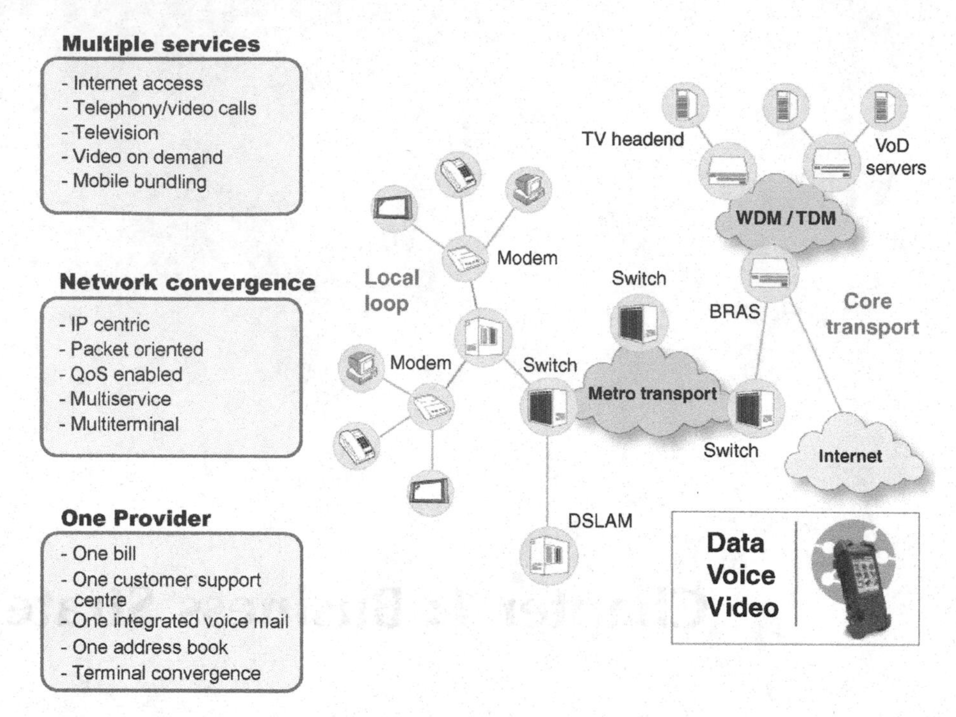

Figure 1-1. Triple Play aims to unify telecommunication services by using a single network to deliver a bundle of multiple applications.

1.1 Expanding Telco Businesses

Voice is still a profitable business with margins of over 50%; however, in the case of traditional fixed telephony, this is rapidly declining (see Figure 1-3). In many cases the fixed phone service is included in the same flat bill with the Internet access and line rental. On the other hand, the growth rate of the cellular telephony business is now less than it was a few

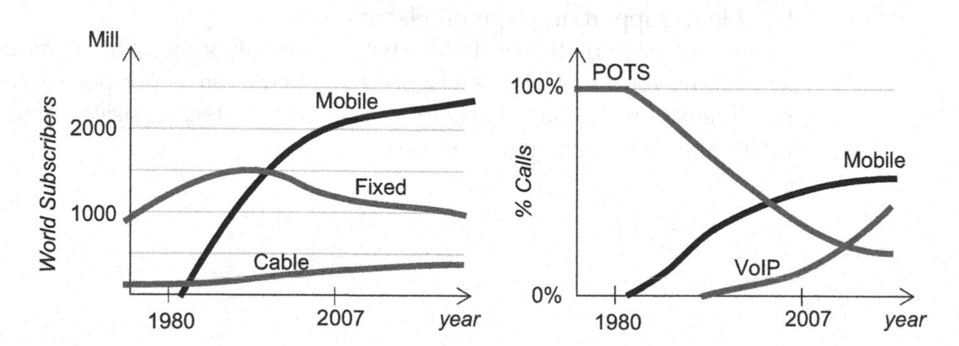

Figure 1-2. Mobile and cable subscribers undermine up to 2% of fixed line subscribers per year. Reproduced by permission of Trend Communications Ltd.

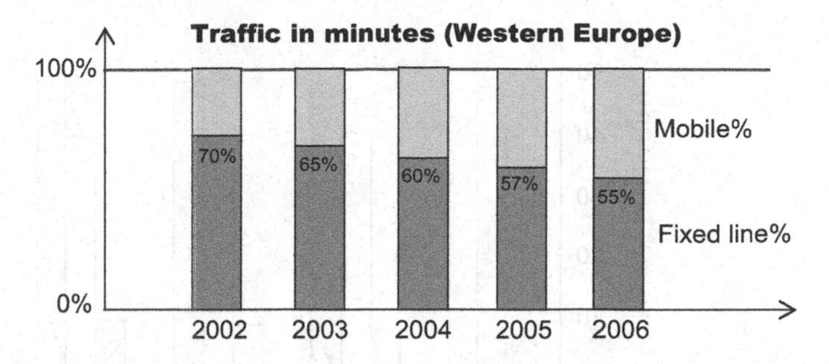

Figure 1-3. Voice revenues based on fixed lines are declining and wireless voice revenues are not growing as in the previous decades. Reproduced by permission of CMT.

years ago and is nearly at saturation point in developed countries. Unfortunately for fixed line operators, alternative services such as broadband access have become a commodity difficult to differentiate, making it impossible to compensate for declining voice revenues.

After several failed attempts the telecoms industry has apparently found a remedy for its continuous headache of offering multiple services as a commercial package that includes fixed line rental and wireless services all in one monthly bill (see Figure 1-4). These multiple services are often referred to as Triple Play. It is too early to know if this is going to be the solution for the telecom crisis that started in 2000 with the 3G licences and the dot.com bubble, or is it just 'another case of mass hysteria in the telecoms industry'? (*The Economist*, 12 October, 2006).

Bundling video with existing data and voice services goes further than a pure marketing campaign; it is essential to keeping telcos in the residential business. It is, in fact, a very

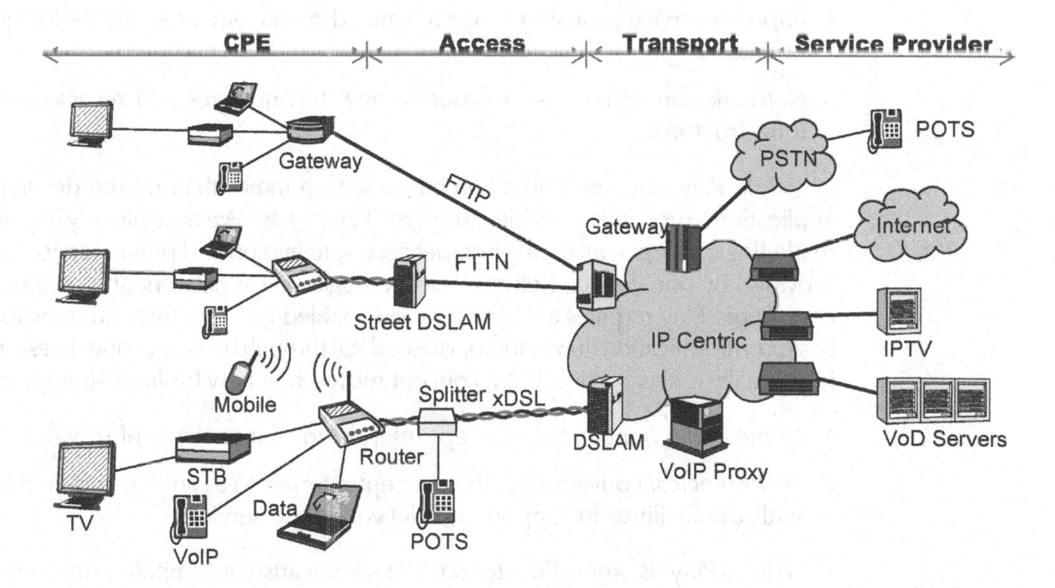

Figure 1-4. Multiple services mean one network, various terminals and many types of access.

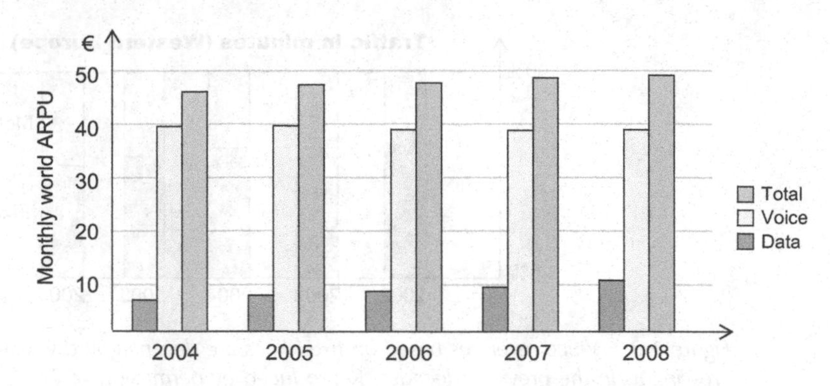

Figure 1-5. Evolution of world ARPU. Reproduced by permission of Telegeography Research.

ambitious strategy with well-defined targets:

- *Reduce churn* – gain customer loyalty with one package that includes all services supplied by one vendor.
- *Minimize costs* – integrate infrastructures and human teams using network convergence.
- *Gain TV customers* – telecom operators should use the same weapons as the cable companies to supply television services.
- *Increase profits* – by using legacy and innovative applications to raise the average revenue per user (ARPU) (see Figure 1-5).
- *Grow the brand name* – cultivate the perception of the company as being able to supply any type of telecommunication service.
- *Improve service provision* – use advanced management solutions for quick and easy provisioning.
- *Network convergence* – incorporate new technologies and recycle existing network infrastructures.

Triple Play is focused on a combined service rather than on the development of new applications, protocols or architectures (see Figure 1-8). We have already explained this is as a marketing concept concerning Internet access, television and phone services. All of them are provided by one vendor, delivered over a single access network and paid with one bill. To offer Triple Play requires a technologically enabled network that can transport all the three basic communication flows (audio, video, data) through the same pipe. In essence Triple Play is not really a new service, but a concept related to a new business strategy on two planes:

1. *Commercial bundle*, the concept referred to as a package of services.
2. *Technological convergence*, the concept referred to as an IP centric network, enriched with the facilities to support and deliver all the services.

Triple Play is appealing to customers because it simplifies the subscription and support of several telecoms applications. The problem for providers is that price, quality

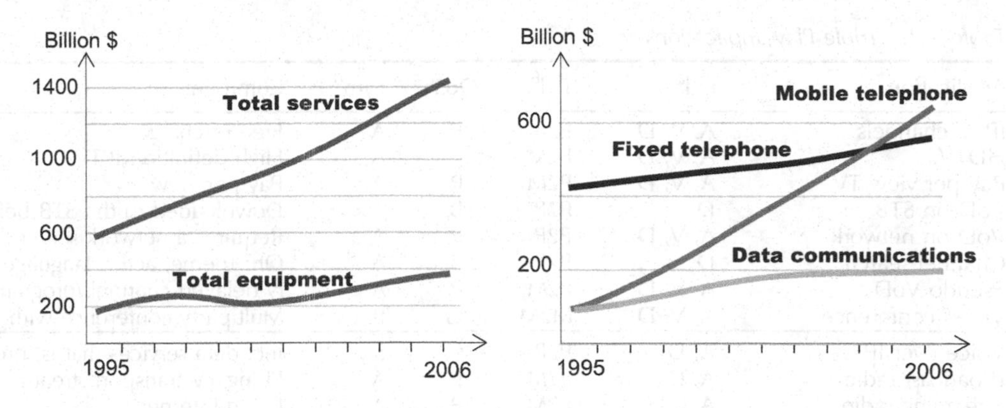

Figure 1-6. Telecom world market revenues in services and equipment. Reproduced by permission of International Telecommunications Union.

and contents are very important, especially as most services are not new at all, except in the format and the interfaces that are used to supply them. Efforts to reduce customer bills would also reduce total market revenues, therefore development of new applications is essential for the business (see Figure 1-6).

1.2 Triple Play Applications

A large number of applications can be designed and supplied over a converged network (see Figure 1-7). Triple Play applications are often a combination of several types of such as data, audio and video, that are managed by a number of parameters such as bandwidth, source/destination relationship, type of routing, QoS and traffic symmetry (see Table 1-1).

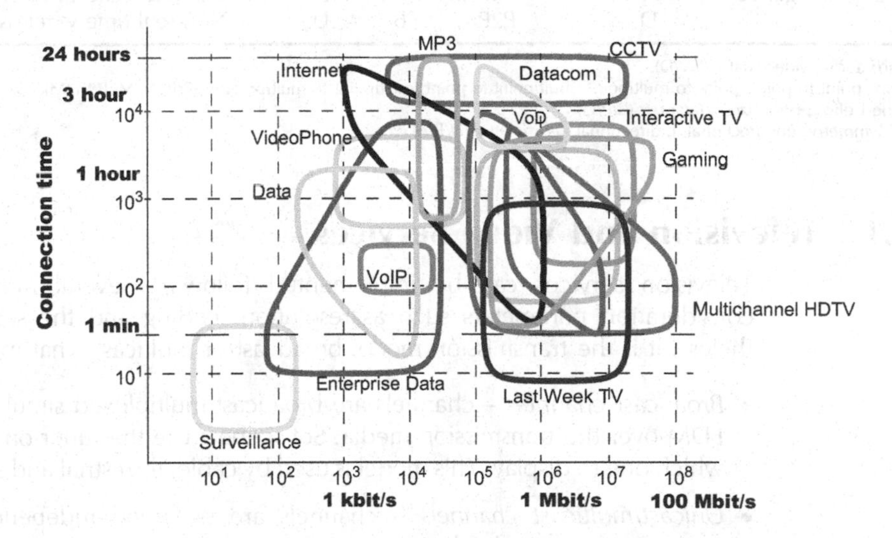

Figure 1-7. Triple Play applications: bandwidth and time requirements.

Table 1-1. Triple Play applications

Application	Pld[a]	Rel[b]	QoS[c]	Sym[d]	Comments
IPTV channels	A, V, D	P2M	P	A	Free of charge
HDTV	A, V, D	P2M	P	A	High definition IPTV
Pay per view TV	A, V, D	P2M	P	A	Pay per view
VoD on STB	D	P2P	B	A	Downloaded to the STB before visualization
VoD on network	A, V, D	P2P	P	A	Requires a network server
Channel search	D	P2P	B	A	On: theme, actor, language...
Pseudo VoD	A, V, D	P2M	P	A	*N* delayed channels/program
Video conference	A, V, D	M2M	G	B	Multiparty conference with image
Voice over IP	A, D	P2P	G	S	Inc. data services that is, presence
Broadcast radio	A, D	P2M	P	A	Using TV transport stream
Streaming radio	A	P2M	P	A	Using Internet
Voice mail	D		P	A	Non-real-time voice messages
Hi Fi audio	A	P2P	P	A	Pay per listening session
Audio downloading	A	P2P	B	A	For MP3 players
Gaming individual	D	P2P	G	A	Individual or group
Gaming group	D	M2M	G	S	Pay per play
Fax	D	P2P	B	A	Fixed to VoIP
e-Commerce	D	M2P	B	A	Web sales applications
VPN services	D	M2M	P	S	Business application
Hi-Speed Internet	D	P2P	B	A	Entertainment, home working
Storage services	D	P2P	G	A	Business application
Surveillance	D	P2P	B	A	Alarms
Home automation	D	P2M	B	A	Remote control, monitoring
Instant messaging (IM)	D	P2M	B	B	Real-time short messages
e-mail	D	P2M	B	U	Non-real-time messaging
www	D	P2P	B	A	Information browsing based on hypertext
File transfer	D	P2P	B	A	Data download and upload from/to a server
UMA	A, V, D	P2P	G	S	Unlicensed mobile access
Mobile convergence	A	P2P	G	S	Call redirection to fixed line
SMS	D	P2P	B	U	Non-real-time wireless short messages

[a]*Payload*: audio, video, data (A/V/D).
[b]*Relation*: point to point, point to multipoint, multipoint to point, multipoint to multipoint (P2P/P2M/M2P/M2M).
[c]*QoS*: best effort, prioritize, guarantee (B/P/G).
[d]*Traffic symmetry*: unidirectional, bidirectional, asymmetric (U/B/A).

1.2.1 Television and Video Services

Television services can be implemented following several models by taking into consideration parameters such as resolution, coding and the service model. Nevertheless, it is the transmission mode, broadcast or multicast what modifies the service:

- *Broadcast channels* – channels are broadcast/multiplexed simultaneously in TDM or FDM over the transmission media. Subscribers use the tuner on the TV box to select which one to display. This model is used by cable, terrestrial and satellite broadcasters.
- *Unicast/multicast channels* – channels are streamed independently to reach the customer premises that have selected the stream previously. This is the model selected by telcos, in principal, because of the lack of bandwidth at the first mile.

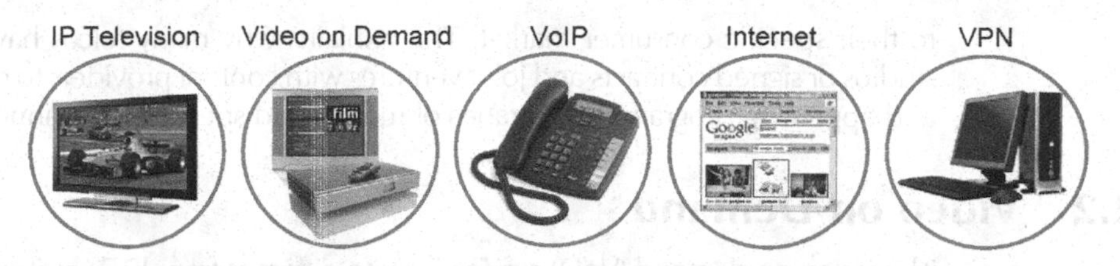

Figure 1-8. Residential customers are focused on applications such as IPTV, VoD, video recording, telephony, Internet access, gaming, hi-fi audio, home automation and mobile bundling. Business customers are focused on connectivity applications such as VPN, broadband access, corporate VoIP and mobile convergence.

Digital video provides a set of interesting possibilities such as metalanguage programmes (one video, several audio signals), customized adverts, pay-per-view or encrypted programmes only for subscribers (see Fig. 1-8). Interaction between the subscribers and the service provider make new capabilities such as games, magazines, voting, competitions, pay-per-view, customized adverts and quizzes. At the end of the day interaction is the key difference between broadcast TV and bidirectional digital platforms.

1.2.1.1 Welcome to the Contents

For most telcos, the television and video business is new. Any previous experience of these services was no more than signals transported in SDH envelopes between the different centres of TV broadcasters, but now that Telcos are also service providers, it is necessary to manage not only the transport and signal distribution, but also the contents; a set of attractive programmes to compete with existing cable operators and broadcasters.

Therefore, telcos must not only acquire new technical and business skills to enter into this already mature market, but should also be involved in the creation of content that is adapted

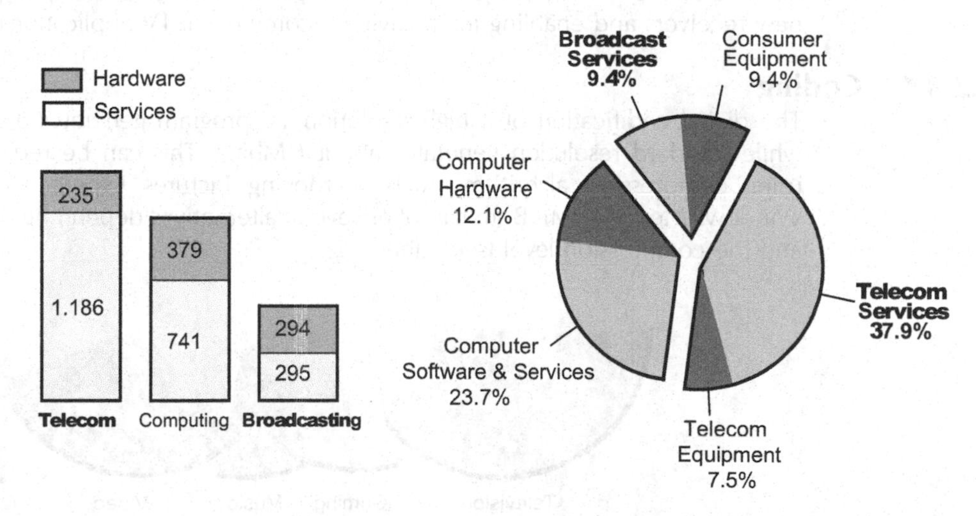

Figure 1-9. The global information and communication technologies market (ICT) indicates that the total broadcasting market is about the 40% of the telecom market; however if we consider only the services segment, the size is only 25%. Reproduced by permission of International Telecommunications Union.

to their specific consumer market. This explains why many telcos have created their own studios or signed contracts and joint ventures with content providers to gain access to suitable and appealing programmes, libraries of movies and specialized channels (see Figure 1-10).

1.2.2 Video on Demand

The video on demand (VoD) service is quite different from IPTV as it enables users to select and watch a video as part of an interactive system. VoD systems have two important features:

1. They enable users to choose the video they want to watch from a digital library selection. Users can control the moment they start watching the video.
2. They provide typical DVD functionality such as pause, fast forward and fast rewind.

For streaming systems this requires more bandwidth on the part of the server, powerful multicast nodes, spare bandwidth and guaranteed QoS control. VoD servers can operate in two ways: streaming or downloading the contents. In both cases this is a point-to-point relation (see Table 1-1). If downloading is being used only a 'best effort' QoS is necessary since the video is recorded onto a network disk, a PC, or a set top box before it is watched.

VoD is one of the killer applications that makes the Triple Play service more attractive when operating over a rich IP network, because cable and satellite operators have more difficulties in implementing this.

1.2.3 New TV Receivers

The migration from analogue and standard definition TV to digital and high definition TV (HDTV) started several years ago. Now mature markets already have many TV receivers enabled to receive digital signals. The new trends are incorporating high definition into new receivers and enabling interactivity in commercial TV applications.

1.2.3.1 Coding

The digital codification of a high-resolution TV program generates a 20 Mbit/s stream while standard resolution generates about 6 Mbit/s. This can be reduced significantly using compression algorithms such as Moving Pictures Experts Group (MPEG) or Windows Media (WM). Both can offer several alternatives depending on the resolution, and the compression level (see Table 1-2).

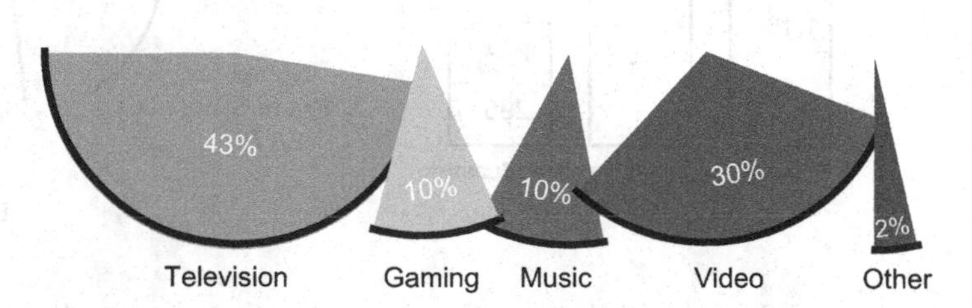

Figure 1-10. Watermelon distribution of the entertainment business. Reproduced by permission of Trend Communications.

Table 1-2. Standard and high-definition bandwidth after compression.

	Lines	Pixels	Broadcast	MPEG-2	MPEG-4	WM9
SDTV	480	704 × 480	6 Mbit/s	3.5 Mbit/s	2–3.2 Mbit/s	2–3.2 Mbit/s
HDTV	1080	1920 × 1080	19.2 Mbit/s	15 Mbit/s	7.5–13 Mbit/s	7.5–13 Mbit/s

The most popular compression family of standards is probably MPEG, defining algorithms based on the discrete cosine transform (DCT) that discards spatially redundant information, and employs movement compensation techniques to minimize temporal redundancy. MPEG-2 is very common, and MPEG-4 is a step-ahead standard that covers small mobile hand sets up to large HDTV receivers. Windows Media 9, a Microsoft development, is also an interesting alternative.

1.2.4 Voice over Internet Protocol

In many aspects Voice over Internet Protocol (VoIP) is also an approach to the Internet world using IP packets to carry the voice signals to incorporate features that would be difficult using traditional phone services. For example, VoIP may allows users to talk for as long they like, subscribers can always be on-line then other users may know about their presence. Depending on how the service has been implemented it is possible to send images, data and videos simultaneously to the people they are talking to. Another interesting aspect is how VoIP phones can use the e-mail address as an identifier and can make calls to an e-mail address as well. Moreover, the phone call list can be made up using a combination of PSTN numbers and e-mail addresses. Access of nomadic users is guaranteed in a similar way to Internet-based mailers, regardless of where in the world the connection to the Internet is established, thanks to proxy servers.

A commercial VoIP service should not be restricted to VoIP phones, but heterogeneous calls between VoIP and ISDN, POTS or GSM phones will be quite normal for a long time. It is not realistic to forecast a full substitution in the short or middle term.

1.2.4.1 Less Expensive Phone Service

The phone service based on VoIP generally costs less than the equivalent based on the traditional Public Switched Telephone Network (PSTN). This fact has been justified by saying that packet-oriented technologies are more efficient that circuit oriented ones because they permit the utilization of a single network to carry voice together with data and video. This hypothesis has been proved, at least according to the 2006 survey carried out by Consumer Reports in the US. The survey said that people who have purchased VoIP service are reportedly saving around $50 per month on their bills.

It is also important to bear in mind that incumbent operators already have a satisfactory quality phone service and in many cases it has already been bundled with ADSL and line rental. That means that the PSTN cost has been deflated and this can explain why legacy operators have little interest in rolling out a new VoIP service that would demand more resources for very low benefits (see Figure 1-12).

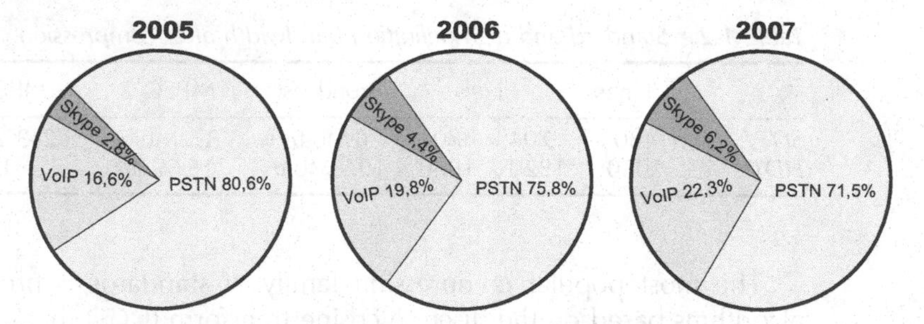

Figure 1-11. International traffic transport. Reproduced by permission of Primetrica and Trend Communications Ltd.

1.2.4.2 Drawbacks

The VoIP phone service relies on the Internet connection so consequently the service can be affected if the quality and bit rate are not appropriate, causing distortions, noise, echoes and unacceptable delays. Most of the drawbacks can be overcome by increasing the bandwidth, minimizing contention, and using more elaborated protocols to prioritize VoIP traffic while controlling delay and packet loss. It may also be worth considering modifying firewalls and adapting network address translation (NAT) tables used at the boundary routers.

1.2.5 VoIP Rollout

VoIP is a very important technological, financial and social change after a century dominated by national operators and a phone service based on circuits (see Figure 1-11).

However, the complete migration, or eventual substitution, will take many more years than *experts* first thought. At the end of the day the installed PSTN base is massive, quality is excellent and customer bills are continuously deflating (see Figure 1-12). Incumbent operators should manage VoIP as a complement whilst maintaining the existing POTS and ISDN services. It is significant that a decade after main manufacturers announced VoIP solutions, the traffic penetration still scores below 20% in developed countries, and this level is thanks to early users like youngsters, travellers, and low income users.

The challenge of VoIP rollout is not insignificant given the complexity of hybrid solutions that also combine traditional PSTN phone calls with the new packet-based

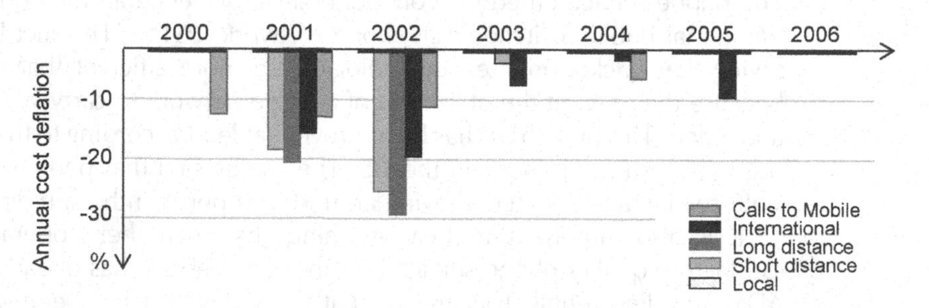

Figure 1-12. Annual deflation of phone calls originated on fixed lines. The Telefonica case. Reproduced by permission of Telefonica.

solution. Just imagine how many protocol translators, gateways and transcodecs that would be necessary to set up a multi-conference between VoIP, POTS and mobile to guarantee compatibility and interconnections.

1.3 Driving Factors of Triple Play

Two main groups of driving factors can be identified: firstly, the necessity to redefine the telecom business and, secondly, the consequence of competitive pressure.

1.3.1 Business Redefinition

Telcos' initial advantage in voice and data access was quickly blurred by new competition forcing them to operate to thin profit margins and even no profits at all. In order to change the shape of the business, telcos have discovered how a DSL broadband router can be combined with a converged network that allows the development of a set of new multimedia applications.

The strategy starts by offering bundled voice and Internet access paving the way for the introduction of television to existing subscribers. IPTV will never completely replace broadcast or cable TV, but will complement them and gain significant market share with an interactive platform built around the IP protocol. Video on demand, pay-per-view (PpV), and video recording services will also need to be more innovative to gain market share.

1.3.1.1 Telcos' Point of View

Above all, telcos feel that they have finally found, in Triple Play, a unique opportunity to redefine the whole business. This strategy requires a high capital expenditure (CAPEX) to develop a converged network, but will allow innovative product creation by combining legacy and new products and will also reduce operational expenses (OPEX) in an integrated management environment.

At the end of the day telcos expect to increase revenues by means of the new services while keeping existing subscribers loyal to the old ones.

1.3.1.2 Consumers Point of View

There is much evidence to suggest that customers are willing to buy a telecom service package, whenever they can get significant savings on their bill for the full service. Customers would also like to reduce the increasing level of complexity in managing their technological devices. Just how many devices, remote controls, configurations and interconnections can there be in modern homes equipped with PCs, satellite, Hi/Fi, digital TV, mobiles, video, gaming stations and surveillance devices? Residential customers are demanding simple, easy-to-use integrated technological devices that offer more automatic features.

1.3.2 Competitive Pressure

Just a few years ago, companies such as cable operators, ISPs and telcos were different businesses and were not competing with each other. Cable operators were focused on video

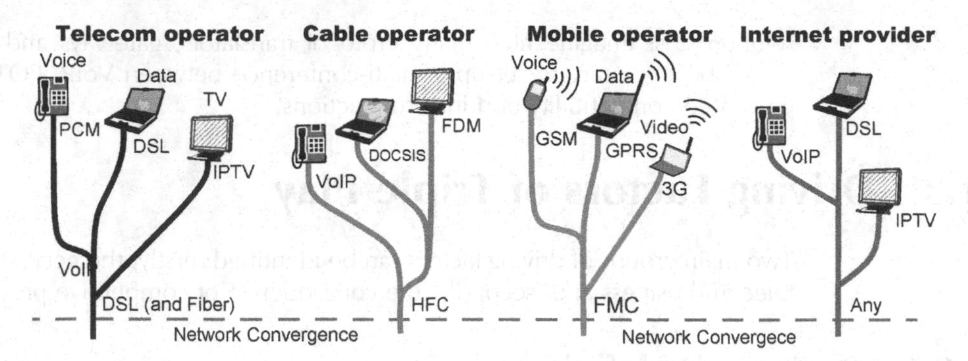

Figure 1-13. Each operator has their own migration strategy for network convergence.

and TV services, ISPs did not go much further than offering IP data services while the main business of telcos was based on voice and datacom services such as T1, E1, FR or DSL.

The advance in technology and new regulations have made ISPs, cable and mobile operators competitors of telcos in voice and data access. So companies that were originally in different markets are all now racing to bundle and offer the same services:

- *Mobile operators*, that are offering wireless technologies have produced a significant social change. Now telephony is a personal service that combines privacy with mobility. It is not a secret that wireless has stolen an important part of fixed line revenues, moreover wireless-only operators are constantly suggesting that all telecom services can be delivered using only wireless technologies. Mobile operators, that have invested in 3G infrastructures, are very keen to bring new applications based on voice, Internet, video and messaging into service. However, it is not very clear what the volume of new revenues will be as multimedia service expectations are not very optimistic (see Figure 1-13).
- *Cable operators*. New regulations have allowed cable operators to grab the traditional telcos market. Cable operators, that started out offering just TV, were the first to offer Triple Play thanks to the Data Over Cable Service Interface Specification (DOCSIS). This technology enabled them to also deliver broadband Internet access to their subscribers which later on opened the door to include VoIP as well (see Figure 1-13).
- *Internet service providers*. ISPs probably are the most important threat to the traditional phone service as inexpensive voice calls can potentially reduce voice revenues to a minimum. ISPs are experts on IP services and do not have to manage large infrastructure assets. This fact has enabled them to build a flexible and innovative model around the Internet. VoIP solutions based on software such as Skype are an excellent example of how ISPs are exploiting this opportunity (see Figure 1-13).

The result is that companies, originally in different markets, are now all racing to bundle multiple services using their own version of a converged network (see Table 1-3). The consequence is that fixed telephony use and revenues are declining after the adoption of mobiles and the penetration of cable operators that also include broadband access and voice services (see Figure 1-14).

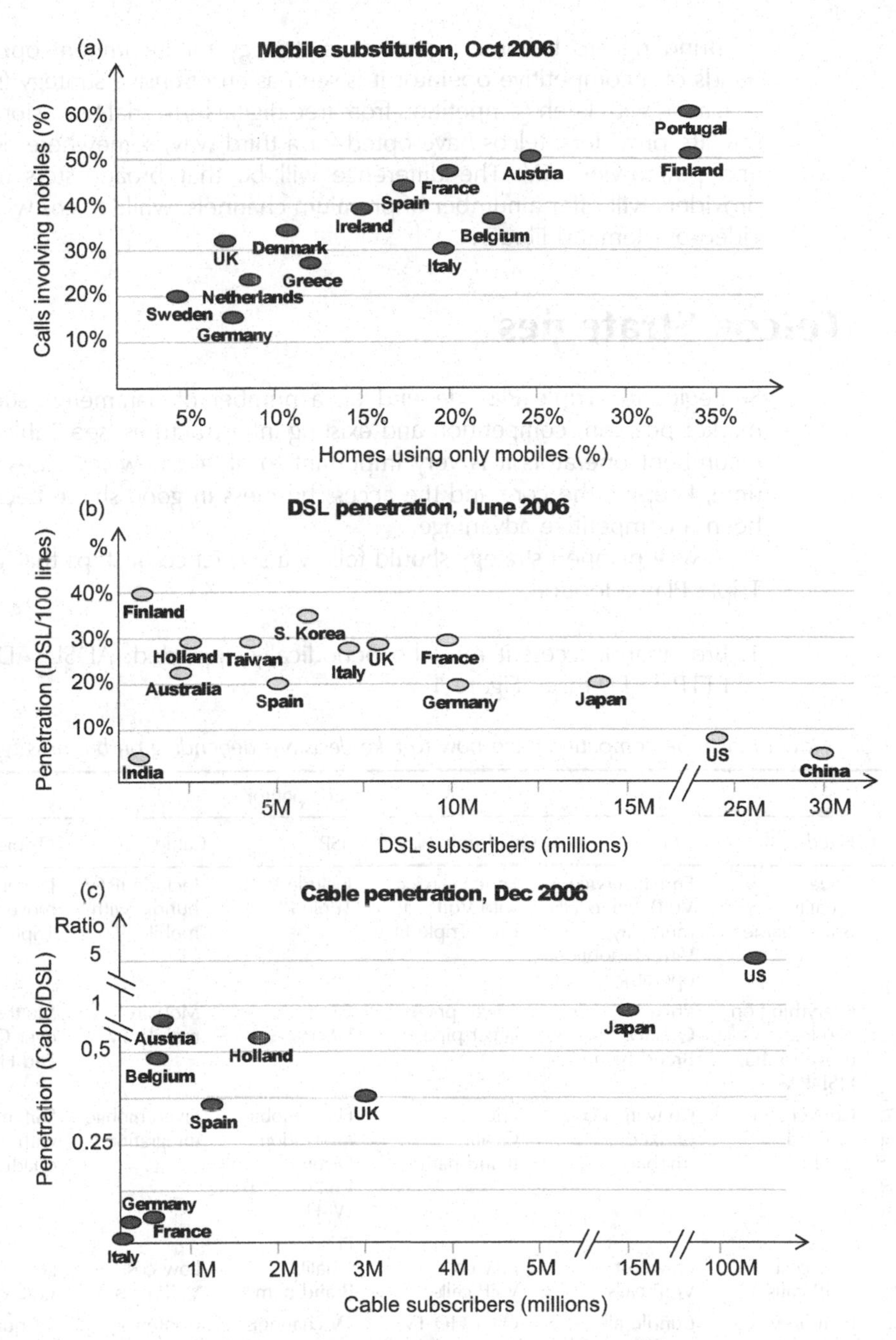

Figure 1-14. Mobile, DSL and cable penetrations. Reproduced by permission of (a) Trend Communications Ltd. (b) Reproduced by permission of DSL Forum; (c) Reproduced by permission of DSL Forum.

Bundling has become a protective strategy for incumbent operators, while in the hands of a competitive operator it is seen as an offensive strategy (see Table 1-3).

Faced with keen competition from free digital terrestrial television and from cable and satellite providers, telcos have opted for a third way, somewhere between free-to-view and pay-to-view TV. The difference will be that broadcasters and pay-to-view TV providers will offer a number of premium channels, while telcos will offer a competitive video-on-demand library.

1.4 Telcos Strategies

Strategies for Triple Play depend on a number of parameters such as business size, market position, competition and existing infrastructures (see Table 1-3). In the case of incumbent operators it is very important to obtain new revenues, while, at the same time, keeping the core and the access business in good shape because this has always been a competitive advantage.

A well planned strategy should follow a sequence of steps that will pave the way to Triple Play adoption:

1. Broadband access it must be periodically upgraded: ADSL, ADSL2+, VDSL2, and FTTP / FTTH (see Figure 1-19).

Table 1-3. How to face the competition and how to take decisions depending on business type.

Operators and Strategies	Competitor						
	Fixed	3G	Quadruple Play	ISP	Cable	Triple Play	Satellite
Fixed	Price Quality Brand name	Enrich service VoIP, videocalls, multiplay Virtual mobile operator	Enrich service with VoIP, video calls, Triple Play	Include VoIP Triple Play	Include IPTV, bundle with mobile	Lower cost, move to Triple Play	TV, rural and isolated regions
3G	Everything on wireless, move to 3G, HSDPA	Price Quality Brand name	Lower price, larger pipe	GPRS, UMTS	Mobility, mobile TV	Multi services over GPRS, 3G and HSDPA	TV, LEO constellation
Quadruple Play	Convergence of fixed + mobile	Convergence of fixed + mobile	Price Quality Brand name	Fixed/mobile integration Divert calls to VoIP fixed using Wi-Fi	Fixed/mobile integration	Differentiate with mobile: quadruple play	TV channels
ISP	Low cost VoIP calls	Low cost VoIP calls	Low cost VoIP calls	Price Quality Brand name	Low cost VoIP calls	Low cost VoIP calls	TV channels
Cable	Include voice and broadband	Bundle all services enrich customization	Offer HD TV, bundle all services	TV channels	Contents Price Brand name	TV quality and quantity of channels	Contents
Triple Play	Multiservice simplify billing, enrich contents	Enrich contents simplify billing, enrich contents	Multiservice simplify billing, enrich contents	Multiservice contents VoD	Focus on Price Support	Contents simultaneous channels Brand name	Multiple and VoD

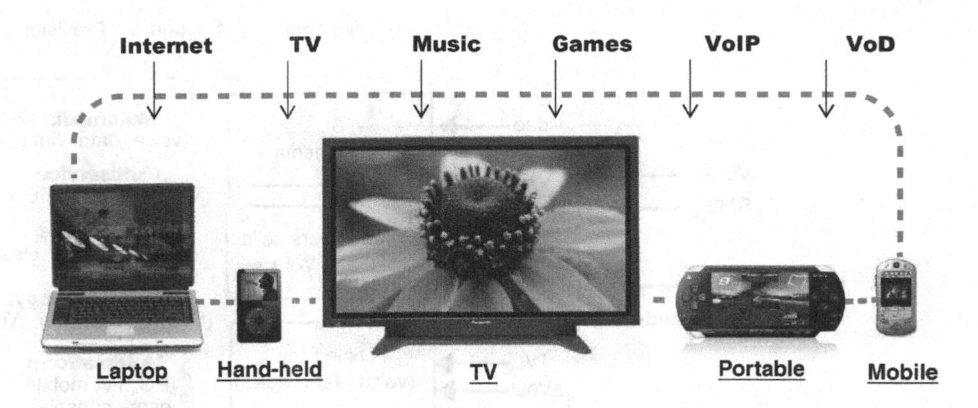

Figure 1-15. Triple Play is not only a matter of multiple information flows but is also about how a wide range of terminals can manage data, audio and video applications.

2. Progressive bundling of local/national/international calls to line rental and adding the Internet access.
3. Converged network rollout, to minimize cost in support and maintenance.
4. Mobile/fixed commercial bundling.
5. Add new applications such as IPTV, VoIP, Video conferencing.
6. Multimedia/multi-terminal services (see Figure 1-15)

1.4.1 Service Bundling and Network Convergence

The bundling process starts as a commercial action unifying in a bill of line rental, DSL service and phone calls. Obviously the first step is the integration of local and short distance calls while the most expensive international calls and calls to mobiles should be last, thus reducing the cost progressively.

Service bundling does not require either technological innovations or network migration. It can be accomplished just by means of market engineering and business alliances where necessary.

On the other hand, the network convergence term refers to the network based on SDH/SONET that is able to support all the Triple Play services simultaneously. It also refers to an IP centric architecture that is able to support different service quality accurately.

At the same time, the existing optical transmission core network has to be migrated into a packet-friendly core layer based on NG-SDH to facilitate the interconnection of the metropolitan that are being deployed based on bridged Ethernet architectures with QoS enablers like VPLS/MPLS.

Regarding the local loops, fibre will be progressively incorporated using FTTx architectures. Fibre will need to be installed in between the customer premises and the central office to reduce the copper span.

Although Triple Play strategies may start only with service bundling, migration to an IP-centric converged network needs to be part of the involved operators' strategy in order to reduce the delivery costs and simplify the management structure (see Figure 1-16).

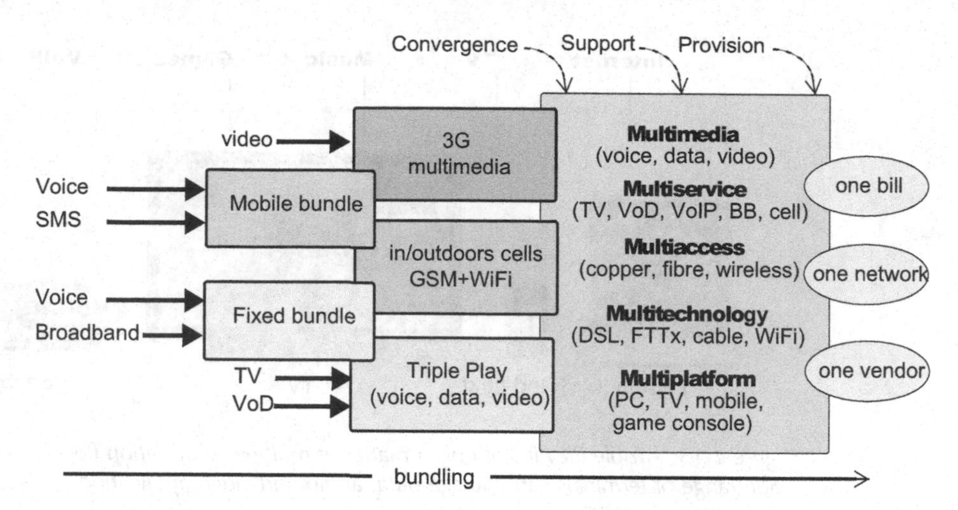

Figure 1-16. Triple Play migration. The big game.

1.4.2 Quadruple Play

When cellular phones are added, the bundle is often called Quadruple Play. However the new bundle is not necessarily a technological convergence, it could be a simple commercial package that only includes bill unification. But it can also be a sophisticated convergence that allows cellular hand sets to connect home routers to the Internet or to redirect calls through the cheaper fixed line. This convergence happens when the mobile is in the wireless hotspot range (wi-fi or bluetooth) of the router. Obviously, mobiles with Wi-Fi capacity are necessary to enable such a level of integration by making a seamless switch from the outdoor cell to the router range.

It is interesting to note that mobiles have their own Triple Play strategy, therefore Quadruple play does not really refer to new applications but to the inclusion of the wireless media and cellular hand sets in the bundle.

Mobile manufacturers and operators have demonstrated that multiple services and technologies can be merged successfully. We find it is quite normal to manage not only phone calls but also text and multimedia messages, agendas, navigation, Internet access, e-mail, radio, video, gaming, photographs, etc., using only our mobile terminal. Wireless operators have proved that it is possible to maintain or even increase revenues in a saturated market by offering innovative services and applications. These are interesting lessons to be learned by the new bundlers, namely, the fixed line operators.

1.4.2.1 Financial Integration

Integral operators, with fixed and mobile divisions, have more possibilities to protect their revenues from single network competitors. This strategy puts more elements into the operator's hands so they can be more aggressive on price than specialized ones. Integral operators can compete much better because they can move forces in both fields simultaneously to complement each other, thus reducing cost and using common staff, and assets. France Telecom and Telefonica are two examples that, by reabsorbing their mobile divisions, have become integral operators (see Figure 1-17).

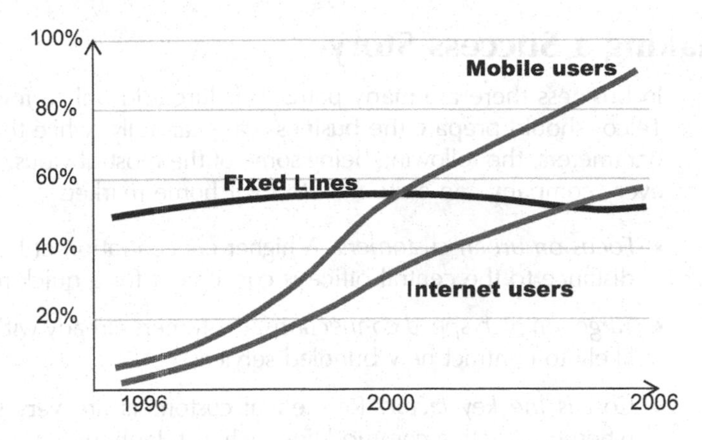

Figure 1-17. Fixed line, mobile and Internet penetration rates in developed countries. Reproduced by permission of International Telecommunications Union.

1.4.2.2 Combined Packages

Combined mobile and fixed services are a natural step forward for integral operators. This can be enhanced with unified features such as call forwarding, unified voice mail, and a common customer service centre. This action reduces mobile churn since it is economically convenient for subscribers and also more difficult for them to break the bundle or to change all fixed, DSL and mobiles services simultaneously, to move to a competitor.

1.4.2.3 Mobile Redirection

Calls originated in a cellular network can be redirected to the owner's fixed lines. To achieve this integration the router and the hand set have to establish a Wi Fi or bluetooth link when they are close enough. Mobiles can also use broadband access to upload/download data, videos, music or to establish low cost VoIP calls over the Internet. This offer would cannibalize some of the revenues, but would also increase the loyalty of existing customers. Such is the case in Unik FT/Orange and BT Fusion, which are bundled services for mobiles that offer low call rates when the mobile user is at home by using the Wi-Fi connection to the router and then the Internet facilities.

1.4.3 VoD: the Key Difference

IP television is the direct way to face competition from cable operators in their core business. Television is a challenge in many aspects, because it requires a new network, a large access pipe, and rich contents to fill up the channels.

VoD can reuse an IPTV network; nevertheless it is a different service that requires new protocols, network elements and terminals to be supported. VoD requires rich libraries of specialized contents like films, documentaries, cartoons, sports, etc.

VoD is a key factor in the Triple Play offer, since broadcasters (cable operators, satellite, digital) have technical difficulties in implementing this service, therefore it can be seen as the front line to gain new customers and to increase the ARPU of existing.

1.4.4 Making a Success Story

In business there are many paths to failure and only a few to building a success story. Telcos should prepare the business case carefully, while the steps will depend on many parameters, the following being some of the most obvious. These are generic trends that every company can customize to their home markets.

- *Focus on urban customers.* A higher concentration of homes, buildings and a shorter distance to the central office is convenient for a quick rollout of the access network.
- *Target on high speed connections.* Customers already with broadband access are more likely to contract new bundled services.
- *Cost is the key factor.* Residential customers are very sensitive to cost, particularly when contracting commodities such as telephony, TV and broadband access.
- *IPTV is a defensive move.* Telcos, in principal, have to see TV services as a purely defensive blueprint to keep cable operators under control.
- *VoD is an offensive move.* Video services should be the front line strategy since only telcos have the most appropiate architecture to support it.
- *Prepare the convergence.* Mobile telephony vs fixed telephony competition is over. Integration of both worlds is strategic.
- *Keep it simple.* One bill, one provider, is probably less important than a service that is reliable, simple to manage by the customer and easy to maintain by the operator.

1.5 Infrastructures

Operators must keep in mind that the control and management of underlying resources are essential. For a successful Triple Play business case it will be necessary to roll out a converged network that is IP centric and QoS enabled. Independent of the access technology (DSL, FTTx, Cable, WiMax or Wi-Fi) and the core network architecture (VPLS, MPLS, Ethernet, NG SDH, or WDM), the converged network must guarantee the QoS to support data, voice and video streams.

Triple Play has different QoS requirements in terms of bandwidth and delays according to the application. For example video conferencing is sensitive to delay and jitter, but non-real-time multimedia applications are less sensitive to delay and packet loss because it makes use of error recovery techniques. While data applications are not sensitive to delay and jitter, packet loss may be a critical factor (see Figure 1-18).

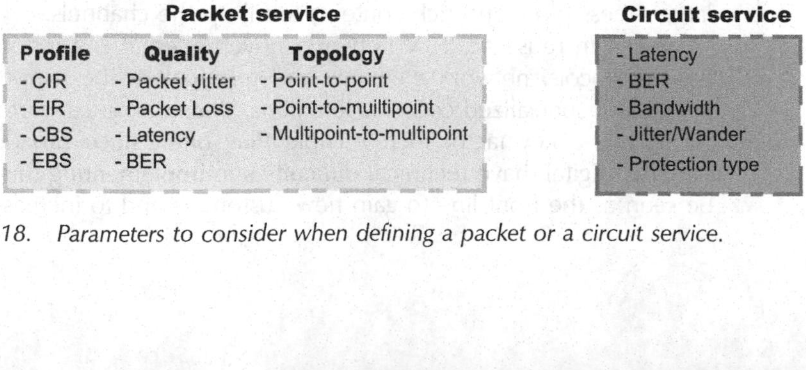

Packet service

Profile	Quality	Topology
- CIR	- Packet Jitter	- Point-to-point
- EIR	- Packet Loss	- Point-to-muiltipoint
- CBS	- Latency	- Multipoint-to-multipoint
- EBS	- BER	

Circuit service

- Latency
- BER
- Bandwidth
- Jitter/Wander
- Protection type

Figure 1-18. Parameters to consider when defining a packet or a circuit service.

There is a number of protocols and architectures to implement QoS in converged networks, including integrated services (IS), differenciated services (DS) and MPLS/VPLS.

1.5.1 CPE Equipment

Customer equipment varies depending on the access technology and manufacturers preferences but, independent of the access technology (copper, fiber, wireless), we need a router which generally has 10/100BASE-T interfaces, and eventually WiFi, to connect PCs, the set-top-box (STB), and IP phones (see Figure 1-4).

1.5.2 The First Mile

During the last decade a lot of progress has been achieved in access technologies. How many are suitable for delivering Triple Play? There are many certainly, based on DSL, cable, Ethernet, wireless and fibre, the cost to bandwidth ratio being the common selection criteria (see Figure 1-21).

Some telcos are installing fibre to the premises (FTTP), which is the best of the possible alternatives available in terms of bandwidth, but the high cost has persuaded many others to continue extracting more bandwidth from their copper pair loops, particularly if they are suitable for the installation of ADSL2+ and VDSL2 (see Figure 1-19). In general, fibre technologies, and specifically the Passive Optical Network (PON), have an associated higher CAPEX because they require fibre deployment to the customer's site, but a lower OPEX because all the elements are optical and passive, being exactly the opposite of copper technologies.

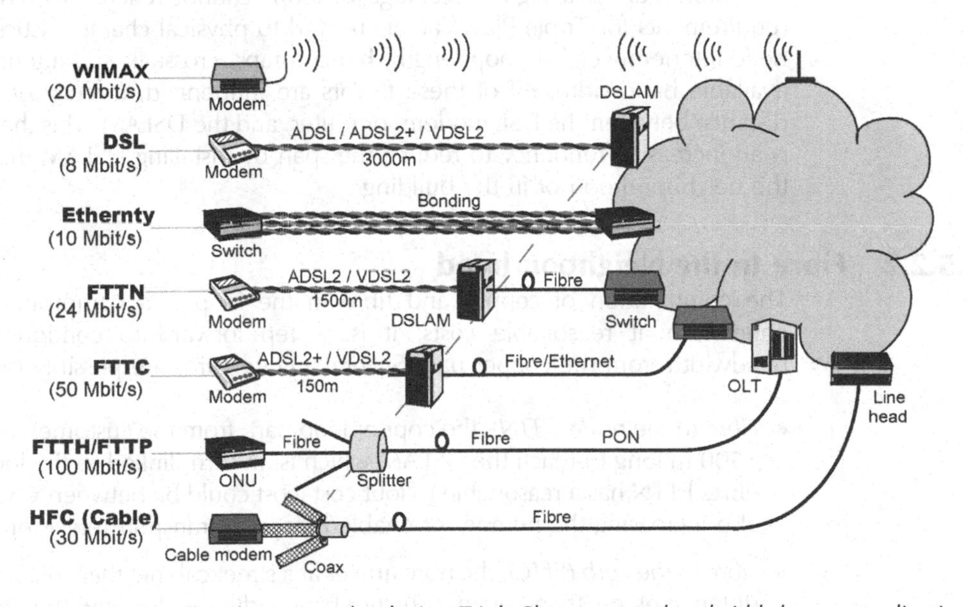

Figure 1-19. Broadband access technologies. Triple Play supports bandwidth hungry applications that require bandwidth of many Mbit/s.

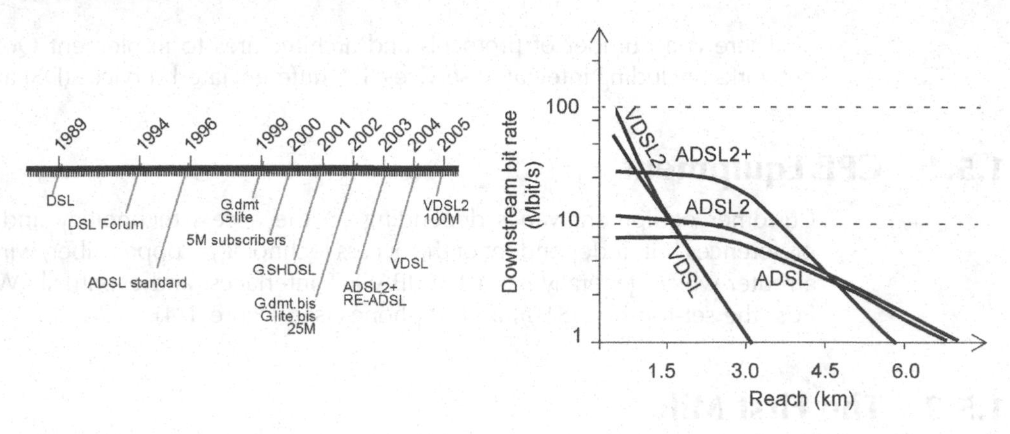

Figure 1-20. DSL has been a real success story in the number of subscribers thanks to its continuous evolution in speed and cost.

Cable operators use hybrid fibre coaxial (HFC), a technology that combines fibre optic and coaxial cable for the last span to the customer premises.

1.5.2.1 Digital Subscriber Loop

Copper loops are available everywhere whilst other media must be newly deployed. That is a competitive advantage for the owners of this infrastructure, who are generally incumbent operators. Several DSL technologies are possible, ADSL2+ and VDSL2 being the main alternatives for residential customers. They must guarantee a minimum of 6 Mbit/s to support one IPTV channel (see Figure 1-20).

Unfortunately, a high percentage of loops cannot reach the minimum bandwidth requirements for Triple Play. Factors related to physical characteristics and electromagnetic interferences like loop length, bridged taps, crosstalk and attenuation limit of the available bandwidth. All of these factors are influenced, directly or indirectly, by the distance between the DSL modem, or router, and the DSLAM. This the reason why there is an increasing tendency to reduce the span by installing DSLAM in street cabinets, in the neighbourhood or in the building.

1.5.2.2 Fibre to the Neighbourhood

The combination of copper and fibre in the loop is a compromise to offer higher bandwidth at reasonable costs. It is a step forward to continue extracting more bandwidth from the copper pair. Several architectures are possible (see Figure 1-19):

- *Fibre to the node (FTTN)*, the copper loop part, from the customer home, can be up to 1500 m long to reach the DSLAM which is, in turn, linked to the local exchange with fibre. FTTN has a reasonable rollout cost (cost could be between €300 and 500) while also increasing the number of enabled loops to transport television channels.
- *Fibre to the curb (FTTC)*, the fibre arrives at a street cabinet that connects homes within a distance of less than 150 m. A higher bandwidth would allow the delivery of more TV

channels simultaneously while the crosstalk, attenuation and noise on the copper are minimized with a shorter wire.

- *Fibre to the building (FTTB),* the fibre arrives at the customer's home connecting the CPE and DSLAM over a distance less than 30 m. Higher bandwidth should be achieved because of a shorter copper section.

1.5.2.3 Optical Access

Fibre is the best media to deliver bandwidth higher than 50 Mbit/s, but it is necessary to figure out how to make it pay. The capacity of optical transmission is the ideal solution to bypass the bandwidth limitation on the last mile, but it is the most expensive. FTTP is based on PON or Active Ethernet and can deliver around 100 Mbit/s to each subscriber.

Fibre optic to the home is often a long-term strategy that is particularly appropriate for green-field installations but difficult to justify when alternative access solutions are already installed (see Figure 1-21). According to several consultancy companies, a home fibre connection can cost a provider anywhere between €600 and 2000 per subscriber, depending on how many difficulties arise. Existing neighbourhoods are the most expensive because it is often necessary to dig a ditch to reach every single home. In new developments it is easier and much cheaper when the fibre is installed together with water, gas, electricity and other utilities.

1.5.2.4 Cable Access

Cable operators use Hybrid Fibre Coaxial (HFC) to deliver dozens of simultaneous TV channels thanks to the big bandwidth pipe this technology can support. DOCSIS architectures enabled Internet access and, later on, VoIP was possible due to a very aggressive strategy to increase the market share of residential consumers of telecom services. However, telcos have used as the answer to deliver TV and VoD. This has made the cable operators business much more difficult.

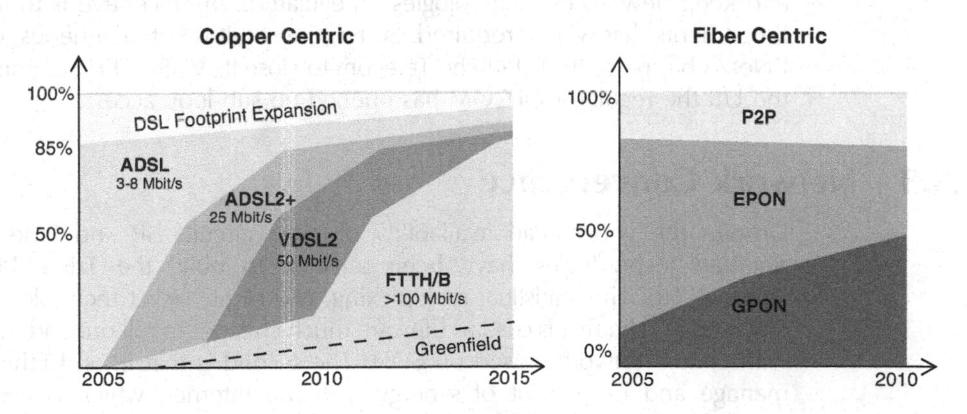

Figure 1-21. Forecast of broadband access technologies. Reproduced by permission of DSL Forum and Trend Communications Ltd.

1.5.2.5 Regulations

Regulatory factors may modify all the access strategies depending on how the unbundling legislation, that was applied to the copper local loops, affects the new access infrastructures:

- Will the regulations for unbundling the local loop be cancelled?
- Will the cable/fibre network be unbundled? This will depend on the unbundled resources: voice frequencies, packet traffic or full capacity.
- Will FTTx, together with wired access, be unbundled?

Obviously owners of fixed lines are applying pressure to prevent the introduction of more unbundling regulations that would allow third parties to use new optical and hybrid infrastructures.

However, there is a significant difference between what is happening on both sides of the Atlantic regarding regulations. In the US recent decisions are backing companies that are investing in infrastructures, while in Europe regulations affecting fibre optics are similar to those affecting the copper local loop.

It is also significant that the US has suppressed some of the unbundling obligations of the owners of cable infrastructures. According to the FCC the intention is to stimulate the investment in new FTTx infrastructures, then new access infrastructures will be offered at a reasonable cost to competitors but not at a regulated price. This decision is supported by the fact that alternative operators have already had time to develop their own business and access networks.

In Europe, new directives will abolish all regulatory distinctions between networks, telephone and the Internet, by placing all services and networks into a single, all-encompassing, regulatory category called 'electronic communications'. The consequence is that some European countries have opened up sub-loop access and optical access as well. These differences explain why the European Commission and National Governments are under pressure from the dominant operators to adopt the FCC model and keep new access technologies unregulated. The incentive is to justify the massive investments that will be required. So far, the result is heterogeneous, while in Germany BNetzA has permitted Deutche Telecom to close its VDSL-FTTN to competitors, while in the UK the regulator OFCOM has opened up sub-loop access.

1.5.3 Network Convergence

Despite the widespread availability of nx64 circuits, IP and Ethernet, two packet-oriented technologies, have been selected to build the Triple Play network (see Figure 1-18). The statistical multiplexing, typical of packet technologies, has important advantages. The first is cost, as they are much cheaper to roll out and maintain compared with circuit networks based on SDH. The second is that IP and Ethernet are easier to manage and have a lot of synergy with the Internet, which is based on the same principles. But probably the ability to implement any new service based on voice, data and video is even more important than the benefits mentioned above.

1.5.3.1 Is the IP Protocol Ready to do it?

IP is considered the best strategy to adopt in the deployment of the new converged services and particularly television, which has for a long time been offered exclusively by cable, terrestrial and satellite operators. The IP protocol differentiates telco portfolios from competitors. Native IP-centric infrastructures that were developed for data transport must be transformed into a multiservice platform also enabled to transport audio and video. To achieve this important investments are required.

The TCP/IP protocol was designed in 1983 and immediately adopted by the US Department of Defense to connect heterogeneous hosts. It has demonstrated itself to be very robust and suitable for managing large and complex topologies. Requiring a minimum of human intervention. The Internet that we know, developed during the 1990s, is an architecture that provides universal connectivity between heterogeneous but open subsystems. TCP/IP protocols are designed to automatically discover topologies and addresses by means of nodes and protocols that are continuously interchanging routing information.

IP networks are robust enough to maintain data flows between hosts but do not necessarily use the shortest or the most efficient route. It means that packet delay is unexpected and can even vary during the time a session is active. In IP networks what really matters is keeping datagrams flowing from source to destination whilst ensuring that they are independent of any events that can affect individual nodes. They are, in some respects, fault tolerant by nature.

Legacy TCP/IP architectures are best effort solutions that are good enough to transport data but are inappropriate for supporting voice or television which demand, not only an accurate QoS control, but also 99.999% availability, high performance and protection mechanisms. To achieve this the layered TCP/IP suite has adopted a number of protocols to emulate the predictable but inefficient circuits based on SDH. The result can be even better and more robust than legacy services when the right architecture is adopted (see Figure 1-22). Elements like SIP, RTP, RTCP, VPLS or GFP are only some of the protocols that can help IP to implement carrier-grade networks to support isochronic applications based on data, and also voice and video/TV.

1.5.3.2 Ethernet at Layer 2

Ethernet has a number of features that have made it the favourite technology for implementing architectures to support Triple Play services. Ethernet scores high in a combination of features like efficiency, simplicity, scalability and cost. It is also important that Ethernet is the technology used in the vast majority of customer premises and service provider installations.

Ethernet is efficient as it is packet-oriented, therefore it obtains the statistical multiplexing gain when transporting independent traffic flows over shared transmission medias. Ethernet is also very simple to set up and maintain, especially when compared with SDH-SONET installations. Other important considerations are the number of engineers and technicians, probably millions, that are confident with Ethernet and all its associated devices and protocols.

Ethernet is designed to be used with many types of optical and metallic media, and different bit rates. Transmission ranges and bandwidths are equivalent to long-haul

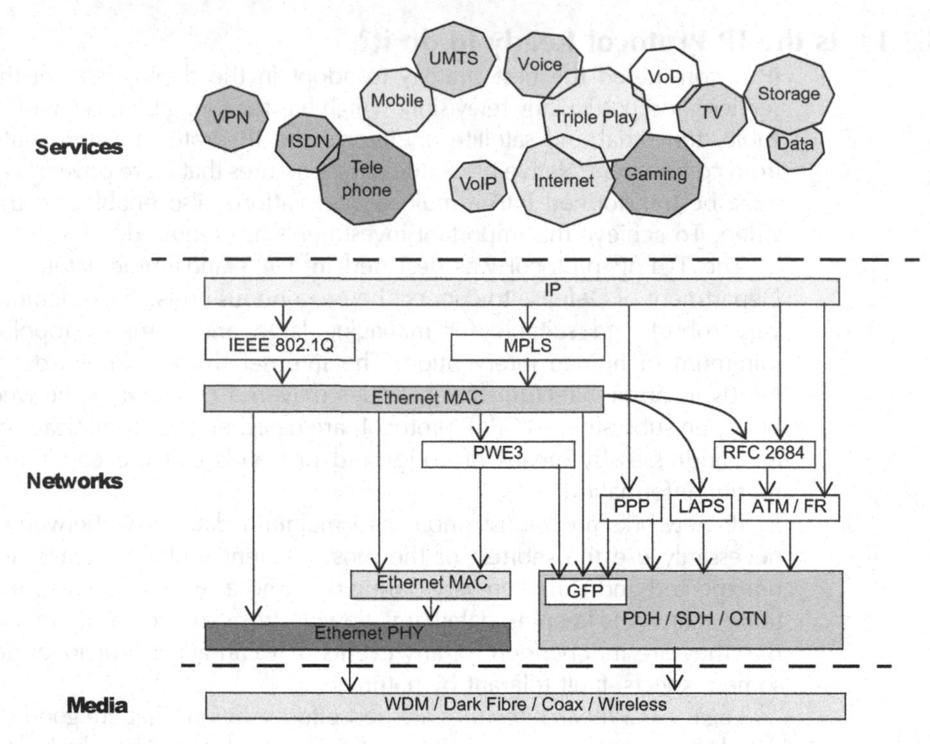

Figure 1-22. Applications, services and protocols.

technologies. Being easily scalable from a few Mbit/s up to many Gbit/s Transmission ranges and bandwidths are equivalent to long-haul technologies (see Table 1-4), it is therefore possible to migrate existing LANs, MANs and WANs to Ethernet using the existing physical media.

1.5.3.3 Ethernet Drawbacks

Unfortunately, native Ethernet lacks some essential functions necessary to supply 'carrier-class' services. Features like reliability, management, rollout, maintenance and QoS are much more demanding in Metro networks supporting Triple Play than in LANs where Ethernet is focused on data transport. Scalability can also be an issue. Ethernet switches work very well when the number of hosts connected is limited and during low traffic conditions, but as soon as the installation grows it tends to degrade in performance, QoS, security and availability.

When Ethernet is extended beyond the LAN, several architectures can fulfil the requirements, including Dark Fibre, DWDM/CWDM, NG-SDH (see Figure 1-23). In principle, all of these architectures are able to support the carrier Ethernet service classes like E-Line, E-LAN and E-Tree, nevertheless some are more appropriate than others.

Depending on the requirements of the applications, budget, customer profile, installed base, optical infrastructure and capacity are more important factors to consider.

Table 1-4. IEEE 802.3 Ethernet versions

	Interface	Media type	Mbps	FDX	Encoding		MFS	Reach
					Data	Symbol		
Ethernet IEEE 802.3a-t	10BASE-2	One 50 Ohm thin coaxial cable	10	H	4B/5B	Manchester	64	<185 m
	10BASE-5	One 50 Ohm thick coaxial cable	10	H	4B/5B	Manchester	64	<500 m
	10Broad36	One 75 Ohm coaxial (CATV)	10	H	4B/5B	Manchester	64	<3600 m
	10BASE-T	Two pairs of UTP 3 (or better)	10	H/F	4B/5B	Manchester	64	<100 m
	10BASE-FP	Two optical MMF passive hub	10	H	4B/5B	Manchester	64	<1000 m
	10BASE-FL	Two optical MMF asyn hub	10	F	4B/5B	Manchester	64	2000 m
	10BASE-FB	Two optical MMF sync hub	10	H	4B/5B	Manchester	64	<2000 m
Fast Ethernet IEEE 802.3u	100BASE-T4	Four pairs of UTP 3 (or better)	100	H	8B/6T	MLT3	64	<100 m
	100BASE-T2	Two pairs of UTP 3 (or better)	100	H/F	PAM5×5	PAM5	64	<100 m
	100BASE-TX	Two pairs of UTP 5 (or better)	100	H/F	4B/5B	MLT3	64	<100 m
	100BASE-TX	Two pairs of STP cables	100	F	4B/5B	MLT3	64	200 m
	100BASE-FX	Two optical MMF	100	F	4B/5B	NRZI	64	2 km
Gigabit Ethernet IEEE 802.3z/ab	1000BASE-CX	Two pairs 150 Ohm STP (twinax)	1000	F	8B/10B	NRZ	416	25 m
	1000BASE-T	Four pair UTP 5 (or better)	1000	H/F	4D-PAM5	PAM5	520	<100 m
	1000BASE-SX	Two MMF, 850 nm	1000	F	8B/10B	NRZ	416	500/750 m
	1000BASE-SX	Two MMF, 850 nm	1000	F	8B/10B	NRZ	416	220/400 m
	1000BASE-LX	Two MMF, 1310 nm	1000	F	8B/10B	NRZ	416	< 2 km
	1000BASE-LX	Two MMF, 1310 nm	1000	F	8B/10B	NRZ	416	<1 km
	1000BASE-LX	Two SMF,1310 nm	1000	F	8B/10B	NRZ	416	5 km
	1000BASE-ZX	Two SMF, 1310 nm	1000	F	8B/10B	NRZ	416	80 km
10G Ethernet IEEE 802.3ae	10GBASE-SR	Two MMF, 850 nm	10000	F	64B/66B	NRZ	N/A	2–300 m
	10GBASE-SW	Two MMF, 850 nm	10000	F	64B/66B	NRZ	N/A	2–33 m
	10GBASE-LX4	Two MMF, 4 × DWM signal	10000	F	8B/10B	NRZ	N/A	300 m
	10GBASE-LX4	Two MMF, 4 × DWM signal	10000	F	8B/10B	NRZ	N/A	300 m
	10GBASE-LX4	Two SMF, 1310 nm, 4 × DWM	10000	F	8B/10B	NRZ	N/A	10 km
	10GBASE-LR	Two SMF, 1310 nm	10000	F	64B/66B	NRZ	N/A	10 km
	10GBASE-LW	Two SMF, 1310 nm	10000	F	64B/66B	NRZ	N/A	10 km
	10GBASE-ER	Two SMF, 1550 nm	10000	F	64B/66B	NRZ	N/A	2–40 km
	10GBASE-EW	Two SMF, 1550 nm	10000	F	64B/66B	NRZ	N/A	2–40 km
Ethernet First Mile IEEE 802.3ah	100BASE-LX10	Two SMF, 1310 nm	100	F	4B/5B	NRZI	N/A	10 km
	100BASE-BX10	One SMF, 1310 (U)/1550 (D)	100	F	4B/5B	NRZI	N/A	10 km
	1000BASE-LX10	Two SMF, 1310 nm	1000	F	8B/10B	NRZ	N/A	10 km
	1000BASE-LX10	Two MMF, 850 nm	1000	F	8B/10B	NRZ	N/A	550 m
	1000BASE-BX10	One SMF 1310 nm (U)/1490 nm (D)	1000	F	8B/10B	NRZ	N/A	10 km
	1000BASE-PX10	One SMF, PON 1310 nm (U)/1490 nm (D)	1000	F	8B/10B	NRZ	N/A	10 km
	1000BASE-PX20	One SMF, PON 1310 nm (U)/1490 nm (D)	1000	F	8B/10B	NRZ	N/A	20 km
	10PASS-TS	One or more telephone pairs	10	F	64/65-octet	T1.424	N/A	750 m
	2BASE-TL	One or more telephone pairs	2	F	64/65-octet	G.991.2	N/A	2.7 km

List of acronyms: H: half-duplex; F: full-duplex MFS, minimum frame size in bytes; N/A, not applicable; MMF, multimode fibre; SMF, single mode fibre; U, up stream; D, down stream

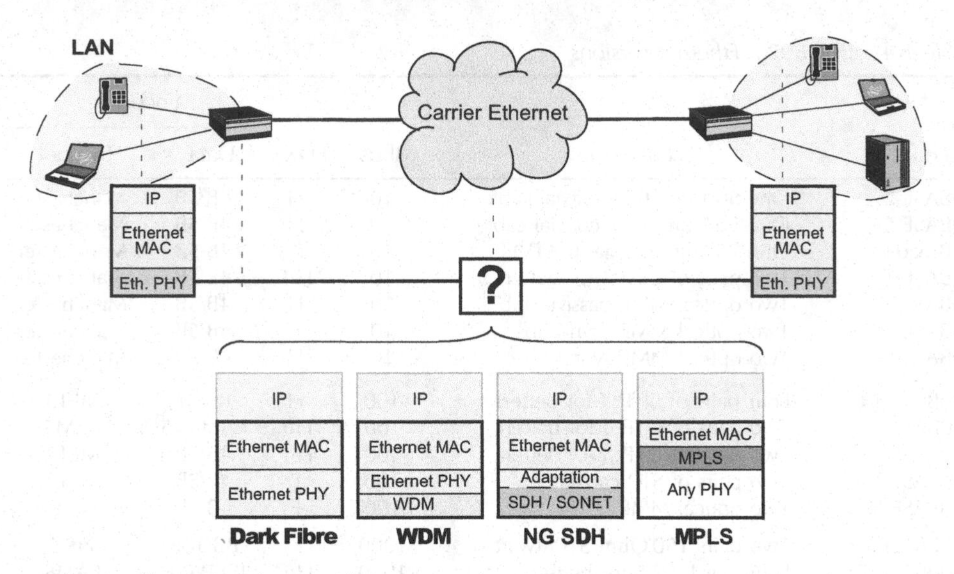

Figure 1-23. Alternatives for providing carrier Ethernet services.

1.6 Triple Play Market

Important growth that can be expected in the residential market for Triple Play services in the US has been calculated at 32 million subscribers annually (Yankee Group August 2006), with an average operator spending rate of about $4000 per subscriber. Pyramid research is not so optimistic, calculating a world market of 35 million subscribers by 2010 (see Figure 1-24). According to Gartner research, the current average combined monthly spend that includes fixed voice, Internet and TV in Europe is €93.70. This roughly matches Fastweb reports in Italy of obtaining an ARPU of €900 a year so the ARPU can be increased by 100% after the adoption of bundled services.

It is necessary to warn that low revenue expectations and the high investments needed for Triple Play could result in significant losses for telecom operators that in some cases can cause an initial cumulative loss higher than €2000 per subscriber (see Figure 1-25).

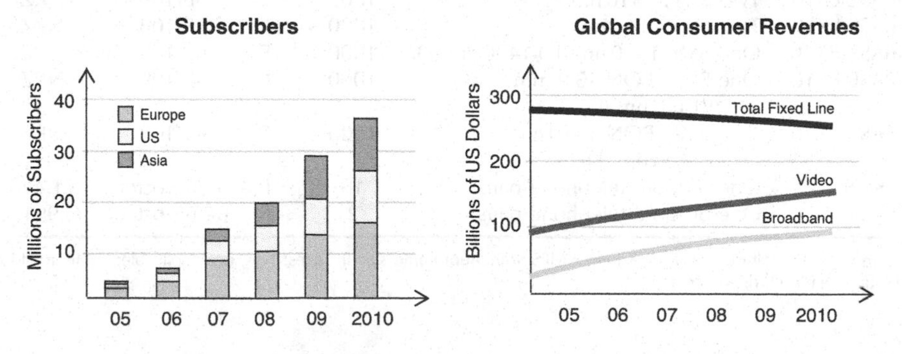

Figure 1-24. Triple Play subscribers and revenues prediction. Reproduced by permission of Trend Communications Ltd.

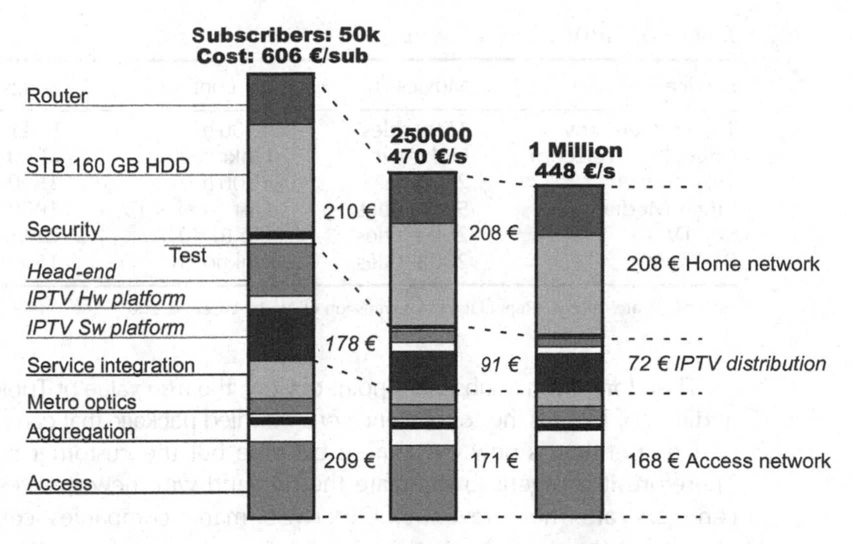

Figure 1-25. Triple Play deployment cost and the scale economies. Reproduced by permission of Alcatel-Lucent © 2007.

Caution is necessary. Several of the most relevant European cable companies providing Triple Play services are not growing quickly, and in some cases, are even losing market share. Put quite simply, many potential subscribers are not interested in having one provider for the three services (*Les Echos* 26 September 2006 and others).

1.6.1 Warning: No Immediate Profit is Expected

1.6.1.1 Providers' View

The size of the broadcast and cable TV market has been calculated to be about 25% of the whole telecom business, and this is a good reason for telcos to go into this business (see Figure 1-9). But the provision of TV has a major impact on telecommunications architectures. Massive investments are required to transform legacy networks into multiservice platforms.

This is the million dollar question. Will service providers get any benefits at all? Triple Play's main message is 'more services for less money' (see Table 1-5). How is it really possible for operators to increase revenues? If the answer is not clear for Triple Play operators it is even more difficult for single service operators. How can they compete against an efficient multiservice network that can deliver any type of application and also integrates mobiles? This is a good point to consider: the cost of not embracing a Triple Play project.

1.6.1.2 Customers' View

The growing enthusiasm for bundling video with existing services demands high investments but customers are not willing to pay a lot for new TV channels since it is a mature service already supplied by broadcast, cable, digital, terrestrial TV and satellite (see Figure 1-10).

Table 1-5. IPTV Customer offering[a]

Service provider	Movies	TV contents	Songs	Other
T-Com Germany	1500 titles	1500 h	Unknown	Sport services
Chunghwa Telecom	Unknown	Unknown	6000	Karaoke
Telefonica	200 titles	1500 h	1600	Sport services
Virgin Media	500+ titles	Last week's TV	1000+	Adult services
Sky TV	200+ titles	Up to 40 h	Unknown	Sport services
BT-Vision	200+ titles	Unknown	150+	15+ games

[a]*Source: Alcatel-Lucent*. (Reproduced by permission of Alcatel-Lucent © 2007)

Therefore, from a subscriber point of view, the true value of Triple Play is not exactly the addition of TV; it is the convenience of a bundled package that can be simpler and cheaper.

Competition is now even more extreme but the customer base remains the same. Therefore it is urgent to stimulate the demand with new services and applications that can generate more revenue. Otherwise many companies could be at risk of not recovering their investments.

1.7 Conclusions

Services bundling and network convergence are different facets of the same strategy, since it is not possible to offer a competitive package, with many varied services, without a network that reduces delivery costs and simplifies the management.

Bundling is the strategy to adopt in order to compete and differentiate the portfolio in mature telecoms markets. Telcos with mobile and fixed networks are moving towards a wholesale business by integrating separate divisions. Single service operators are signing up to alliances with other operators to enrich their services to offer a more comprehensive portfolio that would ideally include wireless, fixed access, voice, video and data. This is also the case for some fixed line operators that may also offer wireless access or those mobile operators that have opened up their network to offer wire based services.

Network convergence is a trend that most operators are having to face (see Figure 1-26). But, will this strategy produce profits? The question is how to bundle multiple services for less money while combining both mobile and fixed networks.

It is still too early to say if Triple Play will be commercially successful. It is a new strategy with many technical challenges and scalability problems ahead. Cable operators are not taking any risks at the moment because they are already one step ahead, as they have already added broadband access and voice to television, their core business. But we should not underestimate the greater variety of services that telcos can offer with the flexible IP-based architectures.

Regarding Triple Play rollout, there are a number of challenges to fulfill. Voice, television, video and broadband access all have different traffic profiles and place different requirements on the network that delivers these services. Voice and television are affected by jitter, while packet loss has a greater effect on video and data services. Consequently, accurate quality of service mechanisms are necessary to develop and control the delivery of Triple Play.

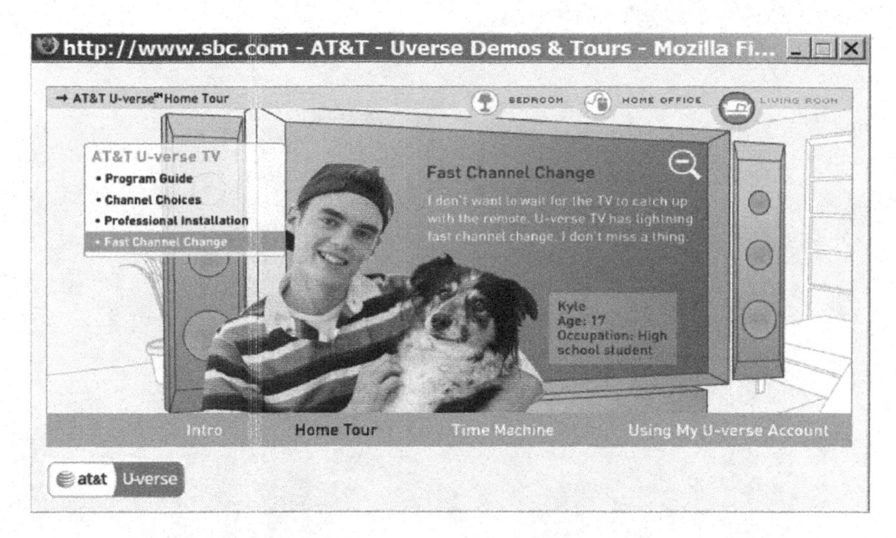

Figure 1 26. Fading phone business pushes companies in a new direction: Triple Play, like AT&T and its U-verse service offers IPTV, VoD, broadband access (wired and wireless), music downloads, gaming, integrated video recording, integrated ISP portals, VoIP, IP video conferencing and file sharing.

In some ways we can find certain similarities with old convergence strategies that happened in the industry during the last few decades. Some of them were successful, like the ISDN convergence of voice and data, but others for example ATM, were not at all able to steal the limelight on the convergence stage. Another recent case of a failed convergence has been 3G, when wireless European operators paid a fortune to operate 3G networks that would also deliver a bundle of voice, data and video services. It must be frustrating when the new 3G services are sold as just offering low-cost calls.

Therefore companies embracing bundling and convergence should take precautions to avoid repeating the same errors.

Selected Bibliography

[1] International Telecommunications Union - Development (ITU -D). Key Global Telecom Indicators for the World Telecommunications Service Sector.

[2] Justin Paul. 2007. How to get ahead in IPTV. A guide to setting up an IPTV Service. Alcatel-Lucent.

[3] TeleGeography Research, a division of PriMetrica, Inc.

[4] *Documentos de Referencia*. Comisión del Mercado de las Telecomunicacions (CMT).

[5] Peter Macaulay. *Broadband Technology Update*. DSL Forum.

[6] Fundacion Telefonica. *La Sociedad de la Informacion 2007*. Editorial Ariel, ISBN: 978-84-08-07798-5.

[7] Vanessa Alvarez. *Triple Play Services Drive Demand for Integrated network testing Solutions, September 2006*. Yankee Group Research, Inc.

[8] Aditya Kishore and Margo De Boer. *Lennar Study 2006*. Yankee Group Research, Inc.

Selected Bibliography

Chapter 2: IP Telephony

Telephony is the technique that enables transmission of human speech over long distances with the help of electromagnetism.

The first days of telephony were dominated by analogue signal transmission and frequency division multiplexing (FDM) of multiple voice signals. Digitalization was a major advance in telephony. By means of digitalization, voice signals, that are continuous by nature, are represented by discrete sequences of symbols belonging to a finite alphabet. Digital signals have many advantages: they can be stored or replicated without limit, they are resistant against noise and they can be efficiently encrypted, compressed and filtered by digital signal processors (DSPs).

The pulse code modulation (PCM) can be used to represent a voice signal by a sequence of ones and zeros. This fact is very important because, historically, PCM was the first modulation used in digital telephony. PCM signals are well suited to time division multiplexing (TDM). The first digital telephony systems, based on PCM and TDM, were introduced in the sixties.

Currently, the world of telephony is facing a new revolution: deterministic TDM is being replaced by statistical TDM which is based on packet switching. The high availability and low cost of IP equipment makes this technology a popular alternative for transporting voice when compared with ATM or FR. Reliability and quality of IP telephony, or VoIP, have improved over the last few years and today it is a fast growing industry (see Figure 2-1).

Triple Play: Building the Converged Network for IP, VoIP and IPTV Francisco J. Hens and José M. Caballero

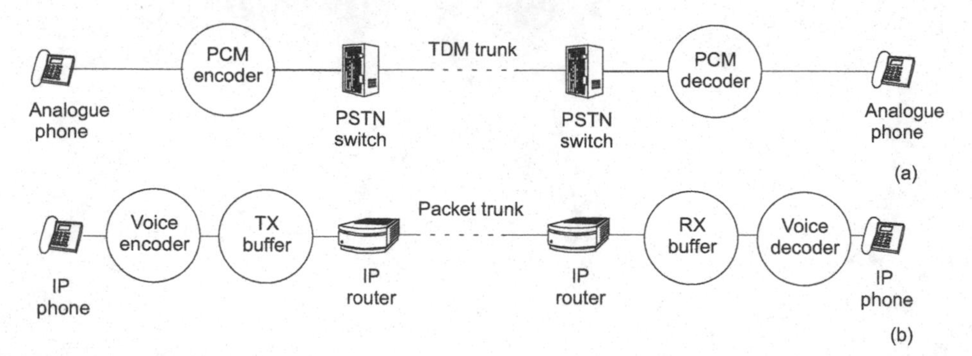

Figure 2-1. Reference models for telephony services. (a) Classical telephony reference model. An analogue signal is generated by a telephone, digitalized with a PCM encoder and transported in TDM trunks. (b) IP telephony reference model. Voice is encoded, digitalized, assembled in packets and transported in packet trunks. Buffering is necessary to deal with variable delay.

IP telephony is a new way to provide a service that is more than one hundred years old. From this point of view, IP telephony alone does not bring anything new to subscribers. They could not even notice migration to IP transport infrastructures. However, IP telephony can potentially bring benefits to subscribers, service providers and carriers:

- Subscribers benefit from lower call costs because operators can avoid paying interconnection charges. IP telephony gives new service providers a competitive edge over incumbent operators because it gives them more and more flexible deployment options for their services. It enables new service providers to avoid many of the regulatory barriers of standard voice provision. Of course, it is thought that subscribers will also benefit from this scenario of increased competition.
- For carriers, IP telephony brings the ability to integrate voice and data over the same network infrastructure. This makes installation and management tasks easier and cheaper.
- It is accepted that VoIP will cut the benefits that service providers will get from telephony but, on the other hand, IP telephony offers new opportunities by means of a new class of services that go beyond classical telephony such as videotelephony, integration of mobile and fixed telephony or telelearning. It is expected that many of the business opportunities for service providers will come from the possibility of bundling classical and innovative services together.

From the technical point of view, the main advantage of IP telephony is that it enables the convergence of data and voice. But convergence is not possible without QoS aware IP networks. During the last few years a lot of effort has been concentrated on designing the architecture of such networks (see Paragraph 7.2). However, most of today's IP networks, including the Internet, remain best-effort networks.

Given the current state of the art of IP technology, integrating time-sensitive and other traffic into Internet backbones is not necessarily a good idea if the carrier is to offer VoIP with quality approaching that of PSTN. Deployment of a dedicated IP network to carry voice is the option chosen by some operators. For them, integration of voice and data is left to a second stage, once the network has been prepared to deal with different traffic classes.

Both integrated and dedicated VoIP networks suffer from QoS problems that were not present in classical telephony. Due to queuing effects, delay in IP networks is higher and far less predictable than in deterministic TDM networks. To correct the effects of variable delay, some traffic has to be buffered by the receiver before sending it to the decoder. This buffer is in fact a dejittering filter that must never become full or empty during the telephone call to avoid losing data or leaving the decoder without voice samples. The need for a dejittering buffer increases the over-all delay experienced by the voice samples. Larger buffers favour jitter absorption but they increase the end-to-end delay. This trade off between delay and jitter filtering capabilities of a VoIP receiver is an important fact in IP telephony. For non interactive applications such as IP radio there is no compromise and dejittering buffers can be larger without any noticeable effect experienced by the end user.

2.1 Coding of Voice Signals

Transmission of digital voice signals involves at least two steps:

1. *Sampling*: the analogue voice signal obtained from a microphone is converted into a sequence of discrete samples.
2. *Coding*: every sample is assigned to a symbol taken from a finite alphabet. Usually, symbols are bit words.

The result from the previous two operations is a bit stream that can be transmitted by a digital channel. To be efficient, the sampling rate of the voice sampler must be as low as possible without compromising the quality of the stream. Something similar happens with the encoding process. The average number of bits generated for every sample must be as small as possible.

Voice sampling and coding take advantage of how the sound is perceived by our auditory organs and how voice is generated. Another point to be taken into account is the required quality level of voice for telephony applications. Telephony users want a clear and easily understandable voice. They will probably value the ability to understand the message and identify the speaker but other features such as high fidelity or stereophonic sound would probably be unnecessary.

Human hearing is able to perceive sounds between about 20 Hz and 20 kHz, but most telephony standards only require the transmission of frequencies between 300 and 3400 Hz. While this is only an small sub-band within the band of all audible sounds, it contains about 80% of the sound information from original voice signals. For transmission systems, adapted for music transmission like FM radio, the whole audible band is required to be transmitted, but for telephony, between 300 and 3400 Hz is considered enough.

Standards for legacy digital telephony systems specify PCM coding of voice signals. PCM is simple and well suited to TDM transmission but things are different in IP technology. Packet transmission can take advantage of variable bit rate codecs. An example of a voice codec for packet transmission is one that only generates data when users are speaking but it does not generate information during silent intervals. The available range of voice codecs for IP telephony is broader than for TDM telephony. Voice codecs for packet transmission can be classified into three different families:

1. *Waveform codecs* encode the voice signal without taking into account how the signal was generated. They try to quantify the voice samples in such a way that the signal decoded by the receiver resembles the original signal as much as possible. They produce high quality output with low coding delay, while maintaining low complexity, but they are not well suited to low bit rate transmission. Another advantage of these types of codec is that they are useful for the transmission of non voice signals such as dual-tone multifrequency (DTMF) digits.
2. *Source codecs* or *vocoders* attempt to emulate the way speech is produced. The parameters of the model are sent to the receiver instead of the waveform itself. Voice is modelled by an excitation passing through a linear predictive filter that models the vocal tract. Human speech sounds can be voiced like an 'a' or unvoiced like an 's'. The way these sounds are generated is fundamentally different. This is the reason why two different excitations are needed. Vocoders break down the voice signal into short chunks of a few milliseconds and then find the parameters of the model that include a flag to choose the excitation type, the parameters of the excitation and the coefficients of the linear filter for each chunk. This usually involves fewer bits than waveform transmission. However, voice synthesized by the receiver sounds non natural and it is difficult to identify the speaker and therefore vocoders are hardly ever used for telephony applications. The linear predictive coding (LPC) is the best-known source codec.
3. *Hybrid codecs* attempt to provide the best of waveform and source codecs. They try to match the waveform of the original signal but at the same time they use models based on how speech is generated by humans. Hybrid codecs are complex and they usually produce a meaningful coding delay. Bit rates obtained from them are a cross between vocoders and waveform codecs but the performance often approaches that of source codecs.

The availability of many voice codecs has the advantage that service providers and subscribers are free to choose the best for their needs. In VoIP applications, more than one codec may be used in the same end-to-end connection since cellular, broadband access and PSTN networks usually require different codecs. These transcoding operations degrade the voice quality. Analysis of distortion caused by transcoding is complex and the penalty paid for the end quality may be appreciable.

2.1.1 Pulse Code Modulation

Pulse code modulation (PCM) is a waveform coding technique that assigns every sample from the signal to a number represented as a binary word. The PCM coding is specified in ITU-T Recommendation G.711 for transporting digital telephony signals in the PSTN.

The chosen sampling frequency for voice applications has been always 8 kHz, slightly more than twice the maximum frequency contained in telephony voice signals (3400 Hz). This is in line with what is required by the Nyquist theorem that states that an analogue

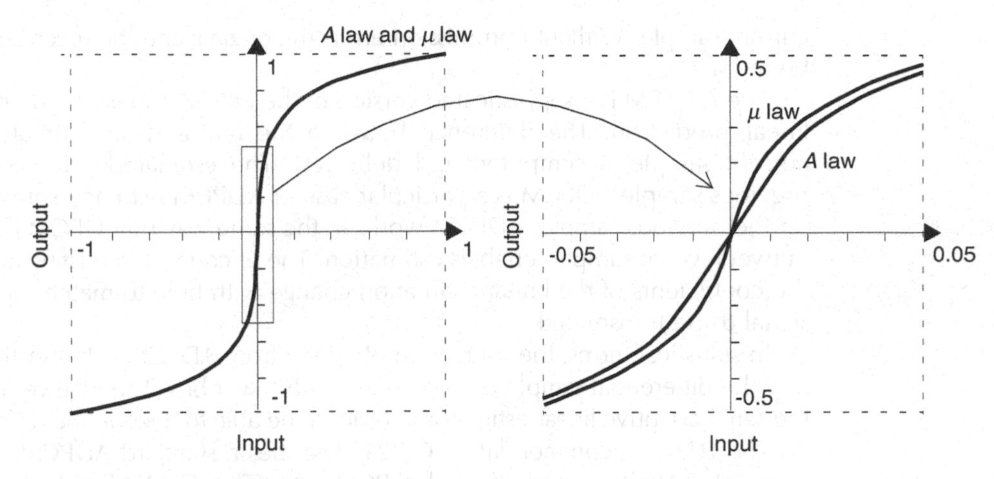

Figure 2-2. Comparison of the A law and the μ law. The comparison curves are nearly identical.

signal of bandwidth B sampled with frequency of at least $2B$ can be recovered without error from the sampled version. On the other hand, every sample of the voice signal is represented by an 8 bit word as specified in G.711. This gives a transmission rate of 64 kbit/s.

The 64 kbit/s is the base for all TDM multiplexing hierarchies and 125 μs, the inverse of the sampling frequency, is the duration of all common TDM frames.

The amplitude of a voice sample is a real number but there are only 256 possible 8 bit words. That means that quantization involves an error that is perceived by the receiver as random noise. The quantifier is chosen to minimize this error. Specifically, G.711 quantizers are not uniform. In the US and Japan the quantifier follows the μ law; in most other countries it follows the A law. The shapes of the μ and A laws are nearly identical and neither of them is uniform (see Figure 2-2).

Statistical analysis shows that smaller amplitude values of voice samples are more likely to occur than larger values. This is the reason because the shapes of the μ and A law quantifiers are designed to give more accurate codifications of samples with small amplitudes. Another reason is that human hearing is more sensitive to variations in weak sounds and it is therefore practical to minimize quantification error in these samples.

2.1.2 Adaptive Differential PCM

The adaptive differential PCM (ADPCM) coding is an evolved version of the basic differential PCM (DPCM). Both ADPCM and DPCM are examples of waveform codecs. With DPCM encoding the difference between two consecutive samples is quantified and delivered rather than individual voice samples as in PCM.

DPCM works because voice signals are not completely random. There is some correlation between samples taken close in time. When computed, the difference between two consecutive samples increases the chance of obtaining a value that is smaller in absolute value than the individual samples. The difference can be quantified with fewer bits than in the PCM case and still maintain good performance. At the receiving end, the previously decoded sample and the received DPCM sample can be used to compute the current sample. Without transmission errors the original and the received sequence must be the same.

The ADPCM is a sophisticated version of the DPCM and is based on the concept of linear prediction. The difference between the real and an estimated value of the current sample is computed and delivered. The estimated value is obtained from previous samples. DPCM is a particular case of ADPCM with an estimation based only on the previous sample. ADPCM works in the same way that DPCM does. Correlation between voice samples enables estimation. The encoder is said to be adaptive because the coefficients of the linear estimation change with time to match the statistics of the signal to be transmitted.

In statistical terms, the estimation obtained from ADPCM is better than from DPCM and the differential sample can be coded with fewer bits. The receiver must implement the same adaptive linear estimator in order to be able to decode the differential samples.

The ITU-T Recommendation G.721 describes a standard ADPCM codec to deliver voice at 32 kbit/s, one-half of the PCM rate. The G.721 has been superseded by Recommendation G.726 that encodes voice signals at 16, 24, 32 or 40 kbit/s.

2.1.3 Code-excited Linear Predictive

Code-excited linear predictive (CELP) codecs are examples of hybrid codecs. Specifically, CELP codecs are examples of analysis by synthesis (AbS) codecs, the most important family of hybrid codecs. They are based on the same modelling principles as LPC. Both have a linear predictive filter to model vocal tract, but models for AbS codecs are more complicated than the model for vocoders because they involve more than two excitations. AbS codecs have a collection of excitations for the linear predictive filter. The encoder finds the output of the filter for all the possible excitations and then computes the distance between the original and the synthetic signal. Finally it sends the coefficients of the filter and a pointer to the excitation that constitutes a better approach to the original signal, to the receiver.

There are many implementations and variations of the CELP codec. Some of them are included in standards for their use in IP telephony applications.

Low-delay CELP (LD-CELP) has been standarized in ITU-T Recommendation G.728. It works with five samples simultaneously that are sampled at 8 kHz. It finds optimum filter coefficients and excitation from a codebook that contains 1024 items. LD-CELP receivers are able to use previously received information to compute the current filter coefficients and, therefore, all that is necessary is to send is a pointer to the excitation signal. This information can be coded in 10 bits since there are 1024 excitations. The resulting bit rate is 16 kbit/s with good voice quality compared with for example, 16 kbit/s ADPCM. LD-CELP introduces a delay of five samples. This result in a delay smaller than 1 ms that is not noticed by the human ear.

Another codec based on a variation of CELP is called conjugate structure CELP (CS-CELP) and is standarized in ITU-T Recommendation G.729. The CS-CELP algorithm works with blocks of 80 samples sampled at 8 kHz. For these blocks the codec computes optimum excitation, filter coefficients and gain parameters and assembles 80 bit frames that are sent to the destination. This results in a bit rate of 8 kbit/s. The coding delay of the G.729 codec is 15 ms.

Table 2-1. *Extensions to the basic G.729 codec*

Standard	Purpose
G.729A	Simplified codec compatible with G.729 but with slightly worse performance
G.729B	Algorithm specifications for voice activity detection (VAD), discontinuous transmission (DTX) and comfort noise generation (CNG). These algorithms are used to reduce the transmission rate during silence periods of speech'
G.729D	Lower bit rate (6.4 kbit/s) version of the basic algorithm for environments with reduced available bandwidth
G.729E	Increased bit rate (11.8 kbit/s) version of the basic codec specially adapted for environments with background music at the input

Recommendation G.729 contains various annexes that further enhances the performance of the basic codec adaptation operation to specific environments (see Table 2-1).

2.1.4 Other Codecs

Although the codecs discussed above are the most common and important, they are just a small selection of the available alternatives (see Table 2-2). An important subset of voice codecs are those defined for digital mobile communications. Mappings into RTP exist for most of these codecs (see Section 4.1). This make codecs for wireless voice available for VoIP applications as well.

2.2 Network Performance Parameters

Quantities such as signal-to-noise ratio (SNR), bit error ratio (BER) or end-to-end delay give information about network performance. However, it is difficult to infer what the users' perception of the network will be when based only on these ratios. The converged network offers heterogeneous services to the subscribers. The perception users have of the network depends on the service but SNR, BER or delay are service-agnostic.

Service-agnostic performance parameters are still useful but their relationship must be investigated together with the perception that users receive from the service. In other words, the values of performance parameters must be interpreted for a given service.

Table 2-2. *Other voice codecs*

Codec	Rates	Main applications
A-CELP (G.723.1)	5.3 kbit/s	VoIP
MP-MLQ (G.723.1)	6.3 kbit/s	VoIP
CDMA Q-CELP	6.2 kbit/s, 13.3 kbit/s, others	IS-95 CDMA wireless systems
GSM FR	13 kbit/s	GSM wireless systems
GSM EFR	12.2 kbit/s	GSM wireless systems
AMR	from 4.75 to 12.2 kbit/s	GSM and 3G wireless systems

For voice, two main QoS parameters are important: packet loss and one-way delay. Delay variation is less important because it can be filtered by the receiver at the price of increased one-way delay. However, delay variation may potentially cause buffer overload or underload at the receiver when the buffering mechanisms are not well dimensioned.

2.2.1 Packet Loss and VoIP

Packet loss (see Section 6.2.3) causes distortion in voice applications that increases as the packet loss ratio increases. When the overall quality of the voice service is represented in terms of the mean opinion score (MOS) (see Section 2.3), the MOS is seen to decrease as packet loss increases (see Figure 2-3).

The ability to keep the quality high with occasional lost packets depends on the following points:

- *The distribution of the lost packets*. Uniform distribution of lost packets tends to be easily absorbed by the decoders. Bursty distributions cause deeper loss of performance.
- *The packet loss concealment (PLC) algorithm in use*. PLC algorithms provide packets for the decoder when the actual packets are lost. A PLC may insert silence or background noise, repeat the last packet received or use interpolation techniques to replace lost packets.

In general terms, packet loss has minimum impact on voice quality if the packet loss ratio is less than 1%. A good PLC helps by keeping quality for loss ratios to about 3%. Higher loss ratios cause severe degradation to the perceived quality. Another factor to take into account is the effect of packet loss depending on the codec. PCM codecs exhibit more tolerance to lost packets than more complex voice compression algorithms. Specifically the G.711 codec is more tolerant to errors than the G.729 and G.723.1 codecs.

2.2.2 Delay and VoIP

As one-way, end-to-end delay increases in the telephony application, switching between the roles of talker and listener becomes more difficult. This undesired effect degrades the

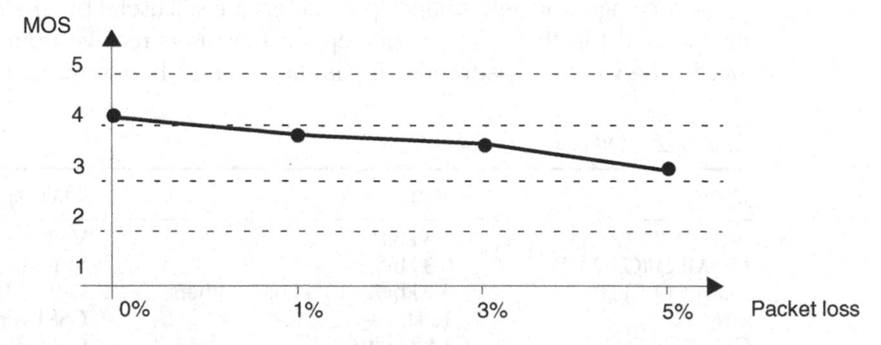

Figure 2-3. Performance degradation of the voice signal expressed in terms of the MOS as a function of packet loss. The considered coding is PCM (G.711 Appendix I) and packet length is 20 ms.

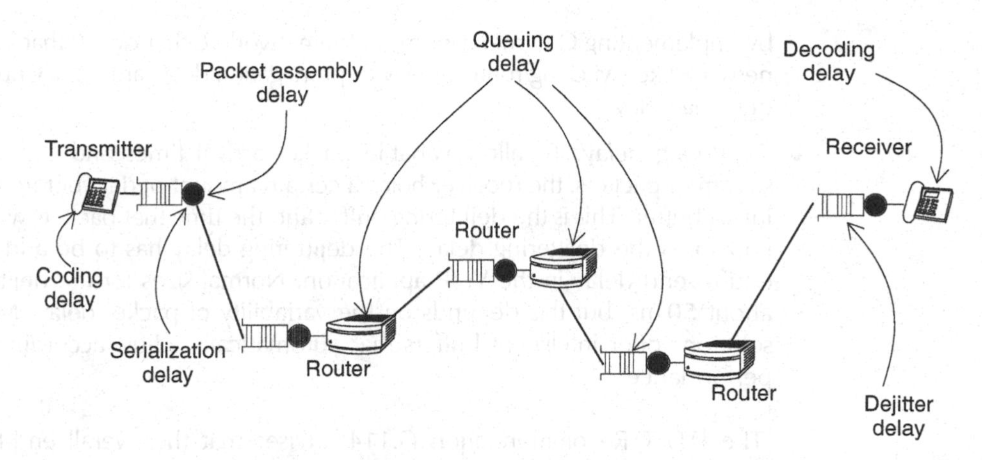

Figure 2-4. Delay sources in the VoIP service.

quality of all types of interactive communications and must therefore be controlled. Dealing with delay is one of the biggest challenges of IP telephony.

The delay experienced in IP telephony uses to be higher than for traditional telephony. The reason is that delay is increased by new elements not present in circuit-switched networks (see Figure 2-4). These are the more meaningful sources of delay for VoIP:

- *Coding/decoding delay*. Sophisticated codecs achieve high compression of the speech data but pay the price of delay. There are two different contributions to the coding delay: the processing delay and the lookahead delay. The former is the time it takes to process a voice frame, the latter is the time the decoder waits for future data to exploit correlation with the frame currently being processed, to achieve better compression. We have to add to the coding delay, the decoding delay, which used to be about one half of the coding delay. The overall coding/decoding delay can be as little as a few tenths of a millisecond. One of the important voice encoders, the G.729 CS-CELP, has a coding delay of 15 ms of which 10 are processing delays and five are lookahead delays. For the same algorithm, the typical decoding delay is 7.5 ms.
- *Packet assembly delay*. Many voice encoders generate very short frames. The G.711 encoder generates an 8 bit frame every 125 μs and the G.729 generates an 80 bit frame every 10 ms. In order to increase the transmission efficiency the transmitter may choose to fill every IP packet with various voice frames. This improves the ratio of user vs overhead bytes but results in increased delay before transmission. For example, the packet assembly delay to generate an IP packet with 5 G.729 frames would be 50 ms.
- *Serialization delay*. This is the time it takes to place one packet on the transmission line. It depends on the size of the packet to be transmitted and the transmission rate in the access link (see Section 6.2.1). For a 100 byte frame and a 64 kbit/s line, the serialization delay would be 12.5 ms.
- *Queuing delay*. Queuing delay appears when the network is experiencing congestion (see Section 6.2.1). Queuing delay is variable and difficult to predict but it can be minimized

by implementing QoS mechanisms in the network. Other delays that happen within the network like switching/routing delay or propagation delay are usually not as critical as the queuing delay.

- *Dejittering delay*. To allow variable packet arrival times and still achieve a steady stream of packets, the receiver holds a certain amount of data before playing them out into a buffer. This is the dejittering buffer and the time that packets wait in the buffer is known as the dejittering delay. The dejittering delay has to be added to the overall end-to-end delay in the VoIP application. Normal sizes for the dejittering buffer are about 50 ms, but this depends on the variability of packet delay. Many commercial solutions offer intelligent buffers that automatically adjust according to the network performance.

The ITU-T Recommendation G.114 advises that the overall end-to-end delay for voice communications can accumulate up to 150 ms without being noticed by the communicating parties. When delay exceeds 150 ms negative effects gradually appear. After calculating all of the possible delay contributions, it seems that achieving the performance level stated in G.114 is not trivial in IP telephony.

The 150 ms delay limit given by G.114 is simple, clear and enables the design of IP telephony systems. However, it has been seen in practice that is difficult to establish a clear limit to acceptable delay in telephony systems. Some authors advise that 200 ms is still a good figure and some others confirm that up to 400 ms can be accepted in long distance calls. An important reason that makes it difficult to agree on the delay limit, is that the degradation effects of delay depend on a new parameter: the echo.

In traditional telephony, the main sources of echo are imperfections caused by hybrids. Hybrids are circuits that adapt four-wire systems used within the telephone and two-wire systems used in the local loop. Hybrids are not needed when all the communicating parties are directly connected to an IP network. In other cases hybrids are still used and may cause echo in VoIP communications but even without hybrids echo may appear caused either by the coupling between the speaker and microphone in the telephone or crosstalk.

It has been seen that echo and delay cause interdependent perception of degradation in voice applications. This interdependence can be summarized as follows: echo sounds louder to speakers as delay between original signal and replication increases. In other words, echo may remain unnoticed when delay is controlled but is perceived by the speaker as a degradation when delay is high enough (see Figure 2-5).

A popular solution to echo, both in traditional and VoIP systems, is echo cancellers based on DSPs that can reduce the strength of echo by up to 25–30 dB.

2.3 Opinion Quality Rating

Opinion quality rates for voice provide information about how the telephony service is operating by taking opinions from users as an input. These ratings have two objectives:

1. *Qualify voice distortion that is due to encoding, decoding and transcoding operations*. To measure codec distortion it is necessary to find a test scenario where any other

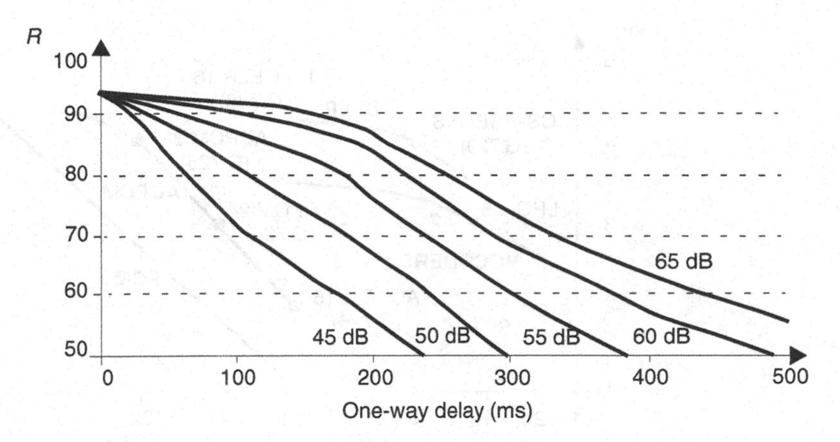

Figure 2-5. Incidence of echo on voice quality expressed in terms of the R factor that measures the quality of the service on a scale of 100 points. The numbers next to the curves represent the Talker Echo Loudness Rating (TELR) measured as the overall two-way attenuation of the voice signal, measured on the path between the transmitter and the receiver.

degradation can be neglected, for example, by directly chaining the encoder and the decoder.

2. *Qualify the service globally*. Thanks to this capacity, opinion rates are often considered very useful when rating voice services. The inconvenience is that they are service-specific. In other words, service performance values for audio may not be useful for video or data applications.

It is worth noting that none of the above mentioned tasks can be achieved with QoS parameters such as delay or packet loss ratio only.

The basic opinion rate for voice is computed by averaging the subjective perception of users. A group of users is requested to rate the quality of the voice in a telephony service as 'excellent', 'good', 'fair', 'poor' and 'bad'. The opinion of each user is mapped into a numeric scale (see Table 2-3), and then the arithmetic mean is computed from all the values obtained from the group. The number given by this methodology is called the mean opinion

Table 2-3. MOS scale

Value	Meaning
5	Excellent
4	Good
3	Fair
2	Poor
1	Bad

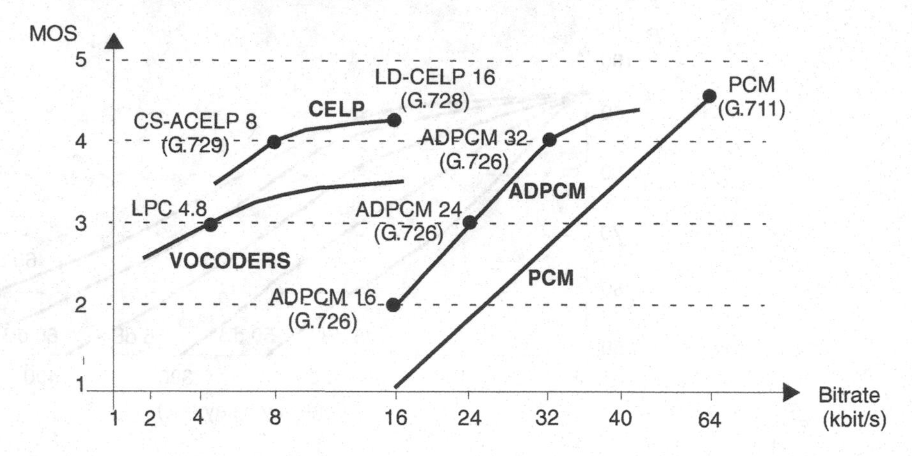

Figure 2-6. Performance of different voice encoders as a function of the bit rate accomplished and the MOS score.

score (MOS). This metric is an attempt to build an objective parameter from many individual, subjective opinions by means of statistical mechanisms.

The MOS is defined under ITU-T Recommendation P.800. It is part of a family called absolute category ratings (ACRs) because quality is rated without a reference. ACRs are opposite to degradation category ratings (DCRs), which are based on the comparison between quality perceived by users in the received voice signal and a reference signal. It is possible to compute a parameter similar to the MOS but that takes degradation ratings as inputs instead of absolute ratings. This new parameter is referred to as degradation MOS (DMOS).

When evaluating the MOS it is useful to distinguish between listening quality and conversational quality. The latter involves evaluation of interactivity in a two-way conversation. Encoding and transmission techniques may be designed to minimize voice distortion, but they may neglect end-to-end delay. If this happens, the listening MOS will not be affected but conversational MOS will show this degradation.

The MOS is very useful to compare the performance of different voice codecs (see Figure 2-6). Over all codecs commonly found in practice, the G.711 for PCM at 64 kbit/s offers the best score. Other codecs such as the G.728 or G.729 have a score that is only slightly worse than the PCM, but they are more efficient in terms of bit rate.

2.4 Objective Quality Assessment

Evaluation of the MOS is complex, time-consuming and expensive. This is the reason why, in practice, direct evaluation of the MOS is replaced by estimations performed by machines or algorithms rather than asking people to rate the service. Algorithmic rating methods are said to be objective because they do not rely on opinions but many of them attempt to estimate the MOS, a parameter obtained from averaging individual opinions.

Depending on the input data they take, objective quality assessment algorithms can be classified into the following groups:

- *Opinion models*: these take as inputs a variety of factors, including delay and other network and terminal performance parameters and produce estimates of conversational MOS. The most important and widely used opinion model is the E-model.
- *Speech-layer models*: require speech signals as inputs and they then generate estimates of listening MOS.
- *Packet-layer models*: in this case the required inputs are IP packets and the output is again an estimate of the listening MOS. Packet-layer models are useful for monitoring speech quality at intermediate network points, where speech signals are not available, but they are necessarily limited because the perceived service performance depends on the network and the terminal features, but the characteristics of the terminal are unknown from intermediate points.

2.4.1 The E-model

The E-model is a computational model that uses transmission parameters to predict the subjective quality of voice. This model was proposed for standarization in the nineties and it is based on an NTT model called OPINE. Today, the E-model has been standarized in the ITU-T Recommendation G.107. It has also been adopted as a network planing tool by the European Telecommunications Standards Institute (ETSI) and the Telecommunications Industry Association (TIA).

It provides results in terms of the *R*-factor that rates the quality of the speech on a scale of between 0 for terrible and 100 for perfect. The *R*-factor values can be theoretically mapped into the MOS scale (see Figure 2-7). Due to its limitations, 94 is the maximum value for *R* attainable by conventional PCM voice over circuit switched networks. For VoIP systems, the normal values for the *R*-factor are usually worse.

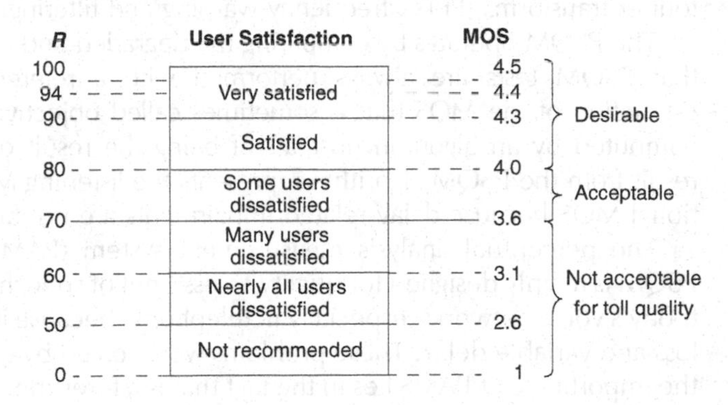

Figure 2-7. Relationship between the R-factor and the MOS, according to G.107 Annex B and G.109.

A basic assumption of the E-model is that the effects of individual degradations can be summed to take account of the psychological perception that all these degradations cause to the user. The *R*-factor is calculated as follows:

$$R = R_0 - I_s - I_d - I_e + A \tag{2.1}$$

where R_0 is based on the SNR, which considers background and electrical noise, I_s captures impairments due to quantizing noise, sidetone and other effects that happen simultaneously with the voice signal generation. The I_d and I_e terms quantify the effects of delay related impairments (including echo) and distortion of voice due to coding and lost packets within the network. Finally, The factor *A*, called the advantage factor, accounts for the willingness of users to accept substandard performance on a new service.

It has been seen that the E-model consider the effects caused by delay and loss of interactivity in voice communications. This is the reason why this model predicts the conversational MOS rather than the listening MOS.

2.4.2 Speech-layer Models

Speech-layer models make estimates of the voice quality perceived by the users taking the voice signal as an input. The first and most basic speech-layer quality evaluation tool is the SNR defined as the ratio between the received voice and noise signal powers. Although simple, the SNR, defined as a ratio of physical quantities, does not consider the psychological effects of impairments. This is the reason why more sophisticated parameters were designed.

The first accurate speech-layer model was the perceptual speech quality measure (PSQM), which is the basis for the ITU-T Recommendation P.861. When it was standarized the PSQM was tested against other algorithms in order to find out which of them provided maximum correlation with opinion ratings.

The idea behind the PSQM is to measure degradation due to distortion in a 'psychoacoustic domain' rather than in the more conventional time or frequency domains. This involves mapping signals into a domain where computing distortion. This mapping needs proper cognitive modelling and involves a sequence of complex operations such as fast fourier transforms (FFTs), frequency warping and filtering.

The PSQM operates by comparing the degraded and the original signals. That means that PSQM tests are always performed with a reference. The algorithm gives an estimation of the MOS that is sometimes called objective MOS (OMOS) because it is computed by an algorithm instead of being the result of subjective perceptions. The result from the PSQM algorithm represents the listening MOS rather than the conversational MOS because delay-related impairments are not taken into account.

The perceptual analysis measurement system (PAMS) goes beyond PSQM. The PSQM is mainly designed for quality assessment of speech codecs but it is unsuitable for today's voice networks, especially IP telephony, because it is unable to deal with packet loss and variable delay. These problems were solved by the PAMS algorithm. Currently the importance of PAMS lies in the fact that is a forerunner of the Perceptual Evaluation of Speech Quality (PESQ) algorithm, the latest and most accurate of the speech-layer objective quality assessment models. In fact, the PESQ combines the excellent time alignment algorithm of PAMS and the psycho-acoustic model of PSQM+ (extension of the PSQM).

The PESQ, as it has been standardized in ITU-T Recommendation P.862, follows this sequence to estimate the performance of the voice signal:

1. It corrects the level of the degraded and the reference signals. After this operation both signals have the same average power.
2. Degraded and reference signals are individually processed by a filter that emulates the receiving device (hand set, loudspeaker or headphones).
3. The two signals are time-aligned to compensate for time shifts.
4. In order to account for the distortions actually perceived by a human listener, degraded and reference signals are transformed into the psycho-acoustic domain by an auditory transformer.
5. Psycho-acoustic representations of both signals are subtracted and the difference is processed by a block that models human hearing (cognitive model).

As a result of this sequence of operations an estimation of perceived voice quality is obtained that can be mapped to the MOS scale.

2.5 Market Segments

The success of IP telephony is favouring the development of commercial solutions specially designed for various market segments from PC to PC voice communications over the Internet, to solutions for service providers intending to replace circuit-switched with packet-switched infrastructures. The main IP telephony market segments are listed below:

- *Single user solutions*: these are the simplest VoIP solutions. They are for users who want to replace or complement costly PSTN calls with low-cost VoIP calls. These are often software-based solutions integrated into multimedia-enabled PCs.
- *Enterprise solutions*: these constitute a market segment directed at companies interested in integrating their voice communications into their IP infrastructure. Most of the currently available solutions for enterprises are based on modified routers with VoIP gateway and Private Branch eXchange (PBX) features.
- *Carrier solutions*: these enable carriers who are providing voice services to a large quantity of residential customers or businesses over a packet based infrastructure. Current VoIP solutions for carriers are based on large residential and trunk VoIP gateways, sophisticated QoS-enabled routers and intelligent signalling modules.

2.5.1 Single User Solutions

There are popular VoIP software solutions that enable voice communications between remote users across IP networks such as the Internet. Some of these software solutions are based on open standards like H.323 or SIP. Furthermore, some sites in the Internet act as SIP proxies or PSTN gateways and offer presence and other useful services at low cost or even free.

The main advantages of these solutions are cost and simplicity. However they do not provide performance in line with a carrier class service like traditional telephony does. Specifically, today's most popular software-based solutions exhibit QoS and availability problems because they use the best-effort Internet as their transport infrastructure.

2.5.2 IP Telephony in Enterprise Networks

IP Telephony products for enterprises exist for companies who want to migrate their infrastructure for voice communications to IP. The new IP infrastructure can be their own Enterprise data network, often based on Ethernet, or WAN data links and circuits provided by an external service provider. Sometimes advantages of VoIP are limited to external calls while internal calls are still delivered through existing POTS or ISDN-based equipment and cabling.

VoIP-enabled IP routers are the main component for some enterprise solutions. One example is the Cisco solution that also uses the CallManager software for signalling and call control of Cisco integrated solutions or third party developments. Another approach to Enterprise VoIP is to upgrade or replace legacy PBXs with VoIP PBXs. The solution-based VoIP router is more focused on integral IP deployments without legacy telephony equipment and cabling. The solution based on VoIP PBXs should be considered, if taking advantage of low cost VoIP calls while maintaining existing internal POTS or ISDN infrastructures is required.

VoIP enabled routers or PBXs must be capable of delivering calls from/to the PSTN, over a virtual private network (VPN) or over the LAN, depending on the destination. For example, IT technicians may choose to deliver internal calls over the existing LAN, calls directed to a remote subsidiary may be delivered to a VPN configured between delegates of the same company and, finally, all other voice calls can be routed to the PSTN (see Figure 2-8). The call routing process must be transparent to users. They do not need to be aware of the underlying transport technology for their calls.

A number of issues appear when voice traffic has to share transmission resources with data traffic generated by other applications. Since VoIP applications are sensitive to delay and jitter, voice performance damage may result. On the other hand, VoIP traffic can potentially degrade performance of data applications. Operation bandwidths are 100 or 1000 Mbit/s in most current LANs but VPN links are built over WAN links, usually with smaller bandwidths per subscriber. That means that it is likely that VPN links can become a bottleneck for the traffic. To deal with this issue, it would be necessary to implement traffic differentiation, customized treatment of traffic classes, resource management and other QoS mechanisms into the corporate network.

Another issue, regarding VoIP communications in Enterprise Networks, is security. By default, voice traffic is not encrypted by most of the applications and may potentially be sniffed and heard by anybody. VPN links are usually encrypted to provide basic confidentiality for data communications and therefore there is no need to perform modifications here, but IT personnel must consider the need to implement security mechanisms in the LAN links in order to provide confidentiality within the local network.

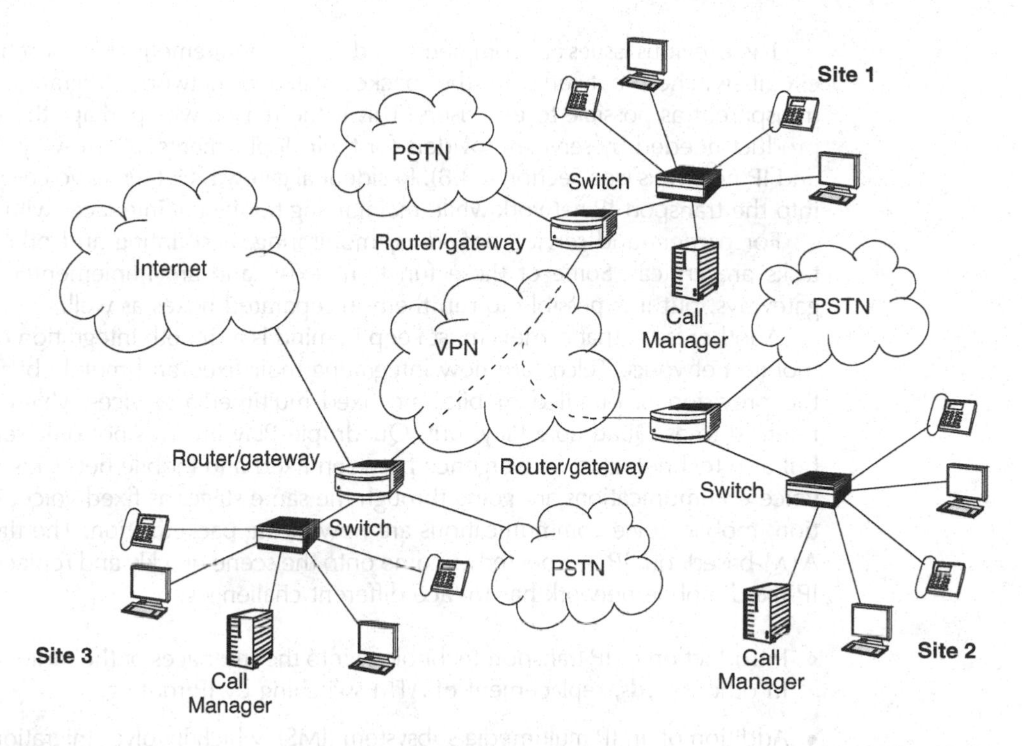

Figure 2-8. VoIP for enterprises. IP telephones are connected to the company IP-based data network. Actual transport of voice calls becomes transparent to users. The system is able to route the calls to the best transport network (IP-based or PSTN) according to the destination and other established rules.

Furthermore, problems arise when the transmission of VoIP calls through a fire-wall is required due to limitations of protocols for transporting packetized voice and signalling (see Section 4.3.7). These limitations must be considered and solutions based on STUN, TURN or other mechanisms must be implemented.

2.5.3 IP Telephony in Service Provider Networks

IP telephony products for carriers are designed to provide voice services to the general public and businesses. There are two important requirements that any carrier-class solution for IP telephony must address:

1. *Interworking*: IP telephony has its own signalling protocols based on SIP or other standards. The implications of interworking between VoIP signalling and traditional signalling must be studied for each particular deployment.
2. *QoS*: the performance provided to subscribers must be at least as good as the performance given by traditional telephony. This involves careful planning and dimensioning, deployment of QoS mechanisms in the transport network or even segregation of the packet-switched voice network and the data network if necessary.

The previous issues are complemented by the requirement of a smooth transition from circuit-switched networks to the packet-switched network. Migration should be as transparent as possible to end users. This is the reason why perhaps the most important product needed by service providers for their deployments, is gate-ways between PSTN and IP networks (see Section 4.3.8). Residential gateways integrate voice communications into the transport IP network while maintaining traditional interfaces with subscribers.

For carriers and service providers, monitoring, accounting and other management tasks are critical. Some of these functions exist, and are implemented in the market gateways, but it is possible to run them in separated boxes as well.

Another issue that carriers must keep in mind is a smooth integration of the fixed and mobile networks. Telcos are now integrating their fixed and mobile businesses to ease the provision of bundled mobile and fixed multimedia services which are sometimes referred to as Quadruple Play. But, Quadruple Play involves not only service bundling, but also technological convergence between fixed and mobile networks. In fact, mobile voice communications are going through the same stages as fixed voice. After digitalization, mobile voice communications are now facing packetization. The first solutions are ATM-based, but IP is expected to come onto the scene quickly and replace ATM. The all-IP fixed/mobile network has to face different challenges:

- Introduction of IP transport technology into the interfaces of the radio access network. In other words, replacement of ATM switching by IP routing.
- Addition of an IP multimedia subsystem (IMS), which involves migration of all services and signalling to packet-based infrastructures. This includes voice, video, data and mobility aspects (for example, handoff).

Selected Bibliography

[1] Minoli D., Minoli E., *Delivering Voice over IP Networks*, 2nd edition, John Wiley and Sons, 2002.

[2] Durkin J. F., *Voice-Enabling the Data Network: H.323, MGCP, SIP, QoS, SLAs, and Security*, Cisco Press, 2003.

[3] Black U., *Voice Over IP: 2/e*, Prentice Hall, 2002.

[4] Hardi W.C., *VoIP Service Quality*, McGraw-Hill, 2003.

[5] Collins D., *Carrier Grade Voice Over IP*, McGraw-Hill, 2003.

[6] Rao K.R., Bojkovic Z. S., Milovanovic D. A., *Multimedia Communication Systems: Techniques, Standards, and Networks*, Prentice Hall, 2002.

[7] Smith, S.W., *The Scientist & Engineer's Guide to Digital Signal Processing*, California Technical, 1997.

[8] Scheets G., Parperis M., Singh R., Voice over the Internet: A Tutorial Discussing Problems and Solutions Associated with Alternative Transport, *IEEE Communications Surveys*, Vol. 6, No. 2, 2004.

[9] Bo Li, Mounir Hamdi, Dongyi Jiang, Xi-Ren Cao, QoS-Enabled Voice Support in the Next-Generation Internet: Issues, Existing Approaches and Challenges, *IEEE Communications Magazine*, April 2000, pp. 54–61.

[10] Hassan M., Nayandoro A., Internet Telephony: Services, Technical Challenges, and Products, *IEEE Communications Magazine*, April 2000, pp. 2–9.

[11] James J. H., Bing Chen, Garrison L., Implementing VoIP: A Voice Transmission Performance Progress Report, *IEEE Communications Magazine*, July 2004, pp. 36–41.
[12] Johnson C. R., Kogan Y., Levy Y., Saheban F., Tarapore P., VoIP Reliability: A Service Provider's Perspective, *IEEE Communications Magazine*, July 2004, pp. 48–54.
[13] Takahashi A., Yoshino H., Kitawaki N., Perceptual QoS Assessment Technologies for VoIP, *IEEE Communications Magazine*, July 2004, pp. 28–34.
[14] Xiaoyuan Gu, Dick M., Kurtisi Z., Noyer U., Wolf L., Network-centric Music Performance: Practice and Experiments, *IEEE Communications Magazine*, June 2005, pp. 86–93.
[15] Nguyen T., Yegenoglu F., Sciuto A., Subbarayan R., Lockheed M., Voice over IP Service and Performance in Satellite Networks, *IEEE Communications Magazine*, March 2001, pp. 164–171.
[16] Baldi M., Risso F., Efficiency of Packet Voice with Deterministic Delay, *IEEE Communications Magazine*, May 2000, pp. 170–177.
[17] Markopoulou A. P., Tobagi F. A., Karam M. J., Assessing the Qouality of Voice Communications Over Internet Backbones, *IEEE Transactions on Networking*, Vol. 11, No. 5, October 2003, pp. 747–760.

Chapter 3: Audiovisual Services

Video and television applications, that delivered the thought technologies the Internet made popular, are seen by telcos as an alternative business to compensate the declining revenues obtained by traditional voice and data applications. Nevertheless, audiovisual services are, so far, new for telcos since previous experiences were no more than transparent transport of TV signals operated by broadcasters. Consequently, telcos are in front of a totally new business case that must be defined to target mainly consumers of entertainment services. Telcos should also learn how to manage contents in order to offer attractive channels that could compete with existing TV operators, including cable, terrestrial and satellite broadcasters.

Audiovisual services are a big market; the demand exists, and therefore it will not be necessary to create it. That is a good news, particularly after the WAP and 3G failures, but the question is, is there room enough for a new type of TV operator? We do not know yet, but it is important to bear in mind three points:

1. First, consumer's willingness to watch TV is declining, the new generation are more captivated by the Internet and digital gaming platforms.
2. Second, TV revenues are not as big as may be supposed; in Western Europe the telecom business is about four times the size of the whole television business.
3. Third, and probably the most important, innovative TV applications that combine television and Internet can create a new market.

Triple Play: Building the Converged Network for IP, VoIP and IPTV Francisco J. Hens and José M. Caballero

IPTV service will only be successful if it is able to differentiate itself from the existing broadcasters in quality and innovative features. Generally talking, IPTV strategies should go for a premium service as it has so many possibilities to define a new television paradigm, binding audiovisual signals with Internet tools to go beyond the traditional TV. An IPTV programme has many possibilities to be enriched by web sites, video-clip retrieval, promotional items, collaterals or chats, which can be perfectly embedded and offered during the programme showtime.

New entrants must find a way to differentiate their service, using the possibilities of the bidirectional features of the IP protocol and the processing capacity of the distribution network. This will be a true IP television (IPTV[1]), a new audiovisual service in which subscribers enjoy the ability to control when and where they receive or download programmes and videos, while combining Internet-like interactions.

3.1 Digital Television

There are two basic models to deliver digital TV, the unidirectional one used by broadcasters and the bidirectional adopted by IPTV operators:

1. *Unidirectional model*. All TV channels are multiplexed and sent simultaneously over the transmission media to receivers. Subscribers have to tune to the programme they want to watch. This is the model adopted by terrestrial, cable and satellite operators.
2. *Bidirectional model*. Each channel is streamed individually in IP packets to those subscribers that have previously requested it. This model, which relies on the Internet suite of protocols, is more complex to deploy, but it is more flexible, and allows the development of the new interactive applications.

3.1.1 The Internet and Television

IPTV is made up of a flexible combination of two major elements, the IP suite of protocols and digital television:

1. The IP suite of protocols is responsible for packaging TV signals that will be routed from the head-end to the customer's site. The IP protocol also provides interactivity features between subscribers, network, service and content providers.
2. Digital television is responsible for the audio, video, data compression and transmission formats according to standards, such as MPEG, that manage image resolution and programme arrangements.

---

[1]When the acronym IPTV is used alone, it generally refers to all the applications that deliver audiovisual programmes by means of IP packets including both live (TV) as well as stored VoD and personal video services (PSV). When IPTV is used in a context together with VoD and PSV, it would refer exclusively to the multicast in real-time television services.

Once we have arrived at this point, it is important to clarify that currently the public Internet cannot support real-time television services for several reasons. *First*, the Internet basically is a best effort network that cannot permanently guarantee the delivery of television with an appropriate quality of service. *Second*, there is not enough bandwidth for standard or high definition TV. *Third*, there are some protocols and multicasting methodologies that are not supported.

The underlying network for IPTV distribution must be a converged network, IP-centric, QoS-enabled, and based on the multicast distribution and delivery of the television signals. This requirement today can only be fulfilled by privately managed IP networks.

3.2 Questioning the IPTV Business Models

Commercial video services delivered over a converged network should match existing services on quality, price and contents because while it is true that a demand for TV exists, this is a very mature and competitive market. Telcos must understand that this is a completely different case than when mobile telephony was launched; cellular phones were a totally new product and customers were willing to pay much higher fees. Mobile telephony was an extraordinary success story, whereas 3G was a complete disaster, at least from a commercial point of view.

In order to minimize the possibility of failure, telcos must define a business model that will help to maximize the chances of writing a success story, by studying the strengths and weaknesses, plus opportunities and threats (see Figure 3-2).

3.2.1 Strengths

IPTV provides a set of interesting possibilities thanks to the combination of digital television and Internet-like features, such as interaction between the subscribers, network and services providers. This iteration is, at the end of the day, the key differentiator from broadcast television.

3.2.1.1 Interaction

The audiovisual services based on IP have the maximum flexibility to combine television with interactivity. IPTV can be a simple replication of broadcast TV and video, but can also add the interaction typical of the Internet. Interactive TV opens an interesting, and still unexplored, dimension that lies in between traditional TV and the Internet.

Imagine how different television could be if there was interaction between groups of subscribers, or between subscribers and content providers during the emission of TV magazines, or pay-per-view programmes, for instance, creating a closed group of users that could interact with each other, sending messages while watching a favourite sport match in different locations.

3.2.1.2 Subscriber Profiles

Operators may build user profiles that can define and change dynamically depending on user preferences. Experiences have demonstrated that there are different profiles of subscribers;

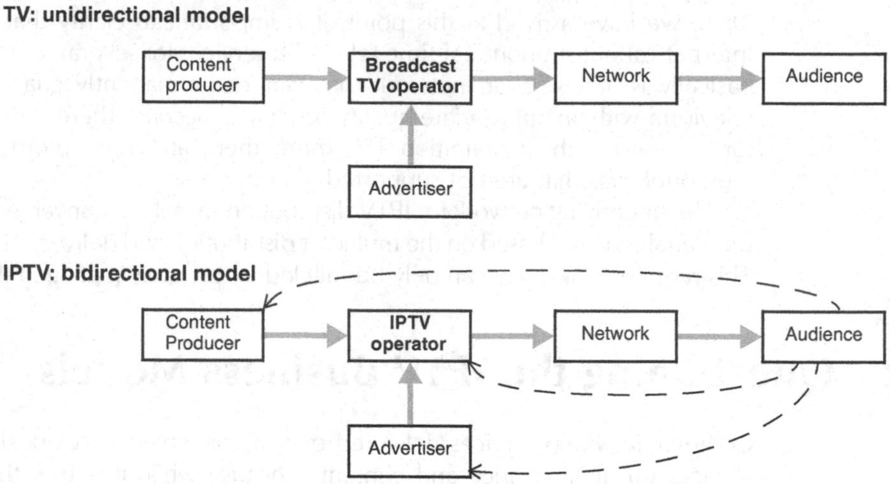

TV: unidirectional model

Content producer → Broadcast TV operator → Network → Audience
Advertiser → Broadcast TV operator

IPTV: bidirectional model

Content Producer → IPTV operator → Network → Audience
Advertiser → IPTV operator

Figure 3-1. In the traditional broadcast TV model there is no direct interaction with the audience, but the IPTV model permits a closer interaction with the audience.

some are passive, others look for specific contents, others enjoy interation. Then, it is possible to establish a direct relationship with the subscribers that would be totally impossible to do in broadcast TV (see Figure 3-1).

Customized information adapted for each profile will allow, for instance, each subscriber to receive reports about specific sport events, films or programmes. It would be also possible to send publicity only to those customers matching specific profiles. This point is particularly interesting for advertisers, since the impact of publicity on TV has been declining.

3.2.2 Opportunities

Bidirectionality is probably the main strength of IPTV, and can be transformed into an opportunity to develop new interactive applications involving all the active agents of the system. Interaction can create additional interest for users and could eventually be the way to capture more audience and increase profits, but this is not a unique opportunity for IPTV (see Table 3-2).

3.2.2.1 Customized Adverts

The interactive IP network enables a large number of opportunities. Moreover, the IPTV industry believes that targeted advertising based on subscriber profiles and interaction will generate a major source of revenue (see Table 3-1).

In IPTV the whole system is able to receive feedback about particular user preferences, which is fundamental to customize adverts and will allow providers to increase publicity fees. It is possible to know exactly the preferences by programme, audience peaks and duration, helping advertisers to target each segment of consumers better. Furthermore, it can be possible to establish direct channels of communications between consumers, advertisers and service providers.

Table 3-1. What do you think will be the principal revenue sources for IPTV?

	Minor	Moderate	Major
Network access fees	45%	40%	15%
Subscription fees for basic content	33%	40%	27%
Subscription fees for premium content	16%	44%	40%
Pay per view	19%	53%	28%
Advertising – mass market	30%	40%	30%
Advertising – targeted	16%	37%	47%

Source: Accenture. (Reproduced by permission of Accenture)

3.2.2.2 Unlimited Channels/Unlimited Videos

Another interesting opportunity that the IP network can offer is that the number of TV channels and copies of the same video that can be offered to the customers are theoretically unlimited. This fact is based on how the channels are distributed and videos are recorded and delivered. We will see how the IPTV distribution of programmes is based on multicast agents, and the videos distribution is based on unicast addressing, depending on the type of application, but a broadcast addressing mode is never used. This strategy permits exploitation of the resources in a very controlled way, maximizing the delivery of the information only to those customers that have subscribed to it.

3.2.2.3 Personal Services

The video and television market is changing; customers are willing to use personal video services (PVS) to save video clips in a network server or to transfer videos between PCs, TV devices or mobiles, and IPTV is perfectly prepared to offer it.

3.2.3 Weaknesses

To create real demand for audiovisual services delivered over IP will not be enough, bundling television with broadband access and phone services. The long-term success requires to a mature business to combine attractive content with the interaction inherent to IP protocol.

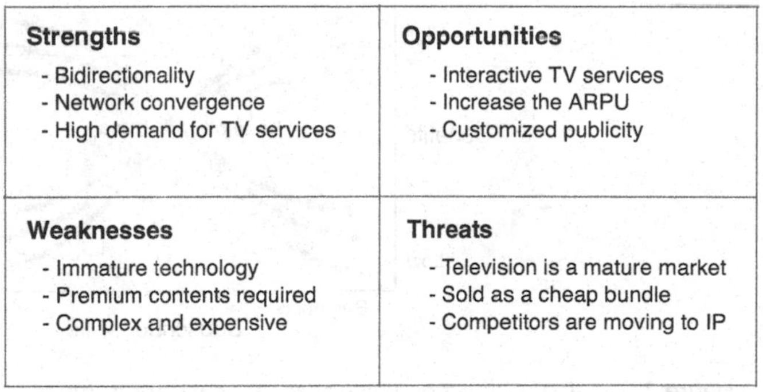

Strengths	**Opportunities**
- Bidirectionality - Network convergence - High demand for TV services	- Interactive TV services - Increase the ARPU - Customized publicity
Weaknesses	**Threats**
- Immature technology - Premium contents required - Complex and expensive	- Television is a mature market - Sold as a cheap bundle - Competitors are moving to IP

Figure 3-2. IPTV business model. Opportunities and threats.

3.2.3.1 Quality Content

Access to quality content is fundamental to entering and staying in such a mature market. Telcos must acquire not only new technical and business skills, but must also be involved in content creation that is adapted to their specific consumer market. This explains why some telcos are creating their own studios or have signed contracts to find attractive programmes, movies and specialized productions from film and programme makers' studios.

3.2.3.2 Complex and Expensive

Interaction increases the complexity of the middleware, and the user interfaces as well; consequently software and the whole platform may become difficult to use and unstable, which would be even worse. In other words, more investment in middleware does not necessarily mean more revenues (see Figure 3-3).

3.2.3.3 High-definition TV

High-definition TV (HDTV) must be planned and be considered a natural evolution of the whole business rather than a differentiator, since cable TV, satellite and terrestrial digital TV are already moving forward in this direction.

3.2.4 Threats

The whole IPTV business is not mature. The main threat may occur if providers do not discover their specific market segment. If eligible subscribers buy IPTV only because it is a cheap bundle of multiple services that also includes television, then the whole triple play service will quickly become a commodity with tiny margins. This would be a similar case to ADSL, where self-installing kits are common to save installation costs.

3.2.4.1 Profitability

Assuming that operators can get an important mass of subscribers and supply them with the proper contents and interactive facilities, it is not clear yet that it can always become a profitable business. Research on the economics of mature cable markets indicates that about a third of cable networks deployed during the last decade are not profitable. How can we ensure that IPTV will be?

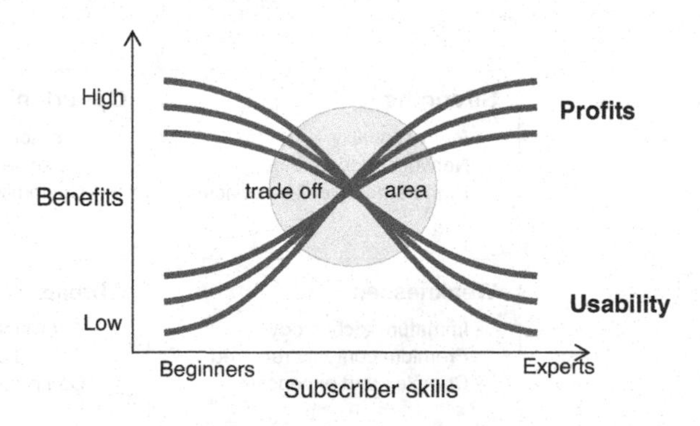

Figure 3-3. IPTV bidirectionality opens up new opportunities to create new applications, whenever the user interface continues to be simple and the middleware stable.

Table 3-2. What are the business benefits your company hopes to achieve through IPTV business? Select up to two benefits.

	Value	Moderate	Major
Reduce churn	13%	Increase broadband	21%
Acquire new customers	28%	Adoption new services	17%
New revenues	43%	Develop new competencies	15%
Drive profit growth	19%	Respond to competition	11%

Source: Accenture. (Reproduced by permission of Accenture)

If IPTV is successful, traditional suppliers will react by offering lower prices, consolidating companies and improving the services in quality and contents.

3.2.4.2 Intellectual Property Rights

The bidirectionality and digital downloading application can also damage the intellectual property rights (IPR). The impact of piracy and person-to-person video transfer can jeopardize the whole business. Content providers, operators and regulator must define the strategy to overcome this issue through controlling customers' use of IPTV facilities according to legal regulations.

3.3 Regulatory Framework

We have explained how IPTV is made up as a combination of television and telecommunications, two areas that in the past have been regulated independently. TV has always been a one-way communication. It was question of radioelectric spectrum, IPR protection and control of several aspects regarding commerce, ethics, political independence, education, language and religious orientation. On the other hand, the Internet still remains very unregulated except on content that can be considered a criminal offence and certain aspects that apply for electronic commerce. Internet regulations focus, in principle, on how to guarantee competition, quality and availability.

So far IPTV is being deployed in a similar way to the Internet, but competitors would prefer similar regulations in line with the TV broadcast. Obviously the IPTV industry is afraid that regulations could compromise the growth of this new technology.

Broadcasters, whoever they are, feel that IPTV has the potential to be an important threat to their business. The worst nightmare of broadcasters is to see subscribers controlling their own TV and video channels, choosing what programmes and events they want to watch and downloading their favourite contents, in a very simple, free and easy way. The question is: can broadcasters suffer similar consequences to those the music industry suffered after Internet success?

In any case, the IPTV industry is beginning to have regulated content. Future regulation may eventually distinguish between linear and non-linear services:

- *Linear*, such as internet-TV – the provider decides when the programme is transmitted, streamed or web-cast.

Table 3-3. Audiovisual service models.

	Resolution	Access	Device	Interaction	Benefits	Issues
Cinema	Maximum	Walking	Camera + wall	No	Latest titles, social event	Cost, time
Video shop	Standard/ high	Walking	DVD + TV	Start/stop/ pause/fwd/ bwd, idiom	Large archive, privacy	Cost, time
Commercial TV	Standard	Wireless	Antenna + TV	No (idiom perhaps)	A lot of broad-, casters free	Ads, poor content
Satellite TV	Standard	Satellite	Dish+ STB + TV	No (idiom perhaps)	Wide area, premium contents,	Installation
Digital TV	Standard/ high	Satellite, cable, TDT	Antenna/ dish/STB + TV	No (idiom perhaps)	Content	Cost
IPTV	Standard/ high	Copper, fibre, FTTx	STB + TV/PC	Internet-like, group of users	Highest flexibility, all possible	Installation, cost
Personal video	Low/ standard	Copper, fibre, FTTx	PC + Camera	Up/ download	Sharing video or archive	Server space
VoD	Standard/ high	Copper, fibre, FTTx	Server +STB + TV/PC	Start/stop/ pause/ fwd/bwd, idiom	Privacy, anytime/ anywhere	Installation, cost, IPR
Mobile TV	Low	Licensed wireless	Mobile terminal	High	Anytime, anywhere, mobility	Not ready
Internet TV	Low	Internet	Internet + PC	Download/ streaming	No specific network	IPR control, QoS
Internet P2P	Standard	Internet	Internet + PC	Start/stop/ pause/ fwd/bwd, idiom	Download to watch, low cost	IPR control, Band-width
Personal video	Low/ standard	Copper, fibre, FTTx	PC + camera	Up/ download	Sharing video, or archive	Server space

- *Non-linear*, such as video on demand – subscribers decide when and where the programme is downloaded.

Linear services may be subjected to regulations that may also apply to broadcasters regarding IPRs, and the sort of content, while non-linear services have similar IPR regulations but may enjoy less control regarding religion, politics, values, violence or adult contents. Nevertheless, non-linear services have increasing legal limitations as, in some markets, many content providers do not allow the storage of their programmes and there exist regulations to control what a personal video recorder can do when connected to, or installed in, an IPTV network.

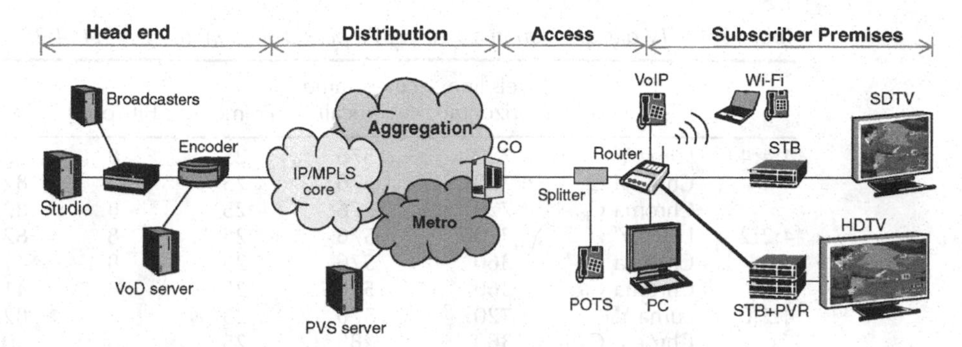

Figure 3-4. Architecture to support audiovisual services including IPTV, VoD and PVS. The IP bidirectionality that allows subscribers interaction with the television network is remarkable.

3.4 Architectural Design

The Triple Play bundle, and particularly IPTV, is a recent achievement thanks to the evolution of IP architectures that can support differentiated QoS. The traditional bandwidth bottleneck at the first mile does not exist anymore, as there is a growing number of alternatives capable of delivering real-time TV. Finally, the evolution of codecs like MPEG, that have made possible more compression and resolution, squeezing the available bandwidth, is also remarkable.

In many aspects IPTV rollout is probably the most important challenge telcos have faced in this decade, because the project requires the deployment of a complete new distribution network, renovating the access together with the installation of new servers and customer devices.

3.4.1 Television Services Rollout

The challenge of rolling out compelling audiovisual services is not an overnight project but a process to be achieved in consecutive steps:

1. *Define the business model.* The analysis of the market and the competitors to determine key parameters such as service charges and size of the market. It will also be convenient to think about how to define service differentiators.
2. *Architectural design.* Defining which nodes and resources are required at each network, including the contribution side, aggregation and access network.
3. *Service definition.* This must be described in terms of picture size, coder and decoder standard, frame formats, encapsulations, and video profiles.

Table 3-4. TV digitalization in a 720 × 576 pixel definition, using 4:4:4, 4:2:2 and 4:2:0 models.

		Pixels/line (horizontal)	Lines/frame (vertical)	Frames/s	Bits/pixel	Mbit/s	Total Mbit/s
4:4:4	Luma Y'	720	576	25	8	82.944	
	Chroma C_B	720	576	25	8	82.944	
	Chroma C_R	720	576	25	8	82.944	248.832
4:2:2	Luma Y'	720	576	25	8	82.944	
	Chroma C_B	360	576	25	8	41.472	
	Chroma C_R	360	576	25	8	41.472	165.888
4:2:0	Luma Y'	720	576	25	8	82.944	
	Chroma C_B	360	288	25	8	20.736	
	Chroma C_R	360	288	25	8	20.736	124.416

4. *Equipment approval*. Once the business model is clear, it is necessary to select the type of servers, network nodes and customer devices that best match the technical requirements and the available budget.
5. *Network commissioning*. The infrastructure rollout requires the verification of IP continuity and network performance that will permit engineers to establish routes, and to define multicast nodes that will control the routing and the quality management.
6. *Service provision*. Video servers and subscriber equipment must be connected, and applications installed to check end-to-end performance and the capability of the network to deliver differentiated video services under real traffic conditions.
7. *Service assurance*. The promised service may be affected by a number of issues that would modify the initial quality conditions. It is necessary to identify and troubleshoot the causes of a bad video quality affecting customers and express it as subscribers' quality of experience.

3.4.2 Business Model Definition

Many of the major operators across the world are deploying converged networks to support video distribution simultaneously with other services like data transport, Internet or telephony. Therefore, the subscriber access network should scale from current bit rates up to a level high enough to support at least one TV channel to target different segments of the TV market, like film, sport, children's channels, live shows, adult content or holiday shows.

However, IPTV requires a new and complex architecture that does not have a unique solution. To achieve good results from a technical and financial point of view it is necessary to plan and deploy a new architecture considering a number of parameters that depend on factors like market conditions, and resource availability, in order to define the business model:

- *Number of programmes*. Each provider has to define the selection of programmes, that will be made available to the subscribers. All of them must be able to reach every subscriber (see Figure 3-5).

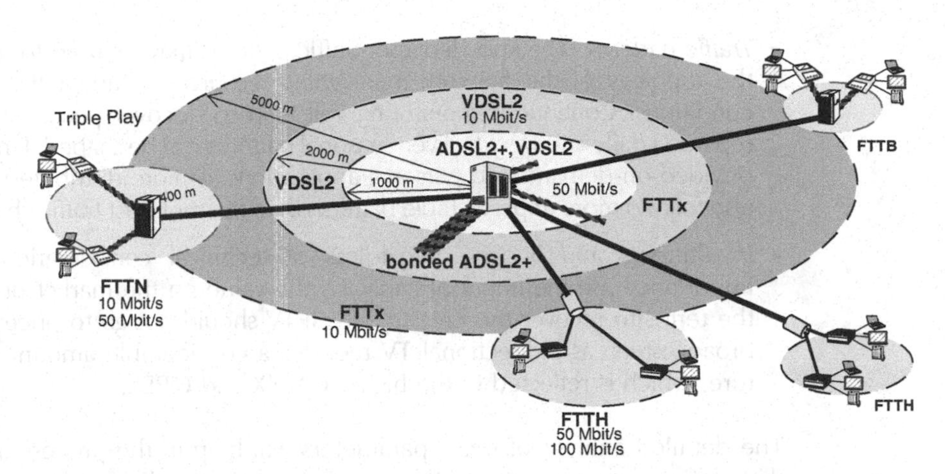

Figure 3-5. Several access alternatives allow bit rate escalation from a few Mbit/s up to 100 Mbit/s for multiroomed HDTV.

- *Image characterization.* Define the quality of the video in terms of number of pixels and frames per second. Generally speaking, the better the quality of the image is, the more bandwidth is required (see Table 3-5). There is a move to HDTV; it is interesting to realize that manufactures are already selling many receivers despite the lack of HD programmes.
- *Coders and decoders.* There are several alternatives to compress and code programmes including the MPEG family of standards, and also the Microsoft alternative. The differences are important; older standards like MPEG-1 are less efficient. Others, such as MPEG2, have an good installed base but are progressively substituted by the latest standards like MPEG-4 and WM-9, which are more flexible and efficient (see Table 3-5).
- *Unicast and multicast traffic.* IPTV generates multicast and prioritized traffic, while video-on-demand services and Internet access generate unicast and best-effort traffic. This means that the network must support a combination of traffic, and depending on what the proportion is, the architecture may be different. Generally the higher proportion of unicast services the more resources and bandwidth the network requires.
- *Number of subscribers.* IPTV requires costly infrastructure and therefore, density of users is a criterion in planning new deployments which are always cheaper in metropolitan areas than in suburban or rural areas. It makes sense to focus first on potential segments that may be less reluctant to contract new services, like those that already have broadband access.

Table 3-5. Bandwidth of standard and high definition TV.

	Screen size	MPEG-2	MPEG-4 Part 10 (H.264)	VC-1 (WM9)
SDTV	704 × 480	3.5 Mbit/s	2–3.2 Mbit/s	2–3.2 Mbit/s
HDTV	1920 × 1080	15 Mbit/s	7.5–13 Mbit/s	7.5–13 Mbit/s

- *Traffic patterns.* The knowledge of traffic patterns may be used to modify dynamically the topology of the network, reassigning resources to adapt the network to traffic conditions. Continuous monitoring will help to discover peaks and valleys of traffic repeated daily and weekly. Even seasonal patterns can be shaped. Unicast traffic, typical of video-on-demand and personal applications, is more distributed across time, but it also follows more unpredictable patterns than the multicast traffic characteristic of IPTV.
- *Investments and budget.* Regardless of technical considerations, always of key importance are the financial capacity, the value on the market of the business, and the terms to recover the investments. IPTV should not go to price battles against TV broadcasters, as bidirectional TV requires a considerable amount of new infrastructure, which is reflected in the higher CAPEX and OPEX.

The detailed analysis of these parameters will help in the process of taking decisions that define many aspects of the business case, including architectural ones like the head-end, distribution network, access network and subscriber premises equipment (see Figure 3-4).

3.4.3 Head-end

The head-end or contribution network is the place where the programmes are coded, compressed, multiplexed and finally streamed in IP packets using the most convenient physical interfaces. Video and audio content is typically compressed using either an MPEG codec to define the presentation parameters, such as pixels per line and frame refreshing, to playback of the video signal, and then streamed through IP networks in multicast or unicast addressing modes.

This scheme may not be exclusive to telcos and new entrants, since there is a trend of using IP packets everywhere, including in satellite and cable networks. Broadcasters are willing to upgrade their networks to two-way capability in order to embrace all the advantages of combining internet concepts with digital television.

3.4.4 Distribution Network

The distribution network has to dispatch the audiovisual signals across regional and metropolitan networks until they reach the customer's premises. A high capacity core, assisted by protocols like RTP or RTSP, must guarantee every IP packet to be multicast or unicast without degrading the quality parameters such as delay or packet lost. IP multicast is a method in which information can be sent to multiple, but not all, nodes or hosts at the same time (see Figure 3-6).

Distribution networks are IP-centric, and generally based on carrier Ethernet, which is responsible for traffic aggregation. Architecture supporting isochronic applications requires excellence in design and management, which is often achieved with engineering tools such as MPLS or VPLS that can transform native best-effort protocols (which is the case for Ethernet and IP) into a quality assured network. Different polices of prioritizing, like Diffserv, can be implemented to maintain the quality of the TV signal.

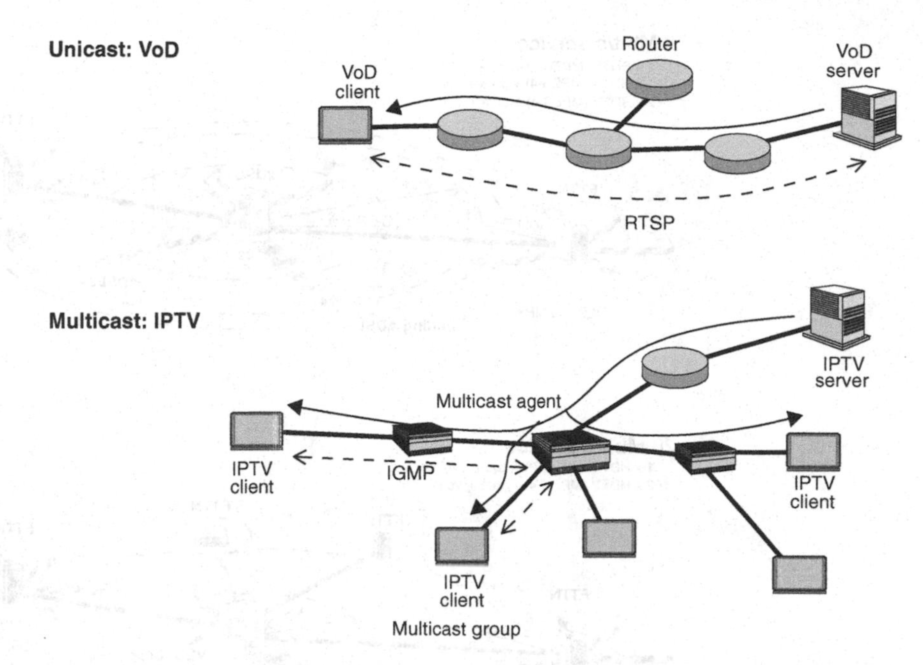

Figure 3-6. Video on Demand is delivered as a unicast service and managed by the RTSP protocol, while IP television is a multicast service managed by the IGMP protocol. This approach has proven to be an efficient and very scalable solution to delivering signals simultaneously to a large number of subscribers.

3.4.4.1 Access Network

The access network spans the first mile, providing a link between the service provider and the subscriber site. Several technologies are available, including ADSL2, ADSL2+, VDSL2, FTTN, FTTH and WiMAX. The selection will depend on parameters like the necessity of bandwidth, the distance and the financial restrictions (see Figure 3-7).

However, despite the great progress in access technologies that has occurred during the last decade, subscribers, due to the bandwidth limitations, are able to receive only two or three channels simultaneously if they have good access; otherwise they will receive only one channel. The zapping function is a request to the network, using the Internet Group Membership Protocol (IGMP), to notify the programme selected to be received, which in technical terms means that the subscriber wants to be part of the multicast group that contains the selected programme.

3.4.5 Subscriber Site

The IP packets finally reach the receiver site as a video stream. The IPTV play-back requires either a personal computer or a set top box (STB) connected to a TV; other gadgets like PDA and mobile hand-sets are also possible. The interaction between the subscriber and the network is done by means of protocols like the IGMP for channel selection in the case of IPTV, and RTSP to manage VoD applications.

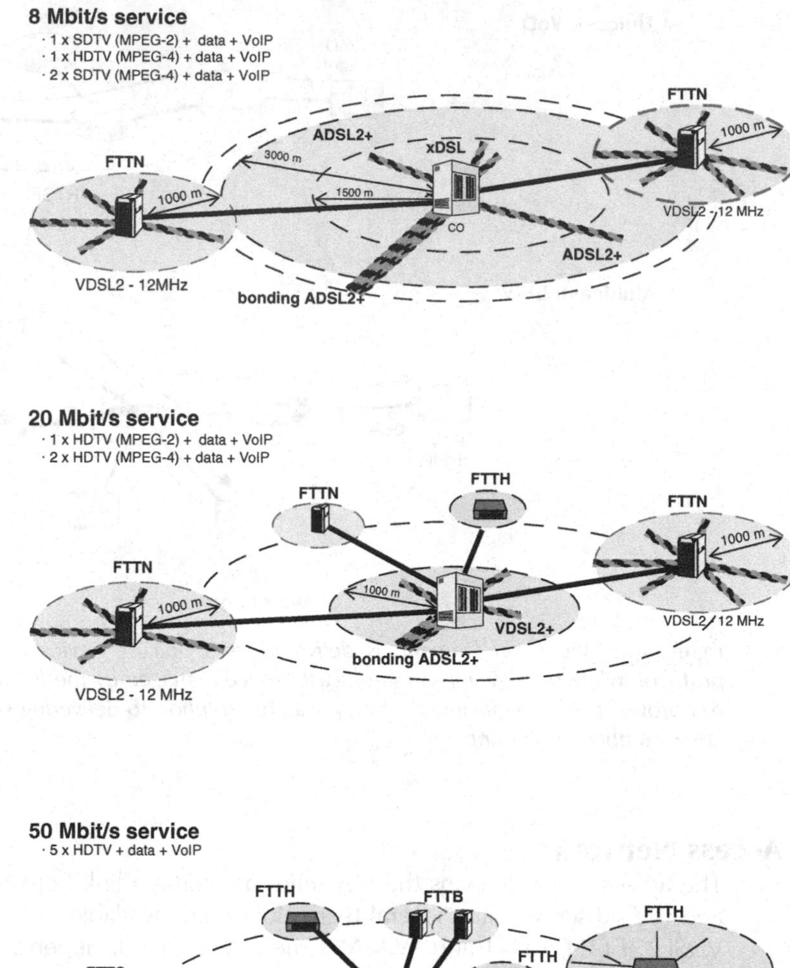

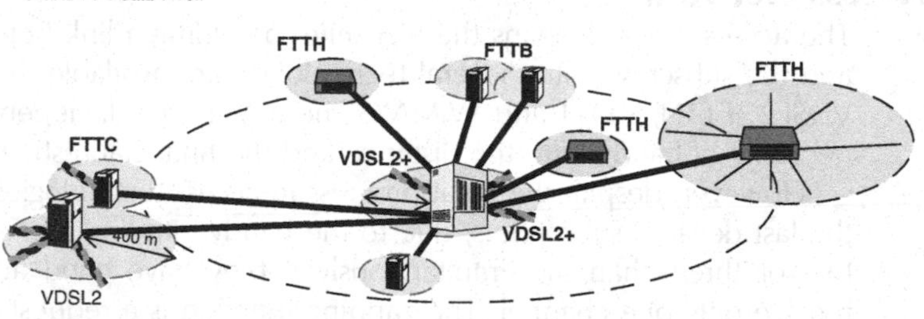

Figure 3-7. The access technology depends on the service to be rolled out, the distance to the central office and the available budget. Copper loop is available everywhere, but has bandwidth limitations compared with fibre optics, which is future proof.

The STB is the central and most relevant element at the customer site, the point where many features must be supported. The STB should be flexible enough to be upgraded as soon as new facilities or standards become available to avoid its replacement, which is always an expensive operation. For instance, some STB support more that one compression algorithm; others have a hard disk for recording programmes; some allow one programme to be watched while recording another; and the best ones can even record several channels simultaneously. In a few words, STB features are so significant for the business development that they must be taken into account in all strategic roadmaps to developing the IPTV business.

3.5 Television and Video Services and Applications

The new IPTV services can be seen as the combination of television with the highly interactive Internet concepts. The result is a number of new applications based on the delivery of audiovisual contents in a bidirectional, customized and controlled way.

3.5.1 IPTV Protocols

The support of multiple data and multimedia applications over common infrastructures is probably the main achievement of converged networks. It is only above layer three or four where we can find protocols devoted only to specific applications (see Figure 3-8). This is the case of IGMP, a protocol intended to support IPTV channel zapping, or the RTSP that provides VCR features (i.e. start/stop/fwd/bwd) specifically devoted to video-on-demand applications.

Regarding the codecs used for video digitalization, compression and transport, several options are available, such as MPEG-4 or VC-1. It is difficult to say which the best option is; it all depends on the business model, strategic decisions or simply the availability of standards at the moment when the service is rolled out.

3.5.2 Video-on-demand Services

The VoD applications are slightly different to live IPTV since they require the establishment of a point-to-point unicast relation between the server and the subscriber to enable

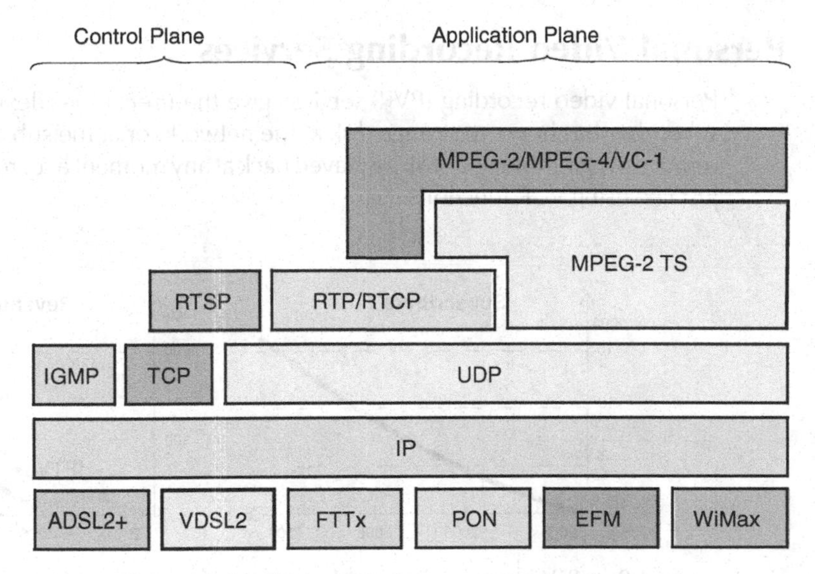

Figure 3-8. Protocols used for audiovisual services over converged networks.

subscribers to select and watch a stored video interactively across the IP network. VoD systems have two important features:

1. Subscribers can choose the video they want to watch from a digital library, deciding the moment and the place where they want to start.
2. Typical VCR/DVD functionality are provided by the RTSP protocol including *start, pause, stop, forward, backwards, fast forward and fast rewind* (see Figure 3-30).

There are many different implementations of VoD, which can be classified in two groups, depending on whether the content is streamed or it is downloaded:

- *Streaming VoD*, real-time service, the video is downloaded directly from the server to the TV; consequently QoS transport requirements are similar to IPTV. A lot of bandwidth and processing capacity is necessary at the head-end and across the distribution network.
- *Downloading VoD*, best-effort service, the video is downloaded and saved before being decoded and displayed in a similar way to those popular peer-to-peer applications on the Internet. The hard disk where video is saved can be at the provider's network or at the customer's site.

Whatever the implementation is, VoD can be the most relevant source of revenue of the Triple Play bundle (see Figure 3-9). The number of stored programmes such as films, video-clips, sport events, etc. can be so massive that traditional video shops cannot match them and the top 10 most popular titles will never become unavailable as they are just downloaded.

Definitely video-on-demand service is one of the killer applications, since cable and satellite broadcasters have more difficulty in implementing this highly interactive service.

3.5.3 Personal Video Recording Services

Personal video recording (PVR) services give the users more flexibility to watch TV. Each subscriber has her own storage disk at the network, or at the subscriber site, to record and remove programmes that will be played back at any moment according to user convenience, just like using VCR functions.

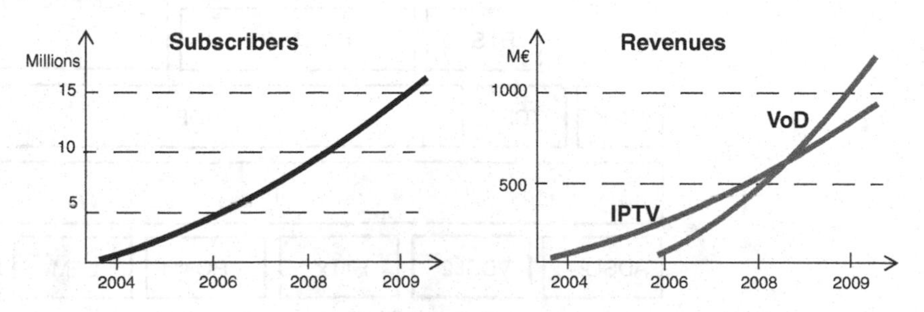

Figure 3-9. IPTV business, some market forecasts for Europe.

The are always programmes and events that subscribers would like to record to replay in the future. Also consider the case when a customer misses a multicast programme, totally or partially, but she would be willing to record it. Think how normal this circumstance is, when the subscriber is watching a live multicast programme and then wants to *pause* for a while, or to *skip* the adverts section, and then *go back* to some point during live transmission.

PVR service includes networked personal video recorder (NPVR), client personal video recorder (CPVR), time shift TV and last week's TV.

3.5.3.1 Networked Personal Video Recorder

NPVR is a consumer service where the real time IPTV stream is captured in a network. When using NPVR subscribers can record programmes in the network-based server, when they want, without needing yet another device or remote control. The recording is performed on the network server so that subscriber's devices are totally independent of the storage process.

It is important to take into account the technological limitations which can result in noticeable delays during playback, and in response to the 'pause', 'rewind' and 'fast forward' features that it shares in common with VCRs and DVD players.

3.5.3.2 Client Personal Video Record

Many people would still prefer to have their own PVR device, as it would allow them to choose exactly what they wanted to record, and eventually to create their own library of videos.

3.5.3.3 Private TV

Groups, enterprises and associations on specific topics can create their own TV, uploading their videos and programmes to be downloaded by VoD or even scheduling the transmission to a closed group of subscribers. This application transfers the control of the contents to the subscribers while the TV operator only provides the architecture and the management of whole system.

3.5.3.4 Last Week's TV

The IPTV operation may offer the complete retrieval of last week's (or any other period of time's) programmes. This service is made available to subscribers who missed the broadcasting schedule.

3.5.3.5 Time Shift TV

Time shift TV is a feature that combines video recording with real-time TV streaming. The subscriber watches the programme with a certain delay, allowing him to pause for a while, skip the advertisement section or go back to some time point during live broadcasting. There is no intention of recording the whole programme, just of doing a kind of buffering of live emissions to allow this flexibility.

3.5.4 Converged Telephony

IPTV can be easily converted into an integrated telephonic system supporting video conferencing that can be a real differentiator. Video conferences have not been very popular despite the effort made by legacy technologies like ISDN, ATM and 3G. Now video conferences have a new opportunity and can finally be successful because they will be low cost and part of a wider bundle that includes many different services.

Supplementary services such as caller id can be integrated on the TV screen and messaging can also be part of the application, which includes TVs, PCs and mobile phones. We should bear in mid that Triple Play is often seen as a defensive strategy to keep mobile and fixed line customers upgrading their service, and integrating it into new platforms.

3.6 Formats and Protocols

3.6.1 Analogue TV

Broadcast analogue TV has been implemented differently across the world. In Europe 625 lines are displayed at a frame rate of 25 Hz, while in the USA and Japan the systems are made up of 525 lines at a frame rate of 30 Hz.

Initially TVs only had luminance, represented by *Y*, to describe the brightness transition from back to white. When colour TV was developed it was necessary to add chrominance by means of two more components, *U* and *V*, resulting in the *YUV* model (see Figure 3-10). The *RGB* (red, green, blue) signals coming from a colour video camera can be transformed directly:

$$\text{Luminance } (Y) = 0.299R + 0.587G + 0.114B$$
$$\text{Chrominance } (U) = 0.493 \times (B - Y)$$
$$\text{Chrominance } (V) = 0.877 \times (R - Y)$$

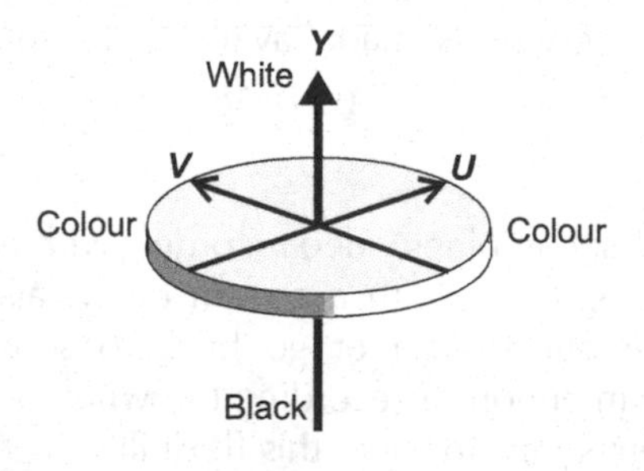

Figure 3-10. The YUV model was invented to separate colour and luminance to permit the compatibility of colour and black-and-white TVs. YUV coding converts the RGB signal to an intensity component (Y) that ranges from black to white plus two other components (U and V) which code the colour. The conversion from RGB to YUV is linear, without loss of information.

The *YUV* model was adopted quickly since it permitted compatibility between the old black and white and the new colour television systems. Note that to represent a colour image in a b&w device it is only necessary to use the luminance while chrominance components are discarded, making $U = V = 0$.

3.6.2 Digital TV

The next milestone in the TV evolution was the digitalization that makes easier the control, storage and distribution of the audiovisual streams, while simplifying the quality management until the signal is delivered.

The ITU-R recommendation BT.601 describes how to digitalize the analogue *YUV* signal into 720 luminance samples and 360 chrominance samples per line. Samples become pixels to create the picture which is refreshed 25 times per second (see Table 3-4). This recommendation indicates how to sample and quantize the signal to obtain a binary sequence of pixels. The digitalization maintains the three digits of the *YUV* model now transformed:

- one number (Y') for the luminance.
- two numbers (C_B, C_R) for the chrominance.

3.6.2.1 Chrominance Subsampling

Human vision is less sensitive to colour than brightness, so chrominance is able to be subsampled at a lower rate than luminance without destroying the image quality. Subsampling results in a significant bandwidth reduction.

The sampling scheme can be expressed with a notation of three digits $A : B : C$, where A indicates the luma sampling reference. In our case A is equal to 4. B represents the chroma horizontal sampling factor relative to A. Finally, C represents the chroma vertical sampling factor relative also to B (see Figure 3-11). Then we have three cases:

- 4:4:4 means that all luma and chroma components are vertically and horizontally sampled at the same rate. If the luminance is digitalized at 720×576 samples, then the chrominance also has 720×576 samples (see Table 3-4).
- 4:2:2 means that the two chroma components are vertically sampled at the same rate of luma rate, while horizontal sampling is done at half the rate. For instance, if the luminance is digitalized at 720×576 samples, the chrominance is at 360×576 samples (see Table 3-4).
- 4:2:0 means that chroma is horizontal and vertically sampled at half the rate of the luma sampling rate. For instance, if the luminance is digitalized at 720×576 samples, the chrominance is at 360×288 samples (see Table 3-4).

3.6.3 Audio and Video Codecs

There are a number of specifications of codecs (coder-decoder) for audiovisual signals that define the coding, compression and streaming of the audiovisual contents. These standards facilitate the independent implementation of interoperable encoders and decoders.

Figure 3-11. The YUV video signals can be digitalized according several schemes to reduce the bandwidth. The 4:4:4 scheme has the same samples for luma and chromas; the 4:2:2 subsamples chromas horizontally, while the 4:2:0 subsamples chromas horizontal and vertically.

The most popular family is MPEG (Moving Pictures Expert Group), defined by the ISO/IEC and ITU-T. Windows Media, which is basically a Microsoft development, is also an interesting codec that is being considered by many of the new IPTV entrants.

- *MPEG-1 (ITU-T H.261)*, published in 1993, was the first digital video standard codec that made possible the evolution from the analogue to the digital dimension regardless of the analogue standards. MPEG-1 compresses and encodes audio and movement at rates around 1.5 Mbit/s, providing a video resolution equivalent to VHS cassettes, targeting the first generation of audio and video CD-ROM applications. MP3, the audio part of MPEG-1, has become very popular thanks to the Internet, digital radio and the new supports for digital audio.
- *MPEG-2 (ITU-T H.262)*, published in 1995, is a superset of MPEG-1, providing a wider range of bit rates from 2 to 20 Mbit/s, several levels of quality and video screen resolutions. MPEG-2 applications have been very popular since the mid 1990s and are being used in satellites, cable, DBS, DVD and early implementations of IPTV.
- *MPEG-4 part 10 (ITU-T AVC/H.264)*, published in 1999, is a very flexible codec which has a range of bit rates from 5 kbit/s to 10 Mbit/s, making it suitable for mobile video, standard definition and high-definition TV. MPEG-4 can save up to 50% of the bandwidth so new entrants to IPTV have selected it, while existing MPEG-2 applications are planned to migrate to MPEG-4.
- *SMPTE VC-1 (WM9V)* is a video codec specification that has been standardized by the Society of Motion Picture and Television Engineers (SMPTE) and implemented by Windows Media 9 (WM-9). It has similar features to MPEG-4, but will probably permit smoother integration with PCs or hybrid PC-TV devices.

3.7 How a Codec Works

The MPEG is a working body within the ISO that is responsible for developing video and audio encoding, compression and standards for digital television delivery and multimedia digital video applications.

A detailed explanation about codecs goes far beyond the objectives of this book. Nevertheless, in order to illustrate the video compression techniques, we have chosen MPEG-2 as it has been a milestone for other codecs such as MPEG-1, MPEG-4 and VC-1, which often use very similar techniques.

3.7.1 MPEG-2 Levels and Profiles

MPEG-2 is an intricate standard intended to support a wide range of applications. It is very flexible, permitting implementations with different resolutions and qualities for making it suitable for DVD, satellite, cable and IP television. However, in order to simplify and to reduce the cost, commercial devices do not need to implement the full MPEG-2 standard but

Table 3-6. Combined constrains, levels, and applications.

		Constrains					
		Picture size		Video streaming			
		Maximum lines/frame	Maximum pixels/line	Buffer size	Maximum frames/s	Maximum Mbit/s	*Applications*
Level	High 1920	1152	1920	9 781 248	60	80	HDTV
	High 1440	1152	1440	7 340 032	60	60	Entry HDTV
	Main	576	720	1 835 008	30	15	SDTV
	Low	288	352	475 136	30	4	Low entry TV

only the subset that matches the requirements of the TV system. Then the selected subset is defined by two descriptors, the profile and the level:

- The *level* describes the constrains on parameters that manage the video presentation and streaming (see Table 3-6). Four different levels have been defined.
- The *profile* defines a subset of algorithmic tools. Five profiles are available to achieve different layers of quality improvements (see Table 3-7).

The exact combination of profile and level exactly describes the MPEG-2 subset of features which encoders and decoders are required to implement.

3.7.1.1 MPEG-2 Levels

Levels describe the constrains on video presentation and video streaming parameters. Constrains are upper limits on what codecs must operate below. The four levels defined are High 1920, High 1440, Main and Low, each one designed to fulfill requirements of different applications in terms of picture size, bandwidth and buffer size limitations (see Table 3-6).

Table 3-7. MPEG-2, combination of profiles and levels.

		Profiles				
		Non - scalable		Multilayer profiles		
		Simple	Main	SNR	Spatial	High
Levels	High 1920		**4:2:0** 80 Mbit/s			**4:2:0/4:2:2** 100/80/25 Mbit/s
	High 1440		**4:2:0** 60 Mbit/s		**4:2:0** 60/40 Mbit/s	**4:2:0/4:2:2** 80/60/25 Mbit/s
	Main	**4:2:0** 15 Mbit/s	**4:2:0** 15 Mbit/s	**4:2:0** 15/10 Mbit/s		**4:2:0/4:2:2** 20/15/4 Mbit/s
	Low		**4:2:0** 4 Mbit/s	**4:2:0** 720 × 576 4/3 Mbit/s		

3.7.1.2 MPEG-2 Profiles

There are five profiles defined. The three more sophisticated are capable of delivering several qualities in a multilayer structure, while the other two are orientated to simpler implementations.

- *Simple*: no-scalable profile suitable for symmetric and time sensitive applications like videoconferencing that requires delays below 150 ms. To achieve that it does not use B-frames (see Section 3.7.2.4) to avoid reordering and buffers, so delays are minimized.
- *Main*: no-scalable profile suitable for asymmetric applications like IPTV, which means that most decoders support the main profile and main level. The main profile manages all types of frames: I, P and B, so it is more efficient but adds about 120 ms delay for frame reordering.
- *SNR*: scalable profile that supports two layers of enhancements. The lower layer transmits the stream to a specific quality while the upper contains video refinements, allowing in this way two qualities simultaneously. This profile has been suggested to be used in terrestrial digital TV.
- *Spatial*: a scalable profile which has been suggested for HDTV emissions compatible with standard resolution receivers. Spatial profile is characterized by images coded and streamed at different resolutions. The lower layer carries low-resolution images while the higher improves with higher resolution.
- *High*: a scalable profile that supports three layers of consecutive enhancements to achieve in this way maximum flexibility and different qualities. The high profile supports the 4:2:2 scheme (see Section 3.6.1) and includes both the SNR and spatial enhancement tools.

3.7.2 MPEG Compression

MPEG reduces the size of the video stream, compressing it to maximize resources such as bandwidth or disks, which are the media for transporting and storing video signals, but without compromising the quality. In some respects, compression is a trade-off between the bandwidth required and the quality obtained (see Figure 3-12). Compression techniques are

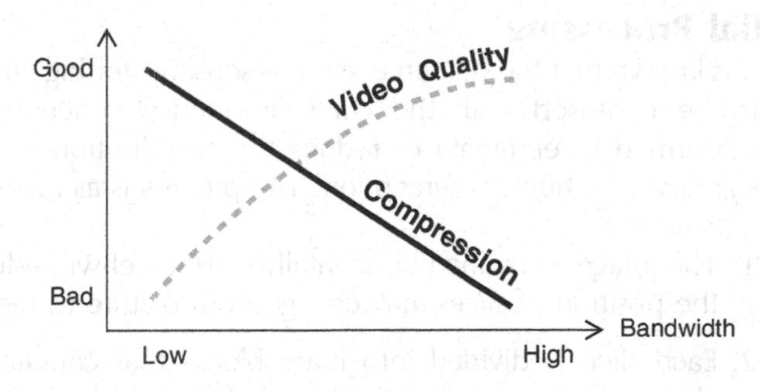

Figure 3-12. The compression level is a trade-off between the video quality and the available bandwidth.

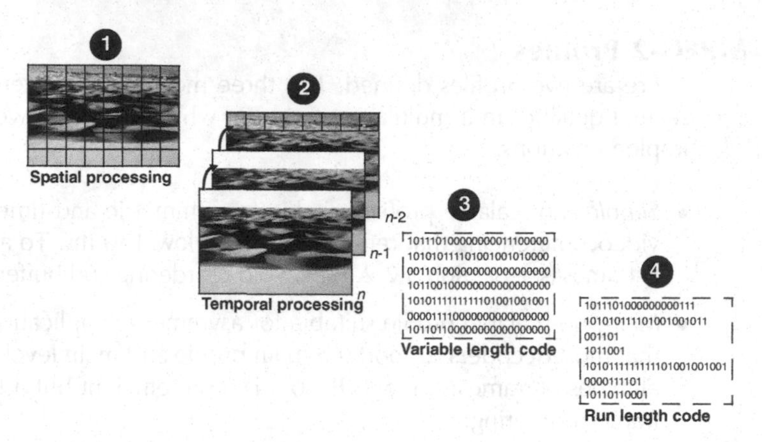

Figure 3-13. The four steps in MPEG-2 compression.

generally based on human perception and mathematical algorithms. In particular MPEG uses the following techniques (see Figure 3-13):

1. *Spatial processing or intraframe coding*, using discrete cosine transform (DCT) to remove shapes the human eye cannot see. This technique is based on the fact that pixels are not independent but correlated in colour and brightness with the neighbourhood.
2. *Temporal processing or interframe coding*, which looks for redundancies between consecutive frames and predicts moving blocks (see Figure 3-13). This compression takes advantage of the fact that many areas of a frame are the same as the previous frame, with some of these areas being predictable, e.g. backgrounds that are repeated across several frames.
3. *Variable length code (VLC)*, which uses shorter encoding to reduce size. This is a mathematical process to reduce the size of the digital code.
4. *Run length encoding*, which replaces large sequences of zeroes to reduce the length of the digital stream; this is another mathematical process.

3.7.2.1 Spatial Processing

It is known that the human eye is less sensitive for high frequencies of picture shapes (not to be confused with the colour-frequency concept); therefore, a DCT process is performed to eliminate or reduce the contribution of those components that are less significant in human perception. The process is as follows (see Figure 3-14):

1. The image is scanned horizontally in 16 pixel-wide slices. Slices must not overlap, but the position of slices may change from picture to picture.
2. Each slice is divided into macroblocks that contain blocks for the luminance and chrominance components. The number of blocks varies depending on the luma/chroma scheme (see Figure 3-11). The 4:2:0 is made from six blocks ($4Y'+2C_B+C_R$); the 4:2:2

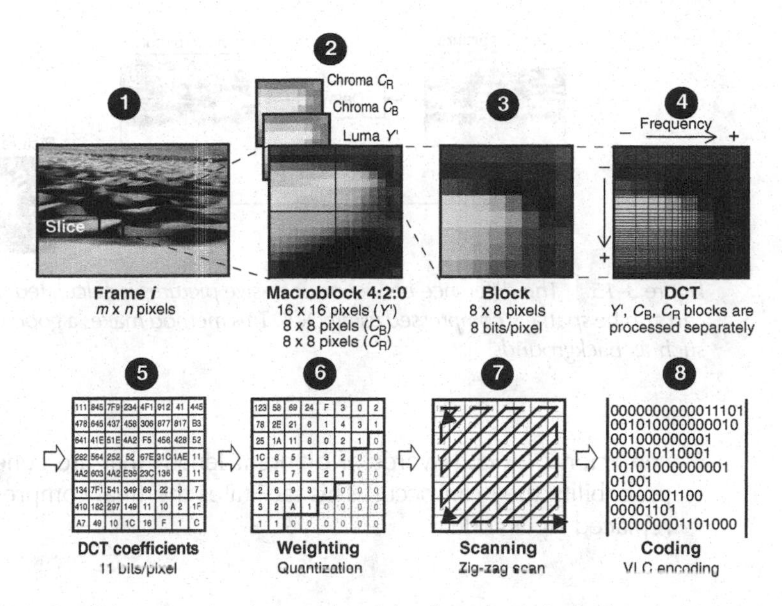

Figure 3-14. Spatial processing is based on the fact that pixels that belong to a block are correlated, so it is possible to remove the correlation. Luma and chroma are processed separately.

is made from eight blocks ($4Y'+2C_B+2C_R$); and the 4:4:4 is made from 12 blocks ($4Y'+4C_B+4C_R$).

3. Each block has 8 × 8 pixels, and each pixel is represented by one byte.
4. A two-dimensional DCT transforms each block into the frequency domain, indicating the contribution of vertical and horizontal components. DCT concentrates the visual energy at the top-left corner, which corresponds to the lower frequency, to which our eyes are most sensitive.
5. DCT transforms images into its frequency components, increasing the coefficient for pixel size from 8 to 11 bits.
6. Weighting is performed according to human perception: higher frequencies are more coarsely quantized, therefore codification of higher frequencies is minimized while zero value is assigned to coefficients that may be below perception thresholds or provide a poor contribution to the image quality. Finally, shorter codes are assigned to most common samples.
7. The non-zero values are then scanned to obtain the sequential binary stream, following a zig-zag scan to get first coefficients in the top-left corner, the most important for human perception, while the bottom-right corner, the least important, is the last to be scanned.
8. The scanning results in a digital stream that tends to put zero coefficients together, then, using Hoffman encoding, a variable length code (VLC) substitutes the non-zero

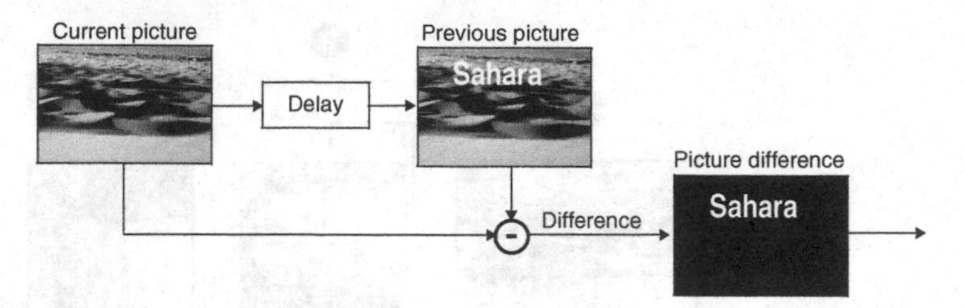

Figure 3-15. The difference between successive pictures is calculated by subtraction. The difference can also be spatially compressed using DCT. This method makes a good prediction for stationary areas such as backgrounds.

coefficients by codewords of different length that are generated depending on the probability they can occur. This facilitates the final compression with zero detection replaced by a code.

3.7.2.2 Temporal Processing

Temporal processing is based on the fact some areas of video are repeated across several frames during a period of time, i.e. stationary backgrounds (see Figure 3-15). Then, instead of coding the same block again and again, the repeated block uses a reference to the place where it has already been represented.

The best interframe predictions are those referred to stationary areas that are repeated across several frames in exactly the same coordinates of the frames. Predictions can be either *backwards* if a previous block is used as reference, or *forwards* if a late block is used as reference (see Figure 3-19).

Slightly more difficult are the predictions of moving areas of the picture, that is, regions that have moved to another place of the picture sequence. For these cases a new algorithm known as motion-compensated predictions is used.

Motion vectors identify translational changes between the pictures. Block matching searches, based on luminance components, identify movements that the encoder will calculate and code into the motion vectors (see Figure 3-16) to predict translations. In the case of *P* frames, motion vectors can refer to previous blocks, and in the case of *B* frames, motion vectors can refer to previous and future frames.

However, predictions are imperfect, because compression introduces distortions that result in important differences at the receiver side compared with the original image. To minimize this effect the coder has a decoder to reconstruct images exactly how the receiver would do, and then it is compared with the original. This operation allows the coder to discover the prediction errors and then it can compensate for them, generating additional information (see Figure 3-17).

3.7.2.3 Codec Structure

To implement the above-mentioned processes to compress the video streams, MPEG codecs have to combine at several elements such as the DCT procedure to remove the

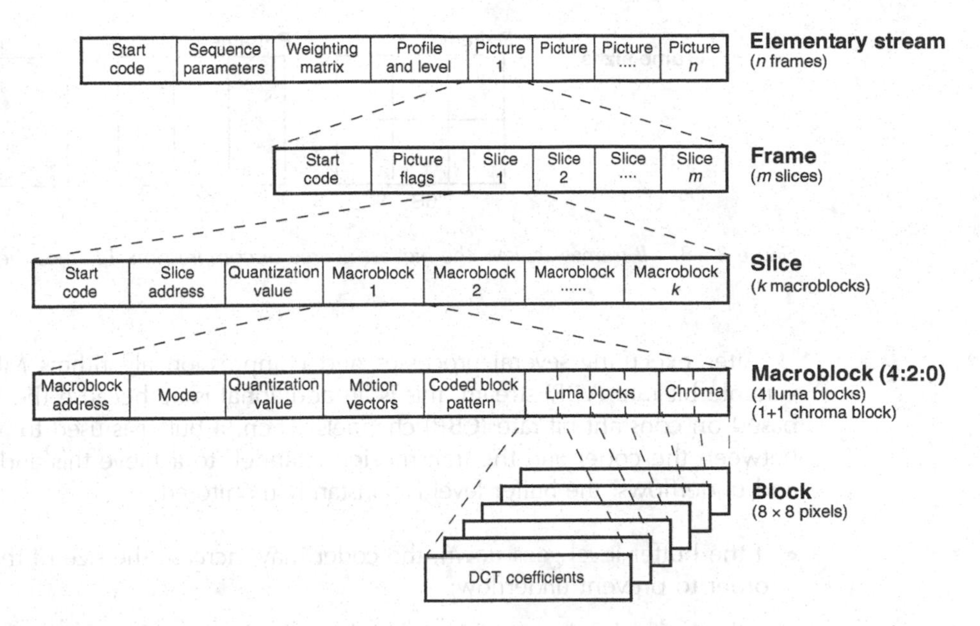

Figure 3-16. Elementary video stream (ES). Each frame has m horizontal slices made with k macroblocks each. This representation corresponds to a 4:2:0 model.

intraframe redundancies, the identification of stationary areas, motion vectors for moving pictures and VLC coding. In addition, the codec also executes the inverse process to identify and compensate compression distortions, resulting in an improved video quality (see Figure 3-17).

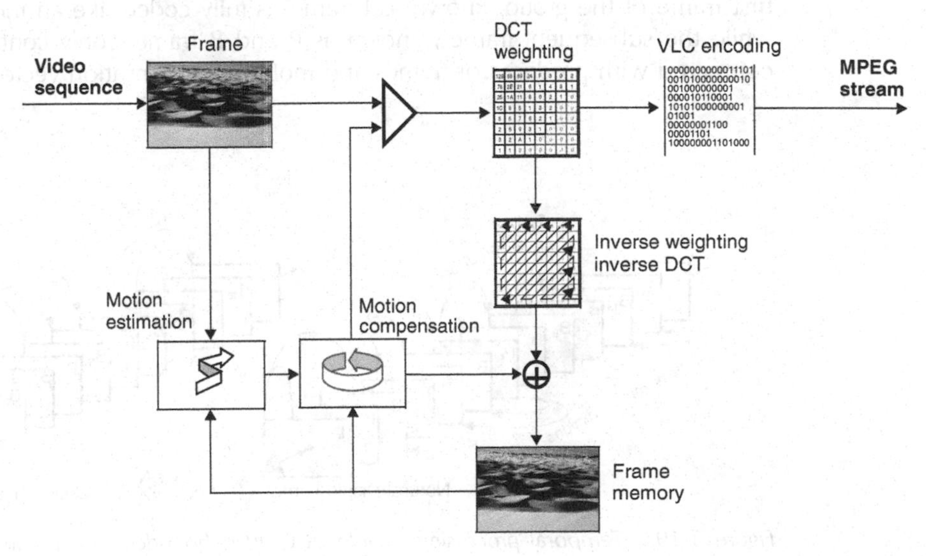

Figure 3-17. Simplified model of MPEG-2 encoder.

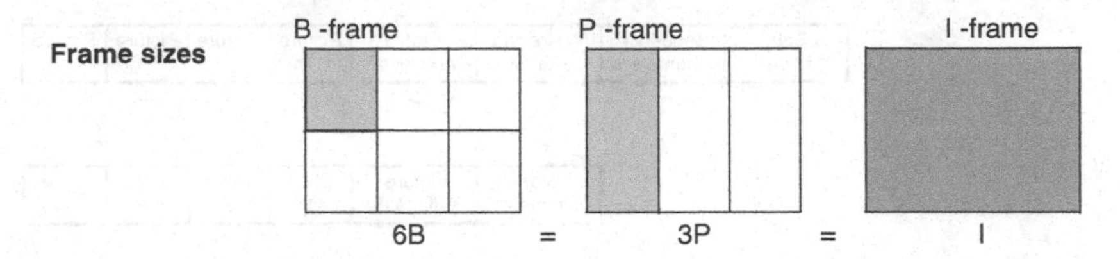

Figure 3-18. B frames achieve the highest degree of compression, I frames the lowest.

After executing several processes and compression algorithms MPEG generates a variable bit rate (VBR) stream, this is an additional issue because the transport is often based on constant bit rate (CBR) channels. Then, a buffer is used to adapt the bit rate between the coder and the transmission channel; to achieve this and avoid overflows and underflows, the buffer level is constantly monitored:

- if the buffer level goes down, the coder may increase the size of the DCT output in order to prevent underflow;
- if the buffer level goes high, the coder may decrease the size of the DCT output in order to prevent overflow.

The result is a nearly constant bit rate output at the average bit rate. At the receiver site the decoder must also have a buffer that will be used to filter the interfamily packet jitter (see Figure 3-28). The size of both buffers should be the same.

3.7.2.4 Type of Frames

The simplest approach to the MPEG video is the sequence of several types of frames. The first frame of the group, known as I frame, is fully coded like an independent picture, while the subsequent frames, known as P and B frames, only contain the differences compared with predecessor frames and motion compensation vectors.

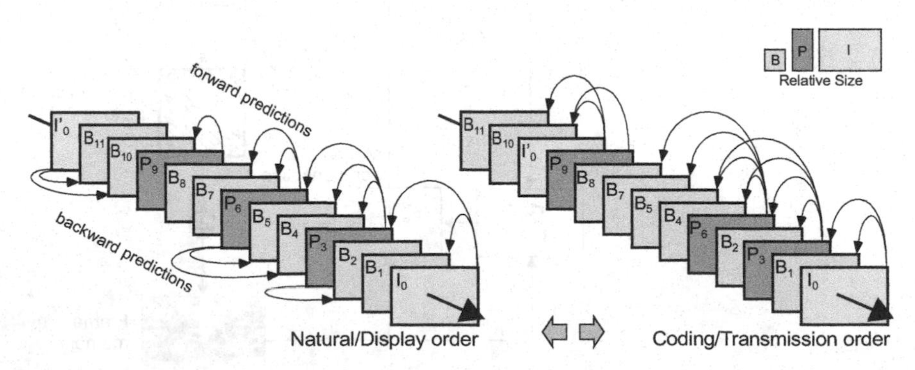

Figure 3-19. Temporal processing sample. A GOP is bounded by I frames 12/15 frames long. Predictions must be made inside the GOP.

Let us analyse in more detail how it all works:

- *I frames* (intra frame) compression is achieved by removing only the spatial redundancy and not the temporal. Therefore I frames do not have references to other frames, and can be reconstructed as a single digital picture, totally independent of any other image.
- *P frames* (prediction frames) contain only the differences from the previous frame. Blocks can either be coded independently, like I frame blocks, or use a reference to the nearest I or P frame previously coded and usually incorporating motion compensation.
- *B frames* (bidirectional-prediction frames) can either be intra-coded or may use backwards or forward references to P and I frames. Forward predictions, or a reference to a future frame, requires a change in the natural frame order before transmission and the receiver must recode. This operation causes a reordering delay (see Figure 3-19).

A group of pictures (GOP) is the sequence of frames embedded between two I frames. In theory, the number of frames in a GOP is unlimited; nevertheless, in practice, typically there are up to 15 P and B frames occurring between two consecutive I frames. This means that an I frame could be received, on average, about every 0.4 s during the normal video showtime.

One should remember the consequences of a lost frame, and realize that the impact is different depending on the type of frame. If one I frame is lost, the whole GOP is affected, since many of the following frames may refer to it, while the loss of a B frame does not impact other frames as it cannot be used as a reference. The consequences of losing a P frame can be similar to those of losing an I frame because it can also be used as a reference.

It is also interesting to analyse the impact of frames on the applications:

- In IPTV the STB should wait to receive the first I frame to start displaying images on the customer's television.
- A video compressed only with I frames would allow stop/forward/backward in every single frame because only I frames are independent. Nevertheless the compression would be minimum.
- A sequence combining I and P frames would achieve better compression but fewer points at which to start/stop the video sequence.
- A sequence combining all three types achieves maximum compression but even fewer points at which to start/stop the video sequence.

3.7.3 MPEG Stream Generation Scheme

Typically a programme has one video stream and one audio stream, and eventually may also have one data stream. Nevertheless, other combinations are also possible, for instance multilingual programmes carry several audio streams and the user choses the idiom. There

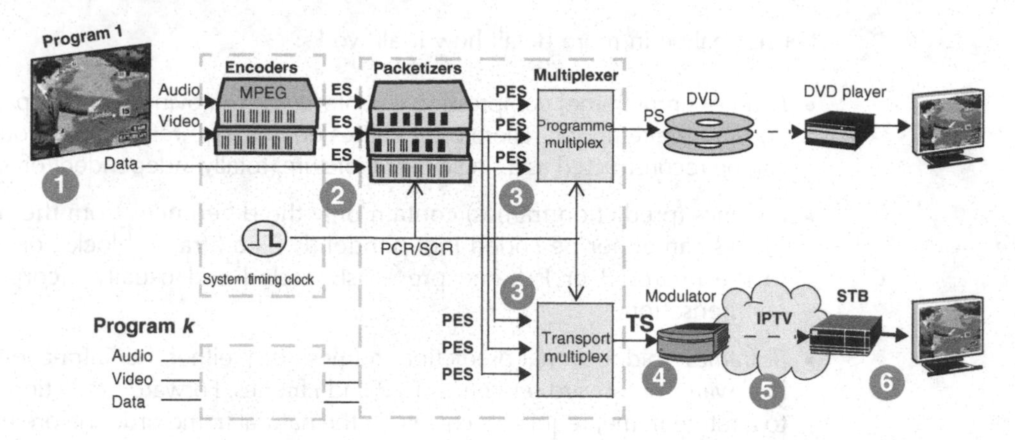

Figure 3-20. Two types of streams can be generated with the same source signal: the programme stream (PS) and a transport stream (TS) are two formats of the same structure. The PS is used in CD-ROM and DVD, while TS is intended for transmission over digital cable, satellite or IP-centric networks.

are also applications that can carry several video streams; a good example is the multicamera TV programmes, where subscribers are invited to select the view they want to display at any time, for instance, during live shows or a sporting event.

Whatever the combination is from the TV studio to the customer television, these are the basics steps of an IPTV system (see Figure 3-20):

1. The video camera output is an analogue multimedia signal made up of the audio and video flows.
2. The encoders digitalize the signals and perform the codification using the MPEG techniques. The outputs of each individual flow for audio, video and data are the elementary streams (ES) (see Figure 3-21).
3. The following step involves packetizers that perform a segmentation of each stream received, adding the packetized elementary stream header (PEH) to identify, synchronize and manage each individual stream (see Table 3-8). The result is a sequence of packetized elementary streams (PES) (see Figure 3-22).
4. The multiplexer combines all the elementary streams defined for each programme in a common transport stream (TS) that accepts transport stream packets (TSP) of 188 bytes (see Figure 3-23). Each TSP is made up of a header of 5 bytes and a payload of 183 bytes in which are dropped the contents of the PES (see Table 3-9).
5. The TS is split to create the IP/UDP packets that can also be mapped in RTP payload. Generally an IP packet holds up to seven TSP in the most common case where standard Ethernet frames are used as a transport layer. After streaming the audiovisual information packets are forwarded and distributed across the IP/carrier-Ethernet network until they reach their destination.
6. Once IP packets are delivered to the STB, the inverse operations must be performed to finally obtain the reconstructed audio and video signals ready to be displayed.

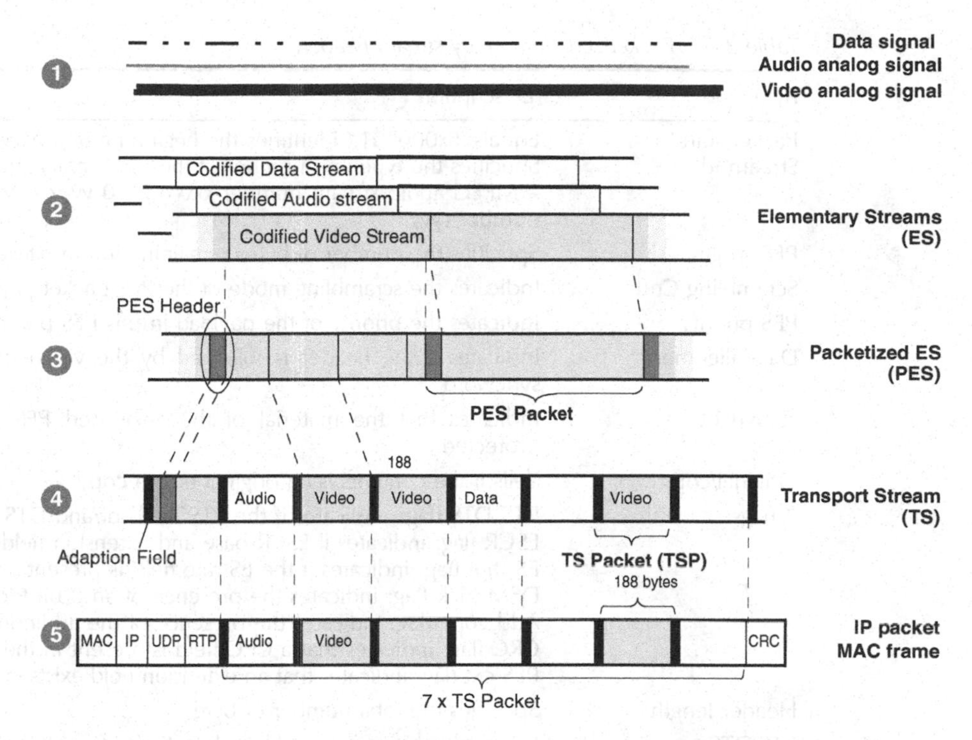

Figure 3-21. Each elementary stream is sent to a processor that creates standard streams called packetized elementary streams (PES) made up of the original inputs of audio, video and data. The PES are finally segmented into fixed-size transport stream packets (TSP) and put together to create the transport stream.

Two type of streams can be generated with the same source signal, the programme stream (PS) and the transport stream (TS) , which have two different formats despite the contents being exactly the same. The PS is used in CD-ROM and DVD, while the TS is intended to be responsible for the multiplexing and transmission of the digital video and audio elements over a transport network such as cable, satellite or IP-centric networks.

3.7.4 The Transport Stream

Regardless of codec being used (MPEG-2, MPEG-4 or WM9), there are several alternatives for the transport of IPTV payloads over IP:

- *MPEG-2 transport stream*: originally designed for MPEG-2 streams, it is also used for carrying MPEG-4 or WM9 streams. The TS main job is multiplexing the digital video and audio stream and synchronizing the receiver. The whole encapsulation is [Payload + TS + UDP + IP].

Table 3-8. Packetized elementary stream header.

Byte	Description
Packet start	Equals '0x000001,' identifies the beginning of a packet
Stream id	Specifies the type and number of the elementary stream, i.e. 110*x xxxx* – MPEG audio stream number *xxxxx* 1110 *yyyy* – MPEG video stream number *yyyy*
PES length	Specifies the number of bytes remaining in the packet after this field
Scrambling Cntl	Indicates the scrambling mode of the PES packet payload
PES priority	Indicates the priority of the payload in this PES packet
Data alignment	Indicates if the header is followed by the video start code or audio syncword
Copyright	Indicates that the material of the associated PES packet payload is protected
Original/copy	Tells if the contents is an original or is a copy
7 flags	PTS DTS flags: indicates if the PTS fields or/and DTS fields are present ESCR flag: indicates if ESCR base and extension fields are present ES rate flag: indicates if the ES rate field is present DSM trick flag: indicates the presence of an 8-bit trick mode field Add copy flag: indicates the presence of the additional copy info CRC flag: indicates that a CRC field is present in the PES packet PES ext flag: indicates that an extension field exists in this packet header
Header length	Specifies the total number of bytes
PTS/DTS	Presentation time stamp/decoding time stamp: indicates the time of presentation, in the decoder of a audio or video unit of the elementary stream
ESCR	The elementary stream clock reference
ES_rate	Specifies the rate at which the system target decoder receives bytes
Trick mode	Indicates which trick mode is applied to the video stream, i.e., fast forward, slow motion, freeze frame, fast reverse, slow reverse
Add copy info	Contains private data relating to copyright information
Prev PES CRC	CRC value of the previous packet
5 flags	Priv data flag: indicates that this packet header contains private data Pack header flag: indicates that a pack header is stored in this packet header PPSC flag: indicates that the program packet sequence counter is present STD buffer flag: indicates that buffer scale and size parameters are present Extension flag: indicates the presence of the PES extension field length
Private data	Field which contains private data
Header length	This field indicates the length of the header
Prog seq count	Optional counter providing functionality similar to a continuity counter
P-STD buffer	Indicates scaling factors to resize the buffer
Extension length	This field specifies the length the data following

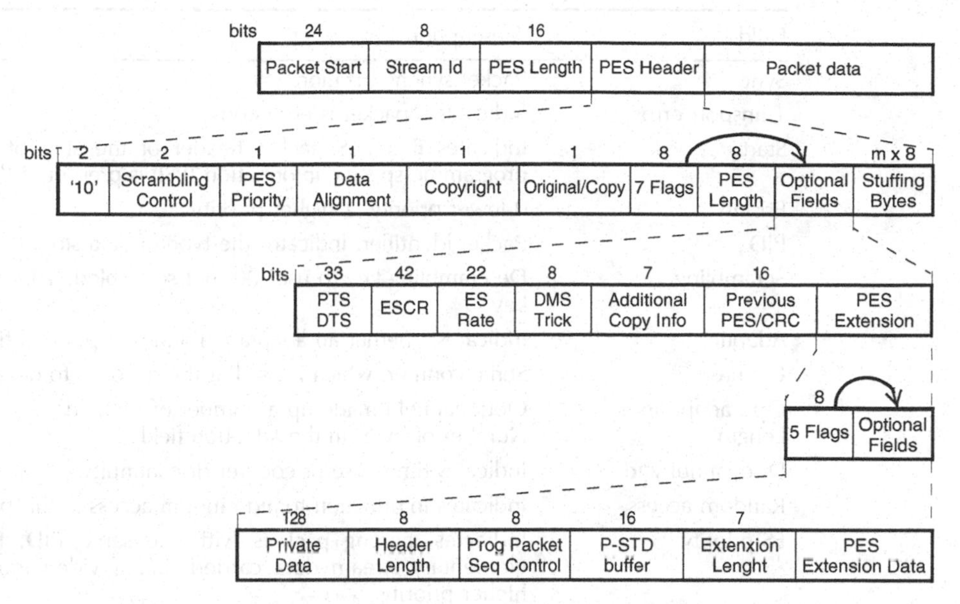

Figure 3-22. Packetized elementary stream (PES) syntax diagram.

- *Real-time protocol (RTP)*: originally developed to carry time sensitive information over IP, it can be used to transport and synchronize IPTV streams that are sent separately because it does not have multiplexing capabilities. A good complement is the RTP control protocol (RTPCP), which is used to synchronize multiple streams, monitoring the quality and providing feedback. This option that includes it is the most efficient since it has has fewer overheads. The whole encapsulation is [Payload + RTP + UDP + IP].

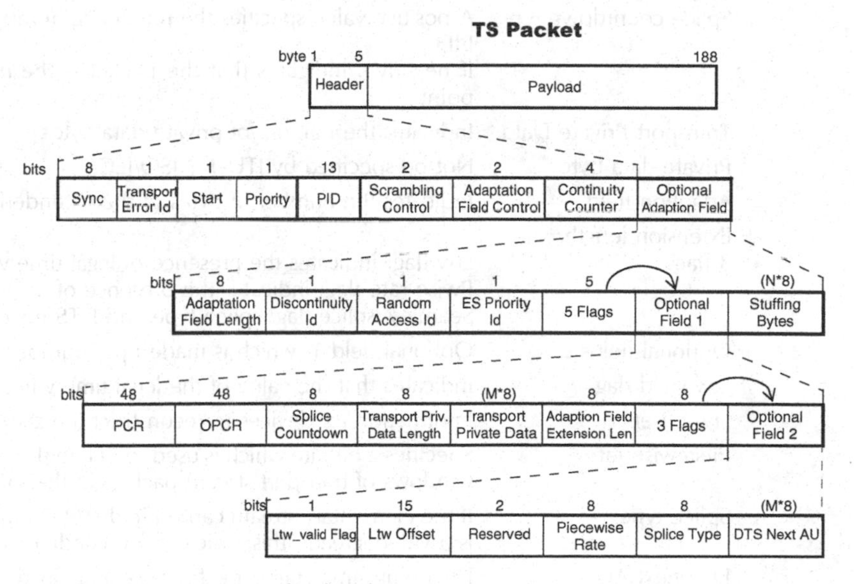

Figure 3-23. Transport stream packet header.

Table 3-9. Transport stream headers.

Field	Description
Sync	Packet synchronization
Transport error	Indicates if packet is erroneous
Start	Indicates if a PES packet header of the start of a table containing programme-specific information (PSI) is present in the payload
Priority	0 lower priority, 1 higher priority
PID	Packet identifier. Indicates the type of data stored
Scrambling	Descrambling key to use: 00, not scrambled; 10, 'even' key; 11, 'odd' key
Adaption	Indicates whether an adaptation and/or a payload field follows this TS
Counter	Serial counter, which is used at the decoder to detect lost packets
Opt. adaption	Optional field made up a number of subfields
Length	Number of bytes in the adaption field
Discontinuity id	Indicates time-base or counter discontinuity for the current TS packet
Random access	Indicates information to aid random access at this point
ES priority	Indicates, among packets with the same PID, the priority of the elementary stream data carried, i.e. if video intra-code slices hold higher priority
5 flags	PCR flag: adaption field contains PCR OPCR flag: adaption field contains OPCR Splicing flag: indicates if a splice countdown field is present Transport private data flag: adaptation field contains private data bytes Extension flag: indicates the presence of an adaptation field extension
Optional fields 1	Optional field 1, which is made up a number of subfields
PCR	The PCR indicates the intended time of arrival
OPCR	OPCR assists in the reconstruction of a single programme TS from another TS
Splice countdown	A positive value specifies the remaining number of packets of the same PID If negative, indicates that the packet is the *n*th following the splicing point
Transport Private Data	Indicates the number of private data bytes
Private data byte	Not be specified by ITU-T | ISO/IEC
Adaption field Extension length	Indicates the number of bytes of the extended adaptation field data
3 flags	Ltw flag: indicates the presence of legal time window field Pwise rate flag: indicates the presence of the piecewise rate field Seamless splice flag: splice type and DTS next AU fields are present
Optional fields 2	Optional field 1, which is made up a number of subfields
Ltw valid flag	Indicates that the value of the legal time window offset shall be valid
Ltw offset	Information for devices to reconstruct the state of the buffers
Piecewise rate	Specifies a bit rate which is used to define the end times of the legal time windows of transport stream packets of the same PID
Splice type	If the elementary stream carried in that PID is an audio, it will be '0'. If it is a video stream, this indicates the conditions for splicing purposes
DTS next AU	Decoding time stamp next access unit: in the case of continuous and periodic decoding through this splicing point, it indicates the decoding time of the first access unit following the splicing point

- A third option, which includes *TS and RTP*, enjoys all the features but is the least efficient as it increases the overheads. The whole encapsulation is [Payload + TS + RTP + UDP + IP].

MPEG-2 TS is today probably the most common transport in digital video broadcast (DVB) and also in IPTV. Many of the pioneers in IPTV started with MPEG-2 and have decide to keep the same TS, even though the payload actually could be MPEG-4 or WM9, because they are confident with it and already have the mechanisms and the experience to monitor the service.

3.7.4.1 MPEG-2 Transport Stream

Transport stream features include the multiplexing of the audio/video/data streams, the description of the TS structure, indications to decode and display the video, the control of transmission error and the synchronization of the receiver to correctly display the video.

The TS is made up of a continuous sequence of transport stream packets (TSPs) that contain the audiovisual information of the programmes (see Figure 3-21). The TSP have a header that describes and manages the packet contents (see Figure 3-23). The packet identifier (PID) is the first field to decode because it indicates the type information contained in the payload, which can be either audiovisual contents or information intended to manage the decoding. Among the fields that manage the decoding there are two essential ones (see Figure 1-23):

1. Programme specific information (PSI) to describe the structure of the TS; and
2. Programme clock reference (PCR), used to synchronize coder and decoder.

3.7.4.2 Programme Specific Information

The programme specific information (PSI) is made of tables that describe the TS structure, providing information about the programme mapping and relationship between the different transport stream packets. Some of the PSI have reserved PID to be distinguished quickly by the decoders (see Figure 3-24). Here we have some of the most relevant PSI tables:

- *The programme association table* (PAT), which has a reserved PID = 0×00, tells the decoder about programmes transported in the TS. The PAT has the pointers to the programme map tables (PMT), which in turn identify the TS packets of video, audio and data that make up the programme (see Figure 3-24).
- *The programme map table* (PMT), which has a PID indicated in the PAT, identifies and describes all the audio, video and data packets that are part of a each programme. Thanks to this information, the receiver can select and combine the different components that are finally decoded and sent to be watched and listened to.
- *The conditional access table* (CAT), which has a reserved PID = 0×01, holds information that is used by the decoder to descramble encrypted programmes in services like pay-per-view or private channels. The table lists the PID value code that provides the entitlement management messages (EMM).

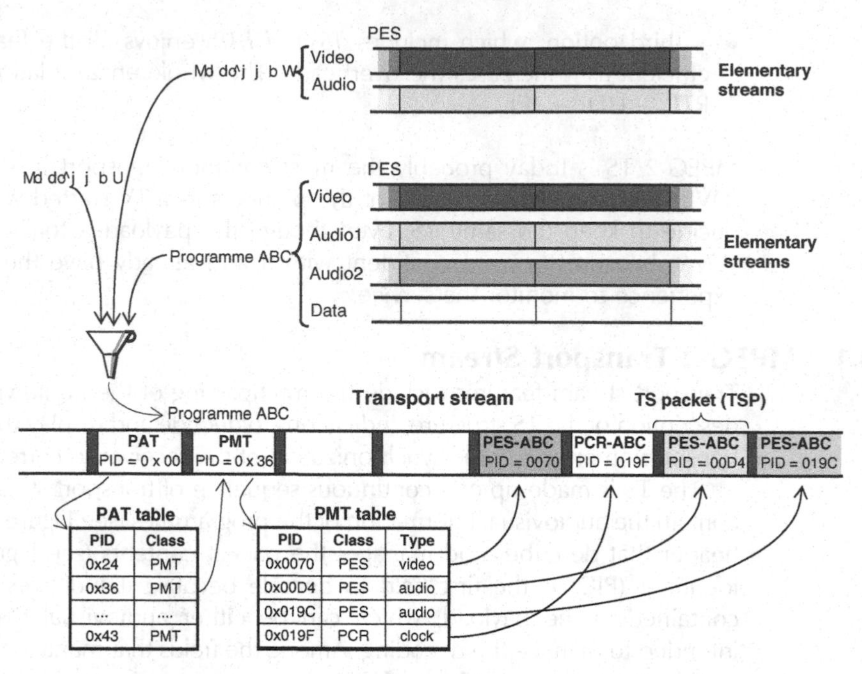

Figure 3-24. The PSI tables describe the structure of the TS, providing the information to retrieve all the contents regarding a programme.

- *The transport stream description table* (TSDT), which has a reserved PID = 0×02, is a private table implementation defined by MPEG, which can be used to provide compatibility with other delivery systems.
- *The network information table* (NIT), has a reserved PID = 0×10. The contents of this table are private, containing information about the physical parameters of the bearer network such as mappings of user services, channel frequencies, satellite transponder and modulation characteristics.
- *The null packet* uses the PID = 0×FF, intended for rate stuffing in order to avoid the bottleneck of insufficient resources. The payload is undefined.

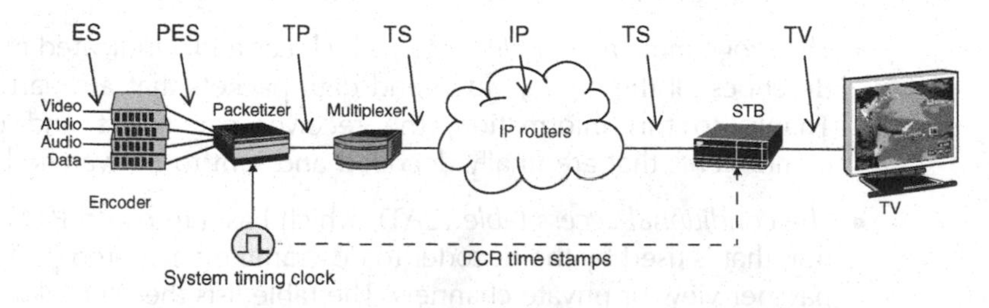

Figure 3-25. IPTV. The PCR is used to synchronize the encoder and decoder.

The first decoding operation the STB has to do is to find the PAT in order to know which programmes are available (see Figure 3-24). Then, once all the PID of the TS packet that make the programme have been identified, the decoder has to capture and drop them to be decoded and finally displayed.

3.7.4.3 Timing and Programme Clock Reference

Video signals are coded according the system timing clock (STC) that runs at 27 MHz, which is used to synchronize the encoder and decoder. The timing information is included on the transport stream as a time stamp (see Figure 3-25) which is known as the programme clock reference (PCR). At the receiver site the PCR time stamps are used to synchronize the internal clock, which is absolutely fundamental to decoding and displaying the television programme.

3.7.5 Packet Distribution and Delivery

Converged IP architectures for IPTV are based on statistical multiplexing to deliver the audiovisual information. In order to distribute IPTV signals across the network until they are delivered to the subscribers, an IP multicast addressing mode is used.

3.7.5.1 Broadcast, Multicast, Unicast

Multicasting is the addressing mode to distribute for IPTV to save bandwidth and processing capacity since the unicast and broadcast alternatives do not scale well in IP networks. Broadcasting every programme to all the nodes would be inefficient and would unnecessarily overload links and nodes. Unicast is used for VoD, which works very much how the Internet works, that is, establishing an point-to-point relationship between server and subscriber.

Table 3-10. VC-1 profiles and levels.

Level	Profile	Bit rate	Resolution
Simple	Low	96 kbit/s	176 × 144
	Medium	384 kbit/s	240 × 176
			352 × 288
Main	Low	2 Mbit/s	320 × 240
	Medium	10 Mbit/s	720 × 480
			720 × 576
	High	20 Mbit/s	1920 × 1080
Advanced	L0	2 Mbit/s	352 × 288
	L1	10 Mbit/s	720 × 480
			720 × 576
	L2	20 Mbit/s	720 × 480
			1280 × 720
	L3	45 Mbit/s	1920 × 1080
			1920 × 1080
			1280 × 720
	L4	135 Mbit/s	1920 × 1080
			2048 × 1536

However, unicast cannot be used for real-time IPTV because server and nodes cannot deliver well when thousands or even millions of subscribers want to watch exactly the same programme at the same time. The IPTV server forwards the television datagrams to a multicast-enabled router, that forwards them to other multicast routers until they reach all hosts associated with the multicast group.

3.7.5.2 The Access Bottleneck

The limited bandwidth available at the access network means that only a few IPTV programmes can be received simultaneously. The number of programmes depend on several factors, including the video resolution and the access bandwidth. The limit can be partially bypassed using more efficient compression algorithms, but in general only a few channels can be delivered simultaneously (see Figure 3-7); then the zapping function has to be implemented at the network side to change the channel.

To tune a channel, subscribers must be part of the multicast group that forwards the signal using the Internet Group Management Protocol (IGMP). When the subscribers want to stop receiving the programme, an IGMP message is sent to leave the group, and the router no longer forwards datagrams.

3.7.5.3 Multicast and IGMP

The IGMP protocol in combination with multicast networks has been designed to make efficient use of the available bandwidth at the access that can only accommodate a limited number of IPTV channels simultaneously. The multicast traffic uses a special set of IP addresses in the 224.0.0.0 to 239.255.255.255 range. Each address is associated with a TV channel.

3.8 Windows Media and VC-1

VC-1 is a video codec specification that has been recently standardized by the Society of Motion Picture and Television Engineers (SMPTE) based on Microsoft developments. The VC-1 is transport and container independent, which means that can be distributed over MPEG-2 TS and RTP as well.

VC-1 supports three profiles (see Table 3-10). The Simple and Main profiles have been ready for several years, and existing implementations have supported the creation and playback of video contents. The completion of the Advanced profile delivers high-definition content, either interlaced or progressive, across multiple medium and devices.

3.8.1 VC-1 Profiles and Levels

VC-1 has a number of profile and level combinations to support the encoding of many types of video. The profile determines the codec features that are available, and determines the required decoder complexity.

3.9 Service Provision

Subscribers of IPTV have at least the same expectations about the quality of the television, as they do of existing TV services. Television is a mature market that has been supplied by many different operators for years. Therefore it is of key importance to match customers' expectations from the first day when the service is launched commercially in such mature markets a correct plan and a detailed verification of the service quality are fundamental before offering it commercially. The trial–error–fix strategies do not work and may have bad consequences for the IPTV business.

The first consideration to take into account is to realize that IPTV is a high bandwidth demanding and also a real-time service, which is more sensitive to impairments than other services based on voice or data.

A bad quality of experience (QoE) by customers has a direct impact on revenues and on the OPEX due to increased calls to customer support centers that have to fix the problem, often having to send engineers to the customer premises, which is always expensive and, in some respects, a defensive reaction. Even more costly could be the impact on the reputation of the service; we already know some IPTV experiences that failed after a bad launch in terms of quality, and that news of the poor quality quickly entred the public domain.

3.9.1 Quality of Experience

Consumers demand high quality for the services, however they cannot describe the IPTV quality using technological terms. Often the description of the experience is only brought to the service provider's attention when the customer complains in plain terms such as 'the image is really bad', 'I can't support such a blurry image' or 'we can't zap unless we wait for ten seconds'.

Quality of experience (QoE) is intended to measure how good the service is from the customer's point of view, rather than from the service point of view as the quality of service (QoS) does. QoE is a subscriber-centric paradigm (see Figure 3-26). For instance, when the IPTV service is degraded the following occur:

- *No service*. No service at all, frequent service interruptions, wrong formatting, very high latency, authentication problems to watch private programmes like pay-per-view.
- *Video degradation*. Frame freezing, blurring, visual noise, loss of colour, edge distortion, pixelation, tiling.
- *Audio degradation*. Drop-out, bad lip synchronization, voice distortion, bad signal-to-noise relation, echo.
- *Interaction*. No zapping capability, loss of VoD functions like start/stop/foreward.

Whatever the case is, the QoE is a subjective perception from the subscriber that operators have to evaluate by measuring an objective combination of parameters. The consequence of a bad service is a miserable video quality which often has the same symptoms, independently of the causes, including a combination of frame freezing, pixelation, loss of colour,

Figure 3-26. Quality of experience is the human perception of how good the service is.

tiling (the picture gets broken up into little blocks), chroma noise or loss of interactivity (see Table 3-11).

3.9.2 Network Impairments

Best-effort networks, like the Internet, have difficulties in delivering differential QoS appropriate for each type of application. This fact is a serious inconvenience for all native IP architectures that, in general, need the aid of additional elements to support IPTV. These features are required to establish admission policies, to control congestion or simply to manage different traffic priorities. The final objective is to guarantee the QoS required for each application and especially the most sensitive, like real-time IPTV. New entrants may be tempted to set up QoS just over provisioning resources whenever necessary; however we think this strategy is wrong, and for the medium and long term could be even more expensive and does not guarantee the result at all.

Quality requirements are different for each application; typically traditional IP data services are happy with best-effort networks, but IPTV or VoIP services require more strict control of

Table 3-11. Events that modify the quality of the IPTV services.

Where	QoS event
Head-end and contents	Coding distortion Server overload Encoder fault Error indication
Transport stream	PCR jitter Continuity error count Synchronization error Interarrival jitter Error on PSI table Unacceptable latency
Distribution network	Network contention IP packet loss, jitter, delay RTP packet loss, jitter TCP Retransmissions
Transaction	IGMP latency (IPTV) RTSP latency (VoD)

Table 3-12. IP network QoS control (ITU-T Y.1541).

Class	Application	Upper bound QoS parameters			
		Delay	Jitter	Loss	Error
0	Real-time, jitter-sensitive, high interaction (VoIP, IPTV, VTC, VoD)	100 ms	50 ms	1×10^{-3}	1×10^{-4}
1	Real-time, jitter-sensitive, interactive (VoIP, IPTV, VTC, VoD)	400 ms	50 ms	1×10^{-3}	1×10^{-4}
2	Transaction data, highly interactive (signalling)	100 ms	Undefined	1×10^{-3}	1×10^{-4}
3	Transaction data, interactive	400 ms	Undefined	1×10^{-3}	1×10^{-4}
4	Low loss only (short transactions, bulk data, video streaming, VoD on local disk)	1 s	Undefined	1×10^{-3}	1×10^{-4}
5	Best-effort IP networks	Undefined	Undefined	Undefined	Undefined

packet loss, delay, jitter and errors. The ITU-T has published the Rec. ITU-T Y.1541 regarding IP Network QoS control (see Table 3-12). The achievement of this level of quality can often be quite challenging. Let us analyse in detail most common misfunctions which can be distinguished into two types of impairments:

1. *Anomalies*: these are the smaller disagreements that can be observed, such as bit error, but anomalies do not cause interruption of the service. Human intervention is, in principal, not required immediately. However, when the density of anomalies is higher, the degradation may affect the quality of the service and then the anomaly can be considered a failure.
2. *Failures*: this is when a fault persists so that the system is unable to perform the expected functionality or even to deliver the service. If the network has been configured as fault-tolerant then a protection mechanism should be triggered to switch to an alternative

Table 3-13. IPTV metrics to evaluate the quality.

Measurement	Description
Video MOS	MOS score in a session
Compression impact	Degradation that can be attributed to the compression
Frame and bit rates	Video rates in frames/s and Mbit/s
Bandwidth use (%)	Percentage of bandwidth used at the access
Packet loss (%)	Percentage of packets that were lost
Max burst length	Maximum period of consecutive packet lost
Out of order	Video packets received out of the sequence
Mean jitter	Mean interarrival packet jitter
Max jitter	Maximum interarrival packets jitter
Unavailable seconds	Time without video service or highly degraded
Number of frames I, B, P	Counter of video frames received

power supply, node, path or subnetwork that would allow delivering the service to be continued with a minimum of interruption. Human intervention is often required.

In the particular case of IPTV, quality delivered to the subscribers can be affected by many different impairments (see Table 3-13). Some factors may depend on traffic conditions; this would be the case for the increment of delays, packet loss or errors. Other factors depend on the design and the quality of the network, for instance an unacceptable IGMP latency or encoder impairments.

3.10 Service Assurance

A systematic approach to the IPTV quality begins with a classifications of the service assurance issues in groups, according the place, the process, or where the event occurs (see Figure 3-27):

- *Content faults*: includes those events generally produced at the contribution network. Monitoring head-end installations permits the assessment of the video quality and determination of the causes of the misfunction.
- *Network impairments*: IPTV applications use a transport and distribution network based on statistical multiplexing; therefore traffic conditions may cause congestion that produces packet loss, delays and jitter.
- *Transaction impairments*: the bidirectionality typical of IPTV rely on protocols like IGMP to manage the IPTV channel zapping and RTSP to control the video streaming on VoD. Both may suffer delay or no functionality at all.
- *Transport events*: the transport, based on MPEG TS or RTP, is fundamental to distributing, synchronizing and decoding the audiovisual streams. Any event at this level can interrupt the ability to decode the programmes contained therein.

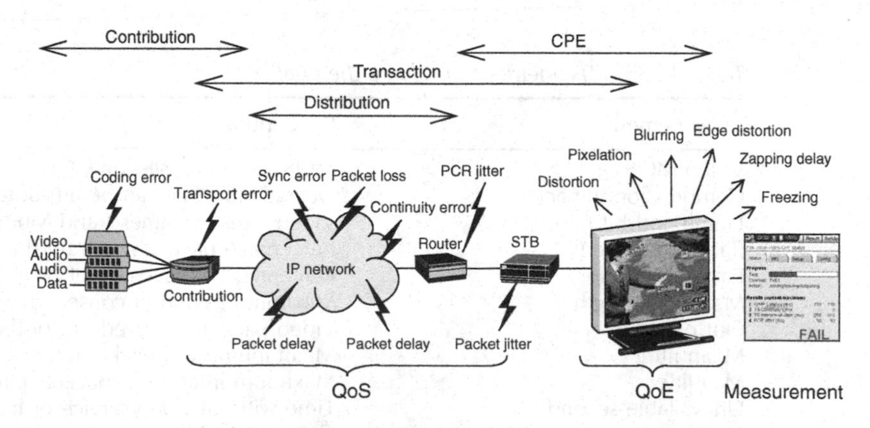

Figure 3-27. Service assurance for an IPTV service.

3.10.1 Content Faults

This group embodies those events occurring before the audiovisual signal is streaming to the network and may include server overloads, audio and video coding distortions and error indications.

3.10.1.1 Coding Distortion

We have seen a number of coding and compression techniques performed by the encoder at the head-end that may cause distortion:

- *interpolative*, like chroma components subsampling;
- *predictive*, like forward predictions that remove redundancies between frames;
- *spatial*, like DCT, which removes internal redundancies of each frame;
- *statistical*, like the Hoffman compression based on mathematical codes.

Compression techniques reduce the size of the audiovisual stream but video loss takes place. At the receive site, the decoder has to perform the inverse processes to play back the video. However the reconstruction of the original is only one approximation because of lossy compressions. This is the coding distortion. The loss factor determines how important the distortion is, and the impact on the quality after the signal has been compressed and decompressed.

3.10.1.2 Error Indicator

This occurs when the encoders detect a failure or corrupted contents of the video source. Therefore the presence of this indication is not related to the IP distribution network but reveals that the contribution network has quality issues that have to be addressed.

3.10.2 Network Impairments

3.10.2.1 Bandwidth Degradation

IPTV applications are delivered by a packet network that may suffer performance degradation caused by traffic conditions, congestion or events such as a link or a node failure. It is important to identify the causes better during the roll-out of the network during the trial period. Once the network is in service the traffic must be monitored continuously, at least at the critical points, to avoid surprises during commercial exploitation.

3.10.2.2 Transport Stream Continuity

The mapping of TS packets into Ethernet/IP packets results in frames that contain seven TSPs. Any out of order, lost or duplication of IP packets are errors that will result in degradation of the video quality. The impact of the error will depend on the nature of the event and the number of packets and the type of frames affected. The loss of I frames is always worse than the loss of B or P frames.

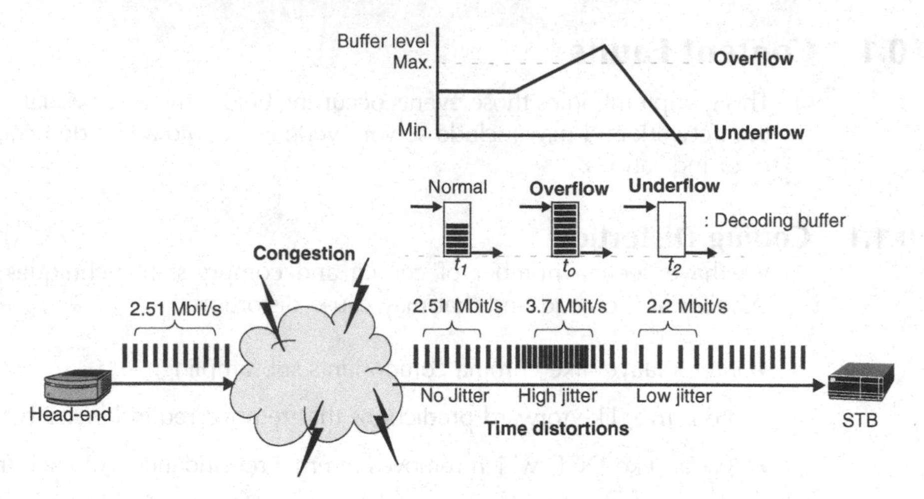

Figure 3-28. Interarrival packet jitter is filtered with the decoding buffer. Higher rates may produce decoder buffer overflow while lower rates may cause buffer underflow. The consequence is, in both cases, bad quality.

3.10.2.3 Interarrival Packet Jitter

There are several issues related to video synchronization. The most common is caused by traffic conditions that produce packet jitter or variation in the delivery of IP packets. This effect generates impairments on recovering the PCR signal, which is used to synchronize the 27 MHz receiver clock.

IP packets transporting the audio and video signals are generated with a specific rate at the ingress node of the network. Jitter occurs when packets are delivered in bursts with inter-packet gaps shorter or larger than when they were generated (see Figure 3-28). Packet jitter is inherent to the IP networks and receivers use buffers to filter it. The problem is that buffers cannot be too big because they increase the delay. The larger the buffers is, the more jitter is filtered, but the packet delay becomes unacceptable so the buffer size is a trade-off between filtering and delay.

There is an obvious limitation of the buffer capacity to filter jitter. Further than this, a buffer overflow or a buffer underflow occurs, causing a service degradation (see Figure 3-28). Peaks of jitter can cause buffer overflow, while jitter valleys cause buffer underflow. In both cases a temporal but severe degradation of the image occurs, either because there are no frames to decode or because some frames have been dropped.

3.10.3 Transaction Impairments

3.10.3.1 Channel Zapping Delay

Channel zapping delay is defined as the delay in changing programmes at the subscriber site. A delay higher than 1 s will produce a bad perception in the customers, this soon will

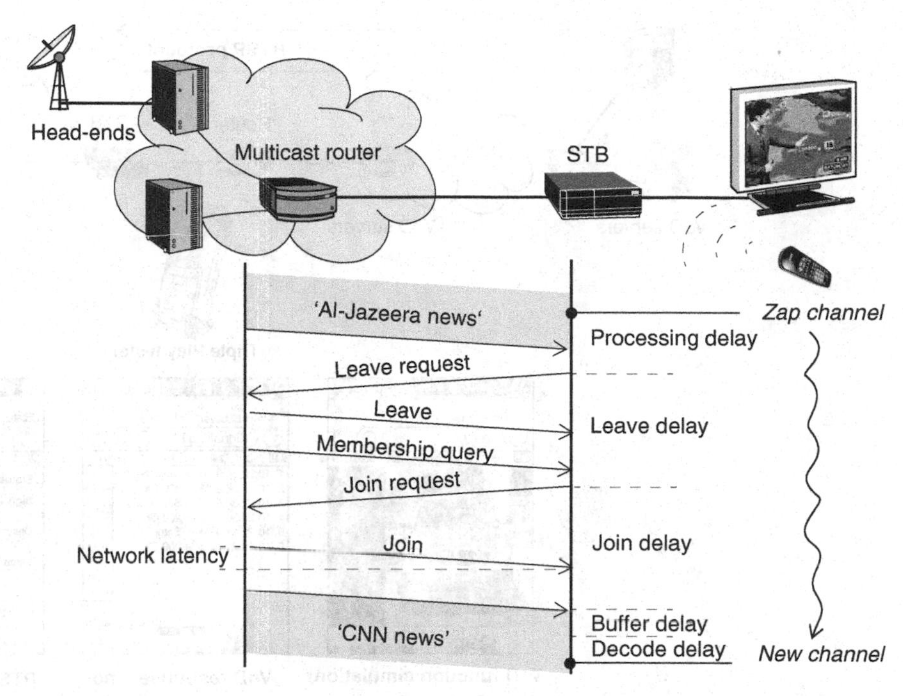

Figure 3-29. In the customer's perception of quality, channel zapping should never take more than 1 or 2 s.

compare it with broadcasters' TV. The time to perform a channel change is not only caused by the IGMP protocol execution; there are also other contributing factors (see Figure 3-29):

- *STB command processing* is the time taken between a channel zap being requested and the IGMP leave message being sent.
- *Network latency* includes all the transmission latencies due to the network.
- *Leave latency* is the time taken to stop the transmission of the old channel.
- *Join latency* is the time taken to start the reception of the new channel.
- *Buffer delay* is caused by the buffer necessary to filter the jitter, but causing delay.
- *STB delay* is the time to find PAT and PMT tables, and the audio and video streams.
- *Decodification delay* is the time to find the first I frame and to decode it.

The IGMP delay could be affected by source server, the network traffic, the multicast flow, congestion on the servers, etc.

3.10.3.2 Real-time Streaming Protocol Verification

The RTSP is an application-level protocol to manage at the delivery a single or several time-synchronized streams over continuous media. RTSP provides a framework to enable

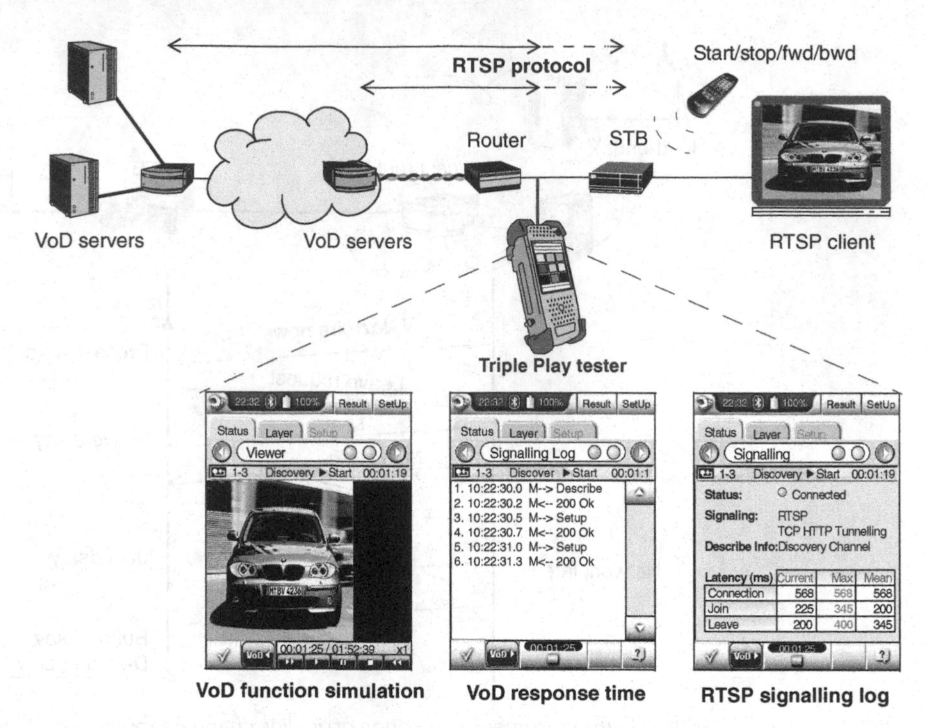

Figure 3-30. Verification of the RTSP performance by means of a Triple Play tester.

on-demand delivery of real-time data, such as audio and video. The RTSP verification test should determine the throughput, the individual and the collective response to functions like Play, Pause, Rewind or Forward (see Figure 3-30).

3.10.4 Transport Impairments

Defined by the DBV organization, the TR 101 290 is a recommendation to help the operators to set up a digital TV service by means of set tests that can be applied to MPEG-2 TS interfaces. This transport stream is particularly important because it is the most common way to transport, not only MPEG-2 coding, but also MPEG-4 and WM-9.

Tests are grouped into three tables according to their importance for monitoring purposes. The first table lists a basic set of parameters which are considered necessary to ensure that the TS can be decoded. The second table lists additional parameters which are recommended for continuous monitoring. The third table lists optional parameters which could be of interest for some applications.

3.10.4.1 First Table of Priority 1

This category classifies the critical elements necessary for decoding of the MPEG signal. The events can be grouped into three types (see Table 3-14):

- *TS sync loss, PAT and PMT error,* which are events that cause service interruption or severe distortion. For instance if the PAT is missing then the decoder can do nothing, no

Table 3-14. *First priority: basic monitoring necessary for decodability.*

Field	Description
TS sync loss	Loss of synchronization with consideration of hysteresis parameters
Sync_byte_error	Synchronization byte not equal 0×47
PAT Error	PID 0×0000 does not occur at least every 0.5 s or a PID 0×0000 does not contain a table_id 0×00
Continuity error	Incorrect packet order, or a packet occurs more than twice, or lost packet
PMT error	Sections with table_id 0×02, do not occur at least every 0.5 s on the PID which is referred to in the PAT
PID error	Referred PID does not occur for a user-specified period

programme is decodable; without a PMT the consequences are similar for the corresponding programme, which cannot be decoded.

- *Continuity count error* is the first indication that the service is degraded, indicating loss of packets, dropped packets, out of order packets or packet duplications.
- *PID error* occurs when the data of a referenced PID is lost for a period of time.

3.10.4.2 Second Table of Priority 2

The following elements should be kept under control to achieve the quality targets (see Table 3-15):

- *CRC error*, indicating corrupted data in the PSI tables;
- *timing events*, indicating difficulties in recovering the PCR, meaning that synchronization problems are likely;
- *CAT error*, an indication that it is necessary to decode a private programme but no CAT table has been found to unscramble it.

Table 3-15. *Second priority: recommended for continuous or periodic monitoring.*

Field	Description
Transport error	Transport_error_indicator in the TS-Header is set to "1"
CRC error	CRC error occurred in CAT, PAT, PMT, NIT, EIT, BAT, SDT or TOT table
PCR error	PCR discontinuity of more than 100 ms occurring without specific indication. Time interval between two consecutive PCR values more than 40 ms
PCR accuracy error	PCR accuracy of selected programme is not within ±500 ns
PTS error	PTS repetition period more than 700 ms
CAT error	A packet indicates that data must be unscrambled but no CAT table is found

Table 3-16. Third priority: application dependent monitoring.

Error	Description
NIT	Id other than 0×40 or 0×41 or 0×72 found on PID = 0×0010 No section with id 0×40 or 0×41 in PID = 0×0010 for more than 10 s
SI repetition	Repetition rate of SI tables outside specified limits
Buffer error	Overflow of transport buffer, or Overflow of multiplexing buffer, or Overflow or underflow of elementary stream buffer, or Overflow or underflow of main buffer, or Overflow of PSI input buffer.
Unreferenced PID	PID not referred to by a PMT within 0.5 s
SDT error	SDT TS not present on PID 0×0011 for more than 2 s
EIT error	EIT-P/F not present on PID 0×0012 for more than 2 s
RST error	Sections with table_id other than 0×71 or 0×72 found on PID 0×0013
TDT error	TDT not present on PID 0×0014 for more than 30 s
Empty buffer	Transport buffer not empty at least once per second
Data delay	Delay of data superior to 1 s

3.10.4.3 Third Table of Priority 3

This is an optional category since many implementations have simplified the structure of the transport stream. Priority 3 table has a reduced interest for IPTV systems except if they have this level of service or require an accurate control of the buffers. The following elements should be kept under control to achieve a high level of quality targets (see Table 3-16):

- *buffer error*: to calculate this indicator a number of buffers of the MPEG-2 reference decoder are checked to determine whether they will have an underflow or an overflow;
- *unreferenced PID*: each non-private programme data stream should have its PID listed in the PMTs;
- *SDT error*: the SDT describes the services available. Without the SDT, the STB is unable to give the viewer a list of what services are available.

3.10.5 Media Delivery Index

The Media Delivery Index (MDI), defined in the RFC 4445, is a measurement for IPTV that gives an indication whether the network is able to transport good quality video independently of the coding scheme. The MDI is generally expressed as two digits XX:YY that refer to two measurements, the delay factor (DF) and media loss rate (MLR):

- DF is the maximum difference between the arrival and the drain of the TS packets. It is a good indicator of the jitter present in the stream and it is useful to calculate the buffer (or the leaky bucket size) necessary to filter the jitter.
- MLR is the count of lost and out-of-order TS packets over a period of time. Out-of-order packets are also included because a lot of consumer equipment does not reorder packets.

MLR is considered a good indicator to evaluate the video quality of experience using a scientific methodology rather than relying on subjective human observations like the video MOS. The MDI is often measured at intermediate points of the network to monitor where the issues that deteriorate the QoE are located. The methodology is simple, based on comparing the DF:MLR components between two or more consecutive points and it is possible to quickly identify the nodes or the links that are causing the degradation.

Selected Bibliography

[1] ITU-R BT.601 Encoding Parameters of Digital Television for Studios (1982–1994).
[2] ITU-T Rec. H.222.0 (2000 E) Information Technology – Generic Coding of Moving Pictures and Associated Audio Information: Systems.
[3] ETR 290 Digital Video Broadcasting (DVB); Measurement Guidelines for DVB Systems, ETSI European Telecommunications Standards Institute.
[4] RFC 2326 Real Time Streaming Protocol (RTSP) 1998.
[5] ISO/IEC 13818-1 Second Edition 2000-12-01 Information technology – Generic Coding of Moving Pictures and Associated Audio Information.
[6] IPTV monitor issue 2, IPTV monitor issue 3. Accenture.
[7] ITU-T Recommendation H.262. Information technology – Generic coding of moving pictures and associated audio information: Video. 2000.
[8] ITU-T Recommendation H.264. Advanced video coding for generic audiovisual services. 2005.

Chapter 4: Signalling

Enabling Triple Play services involves meeting certain challenges. Some of these challenges are related to the network architecture, others to end-to-end service provisioning. Network-architecture-related issues include finding the right QoS framework and the best technologies to provide bandwidth to subscribers, as well as looking for the optimum switching and routing mechanisms. End-to-end service provisioning involves managing multimedia calls between subscribers and adding intelligence to the network.

Signalling is also a service provisioning issue, and currently there are two approaches to it. The first is the ITU-T approach, described in Recommendation H.323. The solution offered by the IETF is to use the Session Initiation Protocol (SIP). These solutions have similar purposes and features, but they follow different design philosophies and are therefore not compatible.

Both H.323 and SIP are essential parts of two different IP multimedia architectures that involve not just one protocol, but many different bits and pieces that can only work properly when they are put together. Some of the protocols that are built on the IP multimedia architecture are the Real-time Transport Protocol (RTP), Real-Time Streaming Protocol (RTSP) and Media Gateway Control Protocol (MGCP) or Megaco. Some of them, such as the RTP, form a part of both SIP and H.323. Others, for example the Session Description Protocol (SDP) or H.245, are specific to just one of them.

Triple Play: Building the Converged Network for IP, VoIP and IPTV Francisco J. Hens and José M. Caballero

4.1 The Real-time Transport Protocol

The RTP is a protocol defined by the IETF to carry time-synchronized unicast or multicast data across IP networks. The typical application for RTP is to carry real-time digital audio or video streams, but it can also be used for other applications, such as storage of continuous data.

The RTP protocol provides payload type identification and sequencing, but its most important feature is *payload timestamping*. Packet networks may delay packets of a data stream for a variable amount of time. The RTP is designed to carry timing information with the data. Marking traffic with timestamps makes it possible to re-time the received data with the source timing and with accuracy good enough for many multimedia streaming applications.

Besides unpredictable delays, packets in IP networks may be lost along the transmission path due to congestion or transmission errors. To deal with this situation, a sequence number is included in the RTP. This number allows the receiving end to calculate how many and which packets are missing in the stream.

Another important aim of the RTP is *payload identification*. There are many alternative ways to transport multimedia streams in packet networks. The RTP provides a common transport framework for all of them. Different payload types may be treated differently at the receiving end. The RTP tells the receiver the type of data that is being transported in the payload and how this data has been encoded.

The RTP is just one component of the IP-based multimedia network framework. It offers some services, such as packet timing and sequencing, but many others are out of the scope of this protocol. For example, the RTP protocol does not offer any mechanisms to guarantee a certain QoS for the data streams it carries. It relies on the QoS mechanisms provided by the network, such as packet marking, shaping and scheduling. Although it is not mandatory, RTP streams are usually delivered by using the User Datagram Protocol (UDP) that offers

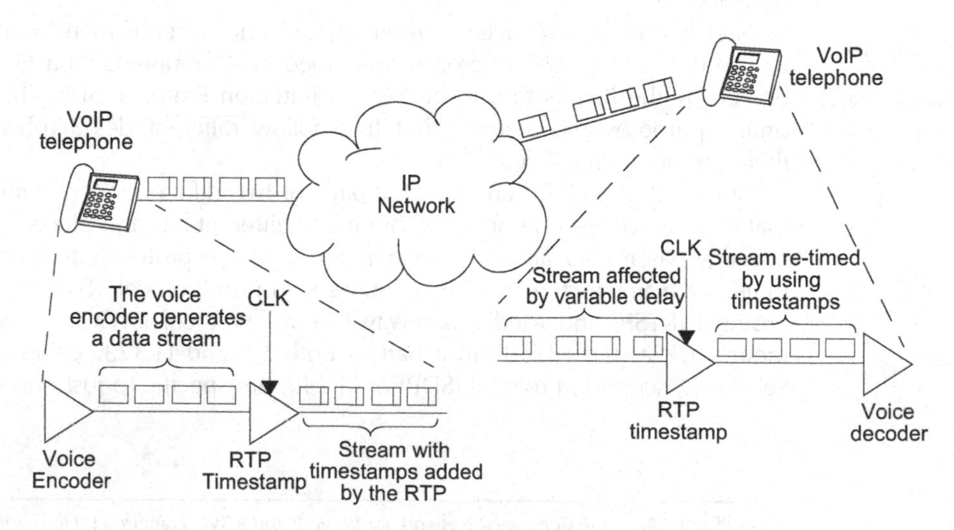

Figure 4-1. Re-timing of a voice data stream performed by the RTP protocol.

non-connection-oriented transport of IP traffic. The UDP provides low overhead and simplified processing, which is convenient for real-time traffic. The UDP is not a reliable transmission protocol in the sense that it does not offer any error recovery mechanisms. In real-time sessions, there is no time to retransmit lost or corrupted packets. If the data is not tolerant to errors or if the error rate is too high, an error recovery scheme based on forward error correction (FEC) must be implemented. Flow control is not supported by UDP either, so some other mechanisms should be used if needed.

4.1.1 Synchronization Sources and Contributing Sources

The RTP streams can be re-timed at the destination, because the generation time of each packet is recorded and delivered with the data. The recorder of the timing information is identified by a 32-bit number and sent with the data as well. This identifier is called the Synchronization Source (SSRC). The same transmitter may generate multiple RTP streams with different SSRCs. For example, two RTP streams with two different SSRCs may be transmitted during a videoconference, one of them for the video, and the other one for the audio stream. Each SSRC, even when coming from the same source, has dedicated sequencing and timestamping.

The simplest case of an RTP stream is communication between two parties; a transmitter and a receiver. A VoIP telephone call to a single person is a good example. In this case, the transmitter encodes the data to be transmitted, and generates an RTP stream. The data is sent directly to the receiver. The transmitter identifies itself with its own IP address, but also with a randomly chosen 32-bit number: the SSRC.

Note that the SSRC is chosen randomly by the transmitters. This would probably not cause a collision, but the parties involved in the transmission must be able to identify and solve this situation.

Often, RTP applications involve more than one receiver or more than one transmitter. Distributing TV signals or multiparty VoIP telephone calls are two important examples of this. When the multi-transmitter situation arises, it might be necessary to combine the

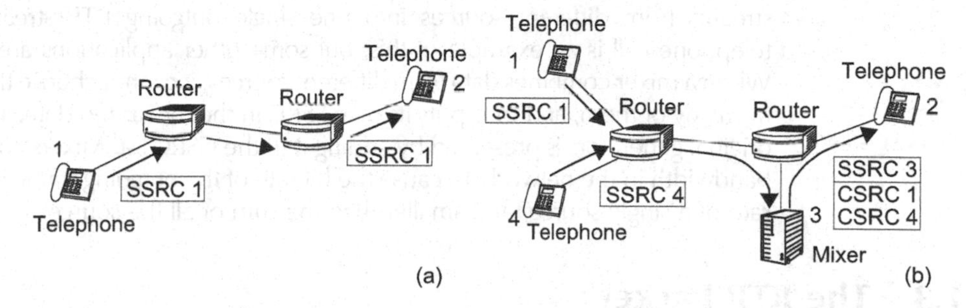

Figure 4-2. Examples of RTP streams: (a) simple RTP session with one transmitter and one receiver. The transmitter identifies itself as the synchronization source for the RTP stream. (b) Multiparty RTP stream. The transmitters deliver the traffic to an intermediate host, a mixer. The streams are combined, re-timed and delivered to the destination by the mixer. The mixer identifies itself as the synchronization source for the new RTP stream, but the original transmitters are still identified as contributing sources.

streams from every transmitter into a single RTP stream before delivery. There is a special device in charge of this task called the mixer.

If a mixer is used, RTP streams are not sent directly to the destination, they are delivered first to the mixer. Thus, the destination IP address of the RTP packets is sent to the IP address of the mixer. The transmitters identify themselves by their IP address and SSRC numbers. Once the packet arrives at the mixer, it is processed and a single RTP stream is generated from the multiple input streams. The mixer sends the new RTP data to the receiver. To do this, it is necessary to set the destination address of the new packets to the IP address of the receiver. The mixer identifies itself by setting the source address of the RTP packets to its own IP address, and setting its own SSRC. The original sources of the RTP stream processed by the mixer are listed in the RTP packets as contributing sources (CSRC). The number of CSRCs in the RTP stream is limited to 15. When the transmission involves more than 15 transmitters, only some of them can be identified.

4.1.2 Translators and Mixers

Translators and mixers are intermediate systems that work at the RTP level. They are in charge of protocol and format translations.

An RTP framework involving only end-user network equipment would be more simple, but some situations arise in practice that make it necessary to use mixers and translators to accommodate the network's resources and data format for different receivers.

Mixers and translators are similar, but the former can do more than the latter:

- *Translators* are intermediate network systems that work at the RTP level. They usually change the format of the incoming RTP packets and forward them to their final destination with their SSRC identifier untouched. The operations performed by translators include downsampling of RTP data for low bandwidth networks, changing the encoding of video or audio streams, and tunnelling of multicast data across firewalls.

- *Mixers* can potentially do the same tasks as translators, but additionally they can combine streams from different sources into one single outgoing RTP stream. A multi-party telephone call is an example of this, but some other applications are possible as well. When a mixer combines data from different sources, it cannot choose the SSRC identifier from any of them, and thus puts its own SSRC in the forwarded data. The identity of the original generators is preserved by adding it to the CSRC list. Mixers contribute to saving bandwidth in the network, because the bit rate of the outgoing traffic is similar to the bit rate of a single source, and smaller than the sum of all the sources.

4.1.3 The RTP Packet

The RTP features are shown in the packet format of the protocol. RTP packets provide a way to define the payload type, use timestamps, sequence numbers, SSRC, CSRCs and so on. The fields of an RTP packet header are the following:

- *Version (V)*: gives information on the version of the RTP. The current version is 2.

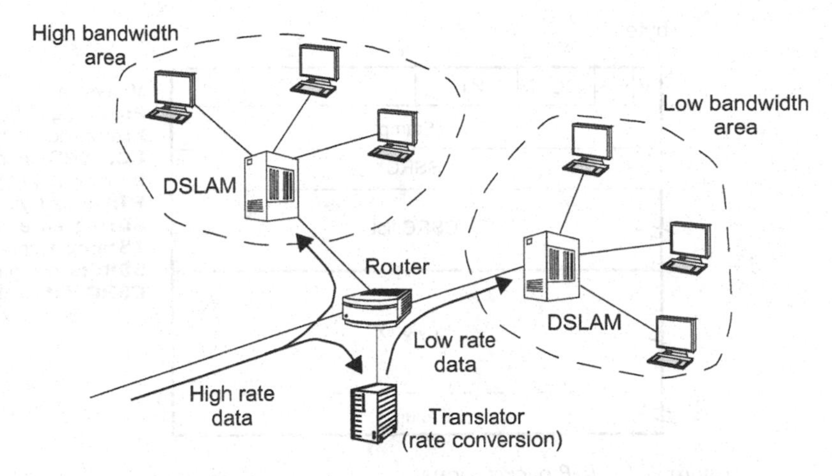

Figure 4-3. One application for translators is rate conversion. They can down-convert the rate of RTP data to be routed to users with low-rate access to the service. This approach makes it possible for the users to take the maximum profit out of their connections.

- *Padding (P)*: this bit is set to 1 to tell the receiver that the packet contains one or more additional padding bytes at the end that do not carry information. The number of bytes to be discarded is carried in the last padding byte.
- *Extension (X)*: this field is set to 1 to tell the receiver that the optional RTP header extension is enabled. Header extensions may be implemented by proprietary versions of the RTP that require additional information to be carried in the RTP header.
- *CSRC Count (CC)*: number of CSRC identifiers for the RTP packet. All CSRC identifiers are listed in the CSRC list field of the RTP header.
- *Marker (M)*: the actual meaning of this bit depends on the profile. It can be used to mark boundaries of higher-level frames, or to signal extra payload types. It may remain unused in some applications.
- *Payload Type (PT)*: PT specifies the payload format and media type carried in the RTP stream. It is used by the receiver to choose the right decoder. Mapping different encoders and payload formats to bit codes depends on the profile. However, there are default mappings for many common media formats.
- *Sequence number (SQ)*: SQ is a number chosen randomly by the receiver when starting to transmit an RTP stream. This number is incremented by one for each packet sent. This makes it possible for the receiver to detect packet loss.
- *Timestamp (TStamp)*: TStamp is a number that represents the sampling instant of the first byte in the RTP packet payload. It allows the receiver to re-time the incoming data according to the clock used to generate the stream. The frequency of this clock may be different for different payload types, and it is identified in the profile or in the payload format specification. The initial value for this field is chosen randomly.

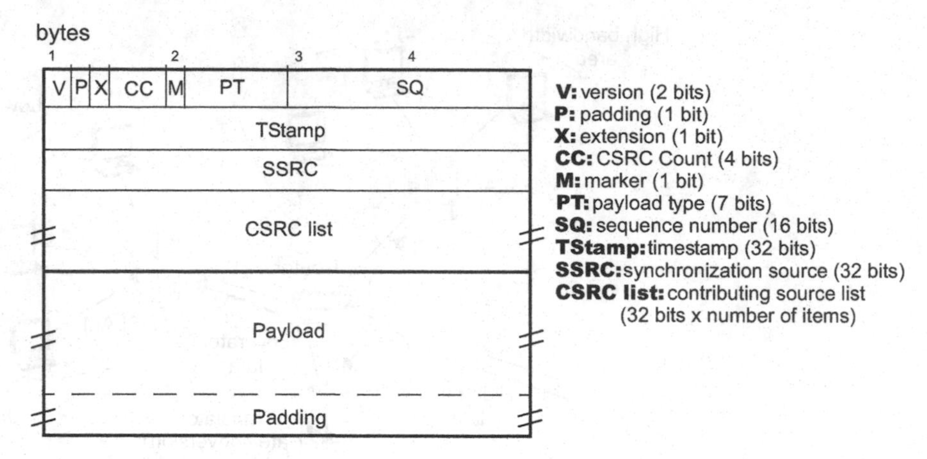

Figure 4-4. RTP packet format.

- *Synchronization Source (SSRC)*: randomly chosen number, identifying the originator of the RTP data stream.

- *Contributing Source list (CSRC list)*: list of contributing sources for the packet payload. The number of items on this list is signalled by the CC header field. The list is limited to 15 items, and a list of 0 items is possible.

4.1.4 Stream Multiplexing

There are different fields in the RTP packet or in the underlying protocols that can be used for stream multiplexing. The SSRC or the PT may be used, as well as the IP address or the transport layer port. Some of the available options are better than others, and some of them must be avoided.

Using only the PT field of the RTP packet has several major inconveniences and must be avoided:

- The format of one or more payloads and their associated PT field values may change during time for the same stream, so it is not a very good idea to use this field to identify the stream.

- Every payload type may require specific sequencing and timestamping and therefore dedicated SSRCs.

- Mixers may have problems in combining incompatible streams. Care must be taken to avoid carrying incompatible payloads under the same SSRC.

Using the SSRC for multiplexing makes it difficult to take profit of some customized IP and UDP layer processing.

- It would be difficult to use different network paths or receive custom QoS treatment for different media types.

Table 4-1. Payload types for RTP defined at RFC 3551

PT	Name	Encoder	Media
0	PCMU	PCM with μ-law scaling as per ITU-T G.711	Audio
3	GSM	GSM standard ETS 300 961	Audio
4	G723	ITU-T Recommendation G.723.1	Audio
5	DVI4	IMA ADPCM wave type, 8000 Hz clock	Audio
6	DVI4	IMA ADPCM wave type, 16000 Hz clock	Audio
7	LPC	Linear predictive encoding contributed by R. Frederick	Audio
8	PCMA	PCM with A-law scaling as per ITU-T G.711	Audio
9	G722	ITU-T Recommendation G.722	Audio
10	L16	Uncompressed audio, 16 bits, 44 100 Hz and 2 channels	Audio
11	L16	Uncompressed audio, 16 bits, 44 100 Hz and 1 channel	Audio
12	QCELP	EIA/TIA standard IS-733	Audio
13	CN	Comfort noise as per RFC 3389	Audio
14	MPA	ISO/IEC 11172-3 and 13818-3 MPEG-1 /MPEG-2 audio	Audio
15	G728	ITU-T Recommendation G.728	Audio
16	DVI4	IMA ADPCM wave type, 11 025 Hz clock	Audio
17	DVI4	IMA ADPCM wave type, 22 050 Hz clock	Audio
18	G729	ITU-T Recommendation G.729	Audio
Dynamic	G726-40	ITU-T Recommendation G.726, 40 kit/s ADPCM	Audio
Dynamic	G726-32	ITU-T Recommendation G.726, 32 kit/s ADPCM	Audio
Dynamic	G726-24	ITU-T Recommendation G.726, 24 kit/s ADPCM	Audio
Dynamic	G726-16	ITU-T Recommendation G.726, 16 kit/s ADPCM	Audio
Dynamic	G729D	ITU-T Recommendation G.729 annex D	Audio
Dynamic	G729E	ITU-T Recommendation G.729 annex E	Audio
Dynamic	GSM-EFR	GSM standard ETS 300 726	Audio
Dynamic	L8	Uncompressed audio, 8 bit samples	Audio
Dynamic	RED	REDundant audio payload as per RFC 2198	Audio
Dynamic	VDVI	Variable rate version of DVI4	Audio
25	CelB	Sun Microsystems CELL-B encoding	Video
26	JPEG	ISO Standards 10918-1 and 10918-2	Video
28	nv	'nv' encoding, version 4 developed at Xerox PARC	Video
31	H261	ITU-T Recommendation H.261	Video
32	MPV	ISO/IEC 11172-3 and 13818-3 MPEG-1/MPEG-2 video	Video
33	MP2T	MPEG-2 transport streams for either audio or video	Audio/Video
34	H263	ITU-T Recommendation H.263 of 1996	Video
Dynamic	H263-1998	ITU-T Recommendation H.263 of 1996	Video

- Networking software makes use of transport layer ports to listen to the network. Two or more processes cannot share the same port, and therefore it is not possible to use customized processes to decode specific media if only SSRC but not the transport layer port is used for multiplexing.

For all these reasons, different RTP media types, for example the audio and video in a videoconference, are normally transmitted with different transport layer ports and/or different IP addresses.

4.1.5 Security

There is a clear need for security in multimedia communications over packet networks. Security in communications is a term that includes different issues:

- *Confidentiality* means that only the intended receiver of the data stream will be able to decode and understand the message. Confidentiality is provided by encrypting the data.
- *Authentication* refers to the ability of the receiver of a data stream to prove the identity of the generating entity. Digital signatures are used for this.
- *Integrity* is achieved when the receiver is able to prove that the message has not been modified by a third party before reception. Here, message-dependent signatures are used that only the transmitter is able to generate.

The RTP protocol supports confidentiality, but authentication and message integrity cannot be provided without a key management and distribution framework that is not natively supported by RTP.

Confidentiality is provided by means of the Data Encryption Standard (DES) algorithm mode. If more options or stronger encryption are needed, it is possible to implement encryption in other protocol layers, for example by means of IP security (IPsec).

4.2 The Real-time Control Protocol

The Real-Time Control Protocol (RTCP) complements the RTP in the following ways:

1. It gives feedback on the quality of transmission. This is, in fact, the primary role of the RTCP. This function is related to the flow and congestion control performed by other transport layer protocols.
2. It carries persistent transport-level identifiers for RTP sources. The SSRC may not be enough for some applications, because it can change when a collision occurs or if the end user equipment is restarted. The RCTP can supply complementary source information.
3. It provides minimum session control. It does not support complex parameter negotiation between the communicating parties, nor membership control. If these advanced features are needed, a session control protocol must be used.

Table 4-2. RTCP packet types

Type	PT id	Purpose
SR	200	Sender report: quality statistics from active senders
RR	201	Receiver report: quality statistics from receivers that are not senders
SDES	202	Source description items: information on the transmitter
BYE	203	Indicates end of participation
APP	204	Application-specific functions

These functions are mandatory for multicast applications, and recommended for all applications. To accomplish these functions, the RTCP packets are multiplexed with the RTP stream. Usually, ports are used for multiplexing; the RTP uses even-numbered ports, and the RTCP uses the contiguous higher odd port number when it is delivered over UDP. The bit rate of RTCP is limited to maintain the performance of the RTP streams that carry user data. For a session, the RTCP band-width is usually no more than 5%.

4.2.1 RTCP Packet Types and Formats

The RTCP packet format depends on the purpose of the packet. The header always contains a 2-bit version field that must match the version field of the RTP header, a padding indicator bit, an 8-bit PT field and a 16-bit length field. Other fields may be present depending on the RTCP packet type.

The receiver knows the RTCP packet type by means of the PT field that is always the second byte of the RTCP packet header. This field may also be used to differentiate RTP and RTCP protocols, because the PT identifiers used by the RTCP are reserved. There are five RTCP packet types: sender report (SR), receiver report (RR), source description (SDES), end of participation (BYE) and application-specific packet (APP) (see Table 4-2).

It is possible for translators and mixers to process and modify RTCP information when it passes across them. For example, several RTCP packets may be combined into a single packet in order to decrease the overhead/payload ratio.

4.2.2 Quality of Service Monitoring

The quality of a packet-based multimedia service can be evaluated with some performance parameters; the most important being delay, jitter and packet loss. The RTCP enables the RTP receiver to generate periodic reports that provide the RTP transmitter or intermediate probes with quality feedback on these three performance parameters.

Statistics are calculated by the communicating parties with the help of specific RTCP fields containing timestamps and packet counts. The reports are generated periodically. If a particular party has transmitted one or more RTP packets since the last report, it will generate a report called SR. Parties that have not sent any RTP packets since the last quality report generate an RR report.

The RR includes information on the sources (identified by their SSRC) from which the receiver has received RTP data packets since the latest report. The number of sources included in the report is given by the Reception Report Count (RC) field of the RTCP packet header. This field has 5 bits, and therefore the maximum number of sources that can be identified

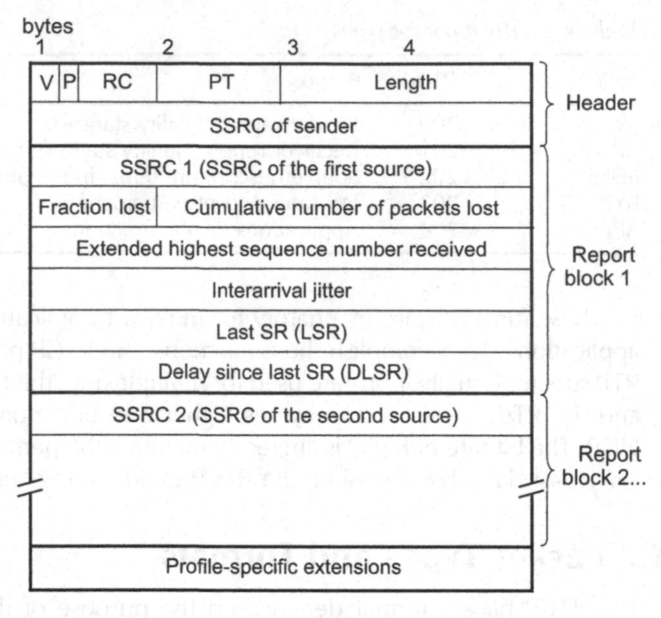

Figure 4-5. RCTP RR packet format.

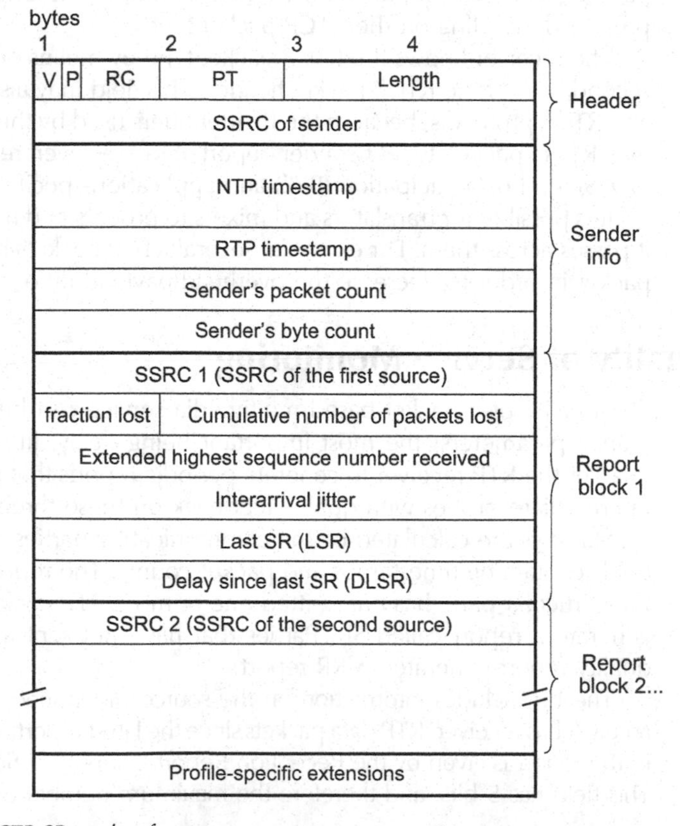

Figure 4-6. RCTP SR packet format.

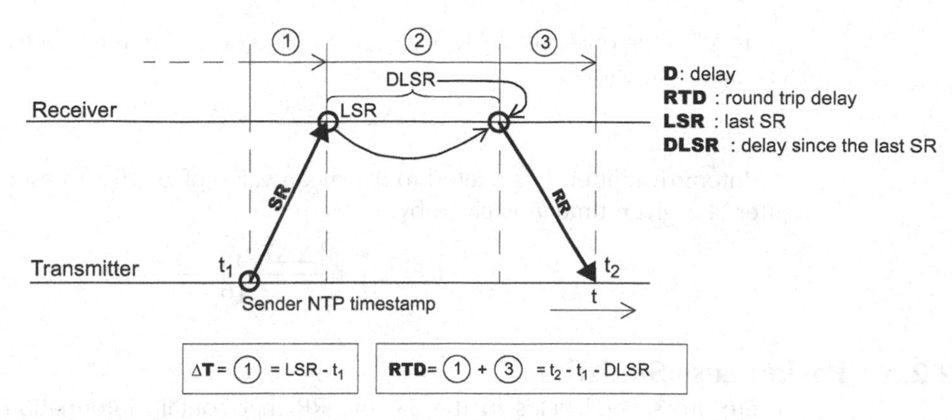

Figure 4.7. Timestamps sent with SR and RR, and estimation of one-way delay and round trip delay.

ranges from 0 to 31. The SR contains similar information as the RR, but the PT field has a different value, and a group of fields includes specific statistics on the RTP transmitter.

4.2.2.1 Delay Statistics

The RTCP enables the transmitters to get an estimation of the Round Trip Delay (RTD), and makes it possible for the receivers to estimate one-way delay.

One-way delay can be calculated as the difference between the originating packet timestamp and the arrival time of the packet. The originating timestamp is recorded by the transmitter and sent to the receiver in a 64-bit field of the SR. The arrival time of the packet is recorded to calculate one-way delay, but it is also sent back to the transmitter within the 32-bit LSR field of the following RR. This makes it possible to calculate the RTD.

The difference between the time the RR arrives to the transmitter and the LSR field of this packet is not the RTD, because it includes the time between receiving the SR and generating the RR. This time does not depend on the transmission performance, but relies on the RTCP protocol itself, which is why it must not be taken into account when calculating the RTD (see Figure 4-7).

For delay estimations, the transmitter and receiver clocks have to be synchronized. The Network Time Protocol (NTP) is used for this purpose. The NTP is a popular protocol used to synchronize the internal time of computers and routers connected to Internet with timing sources that provide an accurate universal time coordinated (UTC) time. The problem is that the accuracy of NTP-based timing is quite unpredictable and ranges from tens of milliseconds to less than one microsecond.

Besides NTP timestamps, an RTP timestamp is also transported in the SR. The RTP timestamp is the same as the NTP timestamp, but measured in RTP units. It is used to synchronize different media of the same source, or to estimate the clock frequency of the source.

4.2.2.2 Jitter Statistics

The receivers provide an estimate of the RTP packet interarrival jitter that is recorded in the corresponding 32-bit field of the RR. Interarrival jitter is defined as the average difference of the delay between two consecutive RTP packets.

If ΔT_i is the one-way delay of a packet (i), the delay difference between two packets is calculated as shown:

$$\Delta\Delta T_{ij} = \Delta T_j - \Delta T_i \tag{4.1}$$

Interarrival jitter, J, is related to the mean value of $\Delta\Delta T_{ij}$. An estimate of interarrival jitter at a given time (i) is given by:

$$J_i = J_{i-1} + \frac{|\Delta\Delta T_{i-1,i}| - J_{i-1}}{16} \tag{4.2}$$

4.2.2.3 Packet Loss Statistics

There are several fields in the SR and RR that contain information that is useful for estimating packet loss. The RTP data source sends the following parameters to the destination:

- The number of packets transmitted since the beginning of the RTP session. This parameter is transported in the *sender's packet count* field of the SR.
- The number of payload bytes transmitted since the beginning of the RTP session. This figure is sent in the *sender's byte count* field of the SR.

An RTP receiver can estimate transmission performance by using the RTP sequence numbers and the SR data received in RTCP packets. Regarding packet loss, the RTP receiver sends the following information back to the transmitter:

- The highest sequence number received in an RTP packet. The difference between the values of this parameter at two different points in time is the expected number of RTP packets in the time interval. The expected number of packets since the beginning of the session is the highest sequence number minus the randomly chosen initial sequence number. This parameter is sent to the transmitter in the *extended highest sequence number received* field of the SR or RR.
- The cumulative number of packets lost since the beginning of the session. This parameter is calculated as the number of expected packets minus the received number of packets. This parameter is included in the *cumulative number of packets lost* field of the SR or RR.
- The fraction of RTP data packets lost since the previous SR or RR. This parameter is calculated as the estimated number of packets lost since the last report, divided by the number of packets expected. This data is delivered to the RTP transmitter within the *fraction lost* field of the SR or RR.

4.2.2.4 Derived Statistics

Delay, jitter and packet loss are important parameters when evaluating the performance of RTP sessions, but there are other possibilities to do this as well. New performance parameters can be derived from the existing ones.

The cumulative number of packets lost at two different points in time can be used to estimate the number of packets lost during that interval. This value is divided by the expected number of packets in that interval, and the result is the fraction of lost packets. This ratio will only match the value of the *fraction lost* field if it is calculated over two consecutive RTCP packets.

By relating the primary performance parameters, the following values can be obtained as well: the rate of lost packets per second (packets/s), the number of received packets, average payload data rate (bytes/s), the average packet rate (packets/s), the average payload size (bytes/packet), etc.

4.2.3 Source Description

The SDES RCTP packets give a description of the RTP source. The SDES packet has a similar structure as the other RCTP packets. It has a 32-bit header followed by source description items for one or more RTP sources identified by their SSRC or CSRC. The terminal RTP equipment generates SDES packets with just one SSRC, but a mixer may contain information for many contributing sources, and the SDES packet may contain many chunks identified by their CSRC. The number of chunks is always identified in the *source count* (SC) field of the header.

All SDES items have similar format:

- an 8-bit identifier showing the type of information included in the SDES item;
- an 8-bit field that includes the length of the SDES item;
- a variable-length field that contains some kind of information about the source.

The information transported in SDES packets is miscellaneous. The most important SDES item is the *canonical end-point identifier* or CNAME. The CNAME is a unique identifier in the RTP session that binds a data source with an alphanumeric identifier. This makes it possible for the destination to identify the source.

Note that the SSRC has a similar objective, except that the SSRC can change during the session if a collision with other SSRC is detected or if the user equipment is restarted. This

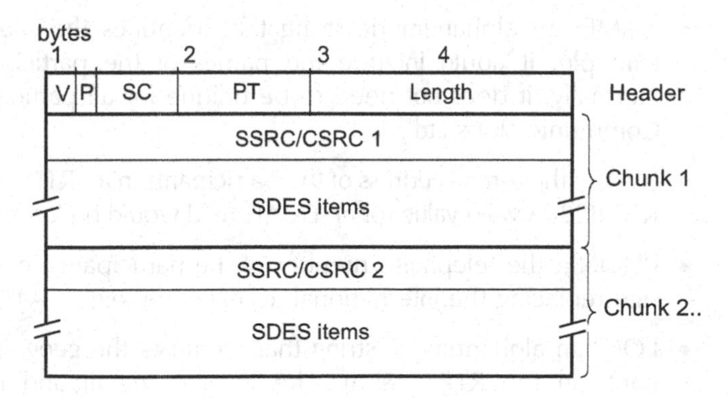

Figure 4-8. RCTP SDES packet format.

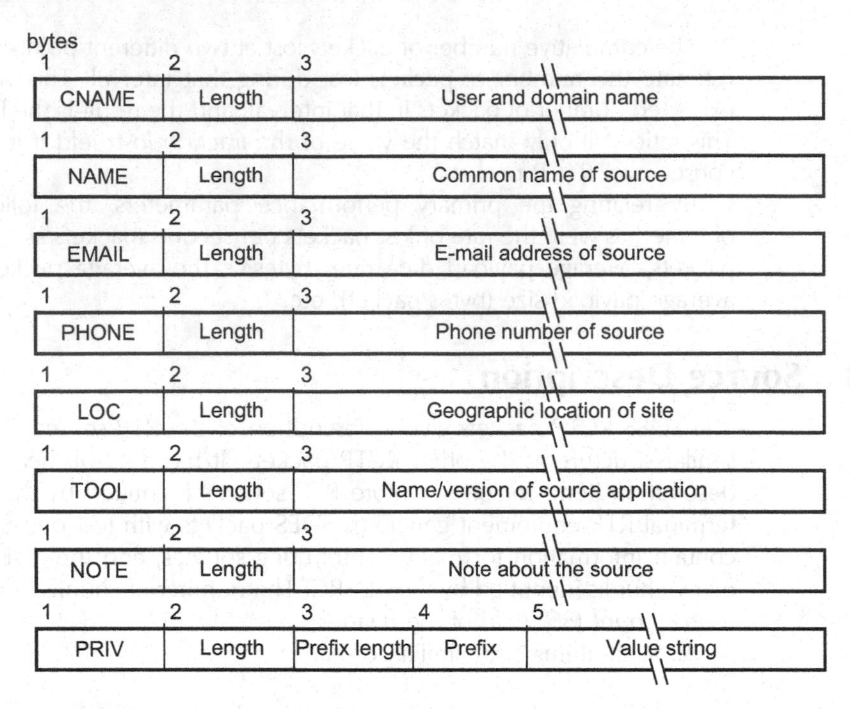

Figure 4-9. Currently defined SDES items.

will not happen with the CNAME. Another interesting property of the CNAME is that it identifies the parties of an RTP session better than the synchronization sources. It could be useful to relate different media coming from the same transmitter but with different SSRCs.

The CNAME has a 'user@host' format, but it is not an e-mail address. The 'user' is normally the login name of a participant in the RTP session and the 'host' identifies the source host. It is normally either the fully qualified domain name of the host, or an ASCII representation of its IP address in dotted decimal format.

There are other SDES item types that can be useful for RTP applications:

- NAME: an alphanumeric string that identifies the source, normally displayed. For example, it could include the names of the participants in an RTP application. Normally, it does not need to be unique for all participants, e.g.'John Smith, Trend Communications Ltd'.
- EMAIL: the e-mail address of the participants in an RTP session formatted as specified in RFC 822. A valid value for an EMAIL field would be 'trend.infoline@trendcomms.com'.
- PHONE: the telephone number of the participants in an RTP session, with the plus sign replacing the international access code, e.g. '+44 1628 503500'.
- LOC: an alphanumeric string that identifies the geographical location of the participants in the RTP session. The level of detail and format may depend on the application, e.g. 'Lower Cookham Road, Maidenhead'.

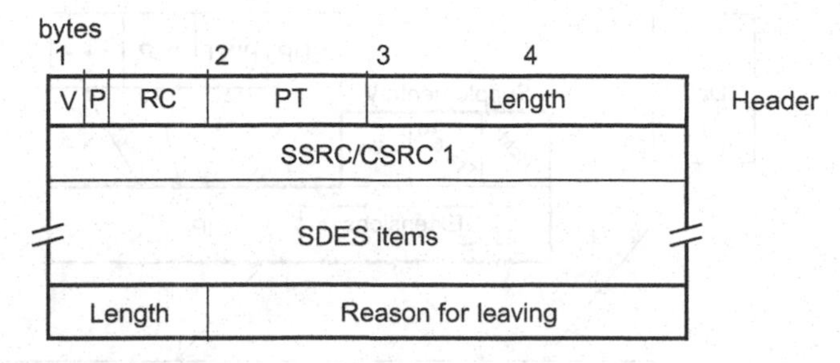

Figure 4-10. BYE packet format.

- TOOL: a string including the name and in some cases the version of the RTP application, e.g., 'Trend Videocom 1.12'.
- NOTE: this SDES item carries transient messages describing the current state of the source, e.g. 'out for lunch' or 'busy'.
- PRIV: application-specific or experimental SDES messages. This SDES item includes a prefix of variable length that may contain a description of the value of this field.

4.2.4 Session Management: The BYE Packet

The RTCP is not a session management protocol. However, it can provide minimum session control in very simple environments. Some SDES messages may be useful to interchange session information between the parties of an RTP communication. Another useful feature is the *goodbye packet* (BYE), which shows that one or more sources are no longer available.

4.3 The Session Initiation Protocol

The Session Initiation Protocol (SIP) is an application layer protocol whose main purpose is to establish, modify and tear down sessions between users in IP networks. The first version of SIP was released by the IETF in 1999, and it is expected to have a similar importance to HTTP for web applications and SMTP for e-mail. The SIP protocol can be used with any kind of session, including telephony, fax, videoconferencing and instant messaging, but an important goal of this protocol is to be able to set up real-time communication sessions across the Internet.

The purpose of SIP can be compared with the Q.931 or ISDN User Part (ISUP) protocols, but while Q.931 and ISUP are restricted to signalling in the PSTN, SIP is designed for Internet. A major difference between SIP and Q.931 or ISUP is that SIP does not make resource reservation. It only finds users in the network and sets up sessions with them.

Besides session establishment and tear-down, the SIP is able to locate users identified by a *uniform resource identifier* (URI). The protocol provides specific SIP URIs for this purpose, with a 'sip:user@domain' structure similar to e-mail addresses. Additionally,

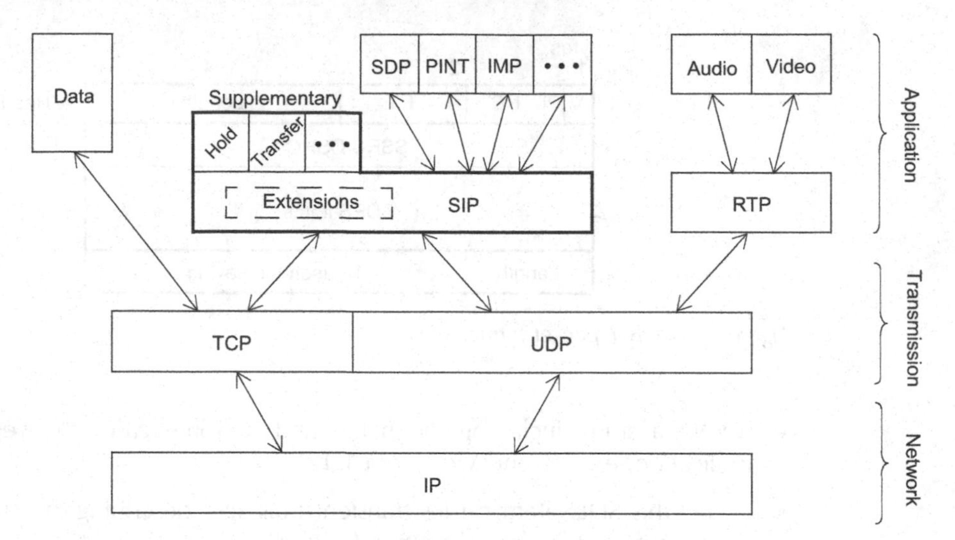

Figure 4-11. IETF multimedia communications protocol architecture.

secure URIs, called SIPS URIs are provided. The SIPS URIs have a 'sips:user@domain' structure, e.g. 'sip:alice@trend.com' and 'sips:7545@pc25.securedomain.org' are valid SIP or SIPS URIs. If a domain name is not available, IP addresses can be used instead, e.g. 'sip:mike@10.0.0.15' is a valid SIP URI. It is possible to establish sessions with remote users by means of other URIs different to the SIP or SIPS URIs as well. For example, the *tel URI* (RFC 2806), to call traditional telephones, is compatible with SIP.

In SIP, Internet telephones, conferencing software and other kinds of end-user equipment are called user agents. User agents are very different to users. URIs identify users rather than user agents. Users are allowed to keep the same identifier (their URI) even if they change their connection points to the network. This is again similar to what happens with e-mail addresses. A particular user agent can be attached to several user agent equipment, or several users may be allowed to use the same user agent equipment. The basic purpose of this approach is to implement mobility management of users in the network.

Unlike other signalling protocols, such as Q.931, ISUP and H.323, the SIP entities communicate between them by means of plain text messages encoded with a syntax similar to HTTP. Text representations of the protocol messages mean additional overhead and processing costs, although they improve readability by humans. It has been shown that call establishment with SIP takes around 1 kB of SIP information. If improved bandwidth efficiency is needed, text compression can be used, but this increases the processing costs higher. SIP messages are often delivered by using the UDP protocol to avoid TCP connection establishment mechanism. This is not mandatory, and TCP can be used as well.

SIP implements a client/server architecture that has made many IETF protocols, such as HTTP, successful. Every SIP transaction takes the form of a request from a SIP client, and one or more replies from a SIP server. The SIP user agent normally implements both the server and the client. It acts as a client when it is the calling party or as a server when being the called party.

4.3.1 Standardization

The SIP was developed by the IETF Multiparty Multimedia Session Control Working Group (MMUSIC WG) and reached the status of a proposed standard in 1999 (RFC 2543). The first standard was replaced by the RFC 3261 in 2002, and it is obsolete today. Version 2 of the SIP protocol was developed by the SIP WG, a group specially devoted to SIP within the IETF.

In addition to the base RFC 3261, there are many complementary RFCs and Internet Drafts for specific applications, supplementary services, interworking and other issues. The standarization effort includes three levels:

1. Maintaining the baseline SIP protocol.
2. Developing SIP extensions to support multiple applications.
3. Developing features and supplementary services for SIP.

As explained, the MMUSIC WG was replaced by the SIP WG. Today, the MMUSIC is in charge of maintaining the Session Description Protocol (SDP).

There are other working groups in charge of the IETF architecture for multimedia communications. The most important of these groups are:

- The SIPPING WG, which identifies and describes the requirements for new SIP extensions. This working group does not implement the extensions, it just identifies and describes them. Then they are delivered to the SIP WG to be included in the SIP RFC or some other standard.
- SIP for Instant Messaging and Presence Leveraging Extensions (SIMPLE) WG devoted to development of Instant Messaging and Presence (IMP) with SIP.
- PSTN to Internet Integration (PINT) WG, which addresses the arrangements through which Internet applications can request telephony services.

4.3.2 Architectural Entities

Direct SIP calls between user agents are possible, but often other entities are needed to take advantage of all the features of the SIP. These extra entities make SIP services intelligent, allowing user location, supplementary service provision or connection to non-SIP devices. There are other architectural SIP entities as well besides user agents: proxies, registrars, redirect servers and location servers. The purpose of every SIP entity is the following:

- *User agents* are the devices that initiate requests, being usually their final destination. Some examples are of these are Internet telephones and software phones known as softphones.
- *Proxy servers* are application-layer routers that forward SIP requests and responses. Their function can be compared with SMTP mail servers.
- *Redirect servers* provide the location of alternative user agents or proxy servers when the original destination cannot be reached.

- *Registrars* keep track of their assigned network domain. They are in charge of authenticating users, and they maintain information about the subscribers available at every moment.

In practice, it is normal to find proxies, redirect servers and registrars in the same physical device, or even implemented in the same software.

4.3.3 SIP Basic Signalling Mechanisms

The communicating parties interchange SIP requests and responses to establish, end or modify sessions. The transaction model for these messages follows the model of the HTTP protocol. SIP client requests invoke methods with specific purposes in the server. The server processes the request and generates one or several responses with success or fail codes.

As an example, we can think of a user, Alice, who wants to start a VoIP session with a remote user, Bob. Alice calls from a softphone installed on her computer that is connected to the Internet and has a *fully qualified domain name* (FQDN) *mkt12.flexanet.com.* Bob will receive phone calls in his VoIP phone at *tec26.trendcomms.com.*

To start the VoIP session, Alice makes use of the INVITE method. INVITE is the most important SIP method. It allows a session to be established with a remote user identified by the user's URI. An INVITE message from Alice with URI *alice@flexanet.com* that wants to call to *bob@trendcomms.com* could look like this:

```
INVITE sip:bob@trendcomms.com SIP/2.0
Via: SIP/2.0/UDP mkt12.flexanet.com;branch=z9hG4bK776asdhds
Max-Forwards: 6
To: Bob <sip:bob@trendcomms.com>
From: Alice <sip:alice@flexanet.com>;tag=1928301774
Call-ID: a84b4c76e66710@mkt12.flexanet.com
CSeq: 314159 INVITE
Contact: <sip:alice@mkt12.flexanet.com>
Content-Type: application/sdp
Content-Length: 142
```

The message starts with the method name, INVITE, the URI of the called party, and the SIP version number. Some header fields follow, with different purposes:

- The *Via* field contains information on the host that is waiting for a response to the INVITE request. It is normally the domain name of this host, and in our example it is the domain name of Alice's PC. The Via field also contains information on how to send the response. Our example message states that SIP version 2 must be used. The transport layer protocol must be UDP. The branch number identifies the transaction.
- The *Max-Forwards* field is decremented by one at each SIP-aware hop. It is used to limit the number of hosts the message can traverse on its way to the destination.
- The *To* field contains the URI of the called party, and sometimes an identifying name (Bob).

- The *From* field is an identifying name of the calling party (Alice) with her SIP or SIPS URI. This field also contains a tag parameter that has a random number for identification purposes.
- The *Call-ID* is a globally unique identifier that groups together series of SIP messages that belong to the same signalling dialogue.
- The *CSeq* field contains a sequence number (314159) and a method name (INVITE). This serves to sort and identify SIP transactions.
- The *Contact* field contains a SIP URI that represents a direct route to the caller. It must contain an FQDN or an IP address for this purpose. The contact field is useful, because while the first transaction in an SIP session is often routed through a chain of proxies, the transactions that follow can be routed directly between the source and the destination.
- *Content-type* contains a description of the SIP message body.
- *Content-length* contains a byte count of the message body.

The INVITE header is not used to negotiate the types of media or codecs to be used. Usually this is done by the SDP protocol that is carried in the SIP message body.

Alice may send the INVITE directly to Bob's phone at *tec26.trendcomms.com* if she knows where to find him. Otherwise, she could use the location services offered by the SIP proxy of her service provider. In this second case, she sends the INVITE to the proxy at *sip.flexanet.com*. It might be interesting for Alice to route the INVITE via *sip.flexanet.com*, even if she knows the IP address of Bob's phone, because in this case she can make use of the ability of SIP to locate users when they change their location. In other words, the SIP proxy can handle mobility. This is not possible with a direct request to the destination.

When *sip.flexanet.com* receives the INVITE from Alice, it tries to find the IP address of Bob's phone or, alternatively, the IP address of another server able to locate Bob. In this example, *sip.flexanet.com* forwards the INVITE to a second SIP proxy: *sip.trendcomms.com*. It adds an additional *Via* field with its own FQDN to the original INVITE request because it wants to receive future notifications about the progress of the call. Then, *sip.flexanet.com* sends a first response back to Alice to inform her that the process of locating Bob has started. SIP responses are similar to HTTP responses. They contain a three-digit code followed by a descriptive sentence. In this case, the code sent to Alice is 100 ('Trying').

When *sip.trendcomms.com* receives the INVITE, it sends a response with code 100 ('Trying') back to *sip.flexanet.com*. It finds Bob's IP address in a database, and it can then forward Bob the INVITE with an extra *Via* field with its own FQDN.

Bob's phone receives the INVITE from Alice and starts ringing. At the same time, Bob's phone sends back a response with code 180 ('Ringing') to Alice through the chain of SIP proxies, thanks to the extra *Via* fields added by them. If Bob's phone is busy or he decides not to answer, an error response will be sent to Alice. Otherwise Bob's phone will generate a 200 ('OK') response to show he has answered the phone and he is ready to talk. The 'OK' response is routed through the chain of proxies, and in this case, it could look as follows:

```
SIP/2.0 200 OK
Via: SIP/2.0/UDP sip.trendcomms.com
```

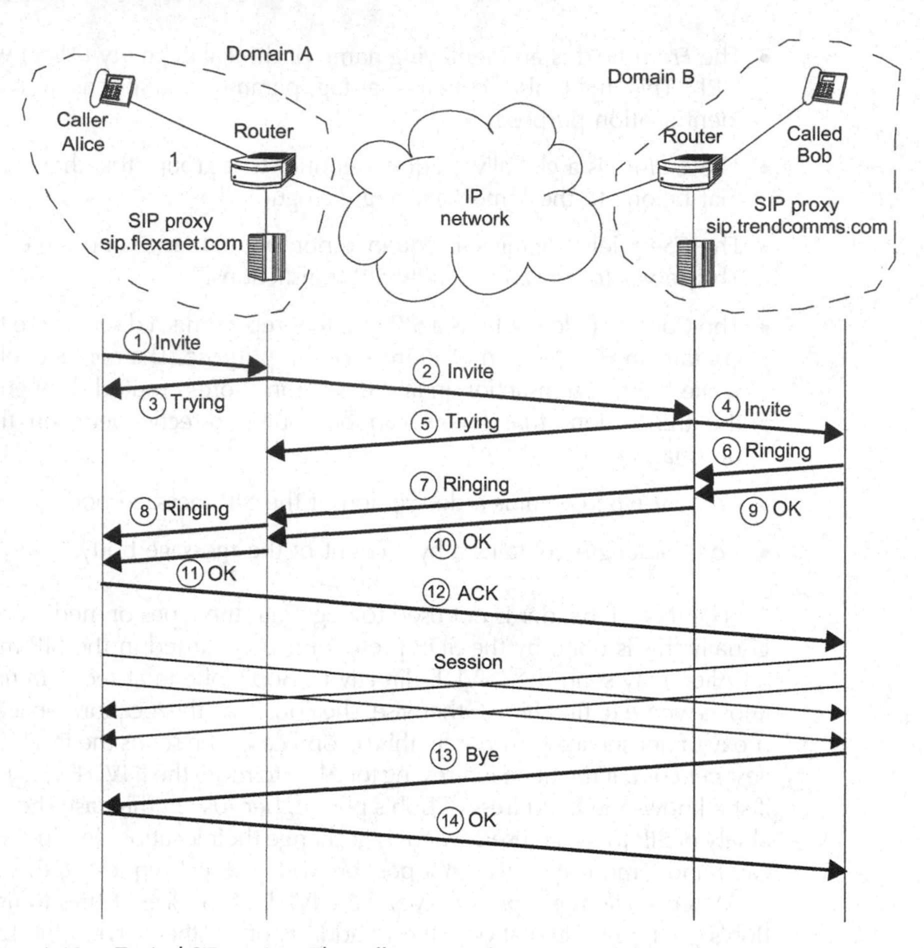

Figure 4-12. Typical SIP session. The calling party does not send the Invite message directly to the called party. It is sent to a SIP proxy that locates the user and starts to negotiate the session parameters.

```
;branch=z9hG4bKnashds8;received=192.0.2.3
Via: SIP/2.0/UDP sip.flexanet.com
;branch=z9hG4bk77ef4c2312983.1;received=192.0.2.2
Via: SIP/2.0/UDP mkt12.flexanet.com
;branch=z9hG4bK776asdhds;received=192.0.2.1
To: Bob <sip:bob@trendcomms.com>;tag=a6c85cf
From: Alice <sip:alice@flexanet.com>;tag=1928301774
Call-ID: a84b4c76e66710@mkt12.flexanet.com
CSeq: 314159 INVITE
Contact: <sip:bob@192.0.2.4>
Content-Type: application/sdp
Content-Length: 131
```

When Alice receives the 'OK' response, she sends an acknowledgement message (ACK) to confirm the reception of code 200 ('OK'). The ACK and all the following transactions in the session are routed directly to Bob's phone, because both Alice and Bob have learned each other's addresses thanks to the *Contact* field of the SIP messages.

After the negotiation phase, the VoIP conversation can start. The SDP handshake usually takes the form of a 'two-way' handshake based on a simple request/reply model. After this, the RTP protocol is used to carry the encoded voice samples. The media formats are those exchanged in the SDP handshake.

Both Alice and Bob may decide to modify the session parameters. They can do this by means of a new INVITE message with the new media description. The VoIP session finishes when one of the communicating parties 'hangs up' the phone. In this case, a BYE message is generated that is answered by an OK, which ends the session.

4.3.4 The Session Description Protocol

The SDP is an IETF application-layer protocol to describe multimedia sessions. It was specified in 1998, and it is described in RFC 2327. The session description includes:

- session name and purpose;
- time the session is active;
- media type comprising the session;
- addresses, ports, formats and other useful information to receive and decode the session.

The SIP protocol does not provide this service, so it needs the help of the SDP. The SIP is suitable for applications like VoIP. An alternative to SIP for multimedia multiparty session invitation, like video broadcast, is the Session Announcement Protocol (SAP). SDP messages can be attached to e-mails and web pages as well. The 'application/sdp' Multipurpose Internet Mail Extension (MIME) content type allows this kind of attachment. The 'application/sdp' is also used by the SIP. Specifically, SDP session descriptions are added to the SIP message body with the 'application/sdp' MIME type.

Like many other IETF application layer protocols, the SDP is text-oriented. This is less efficient than a binary encoding, but easier to understand and generate. All SDP messages have a similar format. They consist of a number of lines with the format 'item'='value'. The SDP items are lower case letters of the English alphabet. Some of them are mandatory, others optional. They are always sorted in the same way to recognize transmission errors.

The SDP items are classified in three groups: session description items (see Table 4-3), time description items (see Table 4-4) and media description items (see Table 4-5).

A session description is made up of one section containing a few session description items, followed by one section with time description items and one or more sections with media description items. A session description might look like:

```
v=0
o=mhandley 28909844526 2890842807 IN IP4 126.16.64.4
s=SDP Seminar
i=A Seminar on the session description protocol
```

Table 4-3. SDP message items for session description

Order	Item	Description	Optional/mandatory
1	v =	Protocol version	Mandatory
2	o =	Owner/creator and session identifier	Mandatory
3	s =	Session name	Mandatory
4	i =	Session information	Optional
5	u =	URL of description	Optional
6	e =	E-mail address	Optional
7	p =	Phone number	Optional
8	c =	Connection information	Optional
9	b =	Bandwidth information	Optional
12	z =	Time zone adjustments	Optional
13	k =	Encryption key	Optional
14	a =	Zero or more session attribute lines	Optional

Table 4-4. SDP message items for time description

Order	Item	Description	Optional/mandatory
10	t=	Time the session is active	Mandatory
11	o=	Zero or more repeat times	Optional

```
u=http://www.cs.ucl.ac.uk/staff/M.Handley/sdp.03.ps
e=mjh@isi.edu (Mark Handley)
c=IN IP4 224.2.17.12/127
t=2873397496 2873404696
m=audio 49170 RTP/AVP 0
m=video 51372 RTP/AVP 31
m=application 32416 udp wb
a=orient:portrait
```

In the example, the 'v=' item shows the SDP protocol version. The 'o=' is a description of the session owner. This is a mandatory field and contains the originating user's login name ('mhandley'), a session identifier ('28909844526') that makes the contents of the 'o=' item globally unique, a version identifier ('2890842807') that can be an NTP timestamp to recognize the most recent announcement, the network type, usually 'IN' for 'Internet', an

Table 4-5. SDP message items for media description

Order	Item	Description	Optional/mandatory
15	m=	Media name and transport address	Mandatory
16	o=	Media title	Optional
17	s=	Connection information	Optional
18	i=	Bandwidth information	Optional
19	u=	Encryption key	Optional
20	e=	Zero or more media attribute lines	Optional

address type identifier (IP4) that can be IP4 or IP6 for IP networks, and the originator's address ('126.16.64.4').

The 's=' and 'i=' items contain the session name and description given by the originator. The 'u=' is an URL that points to a resource with additional information about the session. The 'e=' is the e-mail address of the session responsible that does not need to be the session owner. The 'c=' item contains connection data. It may contain a unicast or a multicast IP address that is available for transmitting and receiving information in the data session. For multicast sessions, a time to live (TTL) is appended to the multicast IP address (127 in the example). The 't=' item includes the start and stop times for the session. The values included in this item are decimal representations of NTP time values in seconds.

In our example, there are three media description fields that specify three media streams: one for audio, one for video and a third one is of 'application' type. All three have the same format. First, the media type is shown. The options available are 'audio', 'video', 'application', 'data' and 'control'. These are followed by the transport layer port to which the media stream will be sent. This is followed by the transport protocol for the stream. Two options are currently available for this: 'udp' if the UDP is to be used, and 'RTP/AVP' to specify the RTP protocol. Finally, the media type (as specified in the RTP profile) is included. In our example, the '0' included for the audio stream means that PCM encoding with μ-law scaling is being used, and the 31 for the video stream stands for the ISO/IEC 11172-3 and 13818-3 MPEG-1/MPEG-2 payload (see Table 4-1). For non-RTP sessions this is not possible. UTP payloads are specified by means of a registered MIME type. In the example, 'wb' is used to specify a payload with MIME type 'application/wb', the correct one for a whiteboard application.

The defined item set for SDP must not be extended. The 'a=' item provides a mechanism to extend and customize the protocol's capabilities. In the example, the 'orient:portrait' is used to define the whiteboard orientation (portrait or landscape) in a webminar. Another interesting application of the 'a=' is the mapping of dynamic RTP payload numbers to customized payload formats. For example, 'a=rtp-map:97 L16/16000/2' would map the RTP payload type 97 to a linear encoded stereo audio stream sampled at 16 kHz.

4.3.5 Security Issues

The SIP header reveals information that the communicating parties may want to keep private. The same can be said about SIP message bodies. This information might be used by a malicious user to access confidential information, change the contents of the session, or fake the identity of someone else. The contents of a SIP message can be secured to avoid these situations. There are mechanisms to ensure SIP message confidentiality, authentication and integrity. Some of them have been specially defined for SIP, while others are of general application in IP networks. These security mechanisms must be combined properly to obtain a trusted service.

The security mechanisms can be classified into end-to-end and hop-by-hop mechanisms. The SIP provides end-to-end security, but hop-by-hop security relies on mechanisms external to the SIP, such as IPSec and the Transport Layer Security (TLS).

The main advantage of IPseq and TLS is that they make it possible to protect the full SIP message header and body, because they work at protocol layers below the application layer.

The drawback of IPsec is that it usually needs special configuration of network nodes. This usually limits the application of this mechanism to servers of the same service provider. TLS is more flexible, but it only works with TCP and is incompatible with UDP.

It must be noticed that, if a user agent uses IPsec or TLS to communicate with a proxy, this does not guarantee that the secure transport will be used on the rest of the end-to-end path. The SIPS URIs provide a way to request the use of TLS along the complete path.

4.3.5.1 Authentication

SIP authentication allows a user agent to prove its identity to another user agent, a proxy server or a registrar. Proxy-to-proxy authentication is not provided, and it must rely on IPSec or TLS.

The SIP authentication is a challenge-based mechanism derived from the HTTP Digest authentication. In challenge-based authentication mechanisms, when a server receives a request, it sends a challenge to the client to prove its identity. The challenge is usually a nonce value uniquely generated and used for one challenge only. The client generates a response value calculated from the nonce, and a secret password shared with the server. The server compares the response with an internally calculated value, and the client is authenticated if the response matches this value.

In challenge-based authentication, the password is never sent as clear text. To be secure, it must be difficult to derive the password from the response generated by the client. Therefore, it is encrypted using some cryptographic hashing algorithms. The default for SIP authentication is the MD5 algorithm.

Some proxies may require authentication. When a proxy server receives a SIP request, it decides whether client authentication is needed or not. If it is, it sends the client a response with code 407 ('Proxy Authentication Required'). The error response is at the same time a challenge, and it contains the nonce and a realm field with information about the registration domain (see Figure 4-13).

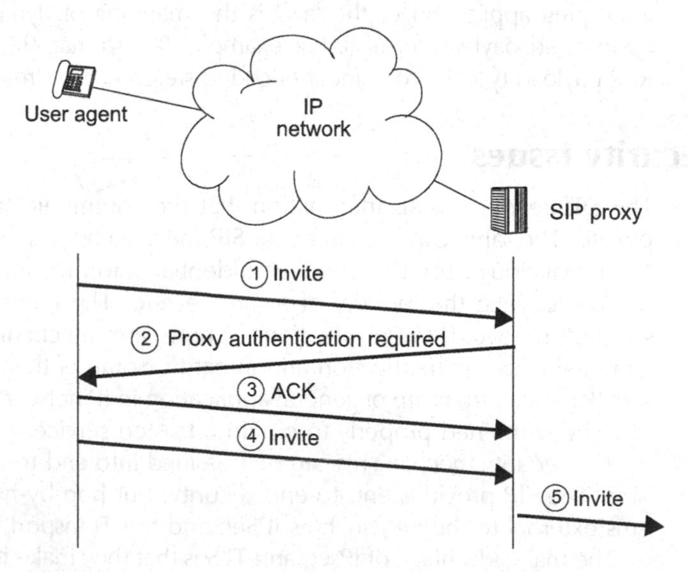

Figure 4-13. Proxy authentication procedure for a user agent.

Another context where authentication may be necessary is when user agents are registered in SIP registrars. The SIP method needed to do this is REGISTER. Some user agents may require authentication as well. The authentication procedure is similar. The difference is that in this case the error code generated by the remote user agent or the registrar is 401 ('Unauthorized').

4.3.5.2 Confidentiality and Integrity

The end-to-end mechanism to ensure confidentiality and integrity in SIP message bodies is based on the E-mail application. SIP messages carry MIME bodies, so S/MIME can be used to protect them.

S/MIME offers a public-key security framework based on certificates. There are two different types of keys in S/MIME: public and private keys. Every private key has a public key associated with it, but it is very difficult to derive the private key from its public equivalent.

Private keys are used to decrypt and sign messages and must be kept secret. Public keys are used to encrypt messages and check signatures. The tasks performed by public keys are not sensible, and therefore there is no security risk in distributing them.

S/MIME is vulnerable to personality theft of the legitimate owner of the public key. This is the reason why public keys are always distributed using certificates. Certificates contain public keys, and they are also a way to ensure that the key belongs to its legitimate owner, because they are cryptographically signed by a trusted entity known by all the communicating parties. This trusted entity is known as the certification authority.

To be able to encrypt or check the integrity of the S/MIME body of a SIP message, one must request a certificate containing the public key of the transmitter.

The main problem with S/MIME is that it only protects the body of the message–the header remains unprotected. There is a solution for this: a SIP message can be tunnelled into an outer SIP message. The original SIP message is sent as an S/MIME body of the outer message. Some of the header fields of the inner message must be copied to the outer message, but the most sensitive data can be hidden within the inner message. The copied fields also provide a way to check the integrity of the outer header fields by comparing them with the inner fields after decryption.

Another inconvenience of S/MIME is that, although it is not very common, some intermediate nodes may change the contents of SIP message bodies. If the contents of the body are hidden by S/MIME encryption, these network nodes will not be able to do their job.

4.3.6 Service Architecture and Protocol Extensions

The basic services provided by SIP core methods and headers include locating resources or communicating parties, as well as sending invitations to sessions. This is complemented by the negotiation of session parameters offered by the SDP protocol. These basic services can be extended to offer a broader set of features.

Some of the additional features of SIP are bundled with the basic services, enhancing and adding intelligence to them. These are called supplementary services, and most of them are directly related to the features provided by circuit-switched telephony, including call transfer, call diversion, call hold and call intrusion. There are some other features that are new applications, not provided by the basic SIP protocol, for example SIP *instant messaging* (IM).

Supplementary services have not been standarized in SIP, as has been done with the H.323 architecture. Specification lists for new services are written by the SIPPING working group. These specifications are analysed by other IETF working groups that evaluate whether it is possible to implement the new service with the baseline protocol, or if a specific protocol extension is needed. Protocol extensions are implemented in SIP by defining new methods, new headers and new response codes. The following sections describe some important SIP extensions.

4.3.6.1 Instant Messaging and Presence

Instant messaging is exchange of information, usually by means of short text messages, between two or more parties in real time or almost real time. The real-time assumption is important to differentiate IM and e-mail; e-mail messages can be delayed by intermediate servers for a long period.

IM also includes presence information and 'buddy lists' that allow users to get information on the connection status of other users (online, offline, busy and so on). The combination of IM and presence information is very powerful, and sometimes it is referred to as *instant messaging and presence* (IMP).

There are SIP extensions that enable IMP applications based on this protocol, for example the MESSAGE method. For presence, there are two methods that form the event communication architecture for SIP, and they are SUBSCRIBE and NOTIFY.

A SIP request with the MESSAGE method would look like this:

```
MESSAGE sip:bob@trendcomms.com SIP/2.0
Via: SIP/2.0/TCP mkt12.flexanet.com;branch=z9hG4bK123dsghds
Max-Forwards: 69
To: Bob <sip:bob@trendcomms.com>
From: Alice <sip:alice@trendcomms>;tag=49583
Call-ID: asd88asd77a@mkt12.flexanet.com
CSeq: 1 MESSAGE
Content-Type: text/plain
Content-Length: 30

Hi Bob, this is an IM example.
```

The header of the MESSAGE request has fields similar to other SIP requests, such as INVITE. The message body can be of any MIME type, including 'message/cpim', thus enabling interoperatibility of all IMP software supporting this format.

SIP MESSAGE transactions follow the same path as other SIP signalling transactions. This is important because, in the past, IMP applications have registered a high volume of usage, and there is a danger of the SIP-based IMP applications interfering with other session signalling. To minimize congestion risk, the size of the messages is limited, and the use of transport protocols (like TCP) with some congestion control mechanism is recommended.

The presence application requires using presence agents that are entities able to store presence information and generate notifications to users when changes occur. Presence agents can obtain presence information from the registrar database, but this is not mandatory. Users interested in receiving presence information about their contacts need to send SUBSCRIBE requests to the presence agent. The presence agent uses the NOTIFY method to inform about andy changes in the 'presence' state of the other users.

4.3.6.2 Extensions for Supplementary Services

Supplementary telephony services, like call transfer or call hold, are currently supported by the PSTN, which is why they are an important requirement for IP telephony, especially for SIP.

A simple example of a supplementary service is call hold. Call hold makes it possible for one of the parties to temporarily interrupt (hold) an active call with the other party. There is no need to define anything new to support call hold with SIP. Call hold can be achieved by a reinvitation: a second INVITE message with a session description set to inactive. The session can be restored after the interruption by sending a third INVITE message with the appropriate SDP description.

Call transfer is another supplementary service that makes it possible to transfer an active session from one user to another (see Figure 4-14). This service is implemented in SIP using the REFER method. A REFER request indicates that the recipient should contact a third party using the contact information provided in the 'Refer-To' header field. For example, the following field included in a REFER request means that the recipient should contact *alice@trendcomms.com:*

```
Refer-To: sip:alice@trendcomms.com
```

If the REFER request is accepted, the URI identified in the 'Refer-To' field is contacted by means of the normal mechanisms provided by SIP, for example INVITE. The NOTIFY method is useful to inform the transferrer on the progress of the transfer. The information is carried in the NOTIFY request body with 'message/sipfrag' MIME type. The body of this messages could be like this:

```
SIP/2.0 100 Trying
```

Many variations of call hold and call transfer can be implemented with SIP. Other supplementary services for SIP are call forwarding, three-way conference, call pickup and automatic redial.

4.3.7 Firewall Traversal

Firewalls isolate trusted networks and hosts from untrusted networks and hosts; in other words they control the traffic between trusted and untrusted hosts. Data packets are allowed to go through depending on the configuration of the firewall.

A typical application scenario for a firewall is to protect a private LAN connected to the Internet. In this case, the firewall is placed between the Internet (external network) and the LAN (internal network). Furthermore, the routers that connect private LANs to the Internet often implement some kind of *network address translation* (NAT) that further enhances security and makes it possible to share just a few public IP addresses among many hosts. These hosts are configured with special private IP addresses that are inexpensive and can be managed by the network administrators of the private site.

NAT replaces internal (private) addresses by external (public) addresses. This means that, for incoming packets, the router replaces the destination address with a private IP

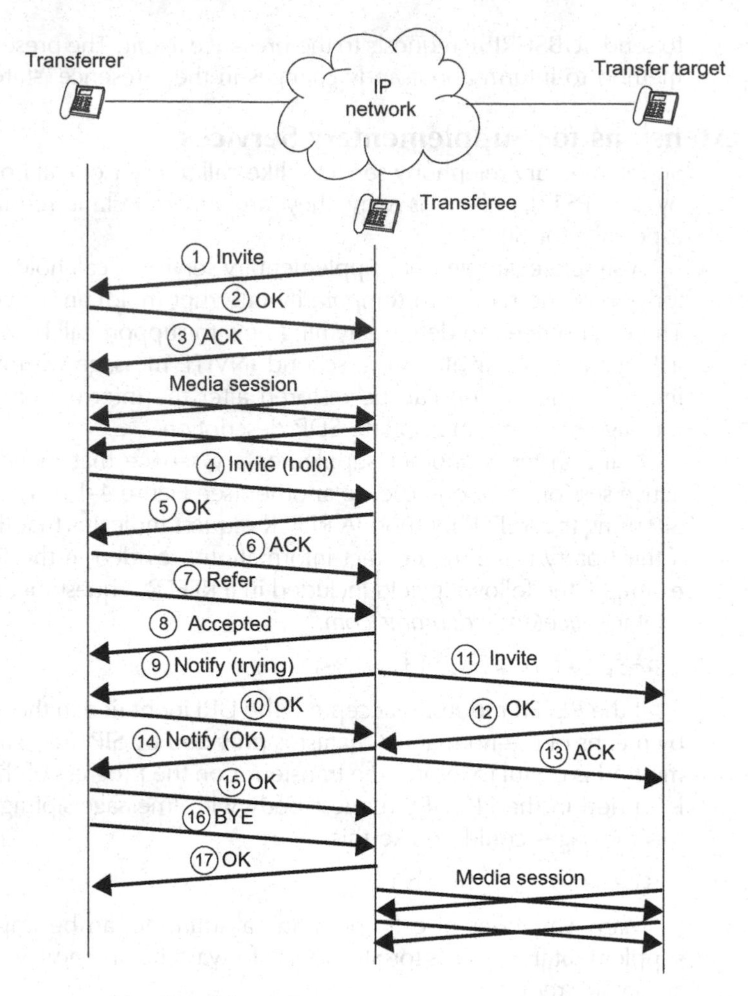

Figure 4-14. Successful call transfer dialogue.

address, and for outgoing packets, the router replaces the source IP address. The router maintains a NAT table that relates external and internal IP addresses (see Figure 4-15).

4.3.7.1 Types of NAT

There are several well-known problems for SIP when crossing firewalls, most of them related to NAT. The complexity of the problem is increased by the fact that there are several types of NATs that exhibit different behaviour with SIP:

- *Full cone NAT*: there is a well-established mapping between internal and external addresses. An external host who wants to send data to an internal host only needs to know the mapping for the address of that host.

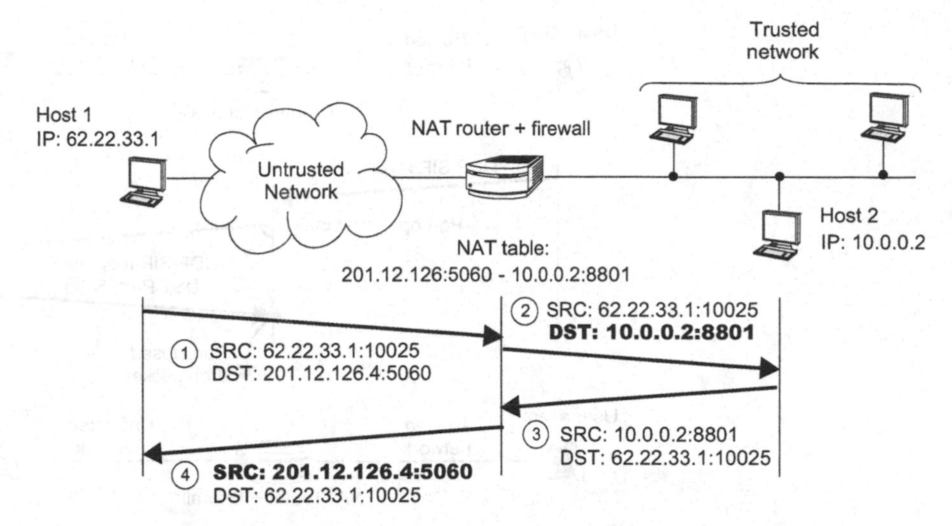

Figure 4-15. A firewall filters traffic between trusted and untrusted networks. NAT further enhances security by hiding private network addresses.

- *Restricted cone NAT*: NAT records are temporary, and they are only created when data is sent from an internal address to an external address. The records are removed from the table after being unused for a certain amount of time. External data can pass through the NAT router if the destination IP address matches an external address in the NAT table. For this reason, NAT connections cannot be started by external hosts in the restricted cone.
- *Port-restricted cone NAT*: the port-restricted cone NAT is similar to the restricted cone NAT. NAT records are temporarily created when data is sent from an internal address. External data can only pass through the NAT router if both the destination IP address and port match an external address in the NAT table.
- *Symmetric NAT*: NAT records contain not only internal and external addresses, but also destination addresses. Packets with different destinations are translated in a different way, and external computers can only respond to the mappings generated for them.

4.3.7.2 Problems

Problems with NAT arise because, with SIP, there is some addressing information that is carried in the application payload. This information is bypassed by devices that only work at layer 3. Addresses and ports carried by the SIP header and payloads are not translated by the NAT server, so the receivers are not able to find the original sources.

This problem affects both SIP signalling and media transport:

- *SIP signalling problems*: SIP responses may fail to find their way back to the originator of the transaction if the 'Via' or 'Contact' fields of the SIP requests cannot be resolved to a

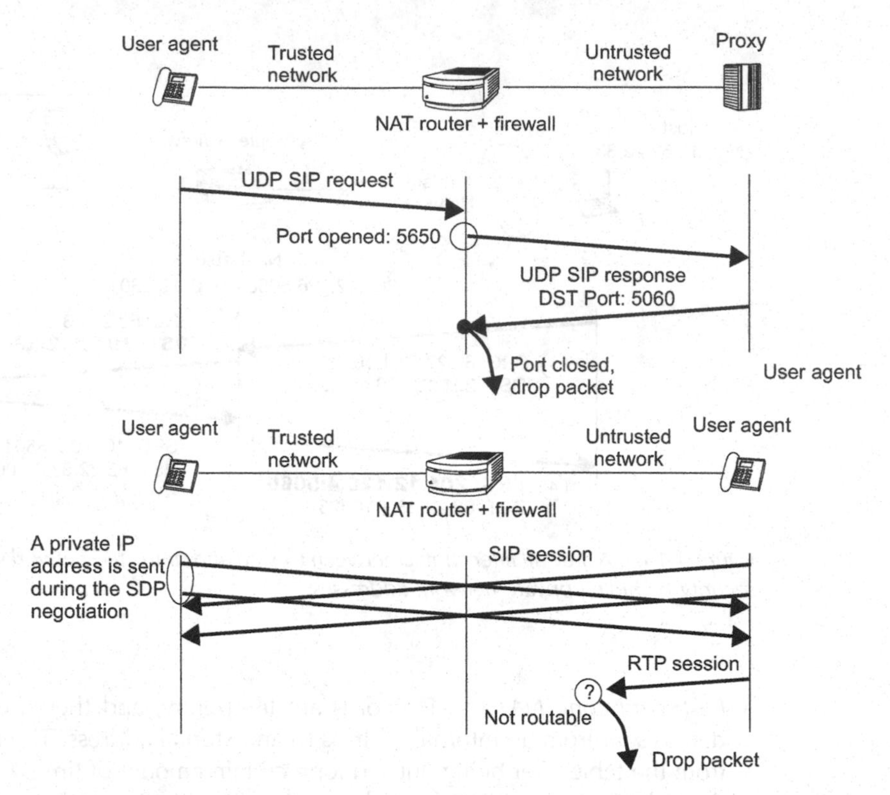

Figure 4-16. NAT translation problems involve both the signalling protocol and the media transport protocol: (a) SIP responses cannot pass through the NAT router, because they are sent to a closed port; (b) RTP packets cannot reach their destination, because the user agent behind the NAT router has sent a private address in the SDP handshake.

public IP address. Even if it is possible to find a route to the source, the reply must be directed to the port opened in the NAT router by the request message. This port may be different from the original source port. Sending responses to the original port or any default port should especially be avoided. Finally, response addresses and ports not always remain open 'forever' after the first transaction. Most of the NAT routers remove ordinary NAT records from the NAT table after an inactivity period that could range from a few seconds to several hours.

- *Media transport problems*: the media transport protocol, usually RTP, may fail to find the participants of a session if they are behind a NAT router. The reason for this is that the destination addresses for the media session are exchanged by the SDP protocol carried in the payload of SIP packets, and therefore not translated by the NAT router.

4.3.7.3 Solutions for NAT Traversal

There are many possible problematic situations when NAT is used. A typical example would be a user agent that wishes to establish a connection with a second user agent without using a SIP proxy. The first user agent is behind a NAT router. If the receiver sends the response to the IP address included in the 'Via' field of the SIP header or to any default port, the communication will surely fail. The best solution is to send the response to the same source address and port received in the request message. These parameters must match the external IP and port of the NAT router.

The problem is more difficult if there is a proxy between the communicating parties. In this case, the session setup cannot continue when the first response arrives to the proxy, because:

1. The source address and the source port of the SIP response are written by the user agent that is in the external network.
2. The connection data of the SIP 'Via' field comes from within the private network, and therefore it is not routable in the external network.
3. All other available data ('Contact' SIP field and SDP connection data) is information from the private network, and it is not useful for the proxy.

The SIP protocol has a solution for this problem. There are header tags that help the proxy to find the way to the source:

1. The 'received' tag is added by the proxy in the 'Via' field when it receives the request from the calling user agent. It shows the IP address of the initiator as it is seen by the proxy. The 'received' tag is still present when the proxy receives the response from the called party, and it provides an IP address where the response is sent (as the original SIP-field IP address is not a public one).
2. The 'rport' tag works similarly to the 'received' tag, but it shows the port to forward responses back to the source.

An INVITE, as it leaves the intermediate SIP proxy, could look like this:

```
INVITE sip:bob@trendcomms.com SIP/2.0
Via: SIP/2.0/UDP sip.trendcomms.com
     ;branch=a71b6d57-507c77f2
Via: SIP/2.0/UDP 10.0.0.2
     ;branch=z9hG4bK776asdhds;received=62.22.33.1;rport=5608
To: Bob <sip:bob@trendcomms.com>;tag=a6c85cf
From: Alice <sip:alice@trendcomms.com>;tag=108bcd14
Call-ID: 4c88fd1e-62bb-4abf-b620-a75659@10.0.0.2
CSeq: 703141 INVITE
Contact: <sip:alice@10.0.0.2>
Content-Type: application/sdp
Content-Length: 131
```

The 'rport' and 'received' tags are a good solution for SIP signalling, but the problem with the media transport protocol remains unsolved. For the VoIP application this may mean that the session startup could be successful, but the users may fail to hear the voice from the remote speaker.

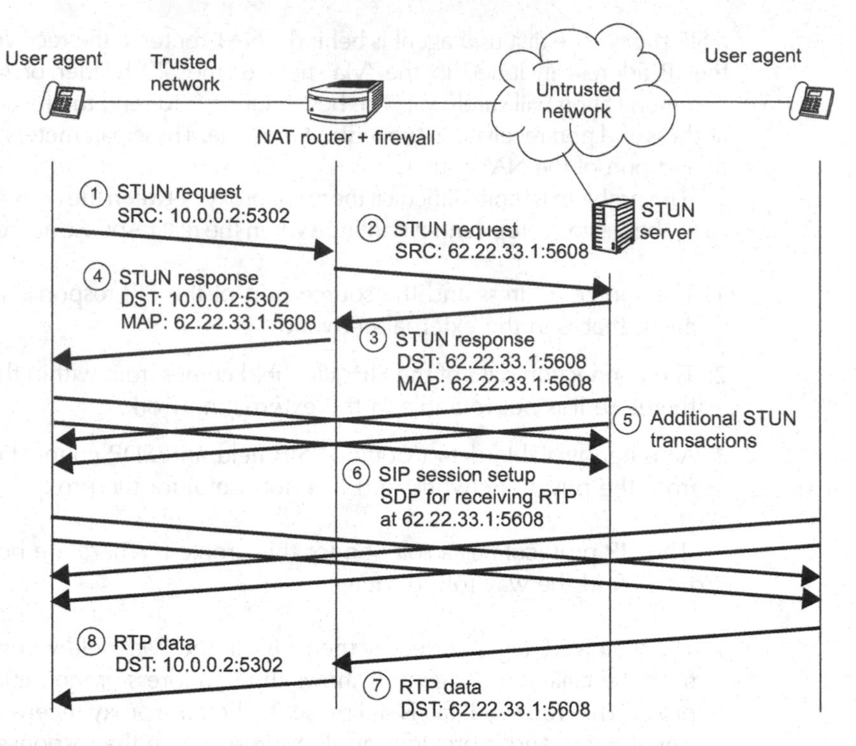

Figure 4-17. STUN servers help to setup SIP sessions across NAT routers. They give information on the external address and port. This information makes it possible to setup SDP fields with the correct port and IP address to receive media data.

Other type of solutions try to replace the private information of the SIP and SDP fields with public information that is meaningful in the external network.

- *Universal plug and play (UPnP)*: the UPnP is a solution that allows users behind a NAT firewall to ask the router what will be the NAT mapping for a particular communication. The information received via UPnP is then inserted into the corresponding fields in the SIP and SDP messages. The UPnP mechanism does not work in cascaded NAT connections, and there are security issues that must be addressed.
- *Simple traversal of UDP through NAT (STUN)*: the STUN is a protocol that allows for external-network queries on how the packets are seen from the outside. The information is sent back to the client, and it may be used in the SIP and SDP messages (see Figure 4-17). The main inconvenience of the STUN protocol is that it does not work with symmetric NAT.
- *Traversal using relay NAT (TURN)*: finding a solution for symmetric NAT is more challenging than finding one for other NATs. The STUN protocol is not valid, because the external mapping seen from the STUN server is different from the address seen from the remote SIP entity. TURN is a solution that can be used for all NAT types. A TURN server provides the calling user agent with an IP address and a port that are written in the SDP

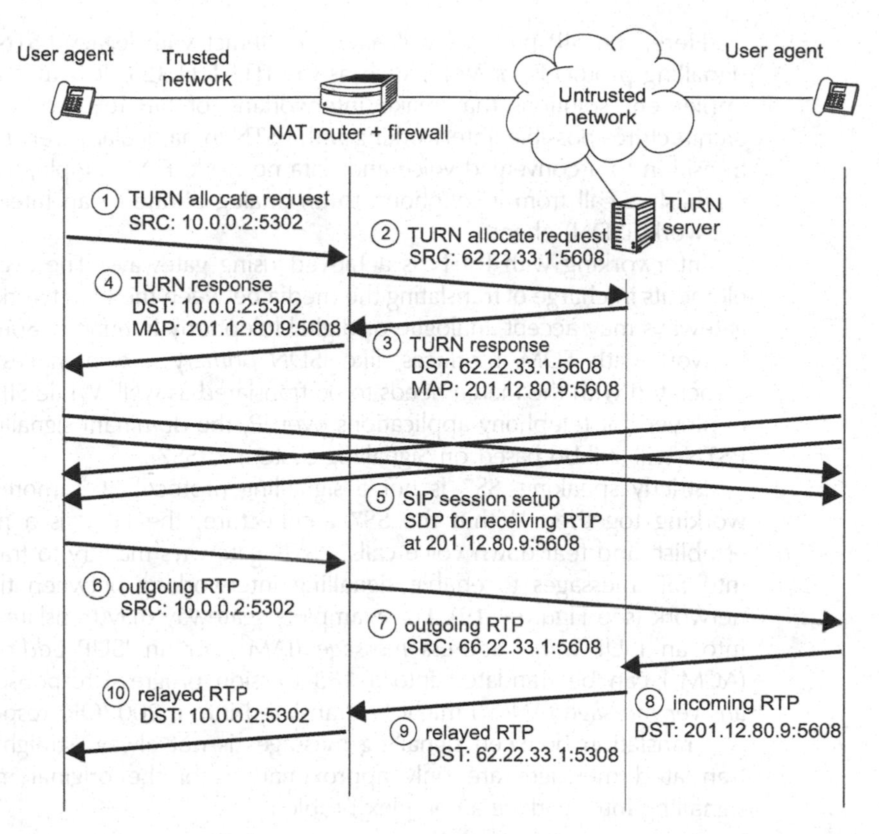

Figure 4-18. TURN servers help to setup SIP sessions across NAT routers. They act as a relay for the incoming media, sent from a remote user agent. They work for all NAT types, but this is not an optimal solution.

payload for receiving RTP data. The SIP session setup can then proceed by using the 'received' and 'rport' tags. After setting the session up, the outgoing RTP data is sent directly to the destination, but the incoming RTP is sent to the TURN server that acts as a relay (see Figure 4-18).

- *Interactive connectivity establishment (ICE)*: TURN works with all NAT types, but it must only be used as the last resort, because it wastes resources on the TURN server. STUN is a good solution, but it is not valid for symmetric NAT. ICE combines both protocols and it gives the best solution for all NAT environments. It gives the client a set of prioritized candidate addresses that can be used for sending and receiving media with the media protocol.

4.3.8 Interworking with the PSTN

Although SIP is a general-purpose protocol for session establishment and tear-down, it has been designed with the VoIP application in mind. It is expected that, in the near future, one of the main SIP applications will be VoIP call signalling.

Here, the SIP protocol will have to interact with legacy PSTN signalling and other signalling protocols for VoIP, such as the ITU-T H.323. It is thus necessary to find and implement solutions that make interworking of SIP telephony and other telephony architectures possible. Interworking with PSTN in particular is very important for a smooth transition to a converged voice and data network. For example, it should be possible to establish a call from a softphone to an analog phone or an Integrated Services Digital Network (ISDN) phone.

Interworking with PSTN is achieved using gateways. These gateways are network elements in charge of translating the media between the IP network and the PSTN. PSTN gateways may accept analogue or digital telephony interfaces. Some of them may need to work with TDM interfaces, like ISDN *primary rate interfaces* (PRI). The signalling associated with the media needs to be translated as well. While SIP is likely to be widely deployed for telephony applications over IP, the dominant signalling framework for the PSTN will still be based on Signalling System 7 (SS7).

Strictly speaking, SS7 is not a signalling protocol. It is more a suite of protocols working together. Within the SS7 architecture, the ISUP is a popular alternative to establish and tear down voice calls. PSTN gateways may try to translate ISUP messages into SIP messages to enable signalling interworking between the PSTN and the IP network (see Figure 4-19). For example, a gateway may translate a SIP INVITE request into an ISUP *initial address message* (IAM[1]), or an ISUP *address complete message* (ACM[2]) can be translated into a 183 (session progress) response code, and an ISUP *answer message* (ANM[3]) might be translated into a 200 (OK) response code.

Translation between signalling messages is not always straightforward. Sometimes, translated messages are only approximations of the original messages. This makes signalling interworking a complex problem.

4.3.8.1 Gateway Decomposition

PSTN gateways must be able to perform complex signal processing operations on large volumes of voice traffic. They must also translate PSTN and IP signalling, and provide services. Signal processing tasks can be handled efficiently by digital signal processors, but this hardware is not well suited to implementing protocol stacks or running general-purpose code.

This is the reason why it has been proposed to divide the PSTN gateway into two separate network elements (see Figure 4-20):

- *media gateways (MGW)* handle protocol, format and encapsulation conversions on the media data;

[1]IAM is the ISUP message that contains the calling number, the called number, the band-width requirements for the communications, the type of caller, and other information.

[2]ACM is the ISUP message that indicates that the call is connected to the destination. This message is a response returned when an IAM is received. It causes a one-way audio to be opened from the destination switch so that the caller can hear a ring-back tone.

[3]ANM is the ISUP message sent towards the originator when the called party answers the phone.

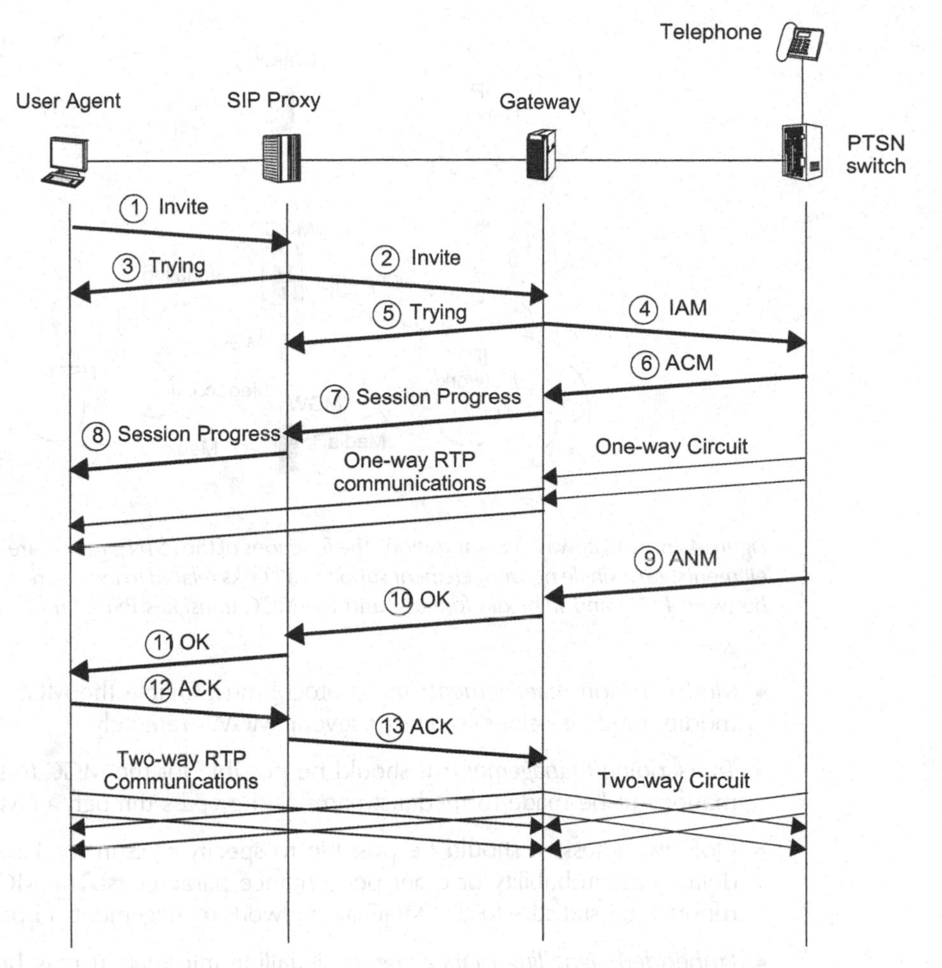

Figure 4-19. The gateway translates signalling messages between the PSTN and the IP signalling protocols.

- *media gateway controllers (MGC)* handle all tasks related to call control and signalling translations.

The gateway decomposition is a step beyond the separation of media transport and signalling transport specified in the IETF VoIP model. Signalling and media may be delivered by different paths in the network and are processed by specialized network elements. It is even possible to send signalling and media across different networks.

Network intelligence is provided by means of signalling, and thus it is normal to think that the MGCs are intelligent network elements. This results in a network model where MGCs are active network elements and MGWs remain mostly passive, controlled by MGCs. A protocol is needed to establish dialogues between MGCs and the MGWs controlled by them. The requirements for such a control protocol are:

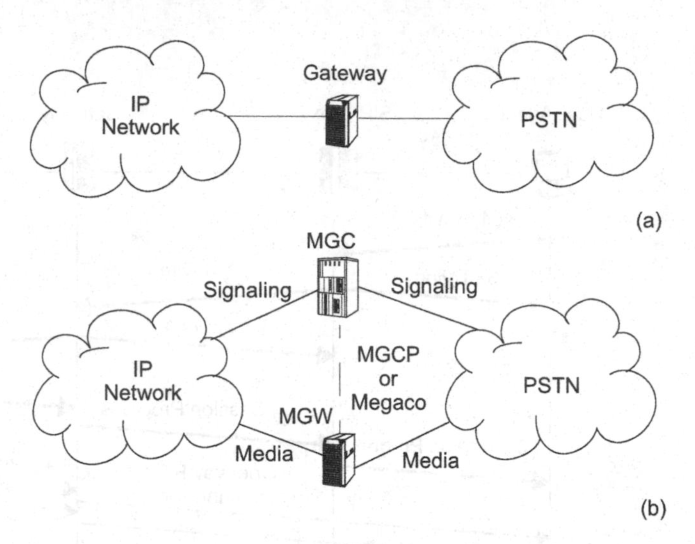

Figure 4-20. Gateway decomposition. The functions of the PSTN gateway are shared by two network elements: (a) a single network element supports all tasks related to interworking; (b) the MG translates between PSTN and IP media formats, and the MGC translates PSTN and IP signalling formats.

- *Media session management*: the protocol must enable the MGC to establish, end or modify media sessions in one or several MGWs remotely.
- *Transcoding management*: it should be possible for the MGC to specify the transformations to be made to media streams as they pass through an MGW.
- *QoS awareness*: it should be possible to specify QoS in media sessions in terms of delay, loss probability or other performance parameters. The MGW must be able to report QoS statistics to the MGC for network management purposes.
- *Embedded signalling management*: signalling information may be embedded in the media stream. An example of this are the *dual tone multi-frequency* (DTMF) tones. It must be possible for the MGW to report the occurrence of this kind of signals. The MGW must also be able to request information about embedded signalling, or be configured to perform certain actions when certain signalling events occur.
- *Generating and managing billing information*: it must be possible to report billing and accounting information. This information may include the start and end times of media sessions, or the transmitted volume of information in the media session.
- *Flexible and scalable MGW management*: it must be possible to assign associations between MGCs and MGWs with flexibility. Certain functions, like back-up MGC designation, are needed for a fault-tolerant system. MGCs must be able to deal with different MGs, as there are many different transmission technologies and heterogeneous implementations by different equipment manufacturers.

The IETF has developed the Media Gateway Control Protocol (MGCP) to deal with MGW/MGC interconnection. This protocol is specified in RFC 2705. There are many commercial solutions based on the MGCP, but the preferred solution for new implementations is the Megaco/H.248 protocol, developed jointly by the IETF and the ITU-T.

4.3.8.2 Converged Telephony

VoIP has revolutionized the world of telephony. It is currently possible to perform voice calls across the Internet at no cost whatsoever. Internet VoIP calls do not meet the minimum requirements of a carrier-class service (high availability, quality and connectivity), but new voice operators are entering the scene anyway, offering VoIP services at very low cost. This new scenario is pushing the traditional telephone operators to adopt IP solutions to transport voice.

MGWs and MGCs are critical for telephone operators to migrate their voice networks to IP while maintaining full connectivity between their subscribers. There are two types of MGs that can be used:

- *Trunking MGWs*: these are MGWs that connect the PSTN with the IP network. They have to deal with SS7 or other type of *network-to-network interface* (NNI) signalling and manage a large number of circuits.
- *Access MGWs*: these MGWs connect user-to-network interfaces to the IP transit network. They may have analogue or ISDN interfaces (PRIs or BRIs) to the user installations.

Access MGWs enable telephone operators to migrate to a packet-based transit telephone network without replacing the analog or ISDN subscriber telephones by VoIP telephones (see Figure 4-21).

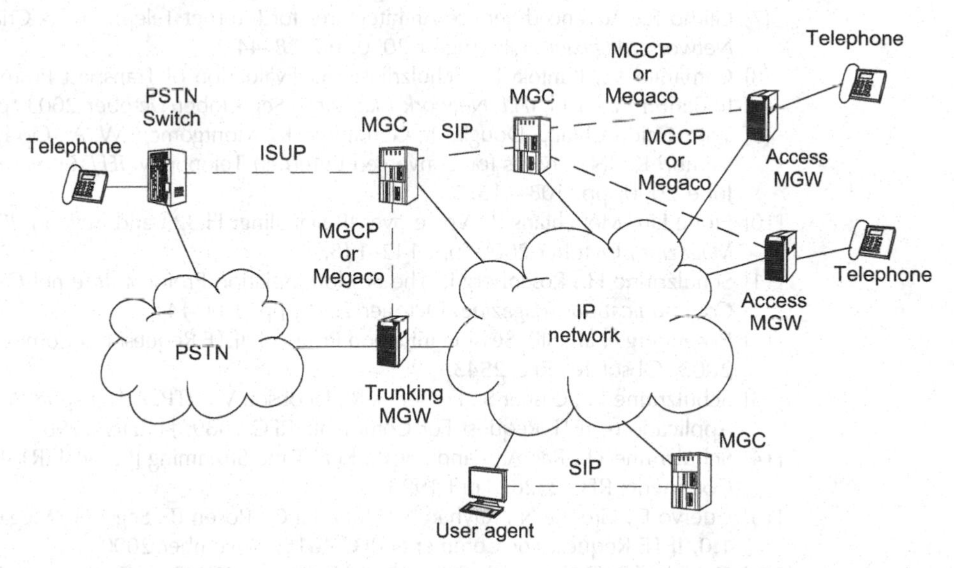

Figure 4-21. Converged VoIP architecture: Telephones are connected to an IP network, thanks to residential MGs that are able to translate voice in a format that can be transmitted over a packet network. PSTN and IP networks can be connected by larger trunking MGWs. All these gateways are controlled by MGCs. Communications between MGWs and MGCs use MGCP or Megaco. SIP is used for communication between different MGCs.

In a network architecture with MGWs, native VoIP user agents can establish voice calls with analogue or ISDN phones. To do this, the VoIP user agent (directly, or with the help of one or several SIP proxies) establishes a SIP session with an MGC. The MGC then forwards the media through an access or trunking MGW. The media may have to travel trough the PSTN. To make this possible, an MGC must translate the SIP signalling into a PSTN signalling language, such as ISUP. In this network architecture, MCGs are seen as normal SIP user agents that establish, modify or end PSTN telephone sessions on demand. Like any other SIP network element, MGCs deliver information between them using the SIP protocol.

Owing to the convergence between PSTN and IP networks, two PSTN telephones could be exchanging information between them without noticing that the voice is actually being delivered via an IP network.

Selected Bibliography

[1] Minoli D., Minoli E., *Delivering Voice over IP Networks*, 2nd Edition, John Wiley and Sons, 2002.

[2] Durkin J. F., *Voice-Enabling the Data Network: H.323, MGCP, SIP, QoS, SLAs, and Security*, Cisco Press, 2003.

[3] Black U., *Voice Over IP: 2/e*, Prentice Hall, 2002.

[4] Hardi W.C., *VoIP Service Quality*, McGraw-Hill, 2003.

[5] Collins D., *Carrier Grade Voice Over IP*, McGraw-Hill, 2003.

[6] Salsano S., Veltri L., Papalilo D., SIP Security Issues: The SIP Authentication Procedure and its Processing Load, *IEEE Network Magazine*, November/December 2002, pp. 2–8.

[7] Glitho R., Advanced Service Architectures for Internet Telephony: A Critical Overview, *IEEE Network Magazine*, July/August 2000, pp. 38–44.

[8] Camarillo G., Kantola R., Schulzrinne H., Evaluation of Transport Protocols for the Session Initiation Protocol, *IEEE Network Magazine*, September/October 2003, pp. 40–46.

[9] Tsun-Chieh Chiang, Douglas J., Gurbani V. K., Montgomery W. A., Opdyke W. F., Reddy J., Vemuri K., IN Services for Converged (Internet) Telephony, *IEEE Communications Magazine*, June 2000, pp. 108–115.

[10] Hong Liu, Mouchtaris P., Voice over IP Signaling: H.323 and Beyond, *IEEE Communications Magazine*, October 2000, pp. 142–148.

[11] Schulzrinne H., Rosenberg J., The Session Initiation Protocol: Internet-Centric Signaling, *IEEE Communications Magazine*, October 2000, pp. 134–141.

[12] Rosenberg *et al.*, SIP: Session Initiation Protocol, IETF Request For Comments RFC 3261, June 2002. Obsoletes RFC 2543.

[13] Schulzrinne H., Casner S., Frederick R., Jacobson V., RTP: A Transport Protocol for Real-Time Applications, IETF Request For Comments RFC 1889, January 1996.

[14] Schulzrinne H., Rao A., Lanphier R., Real Time Streaming Protocol (RTSP), IETF Request For Comments RFC 2326, April 1998.

[15] Cuervo F., Greene N., Rayhan A., Huitema C., Rosen B., Segers J., Megaco Protocol Version 1.0, IETF Request For Comments RFC 3015, November 2000.

[16] Campbell B., Rosenberg J., Schulzrinne H., Huitema C., Gurle D., Session Initiation Protocol (SIP) Extension for Instant Messaging, IETF Request For Comments RFC 3428, December 2002.

[17] Donovan S., The SIP INFO Method, IETF Request For Comments RFC 2976, October 2000.

[18] Roach A., Session Initiation Protocol (SIP) – Specific Event Notification, IETF Request For Comments RFC 3265, June 2002.

[19] Rosenberg J., The Session Initiation Protocol (SIP) UPDATE Method, IETF Requets For Comments RFC 3311, September 2002.
[20] Schulzrinne H., Oran D., Camarillo G., The Reason Header Field for the Session Initiation Protocol (SIP), IETF Request For Comments RFC 3326, December 2002.
[21] Handley M., Jacobson V., SDP: Session Description Protocol, IETF Request For Comments RFC 2327, April 1998.
[22] Mankin A., Bradner S., Mahy R., Willis D., Ott J., Rosen B., Change Process for the Session Initiation Protocol (SIP), IETF Request For Comments RFC 3427, December 2002.
[23] Sparks R., The Session Initiation Protocol (SIP) Refer Method, IETF Request For Comments RFC 3515, April 2003.
[24] Petrack S., Comroy L., The PINT Service Protocol: Extensions to SIP and SDP for IP Access to Telephone Call Services, IETF Request For Comments RFC 2848, June 2000.
[25] Boulton C., Rosenberg J., Camarillo G., Best Current Practices for NAT Traversal for SIP, IETF Draft draft-ietf-sipping-nat-scenarios-03, October 2005.
[26] Greene N., Ramalho M., Rosen B., Media Gateway Control Protocol Architecture and Requirements, IETF Request for Comments RFC 2805, April 2000.
[27] ITU-T Recommendation H.323v4, Packet-Based Multimedia Communications Systems, 2000.
[28] ITU-T Recommendation H.225v4, Packet-Based Multimedia Communications Systems, 2000.
[29] ITU-T Recommendation H.245v8, Packet-Based Multimedia Communications Systems, 2001.
[30] ITU-T Recommendation H.221, Frame Structure for a 64 to 1920 kb/s Channel in Audiovisual Teleservices, 1999.

Chapter 5: IP Multicasting

In the 1990s, the most important Internet application was the World Wide Web (WWW), a unicast client/server application. For the WWW, an application-layer protocol called Hypertext Transfer Protocol (HTTP) is used by the client to request information from a specialized server. This server then sends HTTP answers to the client. Another popular application for the Internet is e-mail, where one sender delivers mail messages to the mailboxes of one or more recipients using an application-layer protocol, the Simple Mail Transfer Protocol (SMTP). The e-mail architecture is quite different from the WWW. E-mail is also a client/server application, but it uses multicasting. In this case, the problem of multicast delivery is solved by SMTP. From the point of view of the network, the paths to the different recipients of an e-mail are seen as different routes by the intermediate routers (see Figure 5-1).

Today, the convergence of circuit-switched and packet-based technologies is making it possible to deliver voice, video and data using one single IP network. Many of the new multimedia applications, such as IPTV or VoIP, are multicast, but the multicast service model of the e-mail application is not adequate for these applications. It does not have the scalability and simplicity needed to deliver real-time voice or video to hundreds or maybe thousands of receivers. To make an IP network support Triple Play, it is necessary to implement multicasting in the network layer of the protocol stack. This means that multicast delivery must be supported by intermediate routers as well, not only by terminal equipment (see Figure 5-2).

The delivery of voice and video makes it necessary to install multicast routing protocols in IP routers. The aim of these protocols is to find routing trees to send multicast data to the

---

Triple Play: Building the Converged Network for IP, VoIP and IPTV Francisco J. Hens and José M. Caballero

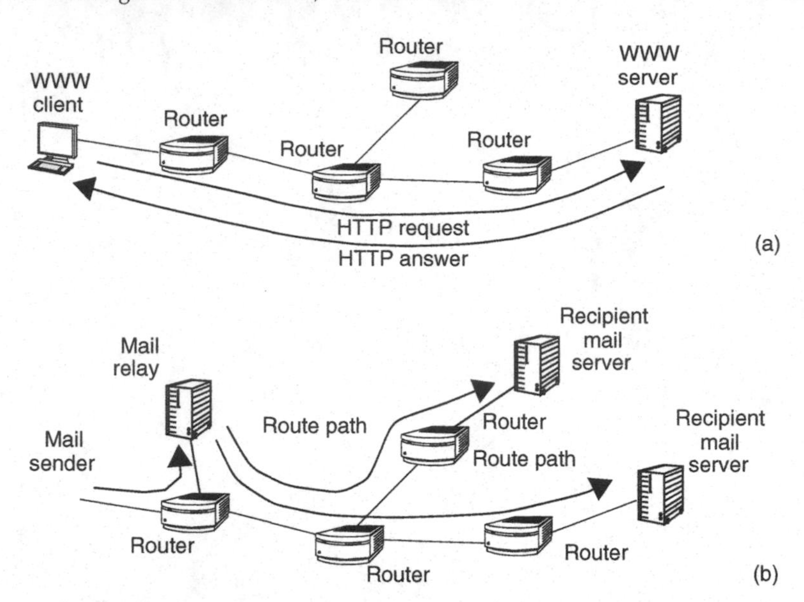

Figure 5-1. Service model for IP data applications: (a) the WWW service model – unicast and client/server-based; (b) the e-mail service model – multicast application; multicast delivery is solved at the application layer.

destination; in other words, choose the correct outgoing interfaces for incoming multicast packets. The second problem to be solved is how to manage multicast groups. This means implementing a mechanism to add and remove recipients dynamically. This is done by using multicast addresses and the Internet Group Management Protocol (IGMP).

5.1 IP Multicast Groups and their Management

In the Internet and in any IP network, multicasting means the transmission of IP datagrams to groups of hosts called multicast groups. Every multicast group is identified by a single IP address. Most of the IP addresses are used to identify hosts in the network, but some of them

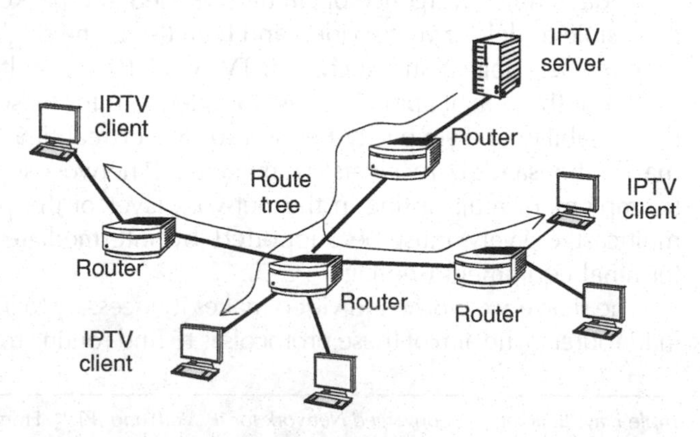

Figure 5-2. Service model for the IPTV application. Multicast delivery is solved at the IP layer and not at the application layer of the protocol stack.

Table 5-1. Internet address classes

Address class	First byte (binary)	Address range (decimal)	Number
Class A	0xxxxxxx	0.0.0.0–127.255.255.255	2 147 483 648
Class B	10xxxxxx	128.0.0.0–191.255.555.555	1 073 741 824
Class C	110xxxxx	192.0.0.0–223.255.255.255	536 870 912
Class D	1110xxxx	224.0.0.0–239.255.255.255	268 435 456
Class E	1111xxxx	240.0.0.0–255.255.255.255	268 435 456

are reserved not for hosts but for groups of hosts. The multicast IP addresses are Class D addresses, which means that they all start with '1110' (see Table 5-1). Expressed as decimals, these addresses fall in the range from 224.0.0.0. to 239.255.255.255.

The Internet Assigned Numbers Authority (IANA) is in charge of the administration of the Internet addressing space, including multicast addresses. Some addresses within the Class D range are reserved by the IANA for specific applications, and others are dynamically assigned to transient multicast groups (see Table 5-2).

Those hosts that need to send data to a multicast group simply put the right multicast address in the destination address field of the IP datagrams sent to this group. There are not many requirements for the transmitter, but IP multicasting must be supported by the network, and the IGMP must be supported by the receiver in order to join or leave multicast groups. Supporting multicasting in IP networks is not mandatory; there are several compliance levels that range from no compliance at all to full compliance.

There is no obligation for the members of a multicast group to be in the same network. However, in a very simple IP multicasting model, all the members of the group are in the same Ethernet network with the sender. In this case, the IGMP is not needed, and it is possible to take advantage of the broadcast nature of Ethernet.

If not all the members of the multicast group are in the same network, a *multicast agent* is needed. This is normally a router that delivers the data to the correct destinations. The routers in the multicast IP network must use a multicast routing protocol to be able to perform this task.

5.1.1 Multicasting in Ethernet Networks

The IANA controls the block of Ethernet MAC addresses between 01-00-5e-00-00-00 and 01-00-5e-7f-ff-ff. When transmitting IP multicast data in an Ethernet network, the destination address can be automatically mapped into a multicast Ethernet address. The

Table 5-2. Some IP multicast addresses reserved by the IANA

Address	Receivers attached to the group
224.0.0.1	All systems on this subnet
224.0.0.2	All routers on this subnet
224.0.0.5	All OSPF routers
224.0.0.6	Designated OSPF routers
224.0.0.9	RIP2 routers
224.0.0.12	DHCP server/relay agent
224.0.1.1	Network time protocol

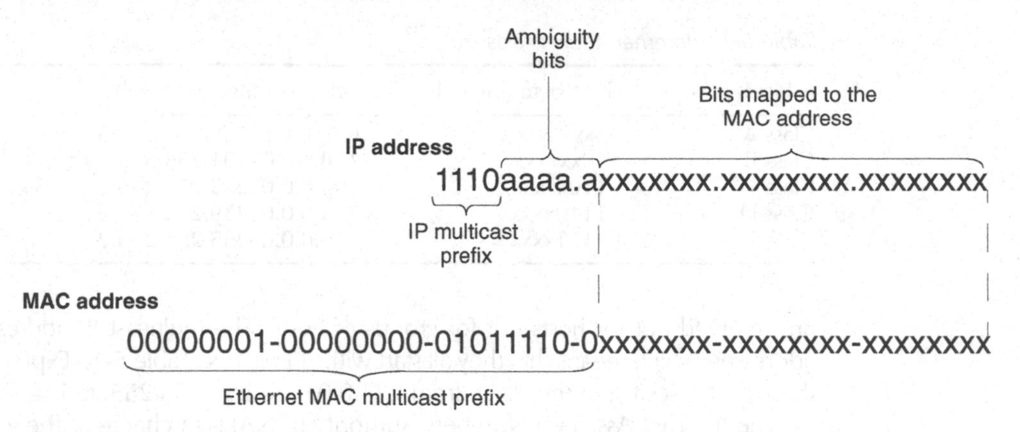

Figure 5-3. Mapping of IP multicast addresses into Ethernet multicast MAC addresses.

28 least significant bits of the IP address are copied to the IANA-managed Ethernet address block explained above. For example, the multicast IP address 239.255.0.1 would be mapped into 01-00-5e-7f-00-01 for Ethernet (see Figure 5-3).

The information arrives to all the hosts in the network, and those belonging to the correct group can recognize the multicast Ethernet address and process the correct frames. Frames arriving at those hosts who do not belong to the group are discarded.

Note that, in the mapping process, five bits of the IP address are lost. This means that there are 32 IP multicast addresses with the same Ethernet address. This does not have to be a problem, if the use of IP multicast addresses is planned carefully.

Ethernet multicasting works in a similar way in broadcast Ethernet networks with shared transmission media and in switched networks with dedicated media. The reason for this is that multicast traffic is broadcast by switches by default. In other words, switched Ethernet behaves the same way as broadcast Ethernet for multicast traffic. This could be a problem, especially in large networks, because multicast Ethernet traffic is directed to hosts that do not request this traffic. In bridged carrier-class networks this may have security implications as well.

It is possible to avoid broadcasting of multicast Ethernet addresses, if IGMP snooping is implemented in the switches. IGMP snooping means that layer-2 switches can check (snoop) layer-3 IGMP messages. This way, the switches know which are the active multicast groups and who belongs to which group, and they can forward multicast data to the correct ports (see Figure 5-4).

5.1.2 Multicasting and the Internet Group Management Protocol

The IGMP is the network protocol that allows hosts to join or leave multicast groups, and therefore it must be implemented by all hosts that wish to receive multicast information.

The multicast agent is an important component of the IP multicasting architecture. This entity is in charge of:

- *Group management*: the multicast agent grants or denies multicast group membership to hosts. It finds and assigns IP addresses and access keys to the groups and removes inactive hosts from them.

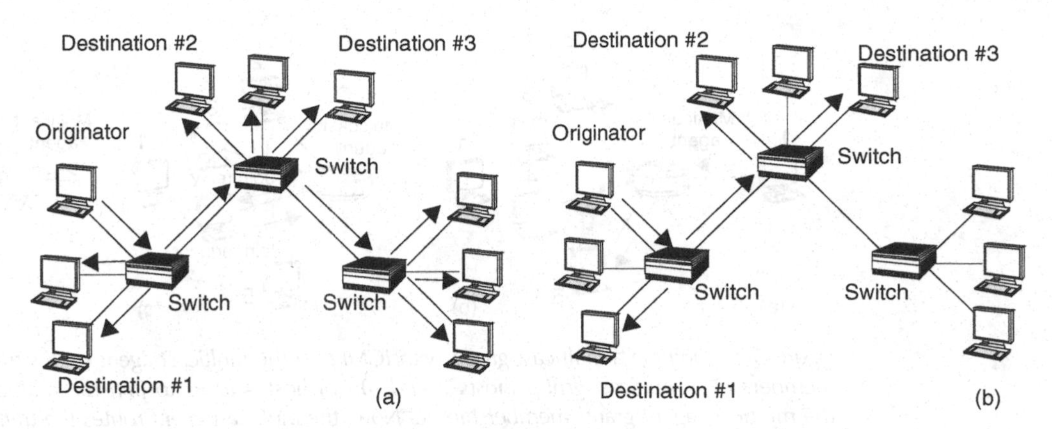

Figure 5-4. (a) Switches not implementing IGMP snooping. (b) Switches implementing IGMP snooping.

- *Multicast routing*: the multicast agent forwards multicast traffic to remote networks. To do this, it associates IP addresses to sets of interfaces. It is not necessary to maintain a full list of members (and their IP addresses) for every group; routing can be performed just by knowing the network interfaces where the information needs to be forwarded to.

IGMP messages are transmitted between hosts and multicast agents that establish a client/server relationship (see Figure 5-5). Hosts are clients and multicast agents are servers. IGMP packets have the same format, defined by the IETF to include the following fields (see Figure 5-6):

- *Type*: describes the purpose of the IGMP packet. IGMP messages can be used to join or leave groups, create new multicast groups, or confirm membership to a particular group.
- *Code*: this field is only meaningful when creating a new group. It shows whether the group is going to be public (value 0) or private (value 1). Public groups can be joined freely, but private groups are protected with an access key. In IGMP replies sent by multicast agents, the Code field shows whether the request is granted, denied or pending.
- *Checksum*: this field is for error detection purposes in IGMP messages. It is the 16-bit one's complement of the one's complement sum of the IGMP message. The Checksum field itself is considered to be zero for this calculation.
- *Identifier*: it is useful to distinguish between the different request messages arriving from the same host. This is why different IGMP messages from the same host have a different Identifier field. When the multicast agent sends a reply, it uses the same identifier as the request message.
- *Group address*: the group address is assigned by the multicast agent; thus, when creating a new group, the address is sent to the host in the Create Group Reply message. When managing group membership, the IP address of the group to be joined or left is sent in the join/leave request message.

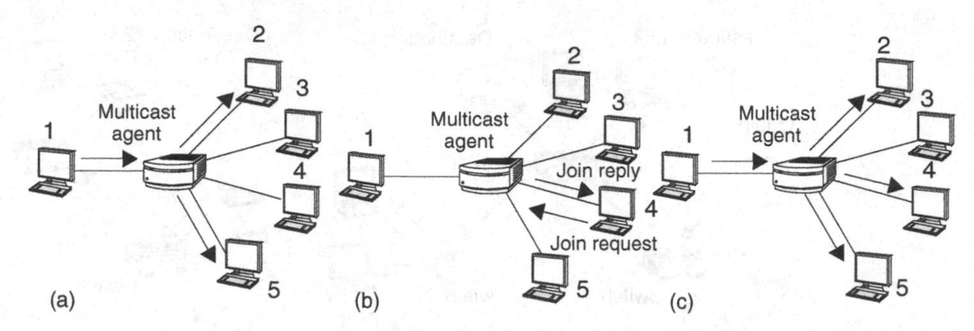

Figure 5-5. Joining a multicast group with IGMP: (a) the multicast agent routes traffic only to the members of a multicast group (hosts 2 and 5); (b) host 4 as us to join the multicast group, and the multicast agent grants membership. (c) Now, the multicast agent routes the traffic to hosts 2, 3 and 5.

- *Access key*: this field is only meaningful when managing private groups. The access key for private groups is assigned by the multicast agent in the Create Group Reply message. The host needs to supply this key in the request message in order to join or leave a private group.

5.2 Multicast Routing

Implementing multicast in broadcast networks such as Ethernet is not problematic, but defining and deploying multicast mechanisms for switched networks such as IP is rather challenging.

IP multicasting across the Internet was first demonstrated in 1992 by using the Multicast Backbone (MBone). The MBone is an overlay network built on the top of the Internet. Several 'islands' supporting multicast routing are connected between them by point-to-point tunnels configured in the unicast Internet to allow long-distance multicast communications.

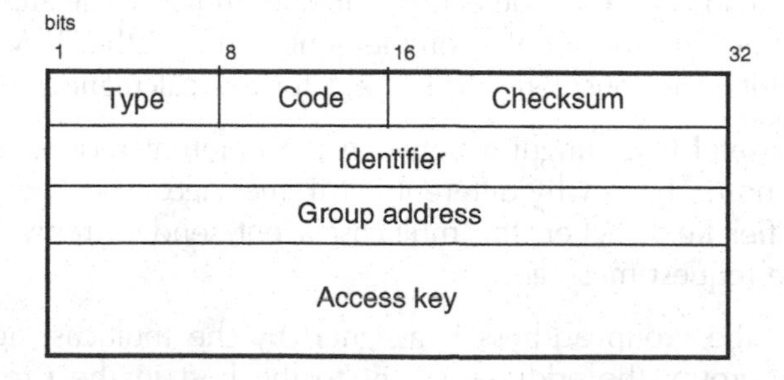

Type	Message
0x01	Create group request
0x02	Create group reply
0x03	Join group request
0x04	Join group reply
0x05	Leave group request
0x06	Leave group reply
0x07	Confirm group request
0x08	Confirm group reply

Figure 5-6. IGMP message format.

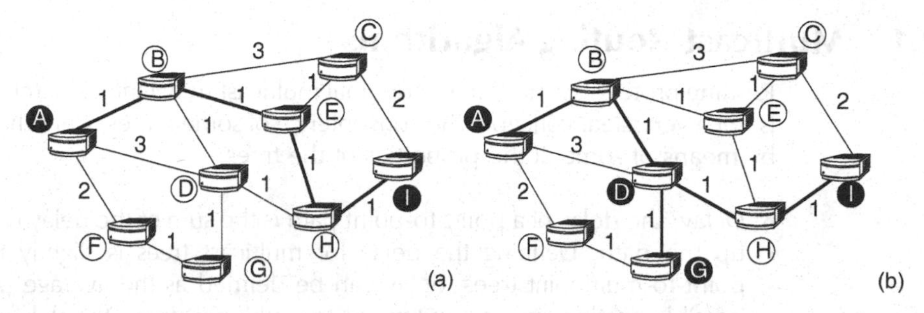

Figure 5-7. Unicast and multicast routing. The 'weight' of the links is identified by numbers: (a) the aim of unicast routing is to find routes between a single source and a single destination across a network; (b) multicast routing tries to find routes to enable communication between multiple transmitters and receivers.

The objective of unicast IP routing is to find routes between a single origin and a single destination in a network. Usually, these routes are paths across several routers. Not all the possible paths between a source and a destination have the same performance. Some paths have less delay than others, and some paths may be more reliable or cheaper. Routes can be manually configured by the network administrator, but for large networks with many routers, it is better to let one or more routing protocols do the job automatically. These protocols make use of routing algorithms to find the best routes. To do this, they need to consider one or more performance parameters (see Chapter 6). Routing algorithms usually define a weight function that uses numbers to represent the convenience of each link (the better the link, the less weight it has). Then they use an optimization process to find the shortest routing paths (those with minimum weight).

Multicast routing has the same problems as unicast routing. Multicast routing protocols make use of special multicast routing algorithms to find the best route. The difference is that the route is not normally a path, because there are always two or more recipients. Multicast routes are defined as trees where the information flows from one or more origins to several destinations. Multicast routing algorithms are in charge of finding the best possible trees (see Figure 5-7).

Routing requirements differ depending on the multicast service to be transported. Multicast services can be classified in source-specific and group-shared services:

- *Source-specific service*: there is only one transmitter and two or more receivers. An example of a source-specific multicast application is IPTV.
- *Group-shared service*: there are several transmitters that can also be receivers. A good example of a service of this type is a multiparty video conference.

Routing trees can also be classified into point-to-multipoint and multipoint-to-multipoint trees. Source-specific trees have a well-defined root node, but group-shared nodes do not have this property. Point-to-multipoint trees are better suited to source-specific services, and multipoint-to-multipoint trees are used for group-shared services.

5.2.1 Multicast Routing Algorithms

Routing protocols try to find the optimal multicast tree. Unfortunately, the word 'optimal' is not a very clear concept. The convenience of some routes over others can be evaluated by means of some of the properties of the trees:

- *Delay*: the delay of a point-to-point path is the sum of the delay of the links that make up the path. Defining the delay for multicast trees is slightly more complex. For point-to-multipoint trees, delay can be defined as the average path delay for every possible destination. For multipoint-to-multipoint trees, the delay is the average of all the possible point-to-multipoint delays when every end-router is considered to be the root of a tree. For example, if in Figure 5-7(b) we consider that the label written at each link is the delay for that link and it is accepted that Router *A* is the root router of the multicast tree, the point-to-multipoint delay for this tree is 3. If the same tree is considered to be a multipoint-to-multipoint tree, the delay is 2.55.
- *Scalability*: the network should support many simultaneous multicast trees, and this places some requirements on the trees. The load of a router in the tree depends on the number of copies to make for every incoming packet to deliver data to the different destinations of the tree. This is the reason why trees that tend to distribute load fairly between routers are better than those that tend to concentrate load on a few routers. Of course, scalability is not only a requirement for multicast trees, but also for the routing protocol. The routing algorithm should be able to compute multicast trees in a reasonable amount of time, and the signalling traffic should remain as small as possible. These properties should scale well with the number of nodes in the network.
- *Support for dynamic multicast groups*: good multicast routing algorithms should allow group members to join or leave the multicast tree in a seamless way, without changing the properties of the tree.
- *Fairness*: a multicast tree should not 'punish' one multicast group member in order to improve the service for other members. For example, in a multiparty teleconference, it is important that all the receivers hear the speaker at the same time, or in a network game; if the delay is not the same for all the players, the game is no longer fair. This means that fair distribution of delay for different receivers is a condition for multicast trees.

These and other properties can be taken into account by multicast routing algorithms. There are many algorithms, and each of them has its own way to calculate routes. Many of the principles for unicast routing algorithms are valid for multicasting as well. It is a common practice to assign a positive number for each link (known as the 'weight' of the link) and then try to find the trees that 'weigh' less. These trees are called *Steiner trees*, and they are the same as the shortest paths for unicast routing algorithms. Specifically, a Steiner tree is a tree that:

1. Spans all the all the nodes to be interconnected.
2. Has the minimum weight. The weight of the tree is the sum of the weights of the links.

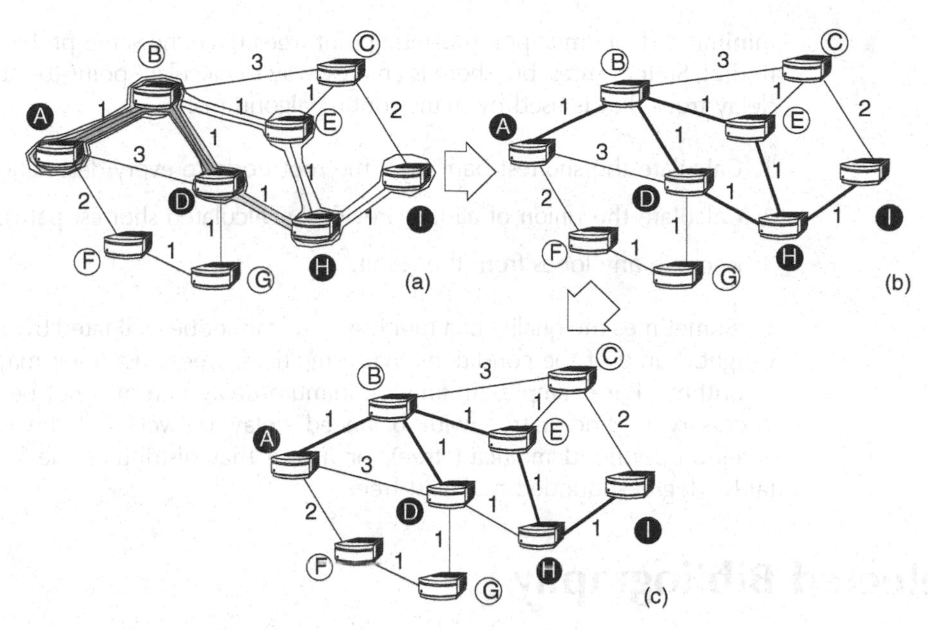

Figure 5-8. Algorithm for finding the minimum-delay point-to-multipoint tree: (a) all the shortest paths are calculated; (b) the union of all the shortest paths is calculated; (c) all the loops are removed to obtain a tree.

The tree shown in Figure 5-7(b) is an example of a Steiner tree that spans nodes *A*, *D*, *G* and *I*. Steiner trees can be used as multipoint-to-multipoint trees to enable group-shared services.

Mathematically, it can be shown that there are no good[1] algorithms for finding Steiner trees. This means that none of the possible algorithms for finding Steiner trees scale well for networks with a high number of nodes, and efficient ways for finding Steiner trees can only be found for some network topologies, as well as for the following cases:

- *Unicast*: there is only one transmitter and one receiver. The problem is how to find the shortest path between two nodes, and it can be solved with the help of the Dijkstra and Bellman–Ford algorithms.
- *Broadcast*: all the nodes in the network belong to the tree. The problem is how to find the minimum-weight spanning tree for the network, and this problem can be solved by means of the Prim and Kruskal algorithms.

Although there are no efficient algorithms for finding Steiner trees, there are some practical approaches that produce good-quality (although not always optimal) trees. Specifically, there are algorithms to find trees with weight that is at worst twice the weight of the optimal Steiner tree. One of these algorithms was proposed by Kou, Markowsky and Berman and is known as the KMB algorithm.

Some other algorithms do not try to find trees similar to the Steiner trees, but take a different approach. Some try to find minimum-delay trees. Algorithms for finding

[1]The word 'good', here, means that the problem of finding Steiner trees cannot be solved in polynomial time by any algorithm.

minimum-delay multipoint-to-multipoint trees have the same problem as algorithms for finding Steiner trees, but there is an easy way to calculate point-to-multipoint minimum-delay trees that is used by many routing algorithms:

1. Calculate the shortest path from the root node to every destination node.
2. Calculate the union of all the previously calculated shortest paths.
3. Remove any loops from the result.

Sometimes the quality of a multicast tree cannot be evaluated by simply calculating its weight. Some of the conditions that a multicast tree must meet may affect the routing algorithms. For example, finding a minimum-delay tree may not be enough; it may be necessary to find a tree with balanced delay between all the destinations (delay-variation-bounded multicast tree), or a tree that distributes the load of every router fairly (degree-bounded multicast tree).

Selected Bibliography

[1] Ramalho M., Multicast Routing Protocols: A Survey and Taxonomy, *IEEE Communications Surveys*, Vol. 3, No. 1, 2000.

[2] Atwood J.W., A Classification of Reliable Multicast Protocols, *IEEE Network Magazine*, May/June 2004, pp. 24–34.

[3] Sahasrabuddhe L. H., Mukherjee B., Multicast Routing Algorithms and Protocols: A Tutorial, *IEEE Network Magazine*, January/February 2000, pp. 90–102.

[4] Bo Li, Jiangchuan Liu, Multirate Video Multicast over the Internet: An Overview, *IEEE Network Magazine*, January/February 2003, pp. 24–29.

[5] Gossain H., De Morais Cordeiro C., Agrawal D. P., Multicast: Wired to Wireless, *IEEE Communications Magazine*, June 2002, pp. 116–123.

[6] Striegel A., Maniwaran G., A Survey of QoS Multicasting Issues, *IEEE Communications Magazine*, June 2002, pp. 82–87.

[7] Smijanic A., Scheduling of Multicast Traffic in HIgh-Capacity Packet Switches, *IEEE Communications Magazine*, November 2002, pp. 72–77.

[8] Dutta A., Chennikara J., Wai Chen, Altintas O., Multicasting Streaming Media to Mobile Users, *IEEE Communications Magazine*, October 2003, pp. 2–10.

[9] Mir N. F., A Survey of Data Multicast Techniques, Architectures, and Algorithms, *IEEE Communications Magazine*, September 2001, pp. 164–170.

[10] Maxemchuk N. F., Reliable Multicast with Delay Guarantees, *IEEE Communications Magazine*, September 2002, pp. 96–102.

[11] Shapiro J. K., Towsley D., Kurose J. Optimization-Based Congestion Control for Multicast Communications, *IEEE Communications Magazine*, September 2002, pp. 90–95.

[12] Deering S. E., Host Extensions for IP Multicasting, IETF Request For Comments RFC 1112, August 1989.

[13] Fenner W., Internet Group Management Protocol, Version 2, IETF Request For Comments RFC 1112, November 1997.

[14] Cain B., Deering S., Kouvelas I., Fenner B., Thyagarajan A., Internet Group Management Protocol, Version 3, IETF Request For Comments RFC 3376, October 2002.

Chapter 6: QoS in Packet Networks

Quality of service (QoS) is the ability of a network to provide services with predictable performance.

Circuit-switched networks have always offered a good QoS when it comes to delay. In particular, the end-to-end delay of these networks is usually small and highly predictable. This makes them very attractive for interactive communication services, such as telephony. The drawback of circuit-switching technologies is that they waste a lot of network transmission capacity, because their resource reservation mechanisms are not very flexible. Packet-switching technologies, however, are very flexible, and they use network infrastructure efficiently, but it is difficult to predict their performance in terms of delay, due to variable queuing and processing times in intermediate nodes. The properties of packet-switched networks have made them a good choice for low-cost data communications.

The trade-off between performance and efficiency in networking technologies arises when a single converged network must support interactive and non-interactive traffic. This is what happens with triple-play applications. An important goal for Triple Play providers is to base their services on a single network that can transport voice, video and data, to generate revenues while reducing capital and operational expenses (CAPEX and OPEX). This network is ideally based on a single networking technology, but it can offer the best of circuit- and packet-switching technologies.

Triple Play: Building the Converged Network for IP, VoIP and IPTV Francisco J. Hens and José M. Caballero

6.1 QoS Basics

Today's converged networks are based on the Internet Protocol (IP). IP is more than 30 years old, but it has a successful history. Most current computer applications, as well as the Internet, are based on IP. The original masterminds of IP designed it as an ordinary packet-based technology in many aspects. IP networks are very flexible and efficient, but they do not have the minimum delay guarantees needed for high-quality audio and video services. In other words, legacy IP networks can only offer best-effort services.

It has been necessary to update IP so that it supports QoS, and this evolution still continues. Future solutions must take these two facts into account:

1. A network with small packets experiences shorter delays and is thus more similar to a circuit-based network. The reason for this is that a packet cannot be sent before it is filled with data. And, obviously, the time needed to fill smaller packets is shorter.
2. Under light load conditions, packet-based networks look more similar to circuit-based networks. There are no variable queuing times, and end-to-end delay becomes a deterministic quantity.

Having small packets in IP networks is not very difficult, because the length of IP packets is variable, and it can be chosen by the transmitter. Another advantage of this variable packet length technology is that packets for data communications can be very long, so that they could achieve a low overhead-to-data ratio, while packets for interactive data and video can be short, to achieve low delay.

Keeping light load in the network is not always easy. Rather than to limit the amount of load, the solution has been to increase the transmission bandwidth to keep network utilization low. This is a good solution when bandwidth is cheap – otherwise it is necessary to find a way to keep delay low and predictable while improving network utilization to the maximum. The current networking technology achieves this by using two features that improve the traditional packet-switching technology:

1. Traffic differentiation
2. Congestion management (control and avoidance)

There are plenty of techniques associated with both traffic differentiation and congestion control and avoidance (see Figure 6-1). The aim, a QoS-aware packet network, can only be accomplished by combining several of these techniques.

6.1.1 Traffic Differentiation

Traffic differentiation separates the bulk traffic load into smaller sets, and treats each set in a customized way. There are two issues related to traffic identification:

1. *Traffic classification*. The traffic is divided into classes or flows. Sometimes it is necessary to explicitly mark the traffic with a class-of-service (CoS) identifier.

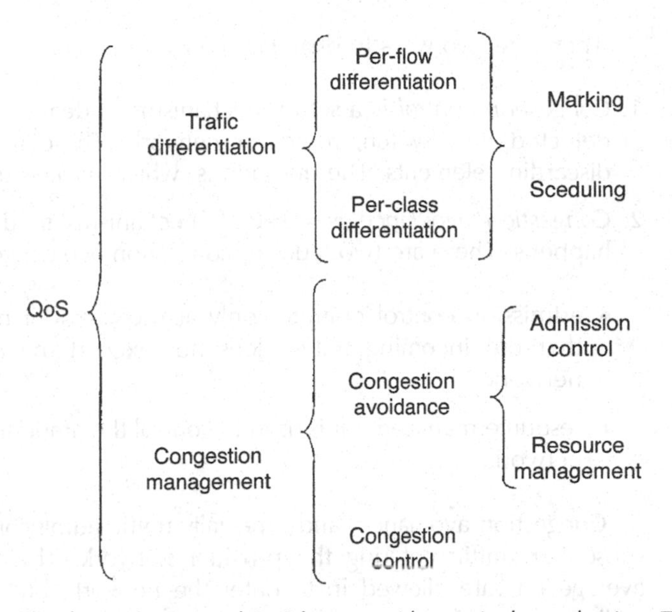

Figure 6-1. It is difficult to achieve good QoS features with one single mechanism. The best way is to mix many elements to get the desired result.

2. *Customized treatment of traffic classes and flows.* Some packets have more privileges than others in network elements. Some may have a higher priority, or there may be resources reserved for their use only.

Traffic differentiation makes it possible to improve performance for certain groups of packets and define new types of services for the packet-switched network.

- *Differentiated services.* We can talk about differentiated services when part of the traffic is treated 'better' than the rest. This way, it is possible to establish some QoS guarantees for the traffic. The QoS defined for differentiated services is also known as soft QoS.
- *Guaranteed services.* Guaranteed services go a step further. They are provided by reserving network resources only for chosen traffic flows. Guaranteed services are more QoS-reliable than differentiated services, but they make efficient bandwidth use difficult. The QoS for guaranteed services is also known as hard QoS.

6.1.2 Congestion Management

Congestion is the degradation of network performance due to excessive traffic load. By efficiently managing network resources, it is possible to maintain performance with higher loads, but congestion will always occur sooner or later. Therefore, when delivering services with QoS, one must always deal with congestion, one way or another.

There are two ways to deal with congestion:

1. *Congestion control* is a set of mechanisms to deal with congestion once it has been detected in a switch, router or network. These mechanisms basically consist of discarding elements. The question is: which packets to discard first?
2. *Congestion avoidance* is a set of mechanisms to deal with congestion before it happens. There are two types of congestion avoidance techniques:

   - admission control operates only at the provider network edge nodes, ensuring that the incoming traffic does not exceed the transmission resources of the network;
   - resource management is used to control the amount of resources being used in the network.

Congestion avoidance, and especially traffic admission, checks the properties of the subscriber traffic entering the provider network. These properties may include the average bit rate allowed in to enter the network, but other parameters are used as well. For example, a network provider may choose to limit the amount of uploaded or downloaded data. Bandwidth profiles are used to specify the subscriber traffic, and the packets that meet the bandwidth profile are called conforming packets.

Non-conforming packets can be delayed, sent without any QoS commitment, or simply dropped. There is a contract between the subscriber and the service provider that specifies the QoS, the bandwidth profile, and how to deal with the traffic that falls outside the bandwidth profile. This contract is known as the service-level agreement (SLA).

6.2 End-to-end Performance Parameters

The first step in offering QoS is to find a set of parameters to quantify and compare the performance of the network. QoS is provided by the network infrastructure, but experienced by the users. This is the reason why QoS is specified by means of end-to-end parameters. Some of the parameters that can be used are end-to-end delay and bit error ratio (BER).

The Internet Protocol Performance Metrics (IPPM) is an IETF working group that defines standard metrics to evaluate network performance. These metrics are intended for network dimensioning and network equipment design, as well as for QoS and SLA verification. Regarding the latter, there are at least four critical metrics to define:

- delay;
- delay variation;
- loss;
- bandwidth.

The IPPM working group has released Requests for Comments (RFC) that deal with these parameters. Other metrics defined by the IPPM include connectivity and packet reordering.

Another goal of the IPPM working group is to define measurement protocols that enable cooperation between measurement equipment. The One-Way Active Measurement Protocol (OWAMP) and the Two-Way Active Measurement Protocol (TWAMP) have been drafted for this purpose.

The IPPM defines atomic metrics: parameters measured from a minimum amount of data. For example, delay is defined for one single packet. However, usually, the information given by a measurement performed in one instance only is not enough to describe the entity under test. A set of related atomic measurements forms a sample measurement. The samples in a sample measurement can be all the possible atomic measurements within the measurement interval, or a more reduced set of atomic measurements obtained at instants chosen statistically over a longer period of time. Finally, there are statistic measurements that summarize the information given by a sample measurement by means of an averaging process.

6.2.1 One-way Delay

The end-to-end *one-way delay* experienced by a packet when it crosses a path in a network is the time it takes to deliver the packet from source to destination. This delay is the sum of delays on each link and node crossed by the packet (see Figure 6-2).

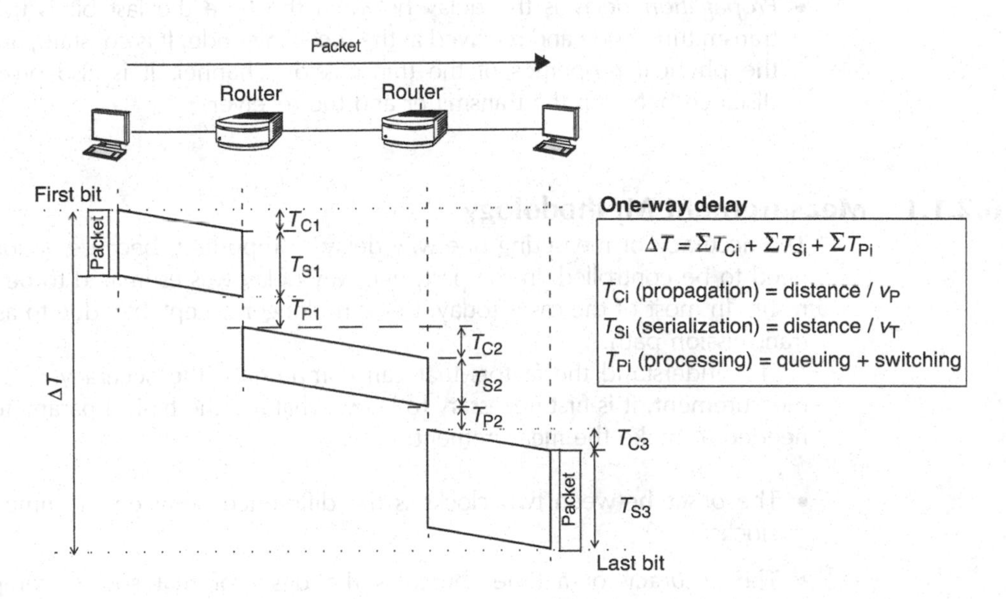

Figure 6-2. One-way delay is the sum of delays on each link and node crossed by a frame.

This definition of one-way delay is atomic, because it is based on a single packet. It is normally more useful to calculate average values of one-way packet delay during a time interval.

The IPPM addresses the definition and measurement of one-way delay in RFC 2679. This RFC deals with IP networks, but the concepts stated there can easily be extended to other packet-switching technologies.

The round trip delay (RTD), or latency, is a parameter related to one-way delay. It is the delay of a packet on its way from the source to the destination and back. RTD is easier to evaluate than other delay parameters, because it can be measured from one end with a single device. Packet timestamping is not required, but a marking mechanism of some kind is needed for packet recognition. The best-known RTD tool is Ping. This tool sends Internet Control Message Protocol (ICMP) echo request messages to a remote host, and receives ICMP echo replay messages from the same host.

There are three types of one-way delay:

- *Processing delay* is the time needed by the switch to choose the output interface when a packet is received (switching delay), and the time that this packet has to wait in a queue when the transmission interface is busy (queuing delay). The switching delay is often much shorter than the queuing delay, and it is sometimes ignored.
- *Serialization delay* is the delay between the transmission time of the first and the last bit of a packet. It depends on the size of the packet. Some intermediate nodes may need to receive the full packet before forwarding it to an output port, but this is not always so. For example, some Ethernet switches are able to forward frames as soon as the destination address is known.
- *Propagation delay* is the delay between the time the last bit is transmitted at the transmitting node and received at the receiving node. It is constant, and it depends on the physical properties of the transmission channel. It is also proportional to the distance between the transmitter and the receiver.

6.2.1.1 Measurement Methodology

The accuracy of measuring one-way delay is important, because various error sources need to be controlled. In the past, one-way delay was estimated to be half of the RTD result. In most of the cases today, this is no longer acceptable, due to asymmetry in the transmission path.

To understand the factors that can compromise the accuracy of a one-way delay measurement, it is first necessary to know what are the typical parameters of the clocks needed to make the measurement:

- The *offset* between two clocks is the difference between the time given by these clocks.
- The *accuracy* of a time source is the offset of that source compared with the UTC.

- The *skew* gives a change rate of a clock's offset or accuracy over time. It can be said, for example, that the skew of a clock is 10 ms per day.
- The *drift* is the rate of change of the skew over time. It depends on the temperature.
- The *resolution* is the minimum time difference that can be distinguished with a clock. In other words, it is the duration of a clock tick.

It is important to note that these parameters are based on the time given by a clock. It is possible to define a different set of parameters based on frequency instead of time. For example, a frequency offset between two clocks is related to the time skew between them.

The errors and uncertainties in one-way delay measurements are the following:

- *Synchronization errors between source and destination clocks*: a one-way delay measurement requires the transmitter and the receiver to be synchronized to maintain the time offset under control. In IP networks, this can be achieved by using the Network Time Protocol (NTP). The performance of this protocol is quite unpredictable, but it is accepted that an error in an NTP-based one-way delay measurement is a few milliseconds at worst. This can be accepted when measuring delays of hundreds of milliseconds. The Global Positioning System (GPS) timing or SDH-derived timing offer performance levels several orders of magnitude better than NTP, but GPS is expensive and requires external antennas to be installed – and SDH is not always available.
- *Errors related to wire times and host times*: bits appear and are detected in the transmission channel at wire times. On the other hand, one-way delay measurements involve packet timestamping. However, packets are timestamped at host times that might not be the same as wire times. The difference between these two times is usually very difficult to predict.
- *Uncertainty related to the resolution of measurement clocks*.

6.2.2 One-way Delay Variation

The one-way delay variation of two consecutively transmitted packets is the one-way delay experienced by the last transmitted packet, minus the one-way delay of the first packet (see Figure 6-3). This is defined by the IPPM in RFC 3393. According to the definition, involving only two packets, one-way delay variation is atomic. More useful results are obtained by averaging several atomic measurements.

There are many parameters related to one-way delay variation:

- *Jitter* is a term widely used in classic telephony. If this term is used for packet-switched networks, its meaning is similar to one-way delay variation. However, while one-way delay variation may be a negative number, packet jitter is normally supposed to be non-negative, and therefore it is often referred to as the absolute value of one-way delay variation.
- *One-way delay variance* is another possible parameter for delay variation. It is a non-negative number and a statistical value. So, unlike one-way delay variation, one-way delay variance is the result of an averaging process.

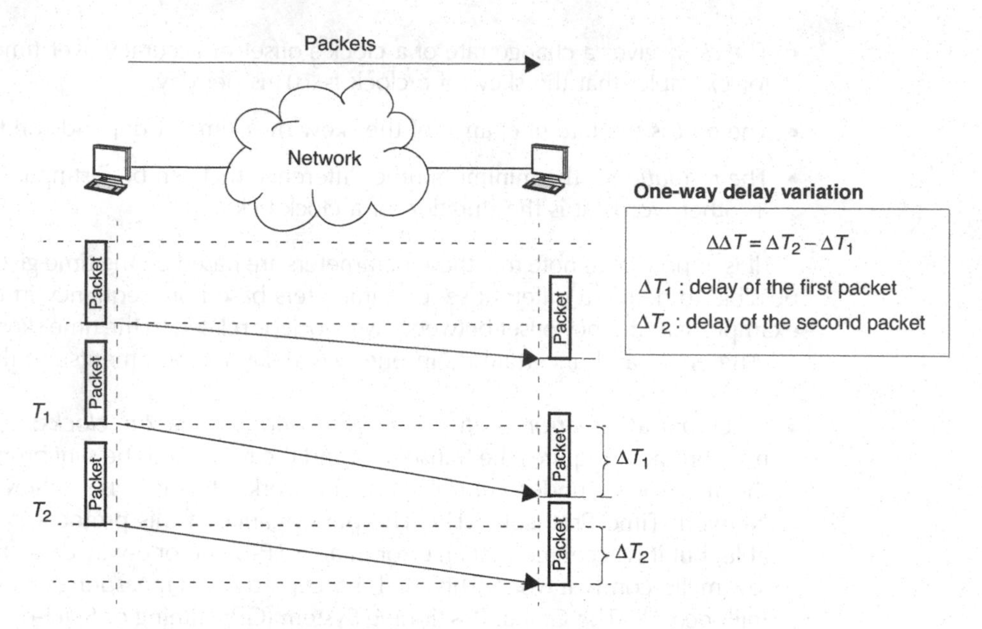

Figure 6-3. One-way delay variation: measurement and impact on data periodicity.

- *Peak-to-peak delay variation* is the difference between the maximum and minimum values of a sequence of delay variation samples.
- The *interarrival time* between two consecutively received packets is the difference between the arrival time of the last and the first packet. Timestamping or sequencing is not required to measure interarrival time, which makes this parameter easier to measure than one-way delay variation. This metric is useful to evaluate the impact of variable delay in packet periodicity but it is not always a good metric for delay variation.

Delay variation is present in both circuit- and packet-switched networks. In circuit-switched networks, delay variation has to do with justification events due to the skew between different clocks. In packet-switched networks, the main sources of delay variation are: variable queuing times in the intermediate network elements, variable serialization and processing time of packets with variable length, and variable route delay when the network implements load-balancing techniques to improve utilization. This makes delay variation in packet networks more remarkable and unpredictable than in circuit switched networks.

6.2.2.1 Measurement Methodology

Evaluating one-way delay variation is easier than measuring one-way delay, because the traffic generator and analyser do not need to be synchronized. The offset between clocks is cancelled, because one-way delay variation is the difference between two delays. Skew usually changes slowly with time and does not affect the results. Although a high level of accuracy is not necessary, one-way delay must be evaluated before one-way delay variation. This is the reason why timestamping of probe packets is required.

The error sources for one-way delay derive from wire and host times and from the uncertainty caused by the limited resolution of the clocks used. The uncertainty of one-way delay is twice as high as that of one-way delay.

6.2.3 Packet Loss

A packet is said to be lost if it does not arrive at its destination. It can be considered that packets that contain errors or arrive too late are also lost. 'Too late' may have different meanings for different applications. In real-time audio and video applications, packets received later than they should have been received are not useful, but it is possible to be less restrictive in many data applications.

The IPPM addresses packet loss evaluation in RFC 2640. The definition given there is based on an application-dependent timeout. The IPPM definition involves only a single packet, and therefore the packet can only take two values: '0' when it is successfully received before the timeout, and '1' when it is lost.

Making packet loss dependent on timeouts may cause problems in practice, because most of the times it requires packet timestamping and synchronization between the analyser and the generator. An exception of this is the round-trip packet loss measurement. In this case timestamping and synchronization are avoided, and only packet sequencing is needed. Ping is a well-known tool for measuring round-trip packet loss.

Nevertheless, there is an alternative to synchronized one-way packet loss, if packets form a part of a sequenced stream: a packet is considered to be lost if a packet is received with a higher sequence number than expected. For networks that allow packet reordering, it may be necessary to let several packets pass with higher sequence numbers before considering a packet to be lost.

A single packet loss event is not meaningful enough to describe the properties of the path between the transmitter and the receiver. There are statistical metrics that give more accurate descriptions. For example, packet loss is often expressed as a ratio of lost and transmitted packets.

Packet loss may occur when transmission errors are registered, but the main reason behind these events is network congestion. Intermediate nodes react to high traffic load conditions by dropping packets and thus generating packet loss. Congestion tends to group loss events, and this harms voice and video decoders that are optimized to work with uniformly distributed loss events.

Loss distance and loss period are metrics that give information on the distribution of loss events (see Figure 6-4).

- *Loss distance* is the difference in the sequence numbers of two consecutively lost packets, separated or not by received packets.
- *Loss period* is the number of packets in a group where all the packets have been lost.

These atomic parameters make it possible to collect statistics that describe the distribution of packet loss events between the transmitter and the receiver.

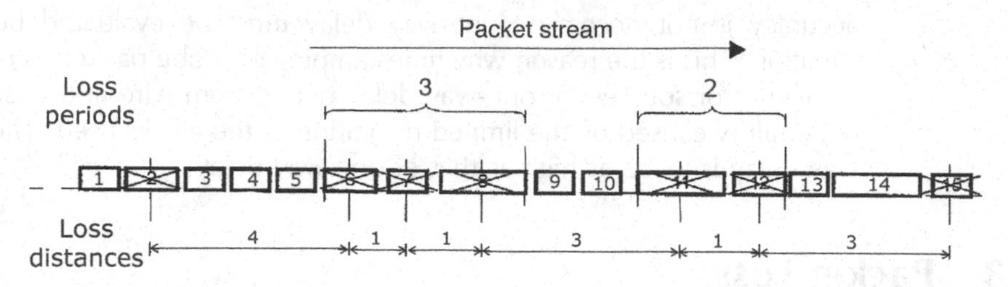

Figure 6-4. One-way delay variation: measurement and impact on data periodicity.

6.2.4 Bandwidth

Bandwidth is a measure of the ability of a link or a network to transfer information during a given period of time. There are different metrics for bandwidth:

- Bandwidth *capacity* defines the maximum rate at which packets can be transmitted by the link.
- *Available bandwidth* measures the bit rate a link or a network can accept for transmission at a given moment.

Bandwidth capacity does not usually change with time, but bandwidth availability depends on the traffic load, and therefore it varies in time. If R_i is the capacity of a link (i), A_i its available bandwidth and u_i a number between 0 and 1 that measures the utilization the link, then:

$$A_i = R_i(1 - u_i) \tag{6.1}$$

Capacity and available bandwidth can be defined for links, or for entire transmission paths composed of various links. The capacity of an end-to-end path is the smallest link capacity of the links that make up the path (see Figure 6-5). The link that limits the end-to-end capacity is called a *narrow link*. A similar definition can be given for the available

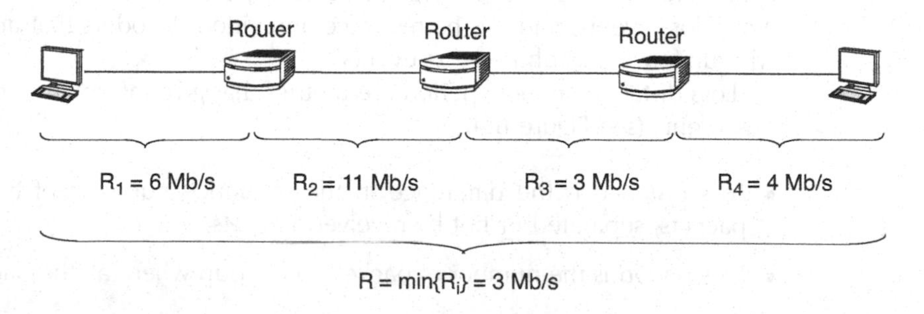

Figure 6-5. The capacity of an end-to-end path is the smallest link capacity of the links that make up the path.

bandwidth of an end-to-end path. In this case, the limiting link is called *tight link*. Both tight and narrow links are sometimes called *bottleneck links*. This term may be confusing, because it does not make distinction between capacity and available bandwidth.

For QoS, the most important bandwidth metric is the available end-to-end capacity, because only end-to-end parameters are relevant when evaluating a service. In RFC 3118, the IPPM defines a very similar metric, namely *bulk transfer capacity* (BTC). This measures the ability of a network to transfer information over a single congestion-aware transport connection, such as the Transfer Control Protocol (TCP).

6.2.4.1 Measurement Methodology

Measuring and estimating bandwidth are challenging tasks. The algorithms available to do this show the connection between bandwidth estimation and delay measurement. The most simple method for measuring bandwidth is the *variable packet size* (VPS) model. This method assumes store-and-forward retransmission in intermediate devices, as well as linear dependency of delay with packet length.

The delay experienced by a single packet in a lightly loaded packet network in the first l links can be approximated in the following way:

$$\Delta T_l = \sum_{i=0}^{l-1}\left(\frac{s}{R_i} + d_i\right) = s\sum_{i=0}^{l-1}\frac{1}{R_i} + D = sk_l + D \tag{6.2}$$

Here, d_i is the transmission delay of the link (i), and s is the packet size. This is the equation of a line with slope, K_l. By delivering packets with different lengths to the destination and to the intermediate nodes, it is possible to collect enough data to calculate link bandwidths, R_i.

The VPS model has many inconveniences. One of them is that it does not consider queuing delay in intermediate nodes. Many network elements have shown significant deviation from the behaviour expected by the algorithm, and this is the reason why there are many versions of this basic method, and other different algorithms have been proposed to measure bandwidth.

6.3 Marking

Marking means classifying packets by means of dedicated header fields, to identify traffic flows or classes. Traffic marking is closely related to traffic and flow identification. Network elements can efficiently mark packets and process marked packets. However, identifying unmarked traffic flows can be difficult and time-consuming, if properties such as the transport layer port, Multipurpose Internet Mail Extension (MIME) type or Uniform Resource Locator (URL) are taken into account for classification.

Sometimes, marking is associated with traffic admission processes. Marks can be set by the subscriber, service provider, or both, for different purposes:

- *To identify the CoS carried within the payload*. For example, VoIP and data traffic have different QoS requirements, and they may be marked with dedicated labels to allow for a customized treatment by the network.

- *To deal with non-conforming packets*. Sometimes, traffic that does not meet the bandwidth profile is not dropped or delayed, but simply delivered without any QoS guarantees. If this is the case, the provider edge node may need to place the correct label on these packets.
- *To enable a customized per-subscriber service provision*. Different customers may have different requirements and SLAs. Marking each customer's packets with a different label makes it easier to provide their customized services.

Traffic marking also has to do with congestion control. A network element that is experiencing congestion in a privileged traffic class may choose to re-classify a part of the traffic to a non-privileged class to free up resources.

Traffic classification can be microscopic (per flow) or macroscopic (per class), but marking is used for both.

6.3.1 Traffic Flows

A *traffic flow* is a unidirectional sequence of packets that are related and have the same QoS requirements. It is commonly understood that a traffic flow results from the activity of a single user. For example, the set of packets that make a HTTP request to a web server is an example of a flow. Another example is the sequence of packetized voice samples from a VoIP application coming from one user. A video conference with both video and audio can be made up of several traffic flows (one for audio and one for video, for instance) and they may have different QoS requirements. Signalling packets associated with this type of multimedia communications usually have special QoS requirements, and they belong to different flows than data packets.

Every flow in a network may have its own QoS requirements, and one possible approach to QoS provisioning is to share network resources on a per-flow basis, assigning every flow its own bandwidth, memory and processor time in network elements. If this solution is implemented, flow identification is a problem that needs to be solved. IP packets include the source and destination address, and they can be used for this purpose. However, this is usually not enough, because IP addresses identify communicating parties, not applications. The transmission port and the MIME type include more details that can help in flow identification, but a better solution is to mark flows according to their specific QoS requirements at layer 3 or layer 2. The IP precedence or the IEEE 802.1p Ethernet CoS fields provide a way to do this.

6.3.2 Traffic Classes

There are traffic flows with similar QoS requirements. All VoIP flows are delay-sensitive, and all web transactions are sensible to transmission errors and losses. These sets of microscopic flows can be grouped into larger entities called *classes*. Traffic classes receive the same treatment in the provider network, and each class may use dedicated resources in the network elements they pass through. It is often a good idea to keep traffic coming from the same application within the same class.

Table 6-1. QoS requirements for different applications.

Traffic type	Bandwidth	Packet loss (max)	Delay (max)	Jitter (max)
Interactive voice (G.711)	12–106 kbit/s	1%	150 ms	30 ms
Streamed video (MPEG-4)	0.005–10 Mbit/s	2%	5000 ms	Insensitive
Streamed audio (MP3)	32–320 kbit/s	2%	5000 ms	Insensitive
Data	Variable	Sensitive	Insensitive	Insensitive

Traffic classes are marked with CoS labels that enable efficient processing. The main packet-based network technologies provide header bytes for marking traffic classes. These technologies include IP, ATM, Frame Relay, MPLS and Ethernet.

By classifying traffic, network resource management can be improved, keeping usage low for critical traffic. Congestion can be controlled so that it will only affect non-critical traffic classes. Another advantage of classification is that it makes it possible to use intelligent packet scheduling in switches and routers. CoS labels are useful when deciding the transmission order for those packets that are queuing for transmission.

6.4 Scheduling

A *scheduler* is a logical component of a network element that chooses the next packet to be sent to an output interface from among all candidates waiting in queues. The decision made by the scheduler can be based on the waiting time or any other characteristic of the traffic. When the scheduler takes traffic flows or class-related characteristics into account, it becomes a key component of traffic differentiation and customized QoS provision.

There are some properties that a good scheduler should always have:

- It may be necessary to provide some *guarantees* to the scheduled traffic, or at least to part of it. For example, some flows may require the queuing time to be shorter than a predefined interval.
- *Fair bandwidth sharing* between some or all of the flows. Privileged flows should not be allowed to consume all the bandwidth most of the time.
- *Processing time* in network elements must be managed carefully. Simple and easy-to-implement schedulers are preferred.

6.4.1 First In, First Out Scheduler

The First In, First Out (FIFO) scheduler provides basic store-and-forward capability. It only considers the order of arrival of the incoming packets. Apart from this, no other characteristics are taken into account. This is a very simple approach, but it has some important drawbacks:

- The QoS-demanding multimedia traffic has to stay in the same queue with bursty data traffic. A short VoIP packet that is in the queue after a long data packet must wait until the

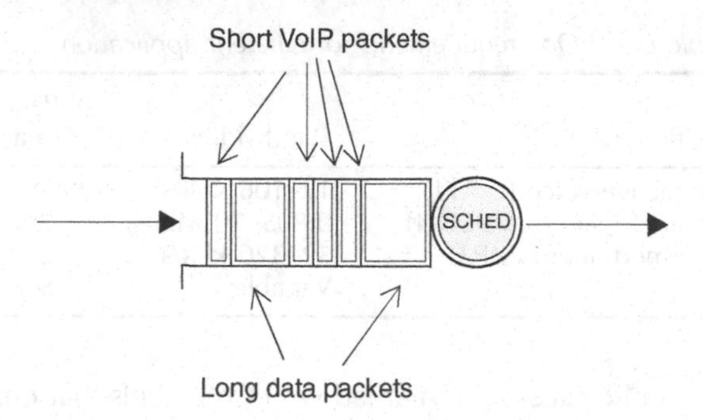

Figure 6-6. VoIP packets have to wait in the FIFO scheduler for the data packets to be served. This disturbs the end-to-end QoS of the VoIP application, even if there are just a few data packets.

whole data packet has been transmitted. This makes queuing delays highly dependent on the bandwidth profile of the bursty data source.

- There is no mechanism to control how the bandwidth is shared between the flows. Those flows that use more bandwidth tend to get most of the resources managed by the scheduler.

6.4.2 Round Robin Scheduler

The Round Robin (RR) scheduler treats traffic quite fairly, with low computational cost. The Round Robin algorithm polls several queues sequentially. Empty queues are skipped. Usually, packets from different flows or classes are stored in separate queues. The available bandwidth is shared fairly between these flows or classes, if all packets have the same length. Otherwise, flows with bigger average packet size are favoured.

The basic RR scheduler assigns the same bandwidth to every flow in those technologies that use fixed-length packets, but it can be modified to work with flows that use customizable bandwidths. The modified RR scheduler is called Weighted Round Robin (WRR), and it calculates a service schedule based on coefficients assigned to each queue. This means that packets in a queue with a coefficient value doubling that of other queues are served twice as fast.

The inconvenience of the RR and WRR algorithms is that they do not consider the length of the packets. In fact, they are not useful for technologies with variable packet lengths. There is a modified version of the basic RR algorithm called Deficit Round Robin (DRR) that deals with unpredictable-length packets. The idea behind the DRR is to keep track of the deficits experienced by a flow in one round, and compensate for these deficits in the next round. To do this, the DRR algorithm assigns a value called *quantum* for every queue. The quantum value is proportional to the level of service assigned for every queue. Additionally, there is a deficit counter on each queue, incremented by a quantum at every round. Packets can only be scheduled if their length is smaller or equal to the deficit counter. Once a packet has been served, the deficit counter is decremented by the packet length, and set to zero when empty queues are found.

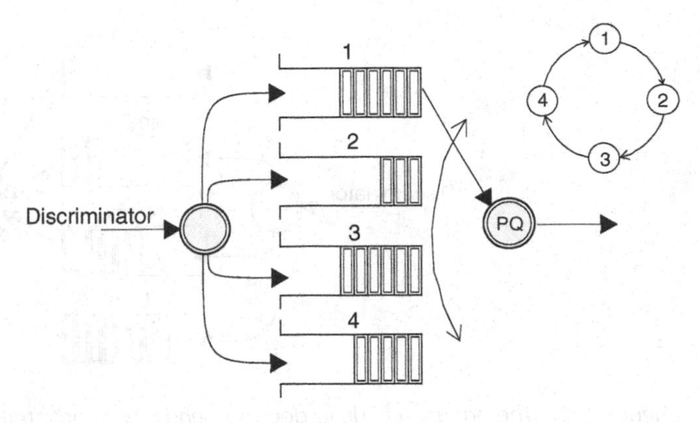

Figure 6-7. The Round Robin algorithm polls several queues sequentially, ignoring empty queues. Traffic flows are treated fairly.

6.4.3 Weighted Fair Queuing

The Weighted Fair Queuing (WFQ) scheduler is an approximation of a theoretical model known as Generalized Processor Sharing (GPS). A GPS scheduler serves traffic queues as if the traffic were a fluid flow. A hypothetical GPS scheduler has several interesting properties: it creates bit-wise fairness for the transmitted traffic, makes it possible to configure a specific bandwidth for each traffic flow, and gives delay and delay variation guarantees. Unfortunately, the GPS scheduler requires packets with infinitesimal size, which is why it cannot be implemented.

The WRR and DRR schedulers are also approximations to the GPS scheduler but they do not treat traffic flows fairly in short time scales or when there are many queues. The WFQ gives better results in these situations, but it has a higher processing load, and it is more expensive to implement.

6.4.4 Priority Scheduler

The Priority scheduler ensures that, the more important the traffic is, the quicker it is handled. Packets with low priority are not delivered if there are packets with higher priority waiting to be sent.

The Priority scheduler is easy to implement, and it does not use up a large amount of resources, but giving absolute preference to high-priority traffic can be inconvenient in practice, because lower-priority traffic may have to wait for a long time before it can be sent, and this is not always desirable. When the available bandwidth needs to be distributed fairly, other more complex schedulers (such as DRR or WFQ) must be used.

6.5 Congestion Avoidance

Congestion avoidance is used to prevent congestion before it occurs. This is done by controlling the amount of traffic in the network as carefully as possible.

Congestion avoidance techniques use open-loop and closed-loop flow control mechanisms. Closed-loop mechanisms rely on feedback messages given by intermediate or end-user

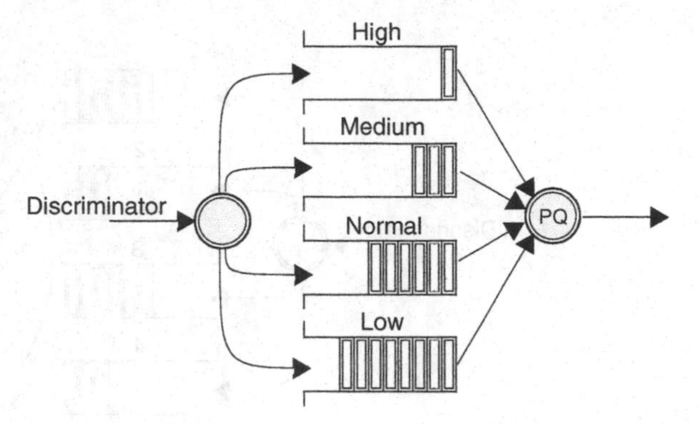

Figure 6-8. The priority scheduler does not send low-priority traffic if there are packets with higher priority waiting to be sent.

equipment to prevent congestion. A good example of closed-loop congestion avoidance is the window-based mechanism provided by TCP in IP networks.

Closed-loop congestion avoidance is well-suited to data networks, but it is slow with multimedia traffic. Open-loop congestion avoidance is the best option for the latter.

There are two open-loop congestion avoidance techniques:

1. *Admission control* filters traffic by discarding, delaying or re-classifying packets or flows, to ensure that the transmission resources of the network are not exceeded.
2. *Resource management* functions reserve resources in the network before starting transmission, and free up resources when transmission is over.

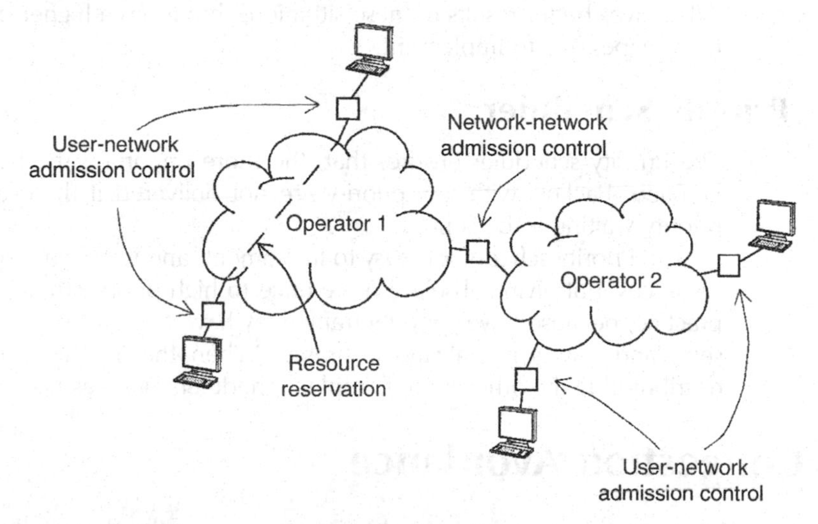

Figure 6-9. Resource management operates within the network, and admission control at the boundaries of the network, but admission control can also be used to split a large network, or to separate networks that belong to different operators.

Resource management is implemented by means of protocols that operate when connection parameters are negotiated or re-negotiated, or when a connection is released. For non-connection-oriented technologies, such as IP, resources are reserved and freed up on a per-flow basis. These protocols are understood by the switches and routers that make up the network, and new connections or flows are accepted by them when there are resources available. Otherwise new connections are rejected.

Admission control is often used at the boundaries of the network, but it can be applied to intermediate network elements as well. Shaping or policing can, for example, be used to separate networks that belong to different operators. Resource management calls for compatibility between reservation protocols in all the network elements used, and operators have to make an agreement on how to reserve resources for each other. These problems may be solved by implementing the necessary shaping and policing functions between networks.

6.5.1 Admission Control

There are two problems related to admission control. First, how to find an algorithm that can decide if a particular packet is conformant or not, and second, what to do with non-conformant packets? Most of the admission control algorithms are based on the token bucket system (see Section 6.5.1.1).

There are different types of filters that can help to classify non-conformant packets, and each of them have different effects on the traffic (see Figure 6-10):

- *Policers* are filters that discard all non-conformant packets. Because packets are dropped, the original information is not fully conserved, but timing is not affected, so, packets are not delayed. Policers are well-suited to those error-tolerant applications that have strict timing constraints, for example VoIP or some interactive video applications.
- *Shapers* work much the same way as policers, but they do not discard packets. Non-conformant traffic is buffered and delayed until it can be sent without violating the SLA agreement or compromising network resources. Shapers conserve all the information that was sent, but they modify timing, so they may cause problems for real-time and interactive communications.
- *Markers* can be used to deal with non-conformant packets (see Paragraph 6.3). Instead of dropping or delaying non-conformant packets, they are delivered with low priority or 'best effort'.

Policers, shapers and markers can be combined to obtain more complex filters that have very interesting properties. For example, there could be an admission control filter that acts as a policer in situations where the SLA agreement is strongly violated, but in the case of eventual or non-persistent violations, the filter would send non-compliant packets – without any QoS guarantees, of course.

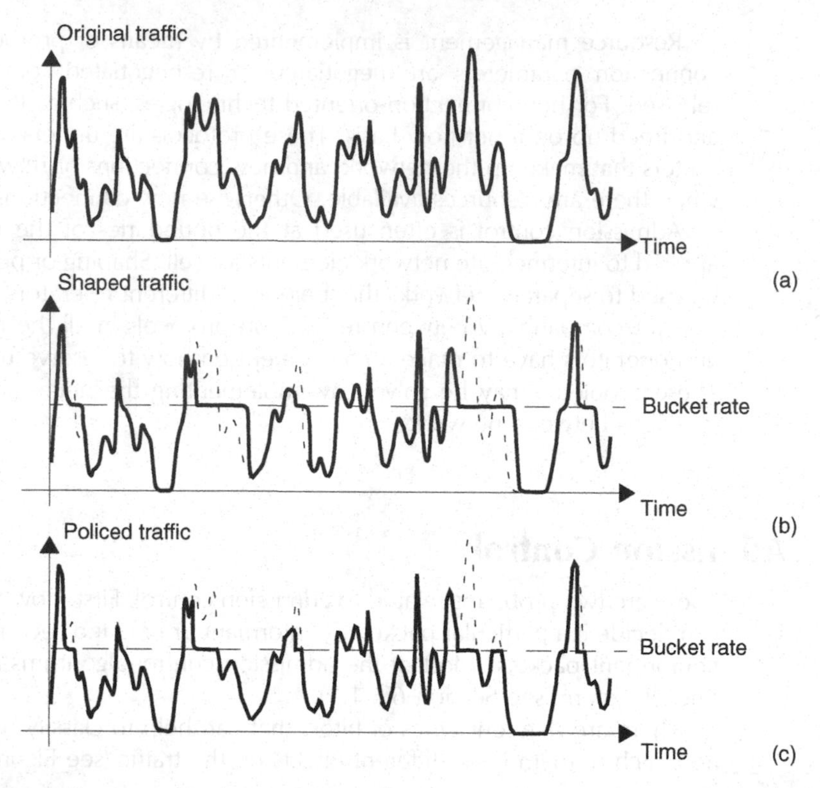

Figure 6-10. Shaping and policing of user traffic. (a) Original traffic. (b) When traffic is shaped, no packets are dropped, but some of them may be delayed. (c) When traffic is policed, it is never delayed, but some packets may be dropped.

6.5.1.1 The Token Bucket

Most of the admission control algorithms are based on the *token bucket* system. These algorithms include a *conceptual bucket*, of size *S*, that is filled with tokens at a configurable rate, *R*. Each packet needs to take a token from the bucket before it can be sent. If there are tokens available, the packet can be delivered, but if the bucket is empty, the packet needs to be treated somehow, and the treatment it gets depends on the type of filter that is being used: policers drop packets when there are no tokens, markers re-classify them, and shapers store them in a buffer until more tokens are available (see Figure 6-11). A bucket can store tokens at maximum. When the bucket is full, it cannot receive any more tokens, so any extra tokens are discarded.

The token bucket algorithm is simple, and it clearly separates conformant and non-conformant packets. It sets limits for the amount of data that enters the network during a specified time interval. Perhaps the most interesting property of this system is that it can be used to make sure that the delay for those flows that are shaped and policed by this algorithm stays within previously defined limits. This way, it is possible to provision guaranteed services in packet-switched networks (see Section 7.2.1.1).

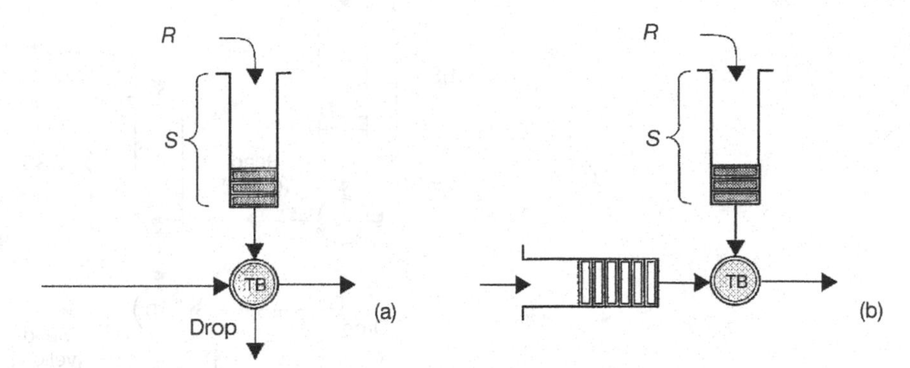

Figure 6-11. *Token buckets used in admission control: (a) token bucket working as a policer; (b) token bucket working as a shaper.*

6.5.1.2 Single-rate Three-colour Marker

The single-rate three-colour marker (srTCM) policer is obtained by chaining two simple token bucket policers (see Figure 6-12). Tokens fill the main bucket until they reach the capacity given by the *committed burst size* (CBS) parameter, at a rate given by the *committed information rate* (CIR) parameter. Any extra tokens in the main bucket are not ignored, but are used to fill a secondary bucket until they reach the capacity given by the *excess burst size* (EBS) parameter.

The traffic that passes through the first bucket (*green traffic*) is delivered with the QoS agreed with the service provider, but any traffic that passes through the secondary bucket (*yellow traffic*) is usually re-classified and delivered as best-effort traffic, or it is given a low priority. Non-conformant traffic (*red traffic*) is dropped.

The srTCM can be considered as a more sophisticated version of the simple token bucket policer. Some non-conformant traffic that would otherwise be dropped can be recovered using srTCM.

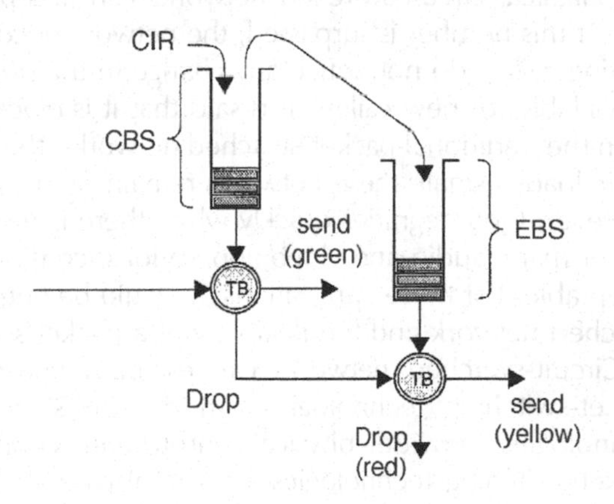

Figure 6-12. *Single-rate three-colour marker policer.*

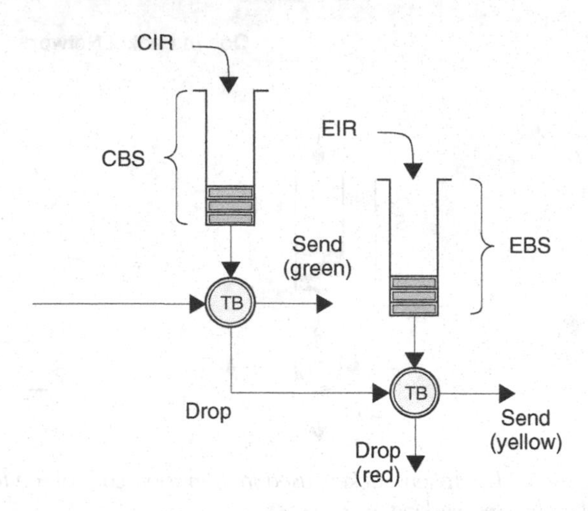

Figure 6-13. Two-rate three-colour marker policer.

6.5.1.3 Two-rate Three-colour Marker

The two-rate three-colour marker (trTCM) algorithm is a modified version of the srTCM that includes a new configuration parameter, the *excess information rate* (EIR). The difference between the two is that, with trTCM, the excess tokens from the main bucket are not used to fill the secondary bucket, but the secondary bucket is filled by a new token flow at a rate given by the EIR parameter (see Figure 6-13).

In practice, the only important difference is that the trTCM allows for a more accurate control over yellow traffic by means of the EIR parameter.

6.5.2 Resource Management

The classical circuit-switched networks can accept a limited number of simultaneous calls. If this number is surpassed, the network becomes unavailable for new callers. The existing callers do not notice any change in the quality of service. When the network is unavailable for new callers, it is said that it is *blocked*.

In the traditional packet-switched networks, the situation is quite different. When the traffic load is small, these networks remain predictable and provide good performance. However, QoS degrades quickly when there is more traffic load.

For many audio and video applications, continuous performance degradation is not acceptable. For these applications, it would be better to have the behaviour of a circuit-switched network and the flexibility of a packet-switched network.

Circuit-switched networks can establish end-to-end connections, although some packet-switching technologies can do the same. However, while circuit-switching technologies provide physical end-to-end connections, the circuits provided by packet-switching technologies are virtual (virtual circuits, VC): they try to emulate the properties of physical circuits.

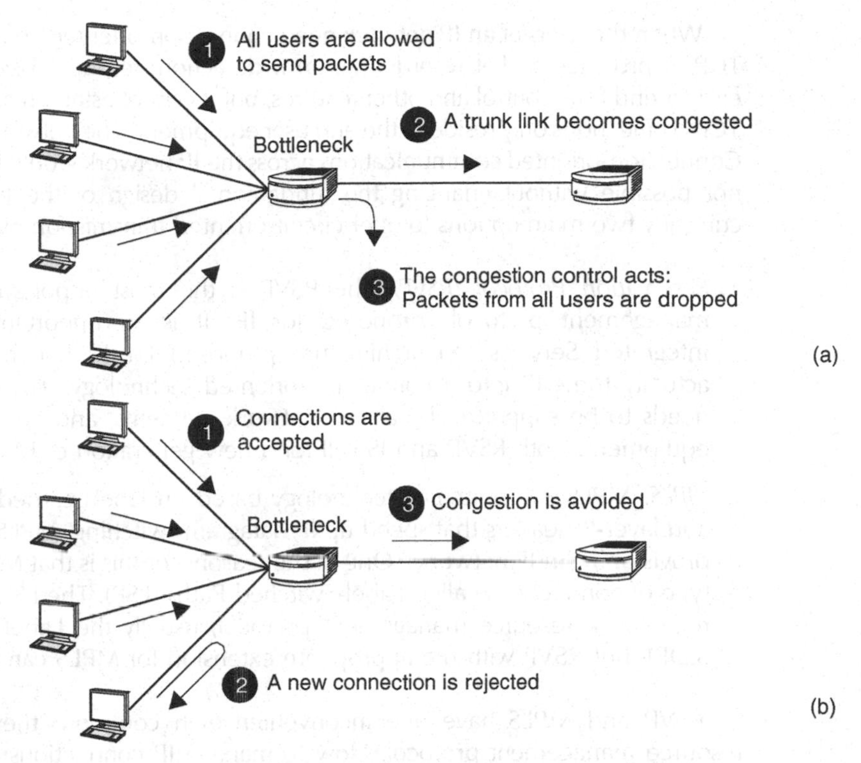

Figure 6-14. How resource management acts: (a) without resource management, all users experience degradation on their applications whenever there is congestion in the network; (b) if congestion management is used, only some subscribers are not allowed to send data, but the others are not affected.

Those technologies that are based on VCs, for example ATM, can potentially provide the same level of service as any other circuit-switched network, while maintaining high flexibility. VCs do have some drawbacks as well, because they need:

- a *resource-management protocol* to establish and release connections: usually these signalling protocols are complex and bandwidth-consuming;
- *to store flow status information* in intermediate nodes: this means that it is necessary to add complexity to network elements by adding more memory and processing time to manage connection information;
- *to waste time establishing connections*: the time wasted in establishing a connection means extra delay.

IP networks are not circuit-oriented for design reasons. These networks deliver packets, and all the information needed for these packets to arrive to their destination is included in the packet header. These packets are called datagrams. Packet-switched networks based on datagrams are simple and scalable, but they are not the best possible solution to transport voice and video.

When the users of an IP network need connection-oriented communications, they use TCP. It provides packet re-ordering, error detection, re-transmission of errored packets, end-to-end flow control and other features, but it cannot establish any VCs in the network. TCP connections only reside in the end user equipment. The network is not aware of them. Connection-oriented communications across the IP network would be desirable, but this is not possible without changing the fundamental design of the IP protocol. There are currently two main options to offer circuit-oriented transmission over IP:

1. *Reservation Protocol (RSVP)*: the RSVP is the most important of all the resource management protocols proposed for IP. It is an important component of the Integrated Services (IS) architecture proposed for IP networks. This architecture actually turns IP into a connection-oriented technology. To be efficient, the RSVP needs to be supported by all the network elements, and not only by the end user equipment. Both RSVP and IS call for a new generation of IP routers.
2. *MPLS*. MPLS is a switching technology based on labels carried between the layer-2 and layer-3 headers that speed up IP datagram switching. MPLS can be used for QoS provisioning in IP networks. One of the reasons for this is that MPLS supports a special type of connections called Label-Switched Paths (LSP). The LSP setup and tear-down relies on a resource management protocol, usually the Label Distribution Protocol (LDP), but RSVP with the appropriate extension for MPLS can be used as well.

RSVP and MPLS have one inconvenience in common: they rely on a complex resource management protocol. How to manage IP connections without a protocol of this type is still an open question.

6.6 Congestion Control and Recovery

Congestion control is the set of techniques that deal with congestion once it has been detected in the network, and its aim is to minimize damage to communications.

In a packet-switched network, a link becomes congested whenever the amount of injected traffic exceeds its transmission capacity. The first consequence of this is that excess packets need to be buffered. This may potentially damage the QoS of real-time multimedia communications. If congestion is severe, buffers may become overloaded and cause packet loss, damaging both data and multimedia communications. Correct packet marking and scheduling help to deal with any damage caused by queuing delays. From this point of view, traffic differentiation is related to congestion control. To deal with the effects of packet loss, it is necessary to implement good Packet Dropping Policies (PDP) for network elements. The ideal features of a good PDP are:

- *Efficiency*: a good PDP should keep network utilization high. Dropping fragmented packets could reduce efficiency. ATM fragments IP datagrams before transmitting them, but there is no mechanism to recover a single lost or errored ATM cell. This makes it necessary to request the transmission of the entire IP datagram at the destination. Fragments that arrive without errors are not usable, but they waste transmission resources. To avoid this problem, the PDP for ATM cells must be aware of the fragmentation of IP packets and drop cells on a per-IP-packet basis rather than on a per-ATM-cell basis. Similar problems occur at flow level. Some flows may not be usable if some packets are lost, so it makes sense to drop entire flows. For the particular case of TCP, a PDP that does not consider TCP connections may cause network underutilization. A single packet loss in a TCP flow would cause the activation of the TCP congestion avoidance mechanism, and TCP would reduce the transmission rate for that connection. If this happens to many TCP connections at the same time, it may lead to underutilization of the network (global synchronization problem). Following the same reasoning, it would make sense to first discard packets that have traversed a smaller number of hops.

- *Fairness*: there should never be more privileged flows than those explicitly allowed by the service provider. Under the same circumstances, all flows should have the same statistical chances of being transmitted without packet loss. If 'unfairness' is decided by the service provider, it should be possible to control the level of 'unfairness'.
- *Simplicity*: as network elements usually have to switch traffic at high rates, it is necessary to keep PDPs as simple as possible. There is a trade-off between efficiency and simplicity. The more efficient and intelligent a PDP is, the more complex it is.
- *Scalability*: PDPs must be prepared for rapid growth of current networks. Scalability tends to be a problem for those PDPs that store connection information.

6.6.1 Drop Tail

The Drop Tail (DT) is the simplest, but not always the best PDP. It accepts packets whenever there is buffer space available, and when there is no space it drops all incoming packets regardless of their type, priority or any other property.

The DT PDP is very scalable, because it is simple, but it has some efficiency and fairness problems. Although DT can be accepted in networks with very light loads, it is not recommended when high utilization level is expected. In fact, it has been demonstrated that in this case the DT can entirely shut down the system.

The DT is biased against flows with a small amount of traffic, and a few high-rate flows tend to get most of the system resources. The DT PDP does not take packet fragmentation or packet flows into account. Long packets have a smaller probability of passing through the policer than short messages when they are fragmented. Furthermore, it has been shown that the TCP protocol exhibits low performance when DT is used.

6.6.2 Partial Packet Discard

The Partial Packet Discard (PPD) PDP has been proposed for ATM networks, to improve the performance of the DT. The improvement lies in that, whenever the PPD has to drop a packet fragment, it also drops all subsequent fragments belonging to the same packet. This saves storage space for other packets, but some fragments that are not usable are still forwarded. On average, one half of the unusable fragments are delivered, which means that it is possible to further improve this PDP.

The PPD needs to store information about packets with previously discarded fragments. This status information adds complexity to this PDP. This is not a big problem for ATM, because the connection table of ATM switches can easily be extended to store the information needed by the PPD.

6.6.3 Early Packet Discard

The Early Packet Discard (EPD) PDP is an alternative improvement for the DT in ATM networks. The buffer with EPD does not accept any more packets when it exceeds a certain configurable threshold. However, fragments from those packets that were already accepted before the threshold was exceeded are queued if there is buffer space available.

The EPD is as complex as the PPD. This PDP needs to store information on all the queued fragments. Performance-wise, the EDP has been demonstrated to be better than PPD in most circumstances. Of course, EPD and PPD can be used at the same time to improve performance even further.

There are many extensions and customizations of the EPD for different purposes; for example, to deal with TCP connections or to improve fairness.

6.6.4 Random Early Detection

The Random Early Detection (RED) PDP is designed to work with IP. It does not need to store any flow status information, and it collaborates with the TCP congestion control mechanism.

The RED splits the storage queue into three regions defined by two thresholds, T_{low} and T_{high}. The RED has different behaviours depending on the queue length (see Figure 6-15).

- *Green region*: the RED operates in this region if the queue is shorter than T_{low}. In this case, all incoming packets are accepted.
- *Yellow region*: this region is defined to be between T_{low} and T_{high}. When RED is operating in this region, an incoming packet is discarded randomly with a probability of P_a.
- *Red region*: RED works in this region if the queue is longer than T_{high}. In this case, all incoming packets are dropped.

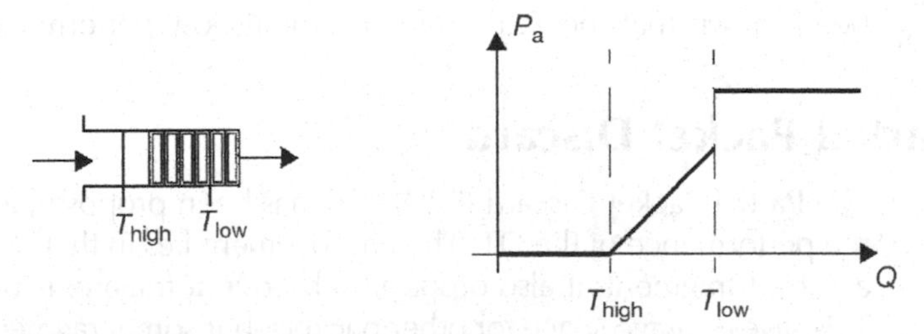

Figure 6-15. Packet drop probability in a queue with RED PDP as a function of the average queue length.

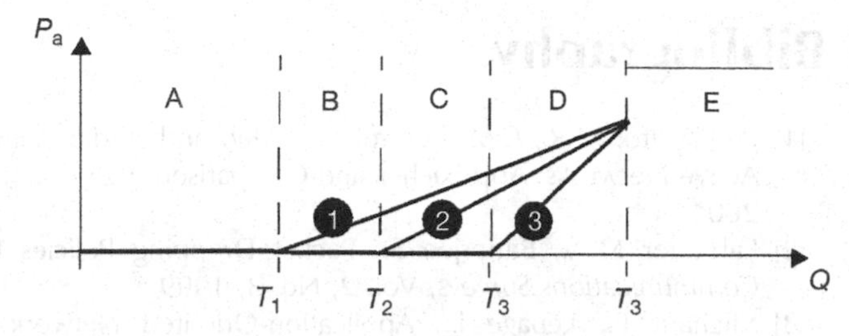

Figure 6-16. A possible implementation of the WRED. When the algorithm is working in region A, all packets are accepted. In region B, packets marked with 1 may be dropped. In region C, packets marked with 2 may be dropped as well. In region D, all three possible classes (1, 2 and 3) can be dropped. Finally, in region E, no new packets are accepted.

Actually, the operating region does not depend directly on the instantaneous value of the queue length. An averaged version, Q, is used for this purpose, to avoid changing the operation region too often. The probability of dropping a packet in the yellow region depends on two factors:

1. *Average queue length*: as the queue length approaches T_{high}, the dropping probability increases.
2. *Counter*: every time a packet is queued, the counter is incremented. This counter is reset when a packet is discarded. High counter values increase the probability of packets getting dropped. In other words, the higher the values are, the more likely it is that packets will be dropped.

The dropping probability is calculated by means of the following formula:

$$P_a = \frac{P_b}{1 - \text{counter}.T_{low}} \tag{6.3}$$

where P_b is:

$$P_b = P_{max} \frac{Q - T_{low}}{T_{high} - T_{low}} \tag{6.4}$$

Choosing the right values for T_{low} and T_{high} is important. T_{high} is related to the maximum delay allowed for a packet, and it is especially important in real-time communications. On the other hand, a value for T_{low} that is too small may affect the quality of bursty data traffic.

The RED avoids dropping packets from many TCP connections at the same time, and also helps to solve the problem of 'underuse', which is caused by the global synchronization problem (see Section 6.6).

The *weighted RED* (WRED) is a variation of the RED algorithm that takes the relative priorities of different packets into account. The algorithm starts dropping low-priority packets before dropping any packets with higher priority.

Selected Bibliography

[1] Bai Y., Ito, M.R., QoS Control for Video and Audio Communication in Conventional and Active Networks: Approaches and Comparison, *IEEE Communications Surveys*, Vol. 6, No. 1, 2004.

[2] Labrador M.A., Banerjee S., Packet Dropping Policies for ATM and IP Networks, *IEEE Communications Surveys*, Vol. 2, No. 3, 1999.

[3] Michaut F., Lepage F., Application-Oriented Network Metrology: Metrics and Active Measurement Tools, *IEEE Communications Surveys*, Vol. 7, No. 2, 2005.

[4] Xi Peng Xiao, Telkamp T., Fineberg V., Cheng Chen, Lionel M. Ni, A Practical Approach for Providing QoS in the Internet Backbone, *IEEE Communications Magazine*, December 2002, pp. 56–62.

[5] Yang Chen, Chunming Qiao, Hamdi M., Tsang D. H. K., Proportional Differentiation: A Scalable QoS Approach, *IEEE Communications Magazine*, June 2003, pp. 52–58.

[6] Adams A., Bu T., Horowitz J., Towsley D., Cáceres R., Duffield N., Lo Presti F., The Use of End-to-End Multicast Measurements for Characterizing Internal Network Behavior, *IEEE Communications Magazine*, May 2000, pp. 152–158.

[7] Christin N., Liebeherr J., A QoS Architecture for Quantitative Service Differentiation, *IEEE Communications Magazine*, June 2003, pp. 38–45.

[8] Almes *et al.*, A One-way Packet Loss Metric for IPPM, IETF Request For Comments RFC 2680, September 1999.

[9] Paxson V., Almes G., Mahdavi J., Mathis M., Framework for IP Performance Metrics, IETF Request for Comments RFC 2330, May 1998.

Chapter 7: QoS Architectures

ATM was the first packet technology that was able to provide custom QoS to subscribers, but most players in the telecommunications market are now abandoning ATM and basing their most innovative QoS solutions on IP.

They have chosen IP because it has become widely available. However, the legacy IP infrastructure is not QoS-aware, and it can only provide best-effort data services, so the IP technology needs to be modified to meet the needs of the market. The lessons learnt from ATM must not be forgotten: the IP QoS must be as simple and efficient as possible.

It is necessary to find a way to achieve a smooth transition from legacy IP to QoS-aware IP, but the task is not easy. QoS provisioning calls for end-to-end support of dedicated signalling protocols, advanced scheduling techniques, as well as innovative congestion management technologies. IP-based QoS can only be provided in small and medium-sized networks – worldwide QoS services over the Internet are not yet possible.

7.1 QoS in ATM Networks

ATM is a versatile technology designed to carry heterogeneous information, including not only data services but also real-time and non-real-time voice and video. In other words, it is designed for a universal converged network.

ATM is a packet-based transmission and switching technology that uses small, fixed-length packets called *cells*. ATM is connection-oriented, which means that, during the process of establishing a connection, switches can reserve resources for new traffic in advance. When reserving resources for users, the network can guarantee a certain QoS level. So that ATM switches can reserve enough resources to satisfy the needs of a new connection, the user must specify the bandwidth profile of the generated traffic and the required QoS level in the connection setup.

The organizations that have created the standards for ATM (ITU-T and the ATM Forum) have also defined the parameters needed for a user to determine the traffic generated, and they have classified all the possible services in categories according to their QoS requirements.

7.1.1 Bandwidth Profile Characterization

The following parameters are defined by the ATM Forum. They are used to describe the bandwidth profile of the traffic that is being transmitted.

- *Peak cell rate* (*PCR*): the maximum rate at which cells can be transmitted in a certain connection.
- *Sustainable cell rate* (*SCR*): the mean transmission rate of a connection, expressed in cells per second or bit/s.
- *Maximum burst size* (*MBS*): the maximum number of cells that a burst transmitted at a rate specified by the PCR may contain.
- *Minimum cell rate* (*MCR*): the minimum transmission rate guaranteed by the network.

In the ATM world, this set of parameters makes up the 'source descriptor'. All of these parameters are not always necessary to describe the source bandwidth profile. It rather depends on the service category.

7.1.2 Negotiated QoS Parameters

When setting up the connection, the originating user supplies the source traffic descriptor and also specifies the desired performance, so that the network could reserve enough resources. The parameters that define the QoS required by the user are the following:

- *Maximum cell transfer delay (max CTD)*: the maximum delay that cells may experience when transmitted across an ATM network, between their generation at the source and reception at the destination.
- *Peak-to-peak cell delay variation (peak-to-peak CDV)*: the maximum interval of variation a delay may undergo. So, if the previous parameter specifies the maximum delay, and if we subtract the value of this parameter, what we obtain is the minimum delay. Cells should experience delays that fall between these two values.
- *Cell loss ratio (CLR)*: the amount of cells lost in a connection as regards the total number of cells transmitted.

7.1.3 ATM Service Categories

The ATM Forum has defined five service categories. Each category is intended for applications with a specific type of traffic and with certain transmission-related QoS requirements. When a connection is established, the service category and the corresponding QoS parameters are negotiated at the same time as the path to the destination is established.

The service categories defined by the ATM Forum are:

1. constant bit rate (CBR);
2. real-time variable bit rate (rt-VBR);
3. non-real-time variable bit rate (nrt-VBR);
4. unspecified bit rate (UBR);
5. available bit rate (ABR).

Some of these service categories are the same as the ATM transfer capacity defined by Recommendation ITU-T I.371, although different names are used (see Table 7-1).

The CBR service is the same as the ITU-T *deterministic bit rate* (DBR), and the VBR service is called *statistical bit rate* (SBR), although at the moment only nrt-SBR is defined. The ABR service has its corresponding transfer capacity, also called ABR, but there is no equivalent for the UBR service. However, the ITU-T defines a transmission capacity called *ATM block transfer* (ABT), and it does not have a corresponding ATM Forum service category.

Each service category has a conformity definition that specifies what happens to out-of-profile traffic. The network will behave differently with excess traffic, depending on the type of service used and the status of the network. Each service has a corresponding quality-parameter priority as well. This is why, if network congestion occurs, the network may prioritize the maintenance of the quality of certain parameters before others, depending on the service used. For example, in a CBR service, the delay and delay variation values will be maintained, although the cell loss rate may increase, whereas in an nrt-VBR connection the situation is the opposite, and the lost cell rate is maintained at the expense of increasing delays.

Table 7-1. ATM Forum service categories and ITU-T ATM transfer capacities.

ATM Forum	ITU-T
CBR	DBR
rt-VBR	—
nrt-VBR	SBR
ABR	ABR
UBR	—
—	ABT

7.1.3.1 CBR Service Category

The CBR service is used in connections that call for constant bandwidth. The transmitted traffic is characterized by the PCR parameter. The source may transmit cells constantly at the rate indicated by PCR, or at lower rates, or it may even remain silent.

All the cells transmitted inside the range specified by PCR have the connection-specific QoS guaranteed. This service has some particular quality requirements concerning transmission delays, delay variations and lost cells. This is why the QoS parameters max CTD, peak-to-peak CDV and CLR must be specified.

7.1.3.2 rt-VBR Service Category

The rt-VBR service is aimed at those applications that have variable rates and high requirements for transmission delays and the variation of these delays. In this case, the traffic transmitted is characterized by PCR, SCR and MBS parameters, as well as the same QoS parameters as the CBR.

7.1.3.3 nrt-VBR Service Category

The nrt-VBR is a service for those applications where there is no temporal relation between the two ends of the connection, and where traffic is variable in time. The same as in the previous case, the traffic is characterized by PCR, SCR and MBS. In this case, however, the quality of service does not have any requirements for delay, but only for the CLR.

7.1.3.4 UBR Service Category

The term UBR designates the best-effort service in ATM networks. The UBR service does not provide guaranteed QoS parameters. Cells are not certified to be transmitted correctly, without any loss, nor is there any assurance of not having delays. In this case, the traffic parameter used, PCR, does not imply any QoS guarantees; the network uses it to obtain information on the maximum bandwidth used. This service is offered by legacy data networks.

7.1.3.5 ABR Service Category

In the ABR service, the transmission rate available for the source varies during the connection time, depending on the network status. This service guarantees a minimum transmission bandwidth, although depending on how loaded the network is, it may provide more bandwidth, up to the maximum indicated by the source. Therefore, the traffic in this service is characterized by MCR (the minimum bandwidth guaranteed) and PCR (the maximum bandwidth provided by the network, if the network is not too loaded). A minimum cell loss rate is guaranteed. However, there is no guarantee concerning delays, which means that this service is not suitable for real-time applications.

7.1.4 SLA in ATM Networks

The SLA contract defines what the client expects from the network, and it is the document the service provider will use to bill the subscriber, and manage its own network. An SLA implies that both the user and the service provider are committed to complying with this agreement; that is, the user will adapt to it and the operator promises to provide the QoS agreed upon.

However, for congestion reasons, due to technical problems or bad network engineering, the operator might end up in a situation where it cannot provide its clients with the transmission they need. The user in his turn should not generate more traffic than agreed, or it could happen that the network, when transmitting this data, ill-treats another user and in this way endangers the previous network engineering.

ATM networks have a set of mechanisms to detect when a client sends more traffic than agreed, and to determine what to do in such circumstances. Once this 'fraud' has been detected, be it intentional or unintentional by the client, the ATM cells violating the contract can be tagged or discarded.

To tag a cell means the following: in the header of an ATM cell there is a bit called the *cell loss priority* (CLP), used to establish a first level of priority among the ATM cells transmitted. If this value is 0, this is an important cell, while if the value is 1, the cell is 'not that important'. Therefore, when a client breaks the contract, the CLP of this client may be put to 1; that is to say, the cell is tagged. In this way, the network will continue carrying this cell, but with its CLP changed. If these cells arrive at a node with congestion problems, they will be among the first cells to be discarded, just like any other cell with CLP = 1.

7.1.5 Resource Management

To provide the performance that users require, in addition to reserving resources during the connection setup, ATM networks implement a series of procedures to control traffic and congestion.

Resource management and traffic control are essential in making an ATM network operative and efficient. These functions try to protect the network and its users from traffic violations, and thus prevent network congestion.

ATM networks implement the following functions to manage and control the resources available:

- *Connection admission control* (CAC): this procedure is carried out when a connection is established, to determine if it can be accepted. This is the first mechanism to protect the network from congestion. The connection will either be accepted or rejected, depending on whether the network has enough resources to meet the user requirements without decreasing the performance of existing connections. This procedure looks for an acceptable path, that is, one with switching nodes and links that can support the new connection.
- *Usage parameter control* (*UPC*): these are functions carried out by the network to supervise and control both the ATM connections used and user traffic. The aim, here, is to detect violations of the traffic parameter values negotiated earlier. The actions to be taken in case of violation depend on network congestion and the type of service used for transmission.
- *Network parameter control* (NPC): the NPC carries out the same functions as the UPC, but for links between networks (*inter-network interface* or INI) instead of those between the user and the network (*user-to-network interface* or UNI).
- *Resource management* (RM): among these functions is the procedure for the network to use special cells to dynamically modify the resources assigned to a certain ATM connection.

7.1.6 The Failure of ATM

ATM was an attempt to bring QoS-aware services to a converged multimedia network, but it failed for several technical and commercial reasons:

- *ATM is more expensive than other technologies, such as Ethernet or IP*. While ATM has become a technology mainly for the backbone, Ethernet has evolved for years as a *local area network* technology, and IP has been used almost everywhere in data applications. The result is that the number of deployed ATM switches is much smaller than the number of deployed Ethernet and IP-based equipment, and these technologies have become cheap and widely available.
- *Insufficient scalability*. ATM interfaces at 155 Mbit/s are common, but equipment for rates higher than this is much more uncommon and expensive.
- *Low efficiency and high complexity*. The advanced features of ATM rely on complex signalling protocols and a reduced 'user data to overhead' ratio. This makes ATM a powerful but difficult and inefficient technology.

When the deployment of ATM began, IP was already being used. In fact, the most important application for ATM is to transport IP traffic. Finally, it seems that IP and Ethernet will replace ATM. To make this possible, Ethernet has evolved to be a carrier-class technology that offers high availability, scalability and advanced *operation, administration and maintenance* (OAM) functions.

7.2 QoS in IP Networks

The birth of the commercial Internet was in 1983 and since then its history has been so successful that today any communication technology must take into account the Internet and its architecture.

The Internet is no more than a mesh of communication devices interconnected between them by means of a very versatile protocol called the Internet Protocol. IP is in the network and inside every host as well. An important reason for the success of the Internet and the IP is that hosts have implemented the IP stack into their operative systems. IP has become the language of the applications. At first, the hosts were all computers, but nowadays many other devices can be connected to the Internet as well, and all of them implement the IP protocol stack.

The success of the Internet is the reason why today there is so much interest in finding a way to base converged communications on IP. This implies that, in the future, an evolved version of the Internet will become a global network for converged communications. The IP technology replaces ATM for this purpose.

As IP is the end-to-end technology of the new converged network, it is quite realistic to think that it will be in charge of QoS provisioning as well. This is the reason why IP is becoming QoS-aware. We are still quite far away from a global converged network, but telecom operators now have the technology needed to offer high-quality voice, video and data services over smaller IP networks.

There are two QoS architectures available for IP:

1. The *integrated services* (IS) architecture provides QoS to traffic flows. It relies on allocation of resources in network elements with the help of a signalling protocol, the Reservation Protocol (RSVP).
2. The *differentiated services* (DS) architecture provides QoS to traffic classes. Packets are classified when they enter the network, and they are marked with DS codepoints. Within the network, they receive custom QoS treatment according to their codepoints only.

The IS architecture is more complex than the DS architecture, but it potentially provides better performance. One of the most important features of the IS approach is the ability to provide absolute delay limits to flows. On the other hand, the DS approach does not rely on a signalling protocol to reserve resources, and does not need to store flow status information in every router of the network. Complex operations involving classifying, marking, policing and shaping are carried out by the edge nodes, while intermediate nodes are only involved in simple forwarding operations. The IS architecture is better suited to small or medium-sized networks, and the more scalable DS approach to large networks.

7.2.1 The Integrated Services Architecture

The IS architecture is an IP-network service model for both best-effort services and other, more sophisticated QoS-aware services (see Figure 7-1).

The IS model assumes that the operation of all IP networks, including the Internet, must be based on:

- A common network infrastructure that relies on a single network element, the router.
- A unified protocol stack. Many protocols can co-exist above and below the network layer, but there should only be one internetwork protocol, IP.

A unified network approach makes it possible use network resources more efficiently, and network administration is easier, but this approach adds complexity to the network. In particular, to make it possible to transport both real-time and non-real-time information efficiently across an IP network, it is necessary to store some flow information in the intermediate network nodes. This is a significant modification to the original design of legacy IP networks that leave all flow processing to end systems while the core remains unaware of end-to-end connections between users. Such a modification implies upgrading or replacing legacy routers by a new generation of connection-aware routers, and this migration may be complex, especially when it comes to the Internet.

Storing the connection status is not useful, if resource reservation is not possible, and the way to reserve bandwidth in packet networks is not very evident. Flow prioritization could be used, but this fails to provide bandwidth guarantees even for high-priority flows. This is the reason why *advanced schedulers* are among the most important components of the IS architecture. For example, the Weighted Fair Queuing (WFQ) scheduler makes it possible to isolate flows and guarantee flexible bandwidth reservation for them.

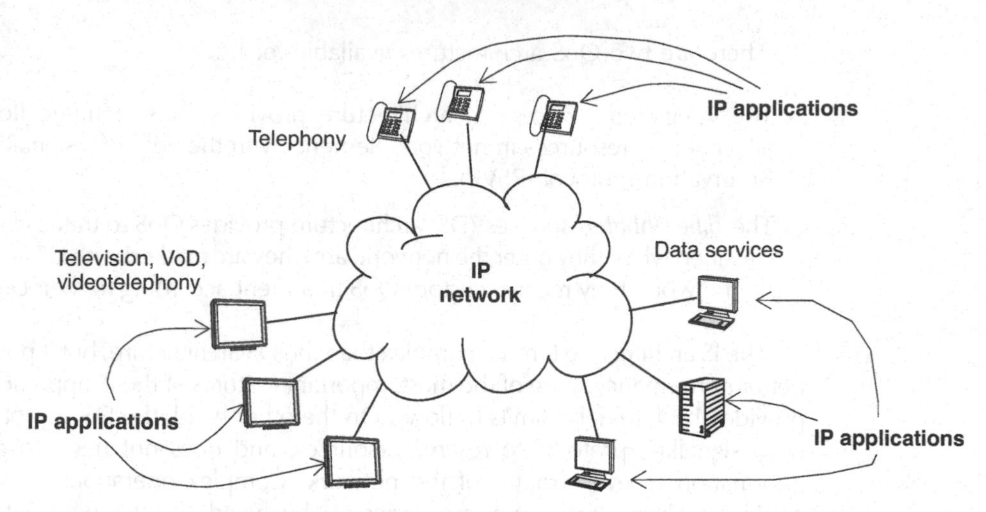

Figure 7-1. The IS architecture redefines legacy IP networks and makes them suitable for converged voice, video and data communications. In the IS architecture, communications applications are based on IP, and IP is the transport mechanism across the network.

Flows are different, and they must receive customized treatment in the network. Flow differentiation and customized end-to-end service types are other key components of the IS architecture. Currently, there are two end-to-end service types: controlled-load services and guaranteed services. The former provides soft QoS, the latter hard QoS.

The final component of the IS architecture is a resource-reservation protocol. This protocol must allow end systems to allocate resources for their flows and choose the right QoS parameters and service types for them.

7.2.1.1 Guaranteed Services

Guaranteed services provide assured end-to-end performance level in terms of bandwidth, delay or packet loss on conforming packets from flows with well-described bandwidth profiles.

It is clear that delay variation is small under light load conditions, due to the small probability of a packet needing to wait in a queue before being forwarded. It is also clear that, under light load, a packet network will never drop packets, because the buffers of network elements will remain empty or almost empty all the time. However it needs to be demonstrated whether it is possible to give absolute limits for the end-to-end performance of a network. In fact, it can be demonstrated that the WFQ scheduler can provide zero loss probability and a deterministic upper limit for the delay with traffic shaped by a token bucket. This is a theoretical result, but with very important practical implications, because it opens the door to a new class of services for packet networks.

It is possible to provide a guaranteed service by reserving B bits of buffer space and R bit/s of bandwidth in a router. If the scheduling algorithm of the routers can be approximated by GPS schedulers, the transmitter will see a dedicated line with bandwidth of R bit/s. In this case, a traffic flow conformant to a token bucket of rate r and depth b will be subject to a theoretical maximum queuing delay of b/R. A more realistic approach to the upper queuing delay limit can be achieved by adding two extra error terms:

$$Q_{\max} = \frac{b}{R} + \frac{C}{R} + D \tag{7.1}$$

where C and D are two constant values. Smaller delays can be achieved, if the peak rate of the flow is limited to p. Taking into account the peak rate of the flow and the effect of packetization (packet size limited to M), the following expression for the queuing delay is obtained:

$$Q_{\max} = \begin{cases} \frac{(b-M)(p-R)}{R(p-r)} + \frac{M+C}{R} + D & r \leq R < p \\ \frac{M+C}{R} + D & r < p \leq R \end{cases} \tag{7.2}$$

The assumptions made are only valid for traffic flows shaped by a token bucket. It is therefore important to perform the corresponding admission control tasks on the traffic that is subject to guaranteed limits. It may also be necessary to reshape the traffic within the network. The network administrator must decide how to deal with non-conforming packets. Often, these packets are sent without performance guarantees, but the network administrator may also decide to drop non-conforming packets.

A guaranteed service is a service for packet networks similar to a dedicated line or a switched TDM circuit. It is also less flexible and needs more resources than other services for packet networks. Guaranteed services are the best option for circuit emulation over IP. They can also be useful for provisioning of delay-sensitive applications, such as certain types of interactive video and audio.

7.2.1.2 Controlled-load Services

Flows associated with controlled-load services are seen by the receiver as if they were best-effort flows on a lightly loaded network, even if the network is not lightly loaded.

Controlled-load services are an example of differentiated services provided on a flow basis rather than on a class basis. The effect of lightly loaded network is accomplished by isolating the flow from the rest of the traffic and devoting resources (in the intermediate routers) specifically to that flow. The amount of resources to be devoted depends on the bandwidth profile of the flow.

Although it can be expected that controlled-load services will behave better than best-effort services, there are no real guarantees for delay or packet loss performance. It is only stated that most of the packets are successfully received with a delay that is not significantly worse than the delay for those packets that do not experience congestion.

Controlled-load traffic requires shaping and policing before it enters the network. Re-shaping and re-classifying may be necessary within the network as well. What happens with non-conforming packets is a choice that has to be made by the network administrator. These packets are sometimes re-classified as best-effort packets. The alternative is to re-shape the whole flow and thus degrade conforming and non-conforming packets equally.

Controlled-load traffic offers performance-levels that fall between best-effort and guaranteed traffic. It can be used to provide interactive and non-interactive media encoded with variable-rate encoders.

7.2.2 The Reservation Protocol

The RSVP enables end-users to reserve, maintain and release resources in IP networks. This protocol dates back to the beginning of the 1990s, and currently it is described in various RFCs, the most important of them being the RFC 2205 released in 1997.

The RSVP has become a key component of the IS architecture. It has two main objectives:

1. To enable the provisioning of services with defined QoS needs that go beyond the best-effort service provided by legacy IP networks.
2. To deal with multicast communications beyond the basic point-to-point model of legacy IP.

The RSVP needs to interact closely with routing protocols to perform its tasks, but it is not a routing protocol. The difference is that routing protocols must look for routes to remote points in a network, and a reservation protocol must be able to manage the resources devoted to sending information in the network across these routes. IP networks change dynamically as transmission conditions change. Routes change accordingly, and this may affect established reservations. The RSVP re-establishes reservations automatically when a route changes.

The RSVP does not need to be transported over UDP or TCP like most other protocols of the IP protocol stack. It can be directly encapsulated into IP datagrams with protocol number 46. Transmission over UDP is also accepted, because some end systems cannot directly map protocols into IP datagrams. All the RSVP messages have a similar format. They are made up of a common header and one or more specific fragments called *objects* that belong to a set of *object classes*. Each class has its own syntax and meaning (see Figure 7-2).

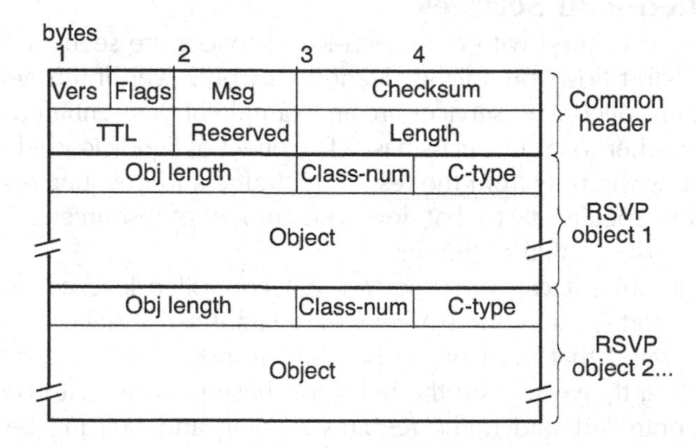

Vers: version, currently 1 (4 bits)
Flags: flags, currently 0 (4 bits)
Msg: message type (8 bits)
1: Path 5: PathTear
2: Resv 6: ResvTear
3: PathErr 7: ResvConf
4: ResvErr

Checksum: error-detection word (8 bits)
TTL: original IP TTL value (4 bits)
Length: total message length (8 bits)
Obj length: object length (16 bits)
Class-num: object class identifier (8 bits)
C-type: class type, unique within class-num (8 bits)

Figure 7-2. RSVP message format.

A particular feature of the RSVP is that those users who make the reservation are not the generators but the receivers of the information. This is a new approach, but it is actually quite logical, because the receivers are the ones who know their own capacity limitations. Furthermore, they are the ones who experience the QoS, and in many cases they are the ones who pay for it. The receiver-initiated reservation paradigm enables heterogeneous receivers in multicast communications. Each receiver can reserve the amount of resources it is able to process. It is difficult for the transmitter of a point-to-multipoint communication to know the capacity of a large and probably dynamic group of receivers and make the reservation for them.

7.2.2.1 PATH and RESV Messages

Resource reservations in a network are made using the RESV message. RESV messages are sent by the receiver to the transmitter, and reservations are made by intermediate routers as these messages pass through them. In certain cases, the amount of resources to reserve is known, but this is not always so. Another type of RSVP message, PATH, makes it easier to know the amount of resources to reserve. PATH messages are delivered to the receiver before sending any data. This can be done after the session negotiation of a VoIP call, for example. PATH messages specify the bandwidth profile of the transmitter, and they may also contain information on the performance of the network.

When routers find PATH messages in the network, they process them. Routers may modify certain fields of the message before forwarding it. Furthermore, intermediate routers store some information from PATH messages that generate path state entries in them. Each path state entry has a clean-up timer associated with it. When the timer expires, the path state is deleted, but if a refresh PATH message is received before the timer expires, the timer is reset. The result is that path state entries need to be periodically refreshed or the path state will be lost.

Something similar happens with reservations; they also need to be refreshed periodically. Implementing this kind of 'soft-state' in the network helps to deal with the ever-changing IP network, and it is well suited to multicast communications with a large number of receivers dynamically joining and leaving communication sessions. Although receivers can free up resources by letting timers expire, they can also use the RESVTEAR message to force a situation where resources must be freed up. The transmitter can clean up the path status without the timer, by using the PATHTEAR message.

So that the RSVP will work correctly, the PATH and RESV messages must follow inverse paths between the transmitter and the receiver (see Figure 7-3). In IP networks, it is normally not possible to make sure that this condition is met. This is why PATH messages are used to store inverse routing information in intermediate routers. This information will make it possible for the RESV messages to find the correct route to the transmitter.

Both the PATH and RESV messages have their own error messages to inform the communicating parties about possible errors. The PATHERR message is sent upstream if the path status cannot be established in the transmission path. The RESVERR message is sent downstream when a reservation fails due to lack of resources, policy setup or for some other reason.

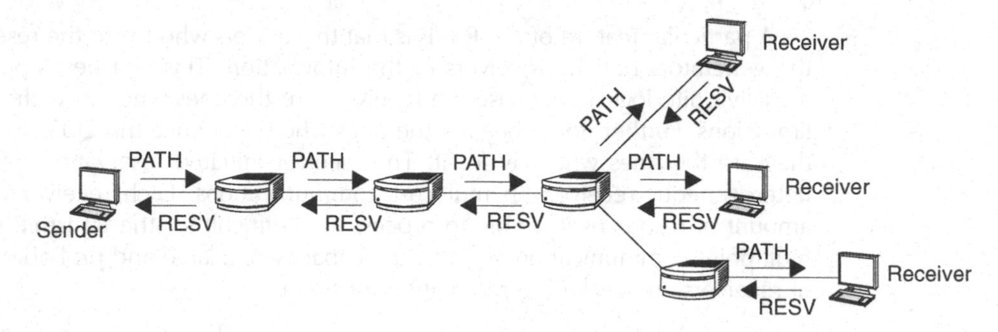

Figure 7-3. Reservations with the RSVP are carried out by the receivers by means of the RESV message, to make it easier to manage multicast sessions. Senders may use the PATH message to let the potential receivers know that there is data to be sent.

7.2.2.2 Making Reservations

Although reservations are made using RESV messages, they cannot be made without a previous PATH message from the source. As explained before, PATH messages have two objectives:

1. They install inverse routing information in the path between the transmitter and the receiver.
2. They help the receiver to know which reservation to make, or force them to make certain reservations.

To do this, the PATH message uses the following objects:

- *Previous hop (Phop)*: this field is upgraded by every router with RSVP support in the transmission path. It gives the address of the last router in the path supporting the RSVP. This field needs to be updated by every router as the packet crosses the network. Thanks to this field, it is possible to know if there is end-to-end QoS support.
- *Sender template*: intermediate routers need information to identify flows and filter them according to the reservations. This information is done by the sender template that contains the IP address and the port of the source. In IP version 6 (IPv6), the source port is replaced by the flow label.
- *Sender traffic specification (sender Tspec)*: contains the bandwidth profile specification of the traffic to be sent.
- Optional *Adspec (advertising specification)*: contains information on the transmission path.

The RSVP specifies bandwidth profiles by means of *Tspecs* (traffic specifications). The Tspec contains the specification of the token bucket that shapes the traffic (see Figure 7-4). The token bucket specification contains the following items:

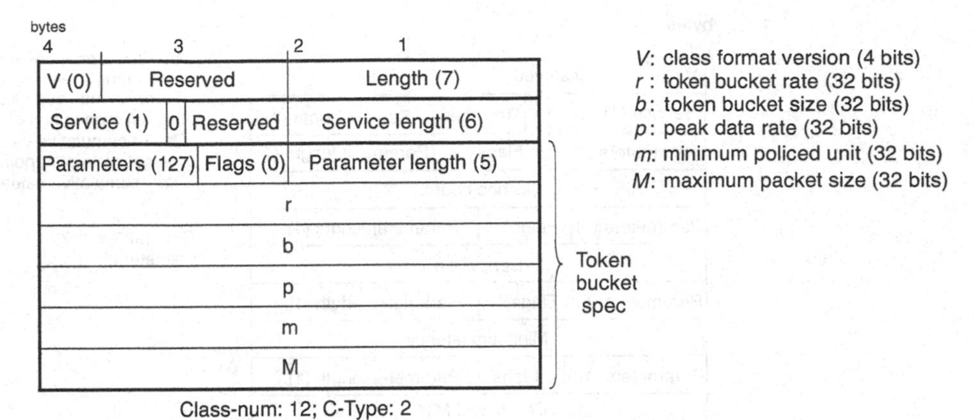

Figure 7-4. Sender Tspec object format.

- *Token bucket rate* (*r*): the maximum sustainable bitrate that is allowed to pass through the shaper.
- *Token bucket depth* (*b*): the size of the bucket in bytes. How much traffic is allowed to pass through the shaper when the transmission rate is above the token bucket rate.
- *Peak rate* (*p*): absolute upper limit of the rate (in bytes per second) that can be delivered through the shaper, even if there are buckets available.
- *Minimum policed unit* (*m*): the minimum packet size (in bytes) needed for allocation and policing. Smaller packets are allowed, but they are considered to consume the same amount of resources as the minimum policed packet. The sender normally sets this parameter to the smallest size possible.
- *Maximum packet size* (*M*): the maximum size (in bytes) for a packet to be delivered. Larger packets might not be sent with QoS guarantees. The sender normally sets this parameter to the largest size possible.

While the sender's Tspec contains information about the transmitter, a PATH message may contain an Adspec object with information on the network, especially on the path between the transmitter and the receiver. The Adspec object contains a fragment with general parameters and at least one fragment with information for a guaranteed service or a controlled-load service (see Figure 7-5). If one of these service fragments is omitted, this means that the service is not available. This can be useful to force the transmitter to choose a reservation type.

The general parameters fragment of the Adspec contains the following information:

- *Global break bit:* bit cleared by the transmitter and set when there are routers without RSVP support along the transmission path. This bit is used to tell the receiver that the Adspec may be invalid.

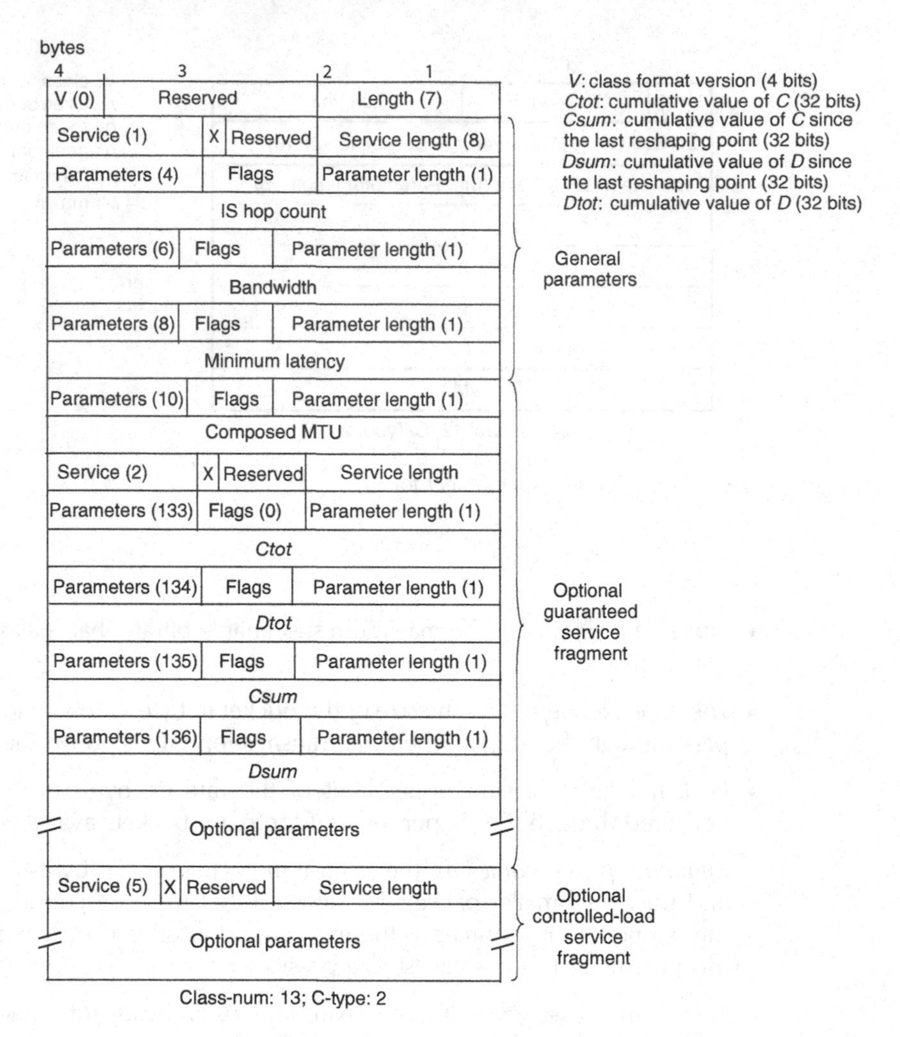

Figure 7-5. The Adspec object format.

- *IS hop count*: the number of routers with RSVP support between the transmitter and the receiver.
- *Bandwidth*: the maximum bandwidth available for transmitting across the path. The smallest individual link bandwidth along the path.
- *Composed maximum transfer unit (MTU)*: the maximum size of a deliverable packet, calculated as the minimum of all MTUs in individual links along the transmission path.
- *Minimum latency*: the sum of individual link latencies. The variable queuing delay is not taken into account, so this value represents the fixed transmission delay along the path.

The guaranteed service fragment includes the delay parameters needed to calculate the bandwidth to be allocated in the routers along the transmission path [see equation (7.2)].

- *Guaranteed service break bit*: bit cleared by the transmitter and set when there are routers without guaranteed service support on the transmission path. This bit is used to tell the receiver that this type of service may not be available.
- *Ctot, Dtot*: end-to-end values of the *C* and *D* parameters of equation (7.2) calculated as the sum of the individual values of every intermediate router.
- *Csum, Dsum*: a composed value of the *C* and *D* parameters of equation (7.2) calculated as the sum of the individual values of all the routers since the last shaping point within the network.

The guaranteed service fragment may contain other optional items that override the values of the general parameters fragment. The controlled-load service fragment of the Adspec does not contain special parameters, but only a break bit for controlled load services, and some optional parameters that override the value of the general parameters fragment.

The Adspec reservation model is known as *One Pass with Advertising* (OPWA). If there is no Adspec available, the receiver can still make reservations with the help of the sender Tspec, but there is no easy way to know the properties of the transmission path in advance. This second model is called *One Pass*.

After reservation advertising with a PATH message, the receiver can allocate resources by sending RESV messages towards the origin. The RESV message contains the following objects:

- *The reservation style*: this object tells the routers how to merge reservations from different receivers in multicast communications. The value of this field depends on the application type.
- *Filter specification* (Filterspec): this object identifies the senders that are allowed to use the reservation made by the receiver. Like the sender template sent with the PATH message, this field includes IP addresses and ports or IPv6 flow labels. Sometimes any host is allowed to use the reservation; if this is the case, this object is not needed and can be omitted.
- *Flow specification (Flowspec)*: this object is used by the routers in the transmission path to calculate the amount of resources to allocate to a particular flow. The format of the Flowspec depends on the service type.
- Optional *reservation confirm (ResvConf)*: this object is used by the receiver to request reservation success confirmation from the last hop.

A receiver that wants to make a reservation has to generate a Flowspec according to its own preferences and sender information. To reserve resources for a controlled-load service, the receiver may simply include the sender traffic specification in the Flowspec sent in the RESV message. In fact, the Flowspec for a controlled-load service has the same format as

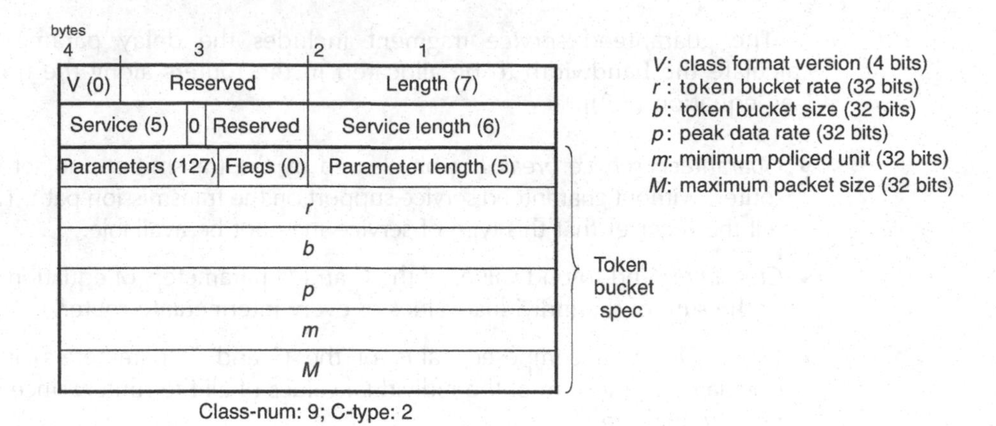

Figure 7-6. The Flowspec object format for a controlled-load service.

the sender Tspec (see Figure 7-6). Usually, the parameter M is taken from the Adspec, if available, and not from the sender Tspec.

Making reservations of guaranteed services is slightly more complicated. The object format of a guaranteed service Flowspec has the same parameters as the controlled-load service Flowspec, but it also includes two new parameters: the required bandwidth (*R*) and the slack term (*S*) (see Figure 7-7).

1. The *token bucket rate* (*r*) gives information on the sustainable bit rate of the source, but to enable a service with bounded delay, it is necessary to allocate a bandwidth higher

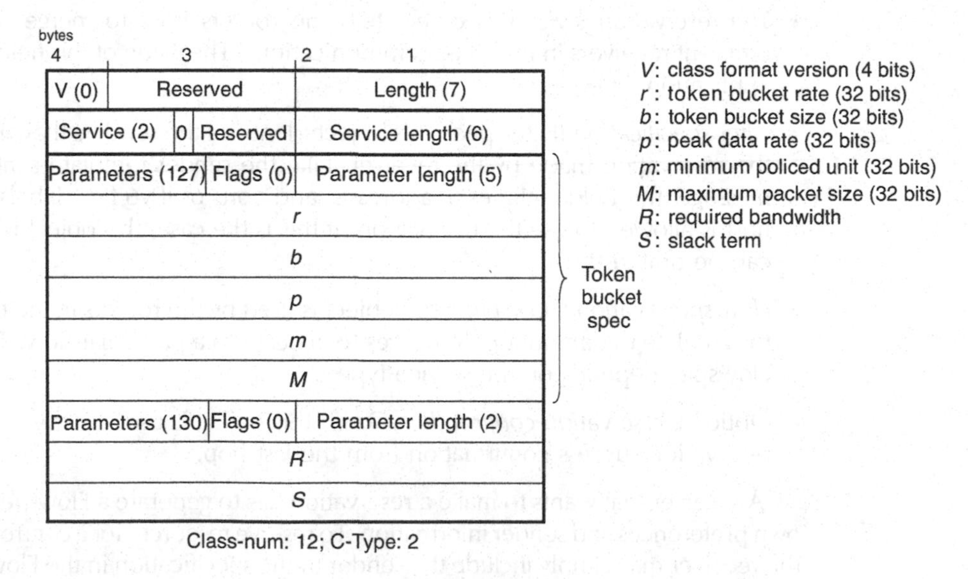

Figure 7-7. Flowspec object format for a guaranteed service.

than r. This bit rate is given by R, and it is calculated using equation (7.2) and with the parameters received in the Adspec.

2. The *slack term* (ms) makes reservations more flexible. The chances of obtaining a successful reservation may be better if R is configured with a value higher than required, and the slack term set to the difference between the configured and the required value. When this is done, the routers along the transmission path may 'waste' more bandwidth than needed, allowing the eventual bottleneck routers to allocate less resources than required, and still meet the delay limit.

7.2.2.3 Dealing with Multiple Senders and Receivers

The RSVP has been designed with multicast communications in mind. This is the reason why reservations are receiver-initiated. However, this is just an example; there are many RSVP-related design decisions that have been made to optimize the operation of multicast communications.

A problem that arises in multicast communications is how to deal with heterogeneous receivers. The amount of resources reserved for the same multicast session may not be the same in different interfaces of the same router. In this situation, routers need to send an RESV message to the transmitter, with the biggest possible reservation. This is to guarantee the required quality to the best receiver. The flow can be left unchanged for this receiver, but it will be shaped for those receivers that do not require high quality (see Figure 7-8). It is worth noting that not all applications may be able to tolerate severe shaping. It may be necessary to re-code the media stream before forwarding it to the receiver. This task is normally carried out by a dedicated host and not by a router.

The second problem involves applications with many senders. Some of these applications do not need to allocate separate resources for each transmitter, because a situation where several transmitters are transmitting simultaneously is very unlikely. The number and identity of these transmitters may also change with time. This is why it is necessary to find

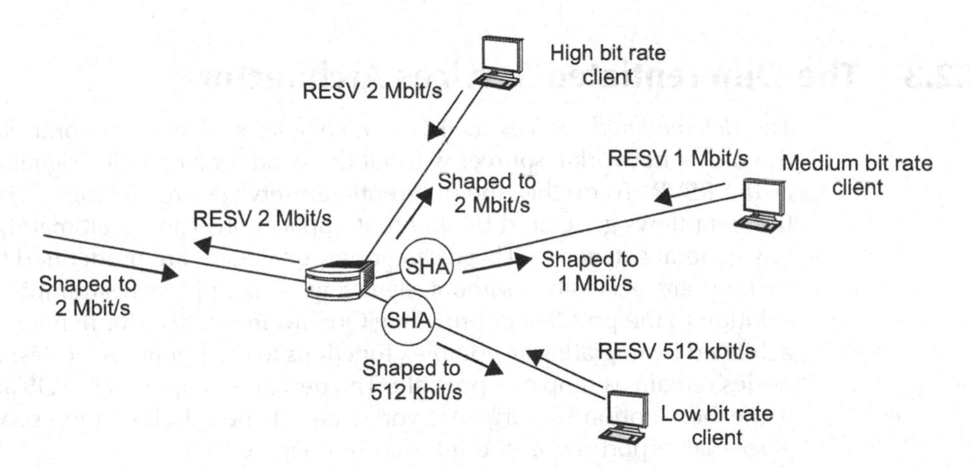

Figure 7-8. Dealing with heterogeneous receivers: the router sends the best reservation to the transmitter, and shapes the flow to meet the quality requirements of all the receivers.

out how to take the maximum profit out of the available bandwidth in this type of application.

The RSVP supports different reservation styles. Each of them has a different approach to the resource-sharing problem between the senders in multisender applications:

- Some share the bandwidth between different senders, and others allocate dedicated bandwidth to each sender.
- Some identify those senders that are allowed to use the reserved resources, but others are open in the sense that any sender can use the reservation.

Currently there are three different reservation styles (see Figure 7-9):

1. *Fixed filter (FF)*: a fixed filter installs dedicated reservations for an identified sender. The reservation cannot be used by other sender, and the receiver must make one resource reservation for every sender. This style is well suited to video applications where many video channels are received from multiple remote sites.
2. *Shared explicit (SE)*: in the SE reservation-style reservations are shared by an identified group of senders. The Filterspec sent upstream is made up of all the senders contained in the RESV messages received from downstream. This reservation style is well suited to audio applications where the senders are well identified and stable. A good example of this reservation style is webcasts and webinars.
3. *Wildcard filter (WF)*: The wildcard filter installs shared reservations available for an unspecified group of senders. A WF reservation is in fact a shared pipe that can be used by a dynamically changing group of transmitters. This reservation style is best suited to audio applications where the number of transmitters and their identity is either unspecified or unstable. Multiparty VoIP calls are a good example of these applications.

7.2.3 The Differentiated Services Architecture

The *differentiated services* (DS) QoS architecture allows for 'controlled unfairness' in the use of network resources without the need for a complex signalling protocol such as the RSVP. To do this, the differentiated services architecture assumes that most of the data flows generated by different applications can be ultimately classified into a few general categories. These categories, or classes, are predefined traffic aggregates, so they are accessible without signalling. This approach attempts to be a scalable solution to the problem of providing QoS-aware services for IP networks. Scalability is achieved by migrating all complex functions to the boundary nodes, while the interior nodes remain as simple as possible. The design principles of the DS architecture make it the best option for large networks, and it should allow QoS service provisioning across large portions of the Internet (see Figure 7-10).

In the DS architecture, the boundary nodes have the following QoS-related primary functions:

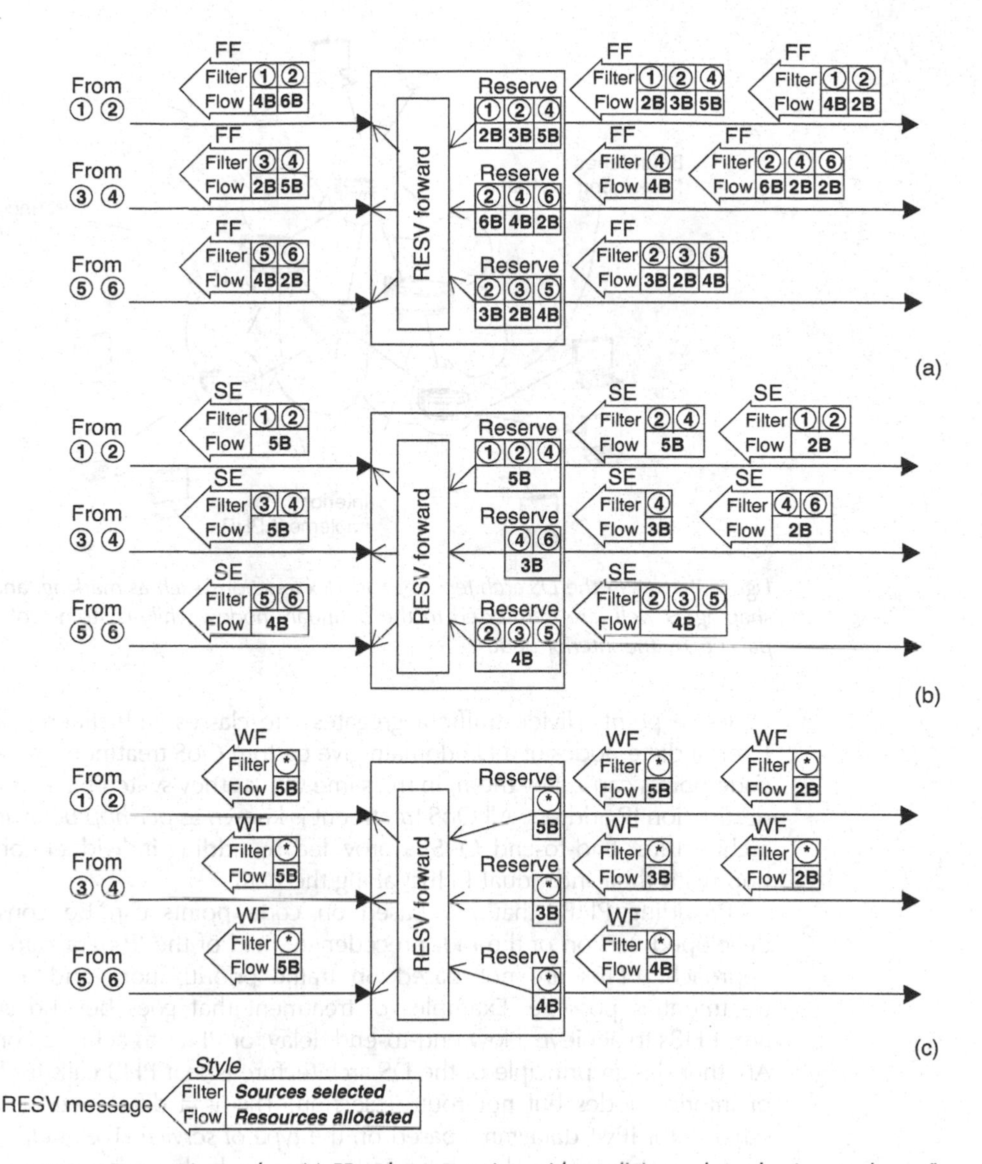

Figure 7-9. Reservation styles: (a) FF-style reservation with explicit sender selection and per-flow dedicated resources; (b) SE-style reservation with explicit sender selection and shared resources; (c) WF-style reservation with wildcard sender selection and shared resources.

- shaping and policing the ingressing traffic to keep congestion under control;
- Marking packets according to the QoS requirements of the associated traffic flow. Marks in the DS architecture are called DS code points, and they are transported inside a special header field of the IPv4 or IPv6 datagram. The DS code points are an example of CoS labels.

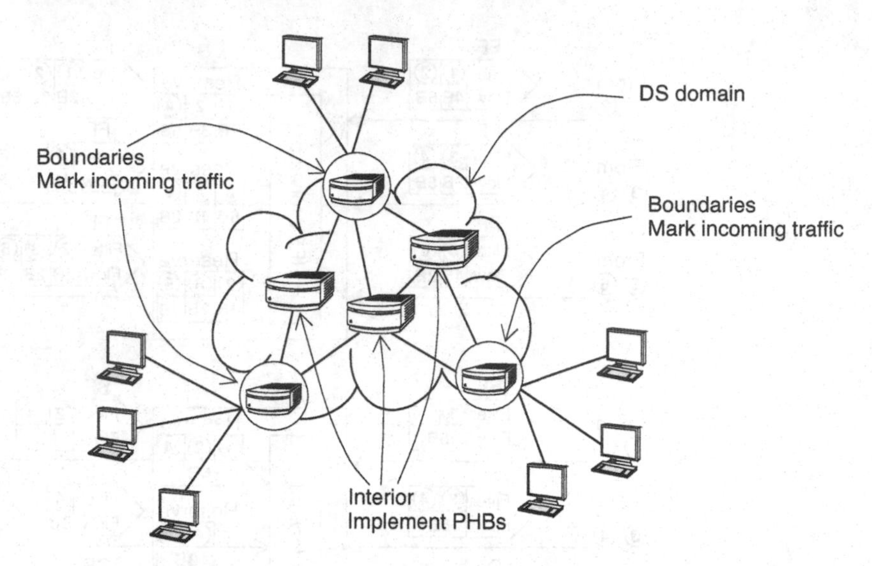

Figure 7-10. In the DS architecture, complex functions such as marking, and maybe policing and shaping as well, are transferred to the boundary nodes, while custom QoS treatment is given to packets by the interior nodes.

Code points divide traffic aggregates into classes with different QoS requirements. Intermediate nodes of a DS domain give custom QoS treatment to packets based on the code points carried by them, in the same way as they switch packets on the basis of their destination IP address. All QoS treatment is known as *per-hop behaviour* (PHB) in the DS architecture. End-to-end QoS is provided by adding individual contributions of every hop (with their individual PHBs) along the path.

Providing PHBs that are based on code points can be considered as a more developed version of the old 'precedence' field of the IP datagram, or a priority-based approach. PHBs are not based on traffic prioritization, and a more sophisticated treatment is possible. Examples of treatment that goes beyond simple prioritization are: PHBs to achieve a low end-to-end delay, or PHPs to achieve controlled packet loss. Another design principle of the DS architecture is that PHB calls for buffer management of interior nodes but not route selection. This is a difference of the service-marking scheme for IPv4 datagrams based on the *type of service* (ToS) field.

Although traffic marking is carried out mainly by boundary nodes, interior nodes may also re-mark traffic. A typical example of this is to re-mark a part of the traffic to a lower priority class under congestion conditions.

7.2.3.1 The DS Field

DS code points are transported in the 6-bit DS field of IP datagrams (see Figure 7-11). In the IPv4 datagram format, the DS field replaces the old ToS bit. The ToS bit has formed a part of the IP specification since the beginning, but it has never been extensively used. The original purpose of the ToS bit was to enhance the performance of selected datagrams, to make it better than best-effort transmission QoS. To do this, a 4-bit field within the ToS bit is defined, and it includes the requirements that this packet needs to meet (see Table 7-2).

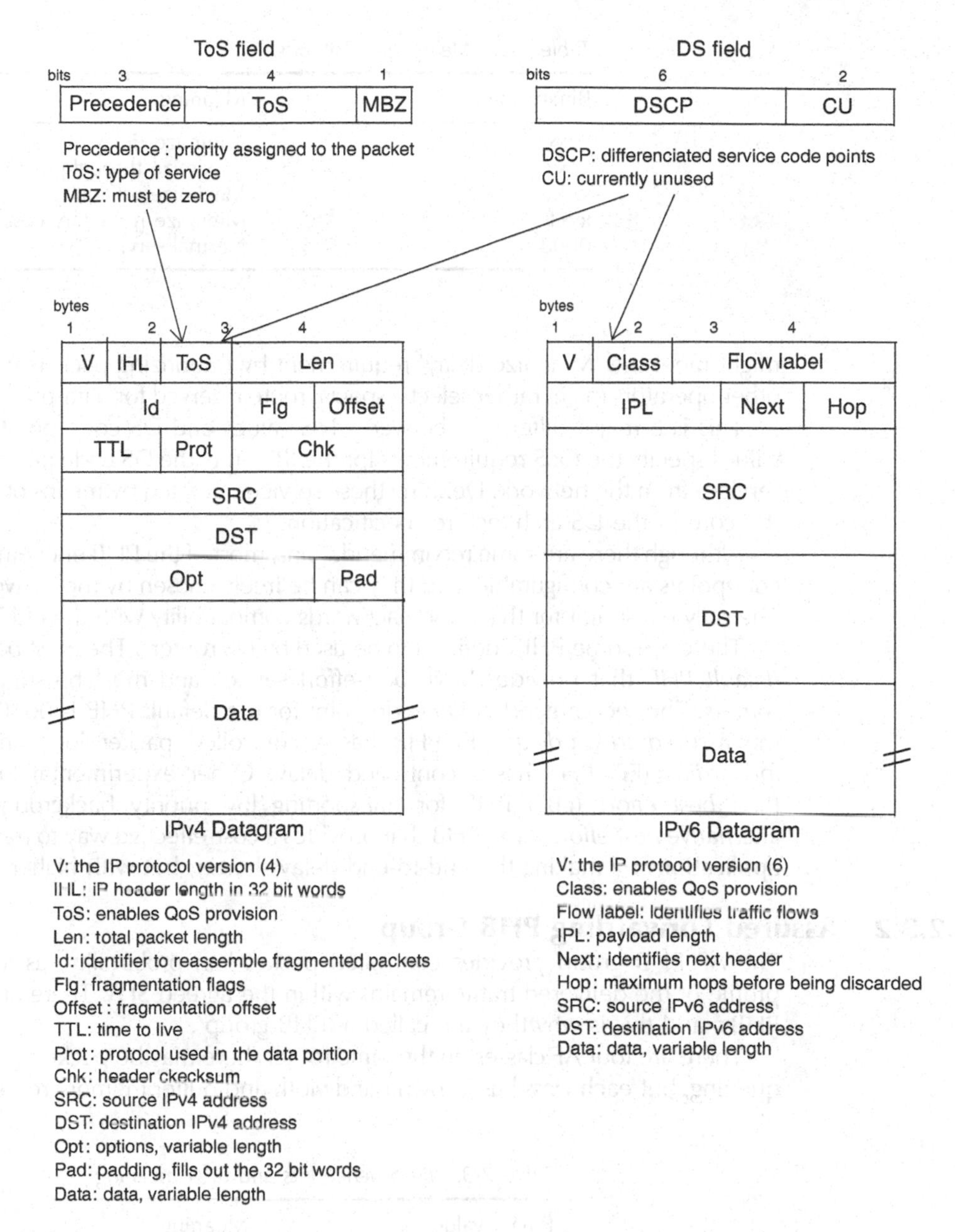

Figure 7-11. IPv4 and IPv6 datagrams and the format of the ToS and DS fields, both related to QoS provisioning.

In addition to the 4-bit field mentioned before, there is a 3-bit precedence field that makes it possible to implement simple priority rules for IP datagrams (see Table 7-3).

The ToS values encode some QoS requirements for the IP datagrams, but the decision on how to deal with these values is left to the network operator. For example, some operators

Table 7-2. Meaning of ToS bits.

Binary value	Meaning
1xxx	Minimize delay
x1xx	Maximize throughput
xx1x	Maximize reliability
xxx1	Minimize monetary cost
0000	Normal service

might meet the 'Minimize delay' requirement by prioritizing packets with this mark, but other operators might rather select a special route reserved for high-priority traffic.

This is a major difference between ToS values and DS code points. While the ToS values specify the QoS requirements for the IP traffic, the DS code points request specific services from the network. Defining these services, created by means of different PHBs, is the core of the DS architecture specification.

Although there are some recommendations, most of the PHB encoding by means of DS codepoints are configurable, and they can be freely chosen by the network administrator. The only constraint for this is the backwards compatibility with the old ToS encodings.

There are some PHBs defined to be used by DS routers. The most basic of them is the *default PHB* that provides basic best-effort service and must be supported by all the routers. The recommended DS code point for the default PHB is 000000. Additionally, the *assured forwarding* (AF) PHB has a controlled packet loss, and the *expedited forwarding* (EF) PHB has a controlled delay. Other experimental PHBs are the *less than best effort* (LBE) PHB for transporting low-priority background traffic, or the *alternative best effort* (ABE) PHB that provides a cost-effective way to transport interactive applications by making the end-to-end delay shorter, but with higher packet loss.

7.2.3.2 Assured Forwarding PHB Group

The AF PHB group provides controlled packet loss probability, as long as the traffic profile of the delivered traffic remains within the agreed SLA. There are 12 different AF PHBs, and this is why they are called a 'PHB group'.

There are four AF classes. In the same router there may be packets from all four classes queuing, but each class has its own bandwidth and buffer memory resources. This means

Table 7-3. Precedence bits and their meaning.

Binary value	Meaning
000	Routine
001	Priority
010	Intermediate
011	Flash
100	Flash override
101	Critic/ECP
110	Internetwork control
111	Network control

Table 7-4. Recommended code points for the AF PHB group.

Recommended code points	Class 1	Class 2	Class 3	Class 4
Low drop precedence	001010 (AF11)	010010 (AF21)	011010 (AF31)	100010 (AF41)
Medium drop precedence	001100 (AF12)	010100 (AF22)	011100 (AF32)	100100 (AF42)
High drop precedence	001110 (AF13)	010110 (AF23)	011110 (AF33)	100110 (AF43)

that the AF classes are isolated, and congestion in one of them does not affect the other three.

Within each AF class, there are three possible drop precedences. If there is congestion, the precedence value of the packet determines its relative importance in the AF class. Routers try to protect packets with a lower drop precedence against loss by discarding packets with higher precedence. All four AF classes together with three drop precedences per class give 12 PHBs. The code points recommended for them are listed in Table 7-4.

The IETF specifies the AF PHB without defining any particular implementations. Network equipment vendors are free to choose the design that best suits them. Figure 7-12 shows one possible implementation for the AF PHB group. A WFQ scheduler can be used to isolate the traffic of different AF classes, but others, such as *weighted round robin* (WRR) may be used as well.

The three drop precedences of every AF class are implemented by means of the Weighted Random Early Discard (WRED) queue management algorithm – although, again, this is not mandatory.

The AF PHB may be used to transport critical data that calls for better transmission parameters than those provided by the default PHB. Another interesting application for the AF is non-real-time video transmission. MPEG-2 video streams have medium tolerance to

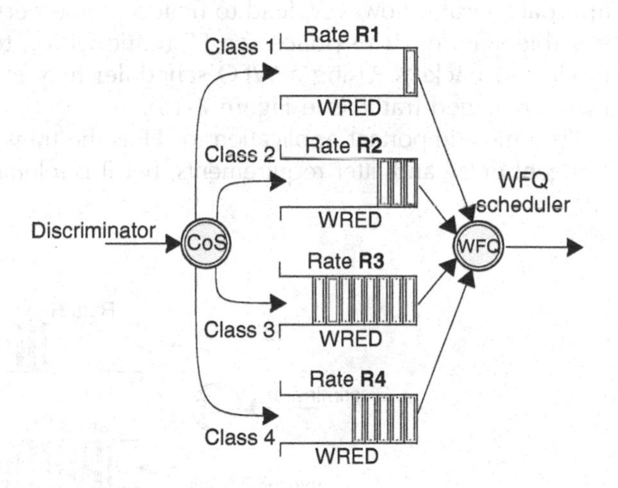

Figure 7-12. One possible implementation of the AF PHB by means of four queues with a WRED PDP and a WFQ scheduler.

jitter and very low tolerance to errored, corrupted or lost packets. Specifically, the drop rate should remain below 10^{-6}. To deal with these QoS needs, video packets may be assigned to one of the AF classes. Within the video AF class, VoD traffic could be assigned to a high drop precedence like the AF12 or AF13 PHB and the IPTV traffic to a low precedence class such as AF11. The reason for this is that the VoD application needs more bandwidth than IPTV, and it is more likely to cause congestion. By assigning IPTV to a low precedence class, this application is protected against congestion caused by VoD streams.

7.2.3.3 Expedited Forwarding PHB

The EF is intended to provide low-delay, low-jitter, low-loss services, similar to a virtual leased line. EF PHB packets receive a treatment that minimizes the queuing delay in the node. This has the following side effects:

- End-to-end delay variation is also minimized, because queuing delay causes end-to-end delay and is the main cause *R* of jitter.
- Packet loss is also minimized, because queues are not very long, and therefore the probability of losing packets is low.

The EF PHB defined in a particular node is specified by a rate, *R*. The rate at which EF packets should be ideally served is *R* or faster.

The EF PHB is similar to the controlled-load service provided by the IS architecture. Both try to emulate the behaviour of a packet-switched network under light load conditions. The difference is that, while the EF PHB emulates this behaviour for an aggregate traffic class, the controlled-load service operates at flow level, offering better performance. The DS code point recommended for the EF PHB is 101110.

The EF PHB can be implemented by means of a priority scheduler. With this solution, EF packets always get privileged access to the outgoing interface. This approach could, however, lead to unacceptable performance for non-EF traffic. One possible solution is to police the EF traffic with a token bucket to limit the rate of privileged packets. Using a WFQ scheduler may also help to protect non-EF traffic from privileged traffic (see Figure 7-13).

The most important application for EF is the transport of VoIP. This application has stringent delay and jitter requirements, but it is tolerant of packet loss.

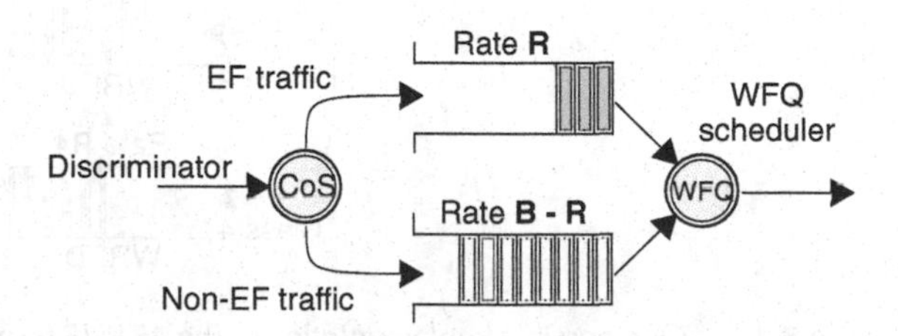

Figure 7-13. A possible implementation of the EF PHB with a WFQ scheduler.

Selected Bibliography

[1] Matrawy A., Lambaradis I., A Survey of Congestion Control Schemes for Multicast Video Applications, *IEEE Communications Surveys*, Vol. 5, No. 2, 2003.

[2] Tryfonas C., Varma A., MPEG-2 Transport over ATM Networks, *IEEE Communications Surveys*, Vol. 2, No. 4, 1999.

[3] Vali D., Plakalis S., Kaloxylos A., A Survey of Internet QoS Signaling, *IEEE Communications Surveys*, Vol. 6, No. 4, 2004.

[4] Marthy L., Edwards C., Hutchison D., The Internet: A Global Telecommunications Solution?, *IEEE Network Magazine*, July/August 2000, pp. 46–57.

[5] Xiao X., Ni L. M., Internet QoS: A Big Picture, *IEEE Network Magazine*, March/April 1999, pp. 8–18.

[6] White, P. P., RSVP and Integrated Services in the Internet: A Tutorial, *IEEE Communications Magazine*, May 1997, pp. 100–106.

[7] Giordano S., Salsano S., Van den Berghe S., Ventre G., Giannakopoulos D., Advanced QoS Provisioning in IP Networks: The European Premium IP Projects, *IEEE Communications Magazine*, January 2003, pp. 2–8.

[8] Mase K., Toward Scalable Admission Control for VoIP Networks, *IEEE Communications Magazine*, July 2004, pp. 42–47.

[9] Welzl M., Franzens L., Mühlhäuser M., Scalability and Quality of Service: A Trade-off?, *IEEE Communications Magazine*, June 2003, pp. 32–36.

[10] Zhang L., Deering S., Estrin D., Shenker S., Zappala D., RSVP: A New Resource Reservation Protocol, *IEEE Network Magazine*, September 1993, Vol. 7, No. 5.

[11] Braden R., Clark D., Shenker S., Integrated Services in the Internet Architecture: an Overview, IETF Request For Comments RFC 1633, June 1994.

[12] Blake S., Black D., Carlson M., Davies E., Wang Z., Weiss W., An architecture for Differentiated Services, IETF Request for Comments RFC 2475, December 1998.

[13] Heinanen J., Baker F., Weiss W., Wrockawski J., Assured Forwarding PHB Group, IETF Request for Comments RFC 2597, June 1999.

[14] Davie B. *et al.*, An Expedited Forwarding PHB (Per-Hop Behavior), IETF Request For Comments RFC 3246, March 2002.

[15] Braden R., Zhang L., Berson S., Herzog S., Jamin S., Resource ReSerVation Protocol (RSVP) – Version 1 Functional Specification, IETF Request For Comments RFC 2205, September 1997.

[16] Wroclawsky J., The use of RSVP with IETF Integrated Services, IETF Request For Comments RFC 2210, September 1997.

[17] Shenker S., Wroclawski J., General Characterization Parameters for Integrated Service Network Elements, IETF Request For Comments RFC 2215, September 1997.

Chapter 8: Broadband Access

The access network connects subscribers to the core network transport facilities, enabling end-to-end service provision. The last (or the first, depending on the point of view) portion of the access network connects the customer premises with a local exchange. This is the part of the network known as the local loop.

The copper pair is still the leading transmission media in the local loop, followed by coaxial, fibre and wireless. Digital subscriber loop (DSL) technologies, designed for digital transmission across existing telephone copper pairs, combine cost effectiveness and acceptable performance. This fact has made DSL a very successful technology for broadband Internet access (see Figure 8-1).

Some of the latest DSLs have been designed for a specific application: video over telephone wires. Unlike Internet access, video applications require several megabits per second for acceptable performance. One MPEG-2 encoded video stream needs about 3.5 Mbit/s. If it is accepted that four is likely to be the minimum number of video channels to be delivered to the customer simultaneously, the bandwidth needed just for video is about 15 Mbit/s per subscriber. If high-speed Internet access and IP telephony services are considered as well, the minimum bandwidth required for each subscriber is about 20 Mbit/s. Furthermore, it is though that, in the future, high-definition TV will require from 60 to 100 Mbit/s of bandwidth per subscriber.

The problem of DSL is that, due to attenuation and crosstalk, it is impossible to achieve long range and high speed simultaneously. While it is possible to deliver a few megabits per

Triple Play: Building the Converged Network for IP, VoIP and IPTV Francisco J. Hens and José M. Caballero

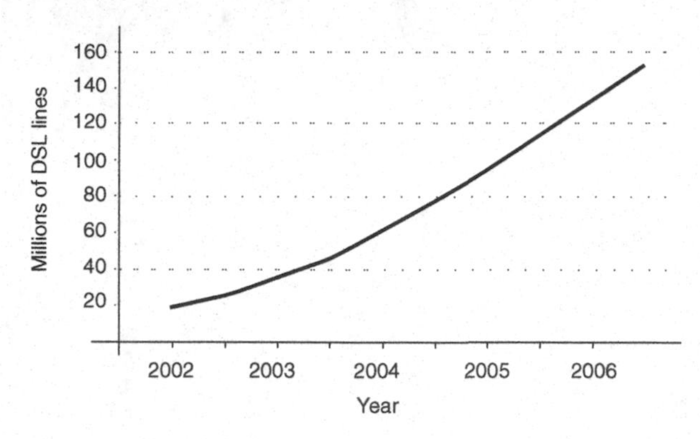

Figure 8-1. Evolution of the worldwide DSL market.

second to customers located several kilometres from the local exchange, delivery of 100 Mbit/s is limited to a few metres. This is the reason why some operators have already started to deploy new access networks based on optical fibres. Only a few of these deployments offer fibre to the home (FTTH). Most of them are (depending on where the optical link is terminated) fibre to the building (FTTB), fibre to the cabinet (FTTCab), etc.

Currently, there are many different options for FTTx (see Figure 8-2). Electrical links can be built with DSL or Ethernet. The ITU-T Recommendation G.993.1 defines the *very-high-bit-rate DSL* (VDSL), a DSL type designed for FTTCab and FTTB architectures. The VDSL technology offers downstream bit rates of around 50 Mbit/s within the range of 300 m. VDSL has been improved in the new ITU-T Recommendation G.993.2. This new technology is known as VDSL2, and it delivers symmetrical 100 Mbit/s bit rate within a range of 300 m. In FTTB architectures, the access network operator may choose to deploy Ethernet over unshielded twisted pair (UTP) cable, if cable lengths are shorter than 100 m. The IEEE 802.3 100BASE-T and 1000BASE-T are likely to be the chosen interfaces. 100BASE-T offers 100 Mbit/s of symmetrical bit rate, and 1000BASE-T 1 Gbit/s of symmetrical bit rate.The range is limited to 100 m for both.

Active Ethernet and *passive optical network* (PON) are the main options for the optical portion of the local loop (see Figure 8-3):

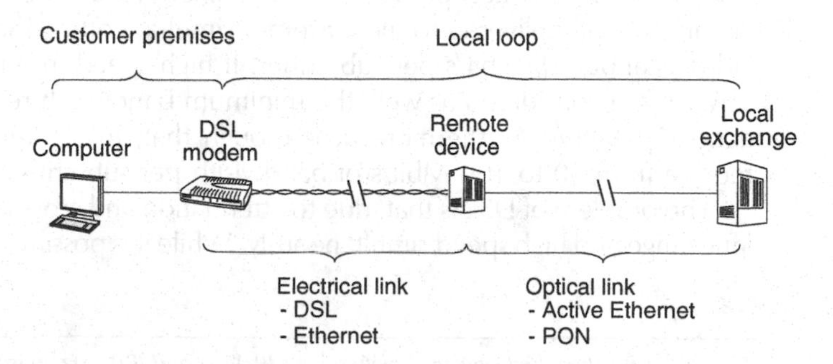

Figure 8-2. FTTx architecture for the local loop.

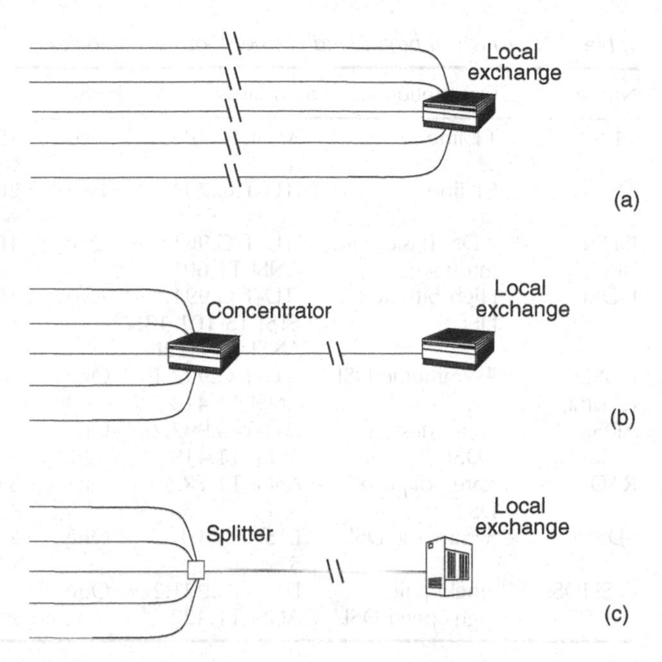

Figure 8-3. Optical fibre installation in the local loop: (a) the point-to-point topology needs a large amount of fibre; (b) with active Ethernet, less fibre is needed, because a switch can be placed close to the subscribers; (c) the PON solution replaces the switch with an inexpensive and passive optical splitter.

- Active Ethernet is made up of point-to-point fibre links between the local exchange and the customer premises. This means that large quantities of optical fibre must be used in the local loop, and this is expensive. However, the use of dedicated fibre links guarantees maximum bandwidth. To reduce the amount of fibre, an Ethernet switch can be installed close to the subscriber, and it acts as a concentrator. Between the switch and the local exchange, it is enough to install a single optical link, or maybe two for redundancy.
- PON has been proposed to avoid installing active elements, such as Ethernet concentrators, in the local loop. Active elements are replaced by simple passive optical splitters, giving as a result a point-to-multipoint topology. PON can be used to offer gigabit-level bandwidth to subscribers. This technology is considered more cost-effective than active Ethernet, and at the same time it is well suited to applications like TV that can be overlapped with data on a different wavelength. The main drawback is the need for complex shared-media access mechanisms to avoid collisions between the traffic of different subscribers.

8.1 Broadband Services Over Copper

Today, the most successful data access technology for residential customers is DSL, with more than 180 million lines installed worldwide in 2007. DSL makes it possible to reuse the existing telephone wires so that they can deliver broadband data services.

Table 8-1. Classical broadband copper loop technologies

Name	Description	Standards	Pairs	Line bit rate	Passband
T1	T1 line	ANSI T1.403	Two	1544 kbit/s symmetric	0–1544 kHz
E1	E1 line	ITU-T G.703	Two	2048 kbit/s symmetric	0–2048 kHz
ISDN BRI	ISDN Basic Rate Interface	ITU-T G.961 ANSI T1.601	One	160 kbit/s symmetric	0–80 kHz
HDSL	High-bit-rate DSL	ITU-T G.991.1 ETSI TS 101 135 ANSI T1.TR.28	Two	1544 kbit/s symmetric	0–370 kHz
ADSL (G.dmt)	Asymmetric DSL	ITU-T G.992.1 ANSI T1.413	One	~1 Mbit/s US ~8 Mbit/s DS	25–138 kHz US 25–1104 kHz DS
ADSL (G.lite)	Splitterless ADSL	ITU-T G.992.2 ANSI T1.419	One	~1 Mbit/s US ~1.5 Mbit/s DS	25–138 kHz US 25–552 kHz DS
RADSL	Rate adaptive DSL	ANSI T1.TR.59	One	512 kbit/s ~8 Mbit/s DS	25–138 kHz US 25–1104 kHz DS
SDSL	Symmetric DSL	ETSI TS 101 524	One	2320 kbit/s symmetric	0–700 kHz
G.SHDSL	Single-pair high-speed DSL	ITU-T G.991.2 ANSI T1.422	One	2320 kbit/s symmetric	0–400 kHz

In order to work properly, DSL uses transmission bandwidth above the vocal frequency band (300–3400 Hz). There are several types of DSL (see Table 8-1), and many of them allow simultaneous transmission of analog telephone signals and data by means of *frequency-division multiplexing* (FDM). DSL signals can use many megahertz above the vocal frequency bandwidth without any noticeable disruption in the telephone service. To achieve this result, voice and data must be separated with the help of a splitter sensitive to frequency, or a similar device (see Figure 8-4).

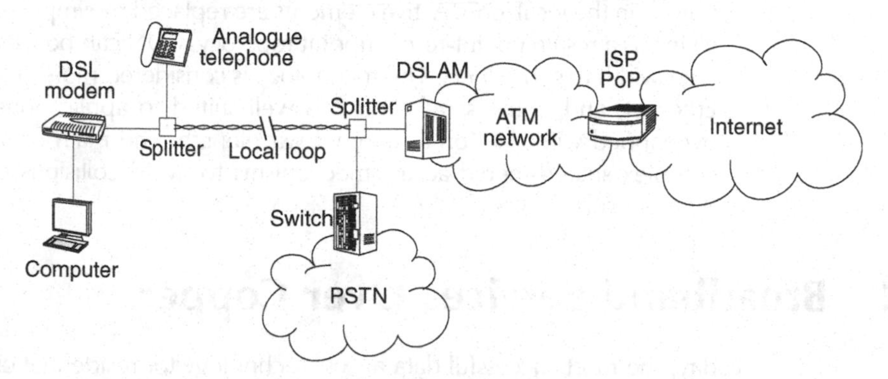

Figure 8-4. Classical access network architecture for Internet over DSL. A splitter is needed to separate the voice and data signals.

The DSL technology dates back to the late 1980s. The first DSL variant which gained wide market acceptance was the *high-bit-rate DSL* (HDSL), still used today to provide E1 (2048 kbit/s) or T1 (1544 kbit/s) to businesses.

Another very popular DSL technology is the *asymmetric DSL* (ADSL), developed by the Stanford University and AT&T Bell Laboratorie. ADSL allocates more transmission bandwidth for the downstream than for the upstream. It was first designed to carry video to the homes across telephone wires. Despite the lack of success of the video application in the early 1990s, ADSL was adopted for Internet access with rates up to 8 Mbit/s downstream and 1 Mbit/s upstream. ADSL took off in the late 1990s, with massive installations in most of the developed countries. To ease and speed up ADSL roll-out, ITU-T released Recommendation G.992.2 for a special version of ADSL called G.lite. ADSL G.lite makes it possible for the subscriber to install the service without an expensive visit from a specialized installer. Today, ADSL standardization and roll-out continue, and new versions such as ADSL2 and ADSL2+ enhance its performance even further.

DSL was developed when some major changes occurred in the telecommunications market. The deregulation of the telecommunications market was intended to increase competition. New network operators, usually referred to as *competitive local exchange carriers* (CLECs), started to emerge, and they offered new services to compete with the *incumbent local exchange carriers* (ILECs) that used to be monopolistic national telecommunication companies.

Unbundling the local loop made it possible for the CLECs to use the ILECs' access facilities. The ILECs were forced to lease their copper pairs to new operators, thus enabling new operators to quickly develop their business.[1] For many CLECs, providing Internet access to homes was the killer application where they put most of their hopes for revenue. DSL was therefore a key technology for them, and unbundling was their opportunity to quickly expand their business with reduced initial investment.

8.1.1 The Limits of Copper Transmission

Despite its many advantages, DSL cannot be seen as the definitive solution for the access network. DSL has been very useful during the past 10–15 years, but this technology alone cannot meet the challenges of future broadband applications such as HDTV, which may need up to 100 Mbit/s. DSL cannot provide long range and high transmission rate simultaneously (see Figure 8-5). Some of the most innovative DSL solutions, like VDSL are designed to achieve very high bit rates, but within a range limited to a couple of hundred metres.

DSL depends on the telephone wires on which it operates, and this means that this technology has some limitations. DSL signals have to suffer many impairments in a transmission channel that was not originally designed to carry them. Two of these limitations are critical:

1. *Attenuation* is caused by progressive loss of the electrical energy of the DSL signal in the transmission line. Attenuation is higher in longer loops, and it also depends on the frequency of the signal being transmitted. The higher the frequency band used for transmission, the more attenuation the signal will suffer. Voice services may work

[1]Sometimes ILECs do not need to lease the copper pair completely; regulations may grant control of the DSL frequency band to the CLEC and let the ILEC control the vocal band for traditional telephony provision.

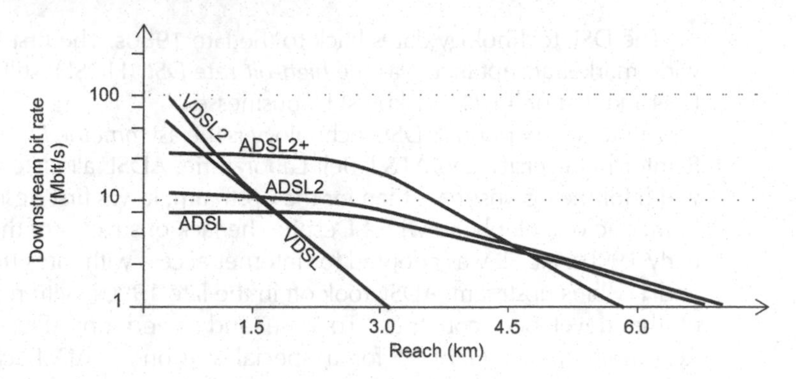

Figure 8-5. Approximate reach achieved with different DSL technologies.

without problems in a copper loop, but the same loop may be unable to transmit DSL signals that use higher frequency bands, due to extra attenuation in these bands.

2. *Crosstalk* is the electromagnetic coupling between transmission lines that are close to one another. In the access network, copper pairs are grouped into binders. One binder may contain dozens or even hundreds of copper pairs, and this is why they are vulnerable to crosstalk. There are two types of crosstalk: *far-end crosstalk* (FEXT) travels in the same direction as the disturbing pair, and *near-end crosstalk* (NEXT) travels in the opposite direction (see Figure 8-6). Both FEXT and NEXT depend on the frequency. Coupling for every disturbing and victim pair is different, and difficult to predict without testing, but generally speaking, FEXT and NEXT increase quickly when the frequency of the disturbing pairs is higher.

Attenuation is being resisted by trying to control the transmission frequency as much as possible. However, for high bit rates there is no choice, and wide portions of the copper transmission bandwidth must be used. Fighting against crosstalk is not as easy as that. Early DSL deployments, such as the first HDSLs, were limited by self-NEXT – crosstalk caused by the same type of DSL at the near end of the cable. This problem was solved by introducing upstream and downstream *frequency-division duplexing* (FDD).

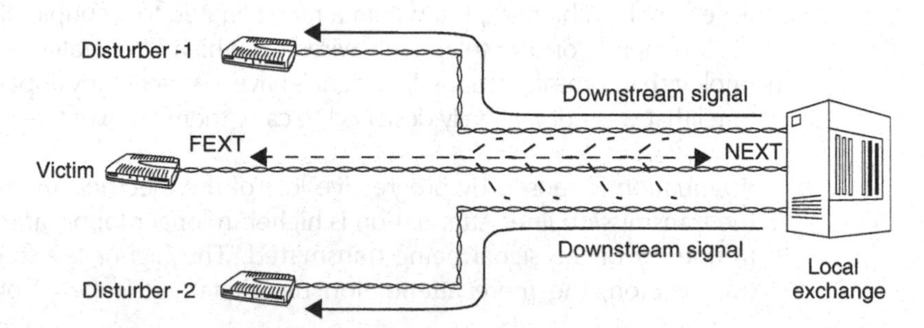

Figure 8-6. Crosstalk between copper pairs. Signals from disturbing lines are coupled to the victim line, damaging communication.

The following DSL generation was limited by FEXT and inter-system NEXT or alien NEXT.

After unbundling took place, the number of signals in the loop started to increase, and crosstalk from some local loop signals could potentially damage other operators' service. Owing to this, the spectral compatibility between copper access technologies had to be studied, and new (national) regulation had to be developed to control the management of the copper loop spectrum.

Another way to fight against crosstalk and external interferences is the use of *discrete multitone* (DMT) modulation. DMT is a multicarrier modulation that spreads the data to be transmitted over many orthogonal carriers instead of using a single carrier. DMT transceivers can slow down carriers affected by narrowband interferences, while unaffected carriers can continue to operate normally. The DMT technology competes with single carrier modulations, such as the simpler *carrierless amplitude modulation* (CAM), or the *quadrature amplitude modulation* (QAM). This competition between modulation technologies led to new standards concerning different and incompatible DSLs. Currently, DMT has more acceptance than CAM and QAM. It is much more widely deployed and preferred by the standards organizations.

8.1.2 ADSL2

The second generation of ADSL (ADSL2), specified in the year 2002 in Recommendations ITU-T G.992.3 and ITU-T G.992.4, is the result of many improvements to the traditional ADSL. The signals that meet the new standard exhibit better behaviour in loops of all lengths, but the results are more impressive in shorter loops: ADSL2 delivers up to 12 Mbit/s downstream and 1 Mbit/s upstream with a range of 2 km.

ADSL2 transceivers are quite complex and they perform many operations on the transmitted and received signals in order to ensure optimal performance (see Figure 8-7). These operations can be grouped as follows:

- *Transmission protocol-specific/transmission convergence* (TPS-TC) processing: it is different depending on the kind of data being transported by DSL.

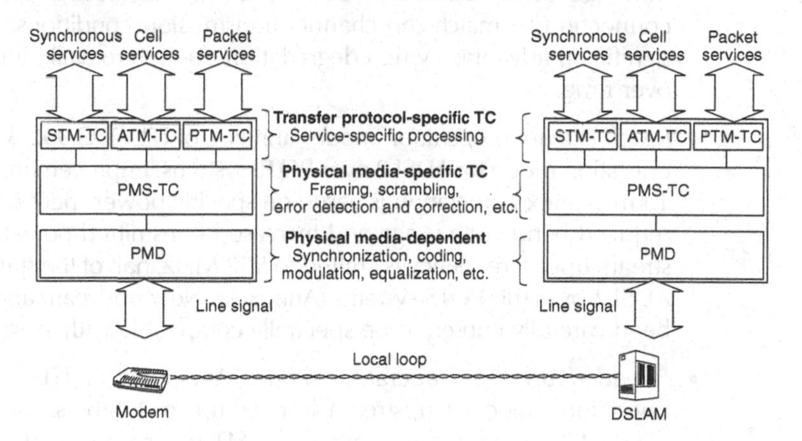

Figure 8-7. Operations involved in ADSL2 transmission and its classification.

- *Physical media specific/transmission convergence* (PMS-TC) processing: it is common for all client data and includes framing, frame synchronization, error detection and correction and scrambling.
- *Physical media-dependent* (PMD) processing: it depends on the physical media, in this case copper pairs, and involves modulation, echo cancellation, equalization, link startup, etc.

ADSL2 has better PMD, PMS-TC and TPS-TC than ADSL. Specifically, ADSL2 includes the following important enhancements:

- *Better support of packet applications*: the new *packet transfer mode* (PTM) enables direct mapping of layer 2 frames like Ethernet over ADSL2 PMS-TC frames. The PTM is added to the *synchronous transfer mode* (STM) and *asynchronous transfer mode* (ATM) already supported by legacy ADSL. Thanks to this new feature, packets do not need to be encapsulated in ATM cells before transmission. ATM in the local loop has always generated discussion because of the large overhead added by cells. Now it can be removed from the loop (see Figure 8-8).
- *All-digital mode*: ADSLwas designed to work simultaneously with *plain old telephone service* (POTS) or *integrated services digital network* (ISDN) voice. Either POTS or ISDN is multiplexed with ADSL with the help of FDM techniques. However, many operators are considering alternative ways of delivering voice. For example, TDM-encoded voice channels can be transmitted in ADSL2 DMT tones. Another solution is to deploy VoIP solutions in the local loop and let voice share DSL bandwidth with existing data applications. Both solutions make it unnecessary to allocate a frequency band for POTS or ISDN in the local loop. ITU-T G.992.3 Annexes I and J define all digital modes for operation without POTS/ISDN. These modes are convenient for local loop with converged voice and data technology. Annex I is designed to minimize crosstalk with POTS and Annex J is spectrally compatible with ISDN.
- *Real-time rate adaptation*: legacy ADSL systems are only able to adjust their transmission rate at initialization, but ADSL2 can seamlessly change the data rate of the connection to match the channel transmission conditions. This new feature offers a significant advantage when degradation due to crosstalk, interferences or noise varies over time.
- *Reach extended operation mode*: annex L of ITU-T G.992.3 defines a reach-extended operation mode for ADSL2 over POTS systems. Improvements in reach may arrive up to 1 km. Annex L operation is based on specific power spectrum transmission masks with reduced bandwidth usage and increased transmitted power. For example, the downstream upper frequency is limited to 552 MHz, half of the standard upper frequency for ADSL2 over the POTS systems (Annex A). New upstream and downstream masks have been carefully chosen to be spectrally compatible with existing signals.
- *Extended upstream operation mode*: Annex M of ITU-T G.992.3 defines a new operation mode for upstream bitrates up to 3 Mbit/s. To achieve this goal, Annex M ADSL2 increases the number of DMT tones dedicated to upstream transmission.

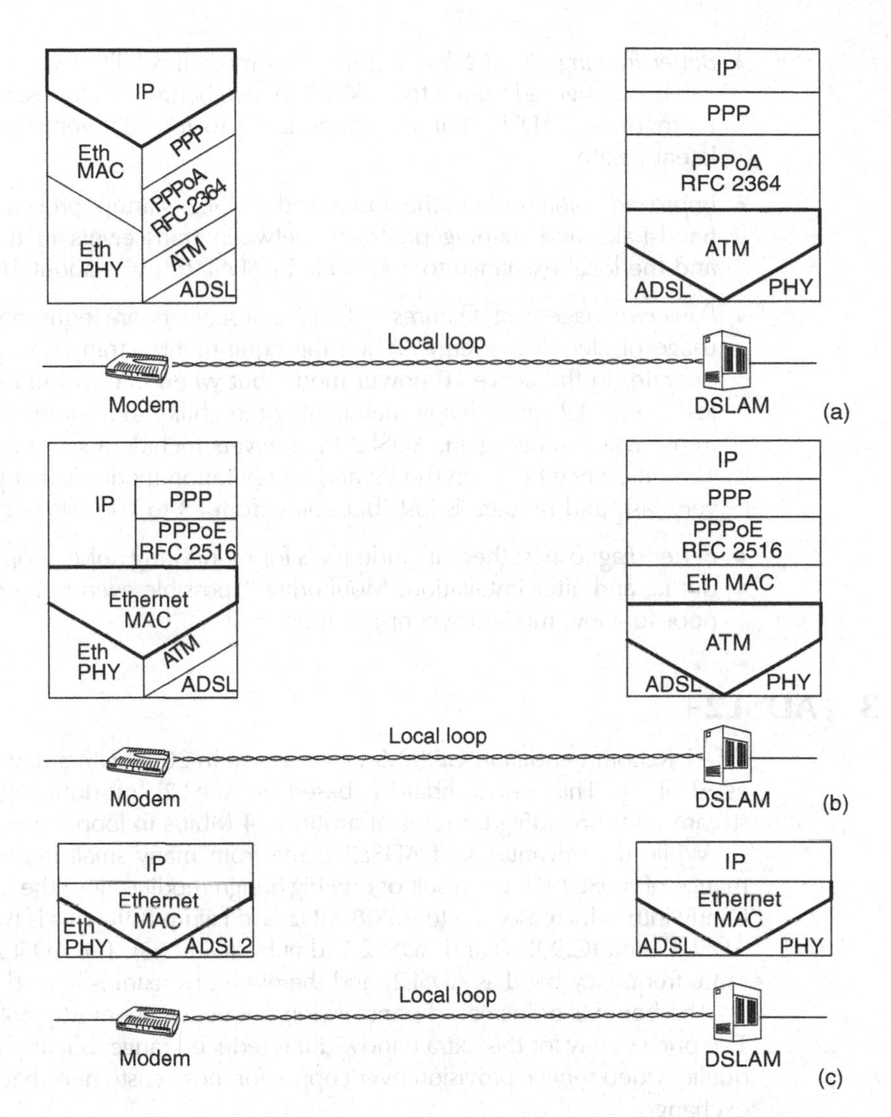

Figure 8-8. Protocol architectures in the local loop with ADSL and ADSL2: (a) routed architecture – the user modem is a router; data is carried in ATM VCs; (b) bridged architecture with ATM – data is encapsulated and carried over an ATM VC; (c) bridged architecture without ATM – uses the ADSL2 PTM to map Ethernet directly in ADSL2 frames.

- *Improved modulation and coding*: a 16-state convolutional Trellis coding is mandatory on ADSL2 DMT carriers. Furthermore, Reed–Solomon (RS) encoding of ADSL2 frames has been improved and now enables higher coding gains. Both Trellis and RS coding add controlled redundancy to the signal, to detect and correct errors without retransmission. These features are specially helpful in longer lines, where the SNR is low and transmission errors are likely to occur. Up to 200 m improvement in range is accomplished, thanks to modulation and coding enhancements in ADSL2.

- *Better framing*: ADSL2 framing provides more flexibility than ADSL. Frame overhead can be configured from 4 to 32 kbit/s to save bandwidth for user data when necessary. Furthermore, ADSL2 framing supports up to four different frame bearers with four latency paths.
- *Improved initialization*: the enhanced ADSL2 startup process reduces the overall handshake and training processes between transceivers in the customer premises and the local exchange to 3 s, while in ADSL this was about 10 s.
- *Power management features*: ADSL2 transceivers are equipped to make optimum usage of electrical energy. When the equipment is transmitting or receiving data, it operates in the active L0 power mode, but when transmission stops, it can go to the low power L2 mode while maintaining the ability to re-enter in L0 mode as soon as transmission starts again. ADSL2 transceivers include a sleep mode called L3 mode. The difference between the L2 and L3 operation modes is that going from L2 to L0 is very fast, and no data is lost, but going from L3 to L0 is slower.
- *Better diagnostics*: these include tools for monitoring noise, loop attenuation and SNR during and after installation. Monitoring is possible even if the channel quality is too poor to allow modem synchronization.

8.1.3 ADSL2+

ITU-T Recommendation G.992.5 was released in 2003 with a new ADSL version known as ADSL2+. This new standard is based on ADSL2, but doubling its maximum downstream rate, providing bit rates of around 24 Mbit/s in loops shorter than 1.5 km.

While the advantages of ADSL2 come from many small improvements, the performance of ADSL2+ is the result of one big design modification: the transmission frequency bandwidth is increased up to 2.208 MHz (see Figure 8-9). This is twice the bandwidth of ADSL G.dmt (G.992.1) and ADSL2 G.dmt.bis (G.992.3). The ADSL2+ upstream uses the same frequency band as ADSL2, and therefore upstream is limited to 1 Mbit/s.

The benefits of increased transmission frequency are mostly noticed in shorter loops. The price to pay for this extra bandwidth is reduced range, but anyway, ADSL2+ enables quality video service provision over copper for those customers that are closer to the local exchange.

8.1.4 Bonded DSL

Bonded DSL provides a single high-speed channel by combining the capacity of two or more low-speed channels. This is done by using inverse multiplexing.

ITU-T Recommendation G.998.1, released in 2005, defines DSL bonding. This Recommendation is based on ATM, which is why it only works with ATM-TC DSL. It is also possible to bond pairs carrying Ethernet with PTM-TC DSL, thanks to the IEEE 802.3ah standard for Ethernet in the first mile (EFM) (see Section 8.3).

Bonded DSL can be used by service providers to deliver high bit rates to subscribers who are several kilometres away from the local exchange. Therefore, the traditional range limitations of this technology are not an issue here. This is why those operators who wish to

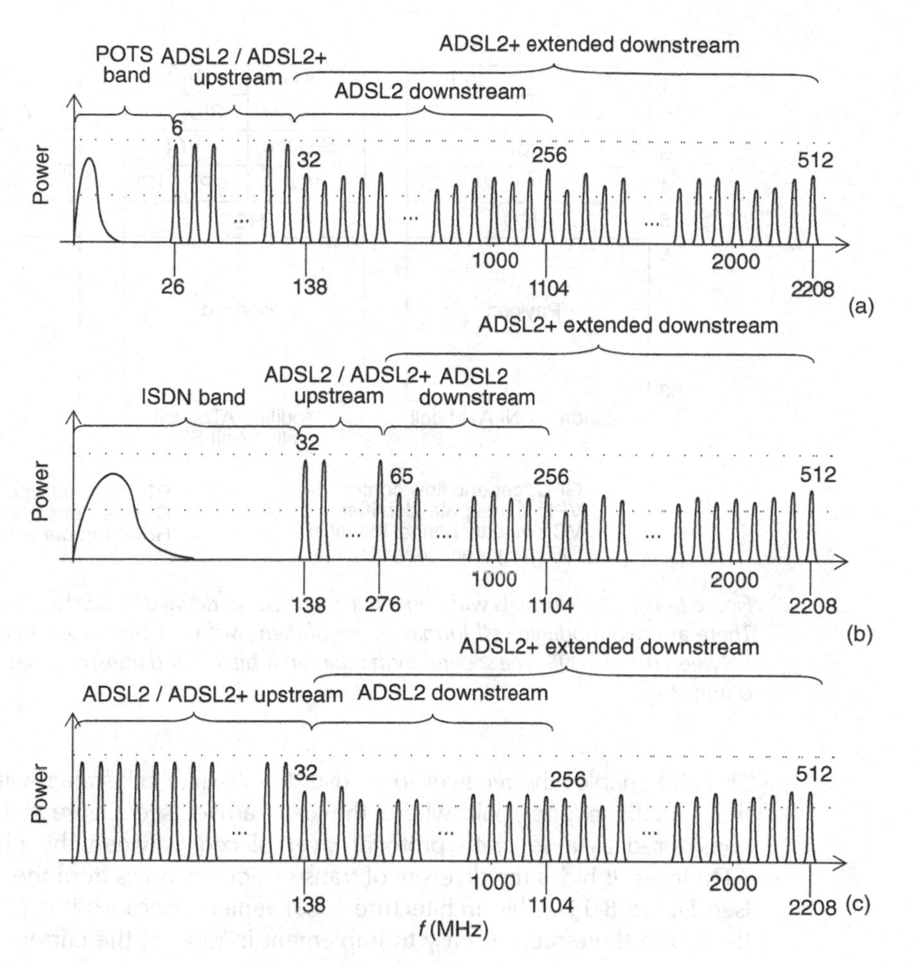

Figure 8-9. Some examples of DMT tone usage in ADSL2/ADSL2+: (a) tone usage for ADSL2/ADSL2 + over POTS as specified in ITU-T G.992.3/G.992.5 Annex A; (b) tone usage for ADSL2/ADSL2+ over ISDN as specified in ITU-T G.992.3/G.992.5 Annex B; (c) all-digital mode as per ITU-T G.992.3/ G.992.5 Annex I.

offer a uniform service bundle to all their potential customers may find bonded links very interesting. For example, customers closer to the local exchange may be served by a single ADSL2/ADSL2+ loop. Those who are further away will have access to the same services, but they will be served by various ADSL/ADSL2+ bonded loops.

Bonding was enabled for the first time in the ADSL2 standard that supported *inverse multiplexing over ATM* (IMA). The main advantage of G.998.1 over IMA is that, while IMA assumes that all the bonded links have the same nominal speed, G.998.1 allows different rates up to a ratio of 4:1 among its pairs.

Unlike IMA, G.998.1 defines a modified cell format with a new field called *sequence identifier* (SID). Cells are sequenced and the sequence number is inserted to the SID field before transmission. The transmitter can freely choose the link in which to put every cell.

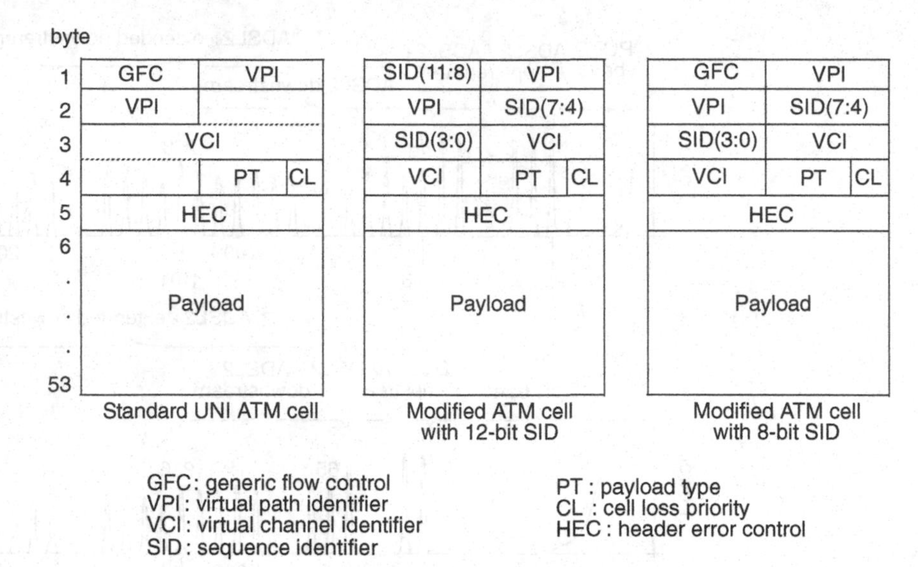

Figure 8-10. Cell formats with sequence numbers, as defined in G.998.1, the bonded DSL standard. There are two modified cell formats. One of them, with a 12-bit SID, can carry sequence numbers between 0 and 4095. The second format has an 8-bit SID and transports sequence numbers between 0 and 255.

The SID enables the receiver to recover the original cell stream without worrying about the specific copper pair where the cells arrive (see Figure 8-10). Bonding can be considered as a separate protocol layer placed between the physical layer and the ATM layer. It hides the diversity of transmission channels from the upper-layer protocols (see Figure 8-11). This architecture is convenient, because it is physical layer-independent, and therefore it is easy to implement in most of the current DSL technologies.

8.1.5 VDSL

VDSL is a DSL technology for high speed (beyond 50 Mbit/s downstream) over short loops (usually a few hundreds of metres) using transmission frequencies up to 12 MHz. Early work on VDSL dates back to 1995, with simultaneous projects at ITU-T, ETSI and ANSI. Later, in 1997, the Full-Service Access Network (FSAN), a forum of telecom operators and vendors, developed the first VDSL specification. The market, however, did not take advantage of this work, due to a disagreement regarding the modulation technology: DMT or CAP. As a result of this, different proprietary implementations of VDSL started to appear, all of them with very limited roll-outs, mainly in Japan and South Korea.

Things did not change until 2003, when 11 major DSL equipment suppliers jointly supported DMT for VDSL, and the ITU-T could finally release Recommendation G.993.1. This Recommendation prioritized DMT – however, QAM was still included in an annex.

VDSL will never be widely deployed, because the second generation of VDSL (VDSL2) was released by the ITU-T in 2005. VDSL2 has better performance than VDSL, so all future improvements will take it as the starting point.

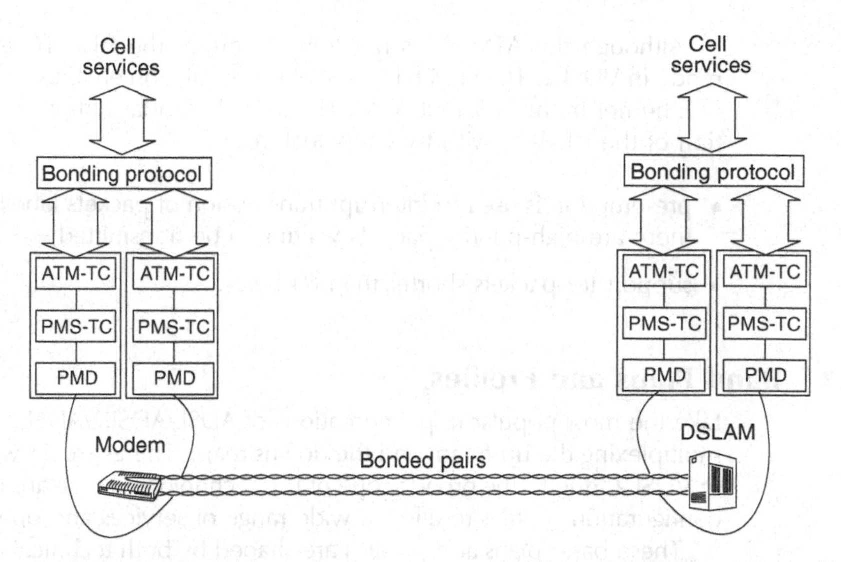

Figure 8-11. ATM cells are transported over two or more bonded copper pairs as if they were a single high-speed stream.

8.1.6 VDSL2

The second generation of VDSL, VDSL2, specified in Recommendation ITU-T G.993.2, is designed to increase bit rate over short and medium-length loops. VDSL2 aims to achieve the following goals:

- 100 Mbit/s symmetric service over loops shorter than 300 m;
- at least 25 Mbit/s in loops shorter than 1000 m.

It would not be possible to provide these rates without opening new frequency bands for digital transmission over copper. Using the band between 12 and 30 MHz increases bit rate. Frequencies between 20 and 138 kHz (optional for VDSL) are used to achieve longer reach than VDSL.

Telcos have been eager to see VDSL2, because this technology enables a new range of services over copper pairs. VDSL2 makes it possible to offer several HDTV channels simultaneously, for example. It is expected that VDSL2 will be key in enabling real competition between telcos and cable operators.

In spite of its name, VDSL2 is not just the successor of VDSL. Thanks to its performance over medium-length loops, it makes sense to migrate some ADSL2+ roll-outs to VDSL2. In other words, VDSL2 can be seen as an evolution of both VDSL and ADSL2+.

VDSL2 equipment can provide both symmetric and asymmetric services by using different configuration profiles. These profiles make it possible to share the spectrum with either POTS or ISDN, but there are also all-digital modes for converged access networks. The QAM modulation, optional for VDSL, is not used, but DMT-based implementations are used instead.

Although the ATM-TC is not fully forgotten, the PTM-TC is the main transmission mode in VDSL2. The EFM 64/65-octet encapsulation enables encoding and transmission of Ethernet frames without ATM. The VDSL2 standard improves the the EFM specification of the PTM-TC with two new features:

- pre-emption is used to interrupt transmission of packets labelled as low priority when there are high-priority packets waiting to be transmitted;
- support for packets shorter than 64 bytes.

8.1.6.1 Band Plans and Profiles

Like the most popular implementations of ADSL/ADSL2/ADSL2+, VDSL2 uses FDD for multiplexing the upstream and the downstream. Things are, however, more complicated in VDSL2, due to the broad scope of this technology. There are multiple band plans and configuration profiles to allow a wide range of services and operation modes.

These band plans and profiles are shaped by both technical and commercial factors:

- Service providers may want to offer symmetric and asymmetric services, custom QoS or other options. The configuration profiles need to accommodate all the possible services that access operators may be interested in.
- Band plans cannot be defined freely, because they are limited by the spectral compatibility between signals. Deploying new signals without considering this degrades the overall quality of both new and legacy services.
- Electronic components have their own limitations. For example, it is difficult (or at least expensive) to transmit very high power across a wide range of frequencies from a few kilohertz to dozens of megahertz. It is therefore necessary to define profiles with specific features: high power, high speed, etc.

The existing VDSL band plans served as the starting point to define band plans for VDSL2. In the year 2000, there were two main band plans for VDSL. The first one, know as Band Plan 998 was better suited for asymmetric services, and the second one, Band Plan 997, was designed for symmetric services. Both Plan 997 and 998 are based on a four-band spectrum with frequencies between 138 and 12 000 kHz used alternatively for transporting upstream and downstream data. The current VDSL specification enables optional transmission in the 25–138 kHz band for upstream or downstream (see Figure 8-12).

The VDSL2 standards maintain spectral compatibility with VDSL band plans below 12 MHz. Many service providers in North America, Europe and Asia adhered to band plans based on the VDSL Plan 998, but some other European operators gave their support to spectrum usage based on Plan 997. As a result, the current version of Recommendation G.992.3 defines different deployment regions (see Figure 8-13):

- In region A (North America), the VDSL2 band plan is based on Plan 998 below 12 MHz. Usage of the spectrum above 12 MHz is currently under study.

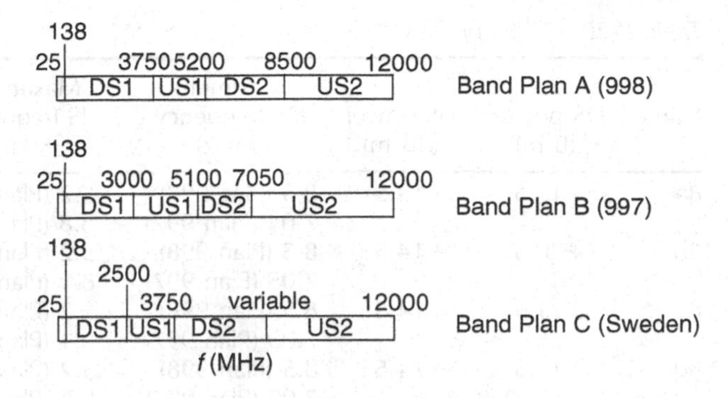

Figure 8-12. VDSL bandplans. Band Plan A (998) is better suited for asymmetric services and Band Plan B (997) is designed for symmetric services. Band Plan C is defined specifically for VDSL in Sweden.

- In region B (Europe), deployments based on both Plan 997 and 998 are accepted. Usage of the spectrum above 12 MHz is to be defined.
- In region C (Japan), deployments will be based on Plan 998. Usage of the band between 12 and 30 MHz is already defined. There are two advanced VDSL2 services of 17 and 30 MHz.

It is worth noting that signals based on Plan 998 and Plan 997 are spectrally incompatible and they cannot coexist in the same plan. Working simultaneously with these signals will cause performance degradation, and as a result, services will stop working as expected. This is more likely to occur in Europe, where two incompatible band plans are accepted.

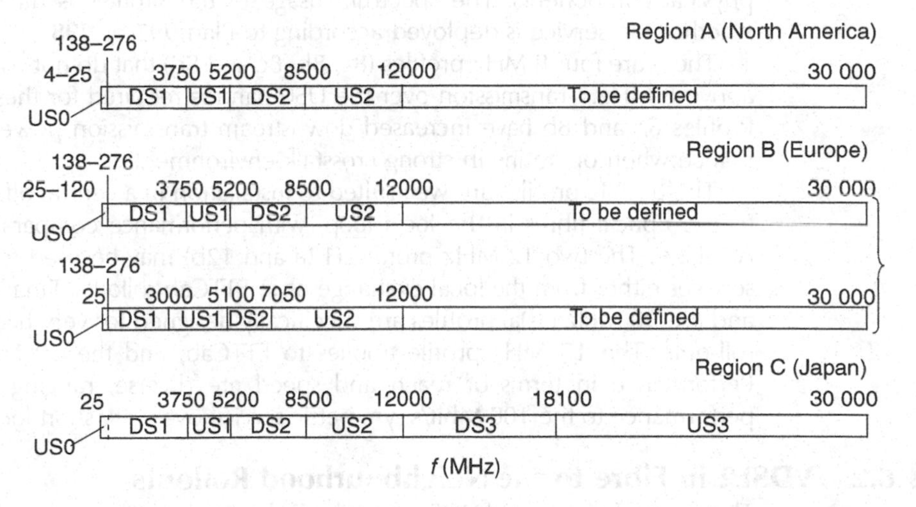

Figure 8-13. VDSL2 band plans for the three deployment regions: North America, Europe and Japan.

Table 8-2. VDSL2 profiles

Name	DS power (dB m)	US power (dB m)	Maximum DS frequency (MHz)	Maximum US frequency (MHz)	US0 usage	Carrier spacing (kHz)
8a	+17.5	+14.5	8.5 (Plan 998) 7.05 (Plan 997)	5.2 (Plan 998) 8.8 (Plan 997)	Required	4.3125
8b	+20.5	+14.5	8.5 (Plan 998) 7.05 (Plan 997)	5.2 (Plan 998) 8.8 (Plan 997)	Required	4.3125
8c	+11.5	+14.5	8.5 (Plan 998) 7.05 (Plan 997)	5.2 (Plan 998) 5.1 (Plan 997)	Required	4.3125
8d	+14.5	+14.5	8.5 (Plan 998) 7.05 (Plan 997)	5.2 (Plan 998) 8.8 (Plan 997)	Required	4.3125
12a	+14.5	+14.5	8.5 (Plan 998) 7.05 (Plan 997)	12	Required	4.3125
12b	+14.5	+14.5	8.5 (Plan 998) 7.05 (Plan 997)	12	Not required	4.3125
17a	+14.5	+14.5	17.7 (Japan)	12 (Japan)	Not required	4.3125
30a	+14.5	+14.5	18.1 (Japan)	30 (Japan)	Not required	8.625

Another issue covered in the definition of the different band plans is the usage of the low frequency band between 20 and 138 Hz (US0 band). This band is currently regulated as optional for VDSL, but it is required in long-reach VDSL2 configuration profiles.

ITU-T Recommendation G.993.2 defines eight different profiles or operation modes for VDSL2 (see Table 8.2). These profiles can be considered as setups of VDSL2 systems, as they contain configurations for the transceivers. The spectrum bands to be used, the maximum and minimum transmitted power and other parameters are given for every profile to achieve specific purposes while taking into account the limitations of the physical components. The spectrum usage of the profiles is different, depending on whether the service is deployed according to Plan 997 or 998.

There are four 8 MHz profiles (8a, 8b, 8c and 8d) that do not use transmission bands above 8 MHz. Transmission over the US0 band is required for these profiles, however. Profiles 8a and 8b have increased downstream transmission power to improve performance when operating in strong crosstalk environments.

The 8 MHz profiles are well suited to installation in a central office without the need to use optical fibres in the local loop, with performance comparable to ADSL/ADSL2/ADSL2+. The two 12 MHz profiles (12a and 12b) may be used to provide broadband services either from the local exchange or in FTTCab rollouts. Finally, the 17 MHz (17a) and the 30 MHz (30a) profiles are specifically designed for very high rates in deep fibre roll-outs. The 17 MHz profile applies to FTTCab, and the 30 MHz profile to FTTB. Performance in terms of reach and speed are diverse, ranging from ADSL2+-alike performance to the 100 Mbit/s symmetric service over very short loops (see Figure 8-14).

8.1.6.2 VDSL2 in Fibre to the Neighbourhood Rollouts

The range of high-speed VDSL2 profiles is shorter than the average distance between the local exchange and the customer premises. This means that VDSL2 cannot achieve its

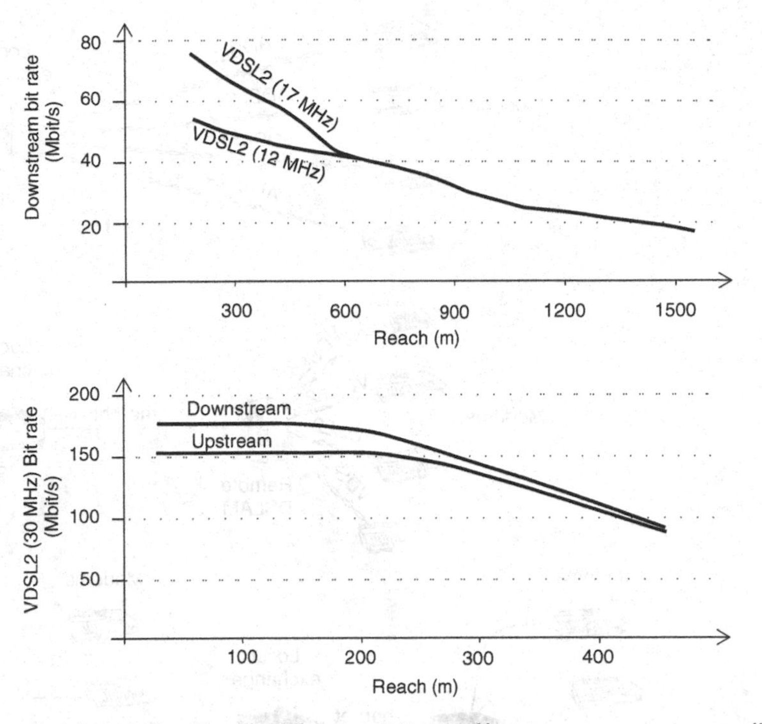

Figure 8-14. Reaches achieved with different VDSL2 profiles. Measurements are collected on a 0.4 mm wire in the presence of 24 VDSL2 disturbers.

maximum performance without moving the *DSL access multiplexer* (DSLAM) out of the exchange.

New DSLAMs, known as remote DSLAMs, can be installed in street cabinets, in the basements of buildings, or in other locations close to the subscribers. By making the distance between the DSLAM and the subscriber shorter, it is possible to increase the bit rate and get the maximum performance from VDSL2. Deploying remote DSLAMs means that optical fibre links must be installed between DSLAMs and the local exchange. Although any optical technology is valid for these links, PON and optical point-to-point Ethernet are the most popular solutions today.

External plant deployment of DSLAM involves several new challenges that need to be solved:

- The equipment must be placed in a suitable place, far from electromagnetic interference. The place must also be secure, to prevent any unauthorized access.
- DSLAMs must be prepared for difficult weather conditions. For example, while in the local exchange typical temperature conditions range from −5 to +40°C, remote DSLAMs must be prepared for temperatures between −40 and +65°C.
- Suitable electrical power sources must be found. Sometimes, remote feeding can be used. In FTTB roll-outs, access to local power sources is an alternative. Using redundant power sources is recommended.

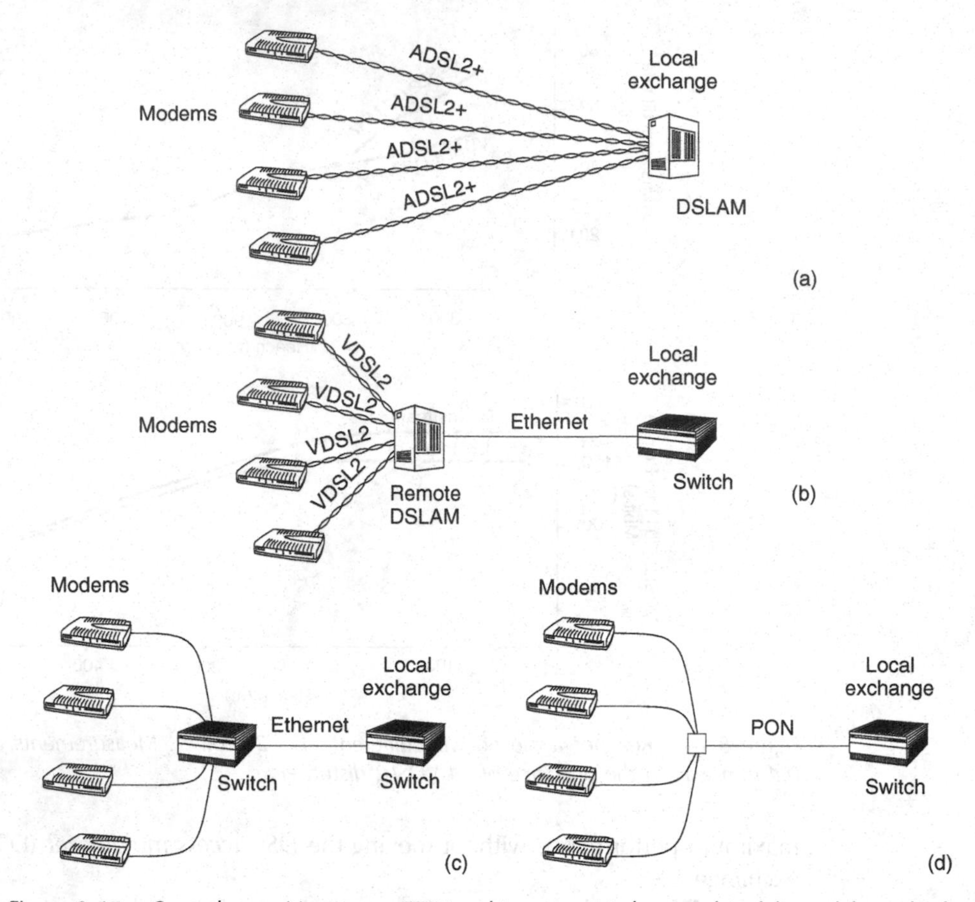

Figure 8-15. Smooth transition to an FTTH architecture: (a) the signal is delivered from the local exchange by means of copper pairs and ADSL2+; (b) remote DSLAMs are used, and now the signal can be delivered at a much higher rate with VDSL2 links; (c) the remote DSLAM is replaced by a switch with optical ports, and the local loop becomes an Ethernet network; (d) the remote DSLAM is replaced by an optical splitter, and subscribers are serviced with PON links.

VDSL2 rollouts with remote DSLAMs are considered by many operators as the first step in the migration path to an access network entirely based on optical communications (see Figure 8-15). Installing optical fibre is expensive, and the cost per subscriber of FTTH deployments is often difficult to justify. Other deep fibre roll-out strategies based on VDSL2 over existing copper wires can be used to reduce the investment per subscriber and simultaneously enable some bandwidth-demanding services such as TV or network gaming.[2]

Existence of VDSL2 transmitters in cabinets may potentially cause disruption in the services transported across local exchange pairs. Crosstalk injected into pairs transporting

[2]Market development may justify massive installation of optical fibre to the customer premises.

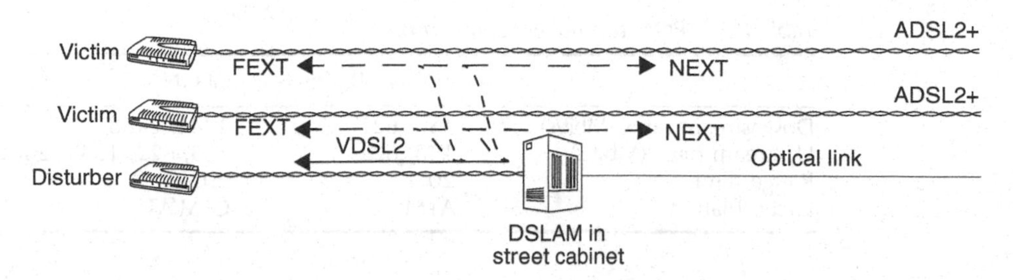

Figure 8-16. Crosstalk injected into ADSL2+ pairs by cabinet signals like VDSL2 is likely to cause damage in victim pairs, even if the signals are spectrally compatible.

attenuated signals will most likely damage them, even if the victim pairs are spectrally compatible, like ADSL2+ and VDSL2 (see Figure 8-16).

To deal with this problem, the power spectrum of VDSL2 signals is shaped by a filter that reduces the power to a convenient level in the frequency band between 138 kHz and 2208 MHz, the band shared by ADSL2+ and VDSL. The amount of VDSL2 attenuation can be programmed: it will be more intense when the remote DSLAM is placed further from the local exchange, and thus victim signals are weaker.

8.2 The Passive Optical Network

The *passive optical network* (PON) is an optical technology for the access network, based only on passive elements such as splitters. In a PON, the transmission medium is shared, and traffic from different stations is multiplexed. Optical transmission increases transmission bandwidth and range dramatically when compared with some copper pair technologies such as DSL. Furthermore, owing to the use of simple and inexpensive transmission elements and shared medium, a PON is a cost-effective solution for the optical access network.

The logical deployment alternative enabling optical communications in the local loop is to replace the copper links by optical fibre links, but this requires a lot of fibre. Installing Ethernet switches acting as traffic concentrators near the customer premises requires less fibre, but massive installation of Ethernet switches has the same inconveniences as remote DSLAMs: suitable placement and power supply must be provided. This is one of the reasons why PON, based only on passive elements that do not need feeding, is a very attractive solution.

Preliminary works on the PON technology date back to the late 1980s, but the first important achievement regarding its standardization was not until 1995. In this year, the *full-service access network* (FSAN) was formed and presented a system specification for *ATM PON* (APON). Later, in 1997, the ITU-T released Recommendation G.983.1 based on the FSAN specification. The APON is known today as *broadband PON* (BPON) to emphasize that, although ATM-based, any broadband service can be provided with this technology.

Since the release of Recommendations G.994.x for *gigabit PON* (GPON) in 2003, APON/BPON is considered a legacy technology. GPON has been specified with the help of the FSAN, and it provides multigigabit bandwidths at lower costs than BPON, while

Table 8-3. PON technology comparison

	APON / BPON	GPON	EPON
Downstream rates (Mbit/s)	155, 622	1244, 2488	1000
Upstream rates (Mbit/s)	155, 622	155, 622, 1244, 2488	1000
Range (km)	20	20	20
Encapsulation	ATM	GEM/ATM	Ethernet

achieving more efficient transporting packetized data with the new lightweight *GPON encapsulation mode* (GEM). The GEM is based on a concept similar to the *generic framing procedure* (GFP), a successful encapsulation for mapping packets in SDH networks.

An alternative approach is the *Ethernet PON* (EPON), released in 2004 as a part of the IEEE 802.3ah standard for Ethernet in access networks. The main innovation of EPON is that it encapsulates data in Ethernet MAC frames for transmission. Today, EPON has become a strong competitor for GPON, and there are supporters and deployments for both technologies (see Table 8-3).

8.2.1 Basic Operation

The physical properties of passive optical splitters make the distribution of optical signals with PON different from other technologies with a shared access to the transmission medium. Ports in optical splitters do not all have the same properties, and thus the network elements connected to them are different:

- The *optical line termination* (OLT) is connected to the uplink port of the optical splitter. Any signal transmitted from the OLT is broadcast to all the other ports of the splitter.
- The *optical network unit* (ONU) is connected to the ordinary ports of the optical splitter. When signals transmitted from the ONT arrive at the splitter, they are retransmitted to the uplink towards the OLT, but not to other ordinary ports where other ONUs could be connected. This makes direct communication between ONUs impossible.

The OLT constitutes the network side of the PON, and it usually resides in the local exchange. The ONUs are the user side. They can be placed in the customer premises in FTTH roll-outs, but they can also be deployed in cabinets, basements of buildings or other locations close to the subscribers. In cases where the ONU is not directly available to the subscribers, the signal is delivered to them by means of other technologies such as DSL or Ethernet. The ONU in FTTH is sometimes referred to as *optical network termination* (ONT).

In the PON technology, the upstream and downstream signals can be multiplexed by physically separating the signals into different optical fibres, or by using *wavelength-division multiplexing* (WDM) mechanisms, i.e. by transmitting the upstream and downstream signals over the same fibre but at different wavelengths.

Signals from two or more ONUs transmitting simultaneously will collide in the uplink, and the OLT will be unable to separate them, unless a bandwidth-sharing mechanism is implemented. WDM appears to be the most natural way to share the transmission media for PON, but it would require either installing tunable lasers in the ONUs or having many different classes of ONUs for transmitting at different wavelengths. The high cost of the first solution and the complexity of the second one make WDM-PON infeasible today, but attractive in the future.

Transmission in current PONs is based on TDM rather than on WDM. TDM allows for a single downstream wavelength, but it relies on complex shared-media access algorithms (see Figure 8-17). Contention based on the *carrier sense media access/collision detection* (CSMA/CD) algorithm was successfully used with broadcast Ethernet, but it cannot be used with PON for the following reasons:

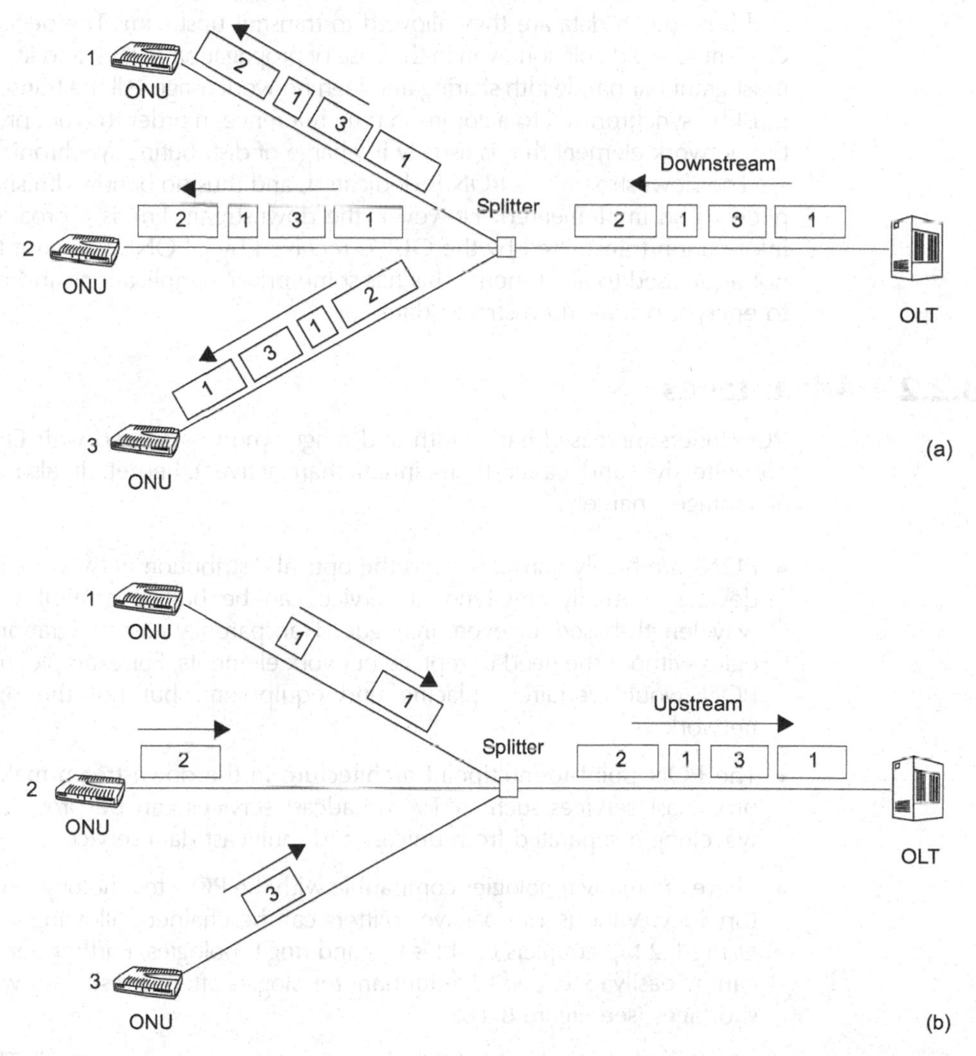

Figure 8-17. Transmission medium sharing a PON: (a) the downstream signal is broadcast to all the ONUs; (b) the upstream signal is point-to point. The section between the splitter and the OLT is shared between all the ONUs.

- ONUs cannot receive signals from other ONUs, and therefore they are unable to detect collisions. The OLT is the only station able to detect collisions between stations.
- The performance of CSMA/CD is closely related to the network size. CS-MA/CD can be used in LAN environments with size limited to hundreds of metres, but it would not work in PONs of dozens of kilometres.

There are bandwidth sharing mechanisms specifically designed for PON. These algorithms take into account that only communications from an ONU to an OLT, but not between ONUs, are possible, and therefore they assign to the OLT controller functions. The OLT decides which ONUs are allowed to transmit, when they are allowed to do so, and how much data are they allowed to transmit upstream. The decisions made by the OLT must avoid collision even in the case of propagation delays, and at the same time they must grant fair bandwidth sharing and high network usage. All the transceivers in the PON must be synchronized to a common time reference in order to work properly. The OLT is the network element that is usually in charge of distributing synchronization.

The downstream of a PON is dedicated, and thus no bandwidth sharing mechanisms need to be implemented. However, the downstream link is a broadcast channel, and information transmitted by the OLT is received by all ONUs, even if this information is not addressed to all of them. This has some privacy implications and makes it necessary to encrypt private downstream data.

8.2.2 Advantages

PON offers increased bandwidth and range when compared with DSL. It is also more cost-effective and easier to maintain than active Ethernet. It also has several other advantages, namely:

- PONs are highly transparent, as the optical distribution network only contains layer-1 devices. Virtually any type of service can be built over PONs, packet, TDM or wavelength-based, or even analogue. Transparency eases migration to new technologies without the need to replace network elements. For example, migration to WDM PON would require replacing end equipment, but not the optical distribution network.
- The PON point-to-multipoint architecture in the downstream makes it easy to offer broadcast services such as TV. Broadcast services can be provided in a dedicated wavelength separated from unicast and multicast data services.
- There are many topologies compatible with the PON technology beyond the basic star topology. Various 1:*N* passive splitters can be chained, allowing for a tree topology. Using 1:2 tap couplers enables bus and ring topologies. Furthermore, basic topologies can be easily extended to redundant topologies offering resiliency when facing service shortages (see Figure 8-18).

On the other hand, using PONs has some inconveniences as well. The most important drawbacks are reduced range and bandwidth when compared with active Ethernet, due to the attenuation introduced by the splitters and the effect of sharing resources.

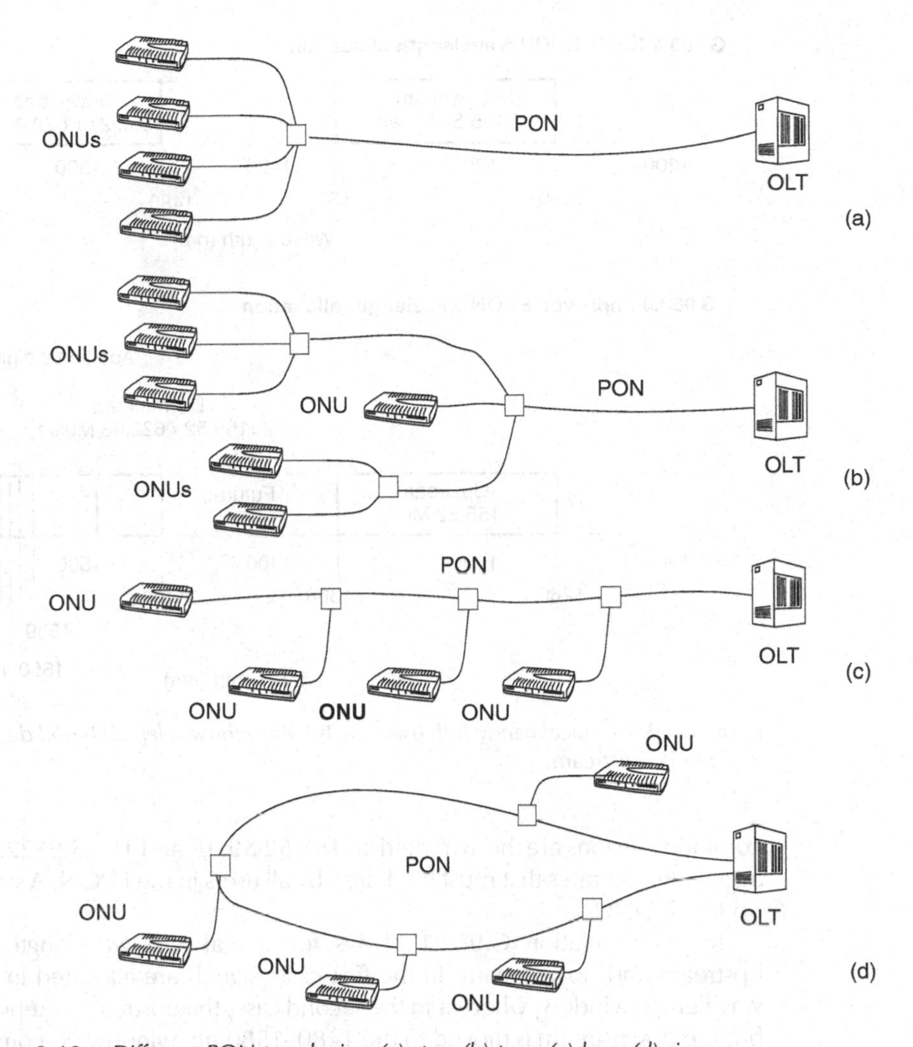

Figure 8-18. Different PON topologies: (a) star; (b) tree; (c) bus; (d) ring.

8.2.3 Broadband PON

Broadband PON or BPON was the first version of a PON to be included in an international standard in the mid-1990s. By that time it was the logical decision to choose an ATM encapsulation for BPON (or APON, as it was known before). BPON was specified by the FSAN, a group of telecommunications companies and telecom equipment vendors. For them, it was natural to choose a technology specifically designed for carrier applications. From the unexpected success of Ethernet and IP arose the need to modify the original design to provide better support of these technologies.

The current version of BPON delivers upstream rates of up to 622.08 Mbit/s, and downstream rates of up to 1244.16 Mbit/s with a range of up to 20 km. However, the most

G.983.1 Basic BPON wavelength allocation

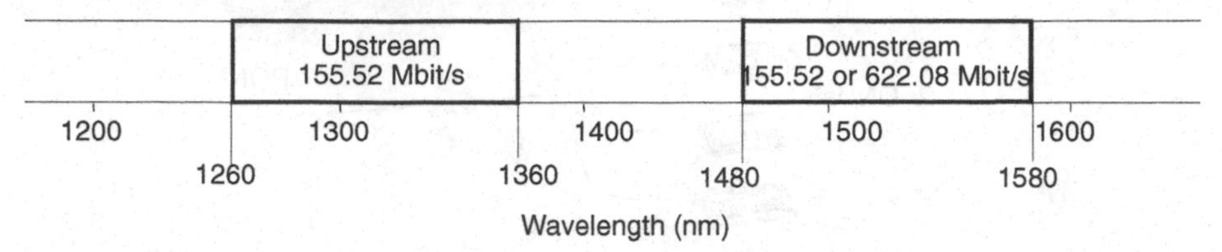

G.983.3 Improved BPON wavelength allocation

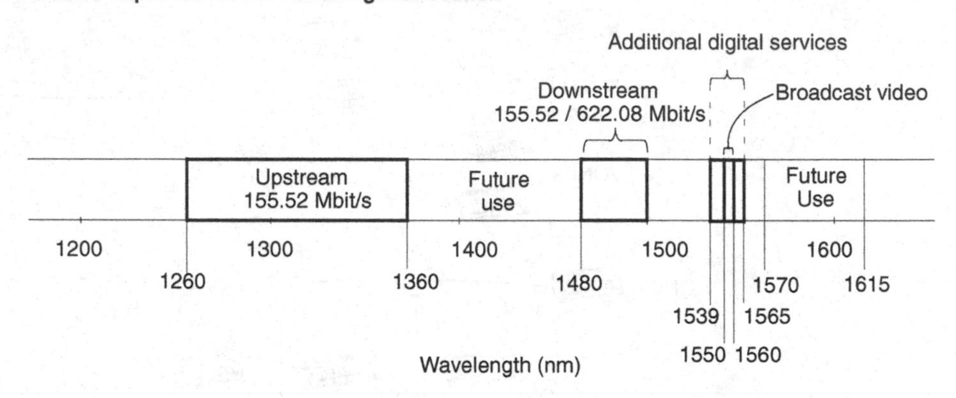

Figure 8-19. Optical bandwidth usage by BPON with wavelength based duplexing of the upstream and the downstream.

common versions are the symmetrical 155.52 Mbit/s and 155.52/622.08 Mbit/s. These are aggregated bit rates that must be shared by all users in the BPON. A single BPON can serve up to 32 ONUs.

Recommendation G.983.1 allows for spatial and wavelength duplexing of both upstream and downstream. In the first case, signals are allocated in the 1260–1360 nm wavelength window, whereas in the second case, the upstream is kept in the same window but the downstream is moved to the 1480–1580 nm window. Recommendation G.983.3 improves this bandwidth allocation by defining two new bands for additional digital services and video broadcasting (see Figure 8-19)

The upstream and downstream in the BPON are continuous streams of cells (see Figure 8-20). The downstream is broadcast and it contains cells for all users. ONUs must extract from the stream the cells addressed to them and discard others. BPON provides downstream privacy by scrambling the contents of every OLT/ONU downstream point-to-point connection with an algorithm that uses a 3 byte key known as churn key. The churn key is provided by the ONUs and updated periodically.

The upstream is shared by all the ONUs using TDM mechanisms. Permission for ONUs to transmit must be granted by the OLT. Grants are inserted downstream in special cells called *physical layer operation, administration and maintenance* (PLOAM) cells. One PLOAM cell is inserted downstream every 28 ATM cells. The OLT grants transmissions to avoid collisions between ONUs and guarantees optimum usage of the transmission channel.

ONUs are subject to transmission delays. The delay may be different for each ONU, depending on the network topology, and the OLT must take this into account to avoid collisions. To do that, the OLT measures the distance to the ONUs, and then requests the ONUs to insert the appropriate delay so that all equivalent ONU/OLT distances are 20 km. This process is called ranging.

8.2.4 Gigabit PON

Gigabit PON or GPON is an ITU-T standard for PON that enables upstream and downstream bitrates above 1 Gbit/s. More specifically, GPON provides 1244.16/1244.16, 1244.16/2488.32 and 2488.32/2488.32 Mbit/s transmission interfaces. Interfaces at 155.52 and 622.08 Mbit/s are also possible for the upstream when higher bitrates are not needed. GPON cannot interoperate with BPON, even if the line rates are the same for the transceivers.

BPON and GPON have similar optical specifications. Both single and double fibre transmission is possible. In the first case, the upstream and downstream are allocated in

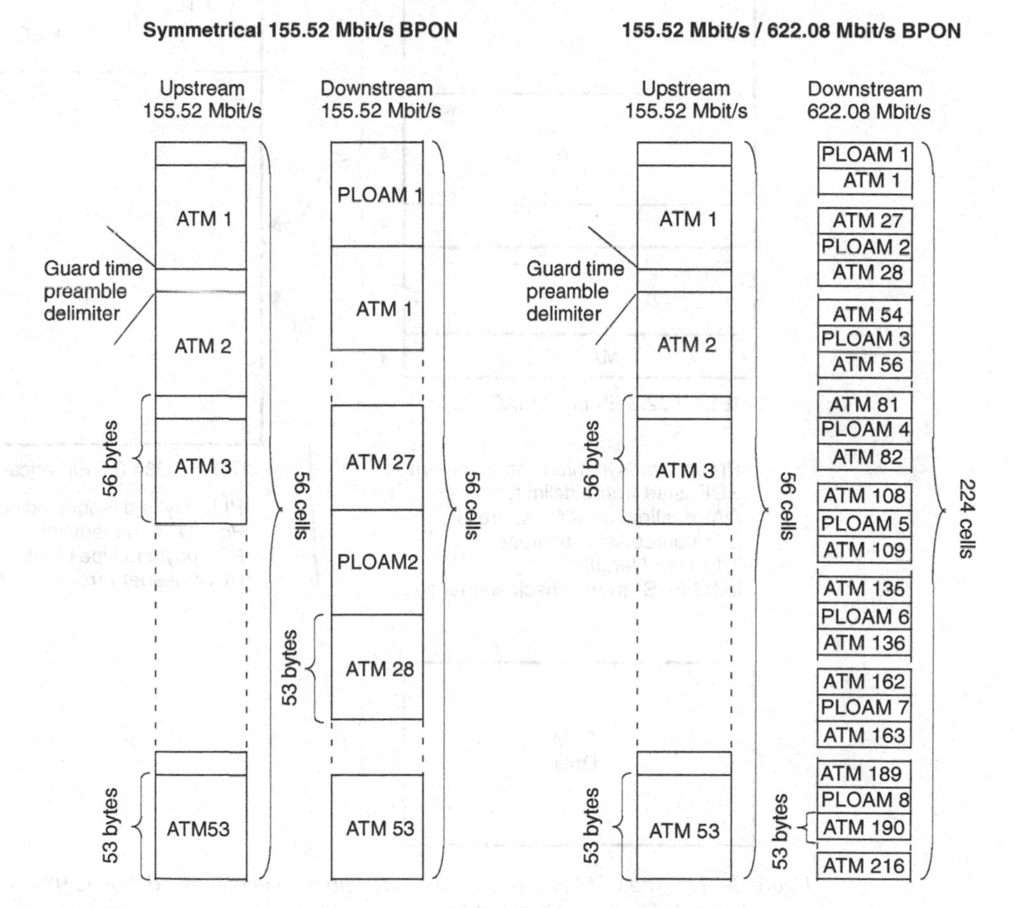

Figure 8-20. Upstream and downstream ATM cell flows in a BPON.

the optical window between 1260 and 1360 nm. In the second case, the downstream is moved to the window between 1480 and 1500 nm. Optical transmission features provide a maximum range of 20 km between the OLT and the ONU.

The main improvement of GPON over the legacy BPON is the new *GPON encapsulation mode* (GEM) that enables transport of TDM and packets like Ethernet without ATM (see Figure 8-21). The GEM is very efficient: it adds just five overhead bytes to encapsulated data, and it can transport variable length packets. Packets can be encapsulated over a single GEM frame, or they can be fragmented and transported over various GEM frames. The GEM has two main objectives:

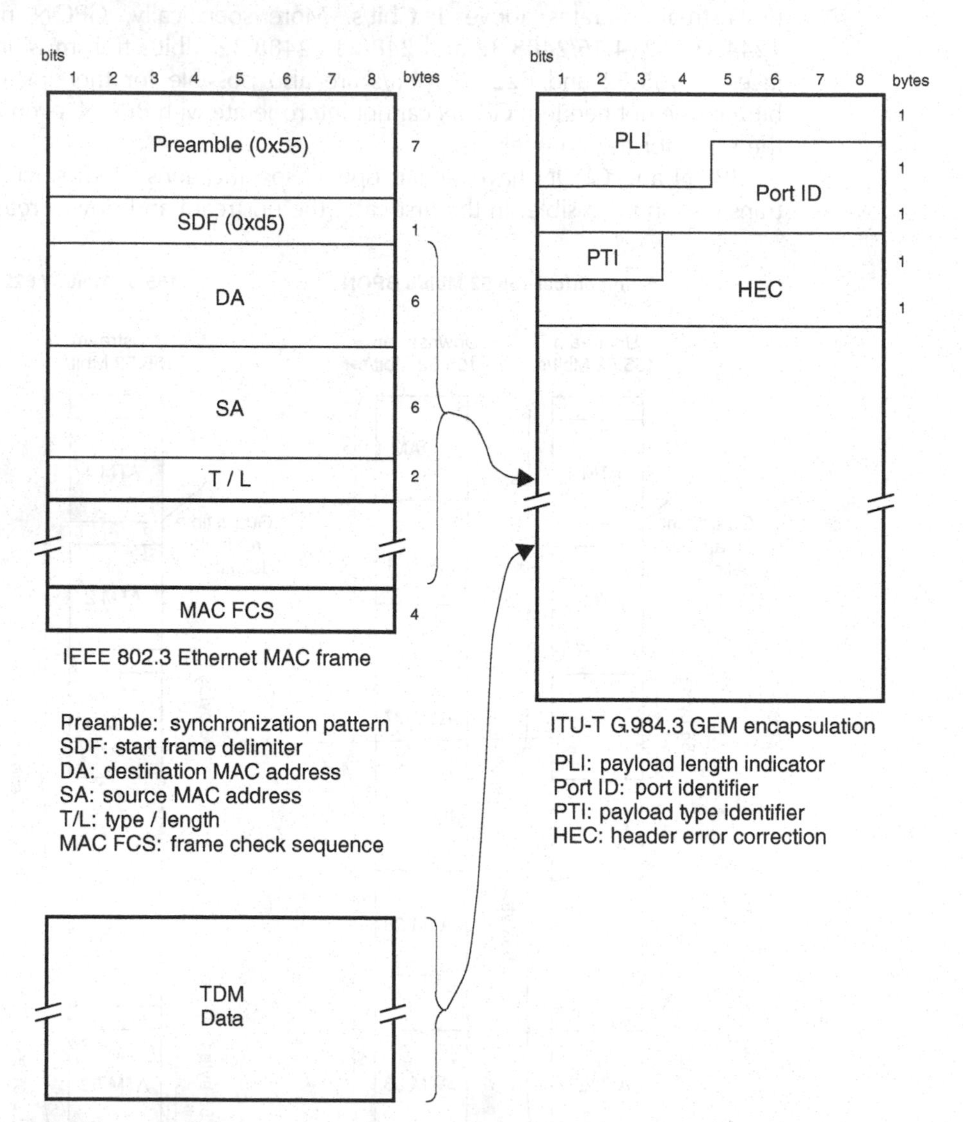

Figure 8-21. The GEM encapsulation, as described in Recommendation G.984.3, enables transport of packetized data as well as TDM.

1. It provides delineation to encapsulated data. To do that, the GEM header contains a 12-bit *payload length indicator* (PLI) that shows to the receiver where the next GEM frame starts. The 15-bit *header error control* (HEC) field detects and corrects errors in the 5-byte GEM header, but it also helps with delineation.

2. It enables multiplexing of various data sources. These data streams are identified by means of a 12-bit label called Port ID.

Although GEM is designed to be the main encapsulation for GPON, ATM is also possible. GPON framing is designed to carry GEM encapsulated data and ATM simultaneously. The GPON downstream is made up of a continuous flow of 125 μs bits frames. A GPON downstream frame includes a *payload control block* (PCBd), an ATM partition that contains ATM cells delivered to the ONUs, and a GEM partition with GEM encapsulated data (see Figure 8-22). The PCB provides different types of overhead for delineation, error detection, PLOAM and other purposes. The PCB also contains a downstream bandwidth map with a pointer list that indicates the time at which each ONU may begin and end its upstream transmission. The upstream is made up of multiplexed transmission bursts from the different ONUs. Every burst starts with a *physical layer overhead* (PLO) that contains fields for delineation, error detection, transmitter identification by means an ONU-ID, and for other purposes. Several contiguous transmissions from the same ONU may be granted. If this happens, the data can be chained in the same burst under the same PLO header.

Besides the PLO, the upstream may contain other overheads by request of the ONUs:

- The PLOAM is a 13-byte overhead containing OAM messages. These messages usually carry alarms, information about the transmitter and encryption passwords and keys.

- The 120-byte *power levelling sequence* (PLS) is inserted upstream to deal with the near–far problem typical of PON. The distance between the OLT and each ONU may be different, and therefore the power received by the OLT from the different ONUs varies. This may affect the ability of the OLT to decode the upstream without errors. The objective of the PLS is to help the OLT to solve this problem.

- The *dynamic bandwidth report* (DBR), may have 1, 2 or 4 bytes of length. Messages sent in this overhead help the OLT to schedule the transmissions sent by the ONU.

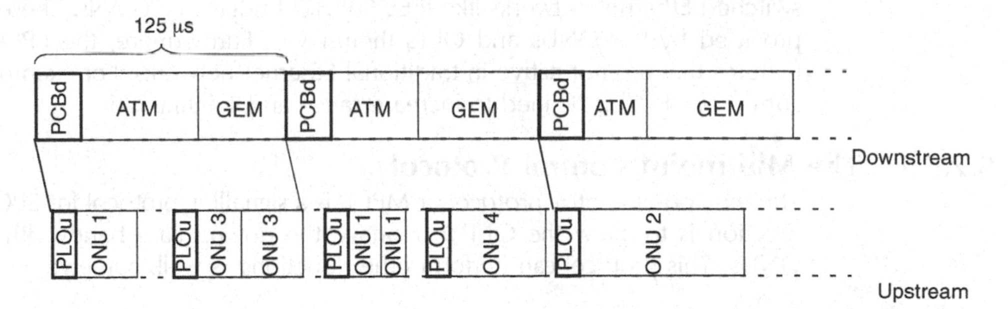

Figure 8-22. Upstream and downstream flows in a GPON. The downstream is composed of a continuous frame stream with ATM and GEM encapsulated data. The upstream contains data bursts from the different ONUs as they have been granted by the OLT in the downstream.

8.2.5 Ethernet PON

The Ethernet PON or EPON is the IEEE alternative for PON. The first version of EPON was released in 2004, which makes this technology the latest PON version to appear at the time of writing. EPON is based on Ethernet, the most successful networking technology specified by the IEEE. In fact, EPON is part of the EFM initiative that attempts to extend Ethernet to the local loop. EPON is a direct competitor of the GPON technology defined by the ITU-T.

There are two alternative interfaces for EPON, known as 1000BASE-PX10 and 1000BASE-PX20. The former has a minimum range of 10 km and the latter 20 km. The typical number of ONUs in an EPON is 16, but alternative splitting ratios are also possible. There is a trade-off between range and splitting ratio, because optical loss increases with both distance and split count. This means that more ONUs can be served if the distance between the ONU and the OLT is shorter.

All the currently defined EPON interfaces are for transmission at 1 Gbit/s, but a 10 Gbit/s EPON standard is expected to be available soon. The EPON upstream and downstream are duplexed in a single SMF fibre. The upstream is transmitted at a nominal wavelength of 1310 nm, and the downstream at 1490 nm. This allows for the EPON to coexist with other services, such as broadcast video or private DWDM transmitted in the 1550 nm window. The signal is encoded with the same 8B/10B code that was specified by most of the Gigabit Ethernet interfaces operating at 1 Gbit/s. This means that the signalling rate for 1 Gbit/s EPON is 1.25 GBd.

The main goal of the 1000BASE-PX physical interfaces is to provide an access point for the connection of MAC entities capable of transmitting standard IEEE 802.3 MAC frames. PON networks are a mixture of a dedicated and shared medium and EPON emulates point-to-point links over this medium. To do that, it extends the traditional Ethernet physical layer by defining:

- a scheduling protocol called *multi-point control protocol* (MPCP) that distributes transmission time among the ONUs to avoid upstream traffic collisions;
- tags known as *logical link identifiers* (LLID) that define point-to-point associations between the ONU and the OLT at physical level.

As a result, the EPON is compatible with most of the advantages provided by switched Ethernet networks like IEEE 802.1D bridging or VLANs. These features can be provided by the ONUs and OLTs themselves. Furthermore, the EPON defines other features that are not native in traditional Ethernet networks. For example, *forward error correction* (FEC) is defined to increase range and splitting ratio.

8.2.5.1 The Multipoint Control Protocol

The *multipoint control protocol* or MPCP is a signalling protocol for EPON, and its main function is to allow the OLT to manage the downstream bandwidth assigned to the ONUs. This protocol can perform other functions as well, namely:

- enable the ONUs to request upstream bandwidth for transmission, and the OLT to assign this bandwidth in such a way that collisions do not occur and network utilization is optimized;

- allow parameter negotiation through the EPON network;
- enable ranging by monitoring the *round trip delay* (RTD) between ONUs and OLT; this feature is important for correctly scheduling upstream transmissions;
- support ONU autodiscovery and registration.

The MPCP is implemented as an extension of the MAC control protocol and therefore MPCP messages are carried over standard Ethernet frames with the Type/Length field set to 0x88-08. There are five MPCP messages currently defined:

- GATE – grants access to the upstream bandwidth for the ONUs for certain periods of time;
- REPORT – used by the ONUs to report local information to the OLT. This information is used by the OLT to decide how the upstream bandwidth is distributed.
- REGISTER, REGISTER_REQUEST and REGISTER_ACK – used for registering ONUs in the network.

The IEEE standards define the protocol for scheduling bandwidth, but equipment manufacturers select the actual scheduling algorithm.

8.2.5.2 Logical Link Identifiers

Logical link identifiers or LLIDs are physical layer link identifiers defined to enable 802.1D bridging over an EPON (see Figure 8-23). The LLID is delivered in EPON Ethernet frames as a 16-bit field that replaces the two last bytes of the frame preamble (see Figure 8-24).

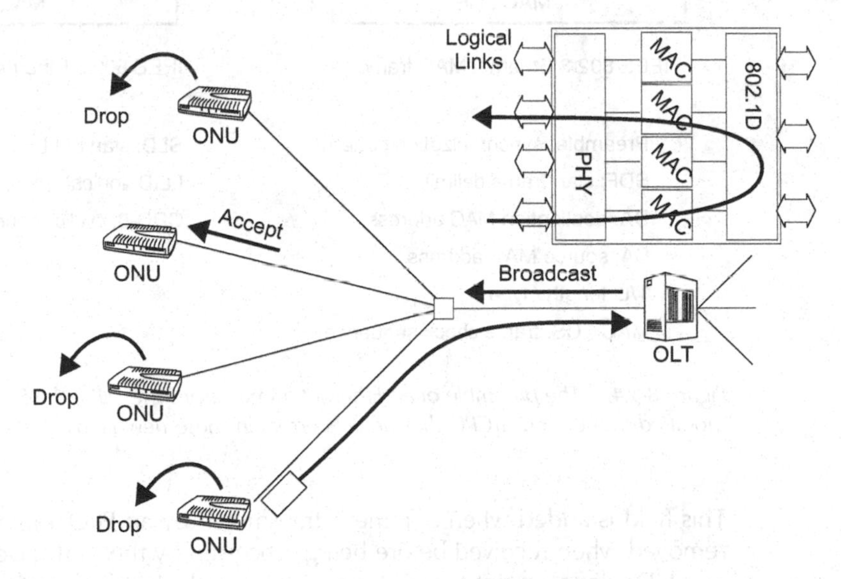

Figure 8-23. ONU-to-ONU bridging would not be possible without LLIDs. ONU-to-OLT associations defined by the LLIDs can be considered as point-to-point logical links. An 802.1D bridge can then perform learning and forwarding operations on the logical links.

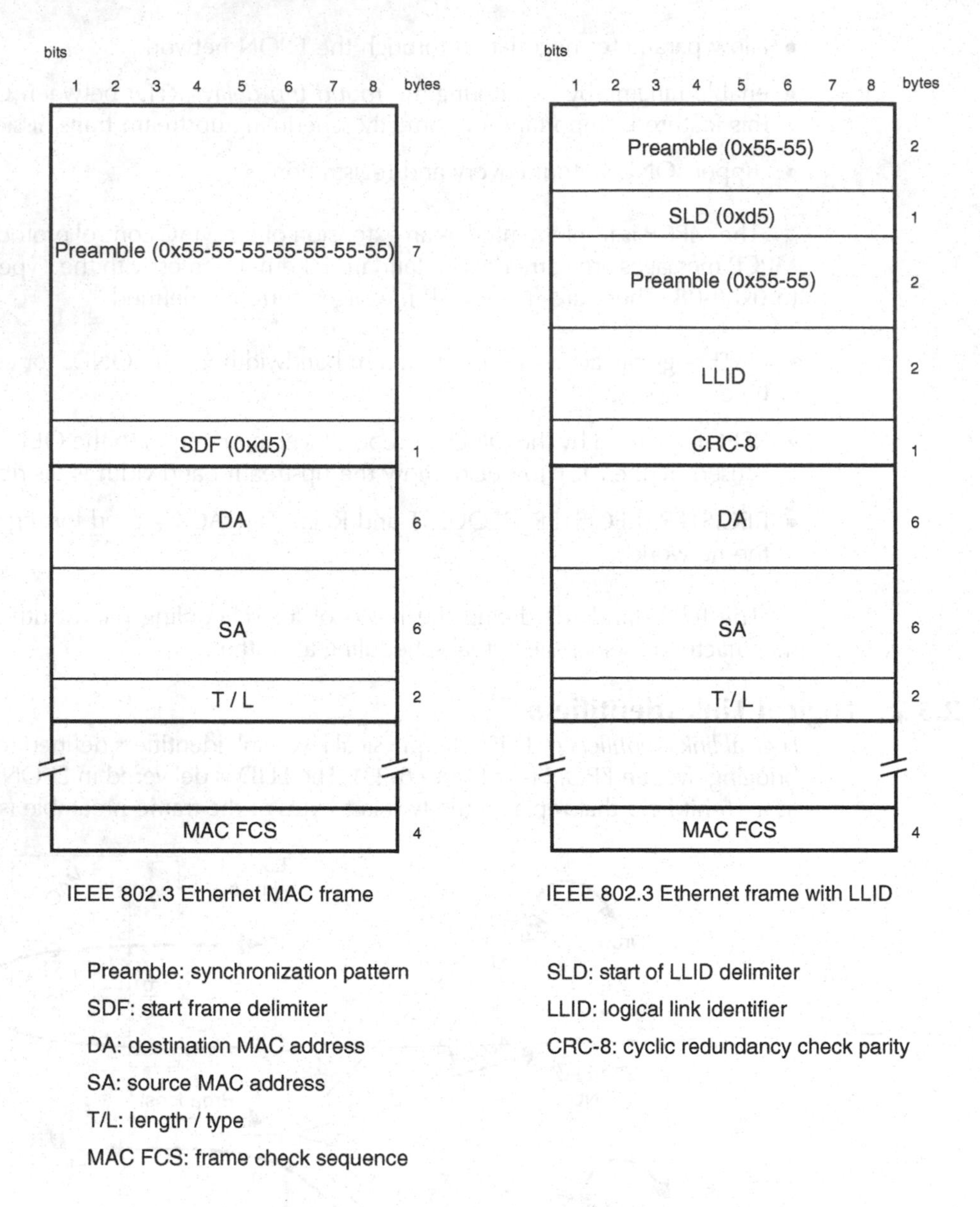

Figure 8-24. The preamble of an Ethernet frame carries the LLID, the SLD that helps processing the modified frame, and a CRC that detects errors in these new fields.

This field is added when a frame is transmitted by an EPON interface and transparently removed when received before being processed by the MAC layer.

LLIDs define point-to-point associations or logical links between the ONU and the OLT. Link identifiers are dynamically assigned when ONUs are registered in OLTs as a part of the initialization process. ONUs and OLTs choose the LLID to put in the delivered

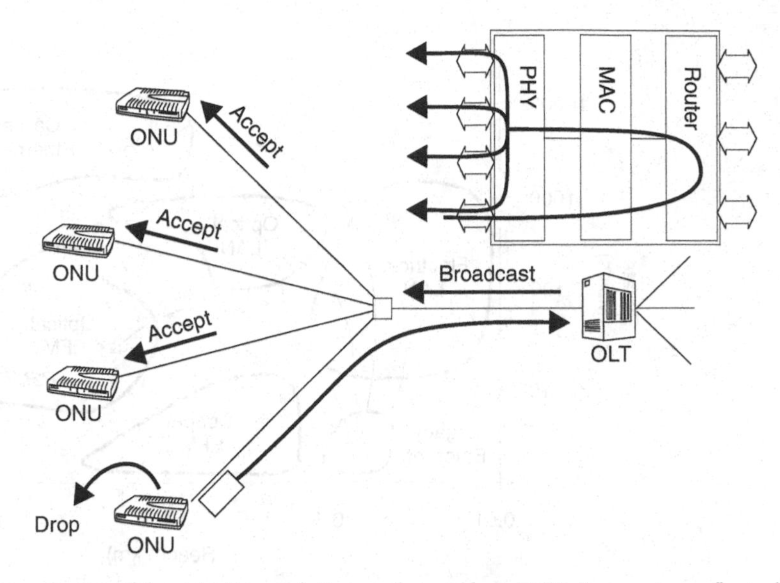

Figure 8-25. Optional shared LAN emulation performed by a PON. Frames are reflected by the OLT and accepted by all the ONUs except for the initial transmitter.

frames depending on the logical link they wish to use. Point-to-point emulation is achieved by following simple filtering rule: if a frame is received by an ONU or OLT with an LLID matching a known link identifier, it is forwarded to the right MAC entity that processes it. Otherwise, the frame is discarded. ONUs need to support a single LLID. They mark outgoing frames with the LLID assigned to them, and they accept frames marked with this LLID. The OLTs are more complex: they need one LLID per connected ONU.

The point-to-point link emulation is the primary operation mode for EPONs, but they may optionally support a shared LAN emulation mode (see Figure 8-25). To do this, it is necessary to modify the LLID filtering rules. The ONU must now accept all the incoming frames, not just the ones marked with their own LLID. The OLT must accept all the incoming frames and reflect them back when they match the defined forwarding rules.

It is also possible to take advantage of the broadcast nature of the downstream by defining a special channel called *single copy broadcast* (SCB) channel. Frames sent by the SBC channel are accepted by all the ONUs.

8.3 Ethernet in the First Mile

The standard IEEE 802.3ah for EFM was released with the aim of extending Ethernet to the local loop for both residential and business customers.

EFM interfaces provide low and medium speeds when compared with the available LAN or WAN standards (see Figure 8-26). The new interfaces, however, are optimized to be profitable in the existing and newly installed provider access networks. The copper EFM

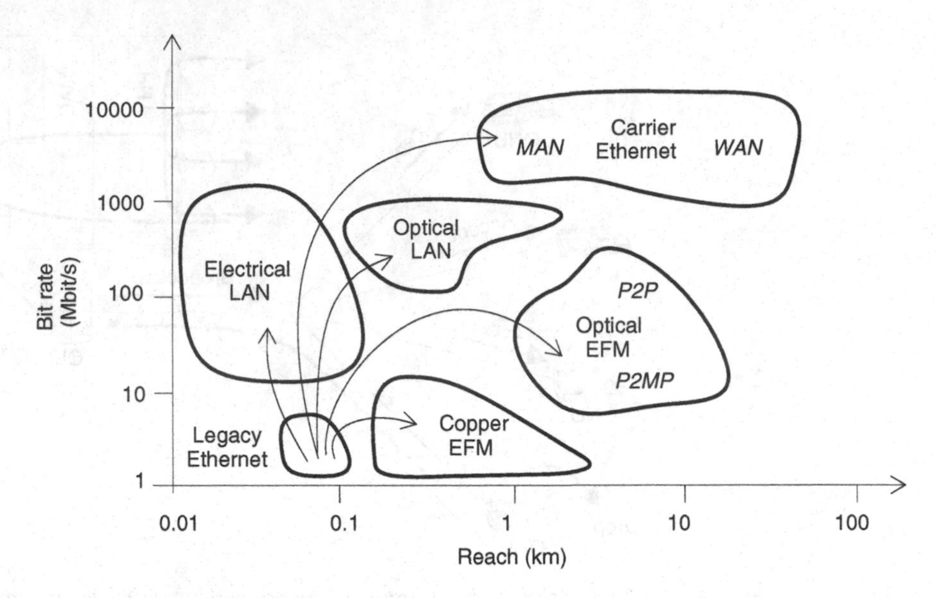

Figure 8-26. Ethernet applications and EFM.

takes advantage of DSL technology for telephone copper pairs, and optical EFM is available for both PON networks and active Ethernet (see Table 8-4).

A major improvement provided by the IEEE 802.3ah is the definition of link *operation, administration and maintenance* (OAM) services. The link OAM enables access network operators to monitor and troubleshoot the Ethernet link between the customer and network operator equipment. This new type of OAM complements OAM signalling at the service level defined in IEEE 802.1ag. The difference between IEEE 802.3ah and IEEE 802.1ag is

Table 8-4. EFM interface summary

Interface	Medium	Wavelength (nm)	Rate (Mbit/s)	Reach (km)
100BASE-LX10	Two single-mode fibres	1310	100	10
100BASE-BX10	One single-mode fibres	1310 (US), 1550 (DS)	100	10
1000BASE-LX10	Two single-mode fibres	1310	1000	10
1000BASE-LX10	Two multimode fibre	1310	1000	0.55
1000BASE-BX10	One single-mode fibre	1310 (US), 1490 (DS)	1000	10
1000BASE-PX10	One single-mode fibre PON	1310 (US), 1490 (DS)	1000	10
1000BASE-PX20	One single-mode fibre PON	1310 (US), 1490 (DS)	1000	20
10PASS-TS	One or more telephone pairs	—	10	0.75
2BASE-TL	One or more telephone pairs	—	2	2.7

that, while the former works at the link level, the latter has been designed for OAM end-to-end signalling.

The functions of the link OAM protocol can be summarized as follows:

- *Discovery* – identifies the devices at each end of the link, along with their OAM capabilities.
- *Link monitoring* – detects and indicates link faults, providing statistics on the registered errors.
- *Remote failure indication* – reports a failure condition detected by the remote peer of a given switch, such as loss of signal in one direction of the link, an unrecoverable error, etc.
- *Remote loopback* – puts the remote peer of a given switch in loopback mode. When a switch is operating in loopback mode, it returns all the traffic it receives back to the origin. The remote loopback mode is very useful for testing purposes.

8.3.1 Ethernet Over Telephone Copper Pairs

The EFM standard defines two interfaces for Ethernet transmission over telephone copper pairs:

- The 2BASE-TL interface is best suited to long-haul applications. It provides a symmetric, full-duplex 2-Mbit/s Ethernet transmission channel with a nominal reach of 2.7 km. It is based on SHDSL as per ITU-T G.991.2. The 2BASE-TL interface is optimized for local exchange applications.
- The 10PASS-TS interface is intended for short-haul applications. It offers a symmetric, full-duplex 10 Mbit/s transmission with a nominal reach of 750 m. It is based on the VDSL (ANSI T1.424) technology and optimized for deep fibre roll-outs like FTTB or FTTCab. It can be combined with EPON or active Ethernet to offer a simple bridged access network. The 10PASS-TS interface is compatible with baseband transmission of analogue voice.

Although the EFM standard is expected to boost the use of Ethernet in provider access networks, Ethernet over copper pairs is really nothing new. The ITU-T ADSL2 standards defined in 2002 a new encapsulation for a *packet transfer mode* (PTM) suitable for Ethernet frame transport. Later the PTM was added to other DSL standards like ADSL2+ and VDSL2. Before the release of ADSL2 and the PTM, Ethernet had to be transported by means the DSL ATM with the mappings defined by the RFC 2684.

The 10PASS-TS and 2BASE-TL are mostly based in existing technology, such as SHDSL and VDSL, mainly for the following reasons:

- Extensive DSL deployments exist and have existed for the past 10 years or so. DSL is a well-known technology, and network operators have a lot of experience with it.

Table 8-5. Copper pair categories

Category	Bandwidth	Common Application
1	—	Telephony, ISDN BRI
2	4 MHz	4 Mbit/s token ring
3	16 MHz	Telephony, 10BASE-T, 100BASE-T4 (four wires)
4	20 MHz	16 Mbit/s token ring
5	100 MHz	100BASE-T, 1000BASE-T (four wires), short haul 155 Mbit/s ATM
5e	100 MHz	100BASE-T, 1000BASE-T (four wires), short haul 155 Mbit/s ATM
6	250 MHz	1000BASE-T (four wires)

- DSL has proven to be efficient, cost-effective and easy to deploy.
- National-level spectrum compatibility standards make it difficult to introduce signals with new spectrum shapes.

One of the challenges of Ethernet over copper is the lack of a strict definition of what is understood by a voice-grade copper pair. The reason for this is that telephone cabling started in the nineteenth century, long before any telecommunication regulations. Most of the current telephone pairs fall into the TIA/EIA categories 1 and 3 (see Table 8-5). Unlike other Ethernet standards, 2BASE-TL and 10PASS-TS are not specified for a transmission media of known features, and therefore the performance of these interfaces remains largely unpredictable in untested cables.

One of the few changes introduced by the IEEE in the DSL specifications was the encapsulation defined for Ethernet. The original ITU-T encapsulation was based on an HDLC framing, but HDLC needs to add extra overhead bytes to avoid false alignment sequences within the frame. As a result, the length of an HDLC frame depends on its contents, and this makes it difficult to provide deterministic data bitrate. The HDLC encapsulation was replaced by the new 64/65-octet encapsulation. Finally, the 64/65-octet encapsulation has been accepted by the ITU-T, and it seems that it will be the main encapsulation for Ethernet over DSL in the near future.

The 64/65-octet encapsulation operates on 64-byte traffic fragments. It encodes both data and idle times. The traffic to be encoded is fragmented in 64-byte blocks. A single sync byte is added to the 64-byte fragments, producing a continuous stream of 65-byte blocks.

Another important feature of the EFM interface for copper is the *bonding function*. This feature is useful in providing Ethernet services over copper without the severe rate limitations. It can replace fibre in places where fibre is not available.

Ethernet bonding allows up to 32 pairs to be grouped together to make them appear as a single-capacity transmission channel (see Figure 8-27). Ethernet bonding works on heterogeneous links, since it is independent of the physical layer.

Bonding is possible even if the links operate at different speed. The IEEE 802.3ah bonding constitutes an Ethernet-optimized alternative to ATM-based bonding (see Section 8.1.4).

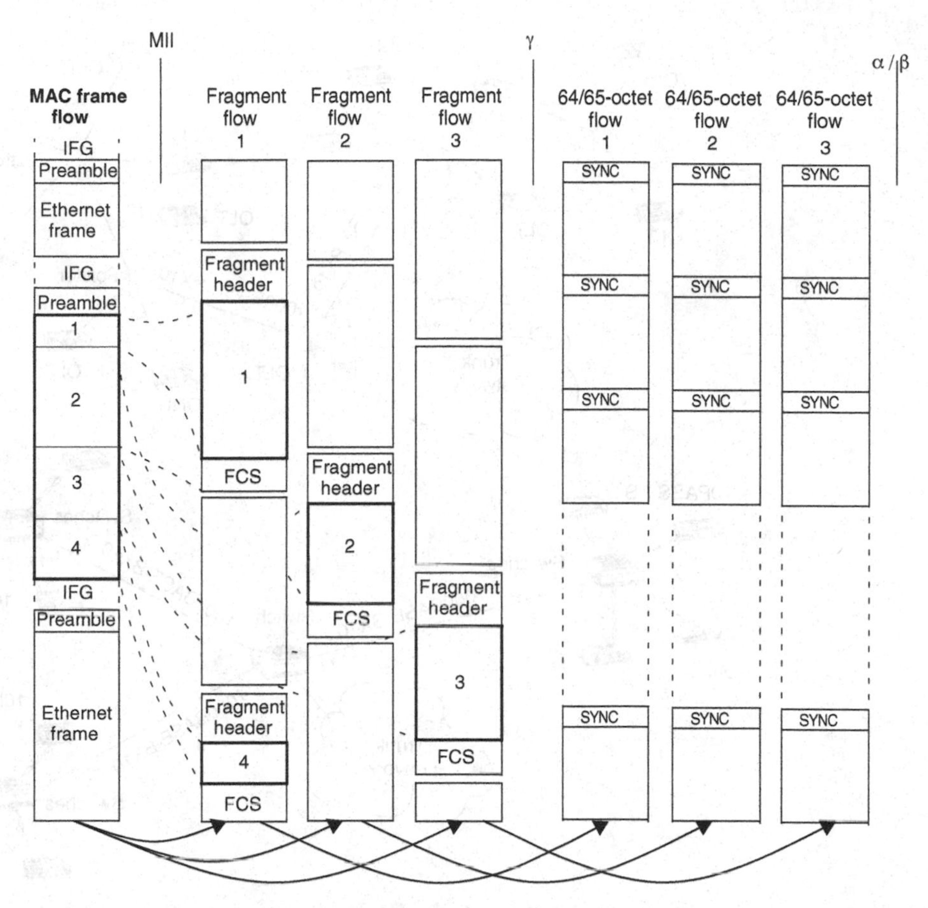

Figure 8-27. 64/65-octet coding of an Ethernet-bonded interface made up of three physical links.

8.3.2 Ethernet in Optical Access Networks

Optical EFM interfaces provide better performance than copper EFM in terms of reach and bit rate, but they require optical fibre. These interfaces have been specially developed for deep-fibre rollouts based on *point-to-point* (P2P) and *point-to-multipoint* (P2MP) architectures.

In the case of the P2MP architecture, the EFM interface offers EPON (see Section 8.2.5). For P2P, the EFM adapts the available Ethernet interfaces so that they operate in the access network. For example, bidirectional interfaces take advantage of the WDM technology to duplex the upstream and downstream in a single fibre. This makes it unnecessary to install two fibres per customer. Extended temperature operation is another improvement that is important for external plant operation.

The existing EPONs provide 1 Gbit/s symmetrical capacity, typically to be shared by 16 subscribers. This means that the minimum guaranteed bandwidth in FTTH is around 60 Mbit/s per subscriber, but depending on the network load, it could increase up to several

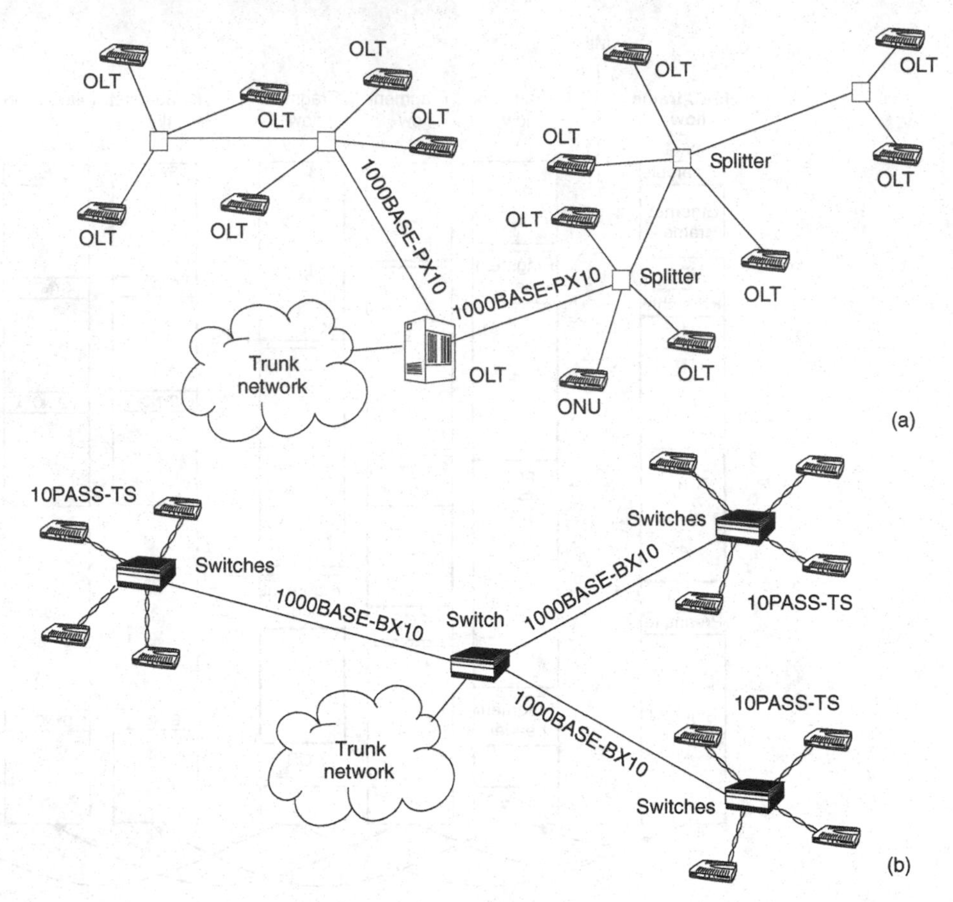

Figure 8-28. Two rollout alternatives for residential customers based on EFM interfaces: (a) deployment based on a shared EPON; (b) deployment based on dedicated P2P optical links and short haul copper links.

hundreds of megabits per second. The P2P interfaces for active Ethernet roll-outs provide 100 Mbit/s per customer in FTTH. Gigabit interfaces also exist, but these are typically used for backhaul in fibre-to-the-neighbourhood applications, or they may be combined with copper in FTTB deployments (see Figure 8-28).

8.4 Service Provisioning

The legacy model for offering Internet access in DSL networks is quite simple. Connections are managed by the *point-to-point protocol* (PPP). This protocol performs user authentication and IP address assignment. PPP connections are usually terminated in the *remote access server* (RAS), a router managed by an ISP. In the local loop, IP data is encapsulated in an ATM *permanent virtual circuit* (PVC). The metropolitan network is usually based on ATM or FR. Subscriber PVCs are forwarded to the ISP *points of presence* (PoP) by this network.

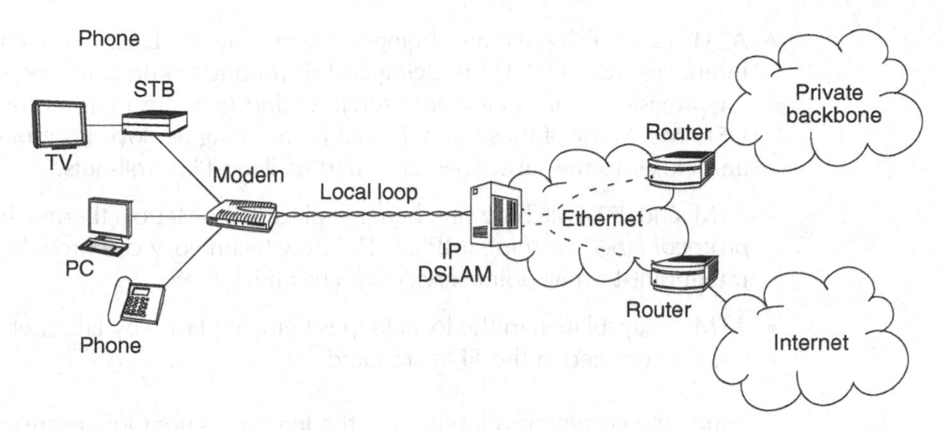

Figure 8-29. Network architecture for Triple Play service to residential subscribers. ATM and FR are replaced by Ethernet and IP.

The evolution of the access network makes it appear similar to a big LAN with multimedia capabilities (see Figure 8-29). This affects the service provisioning model in several ways, namely:

- The *dynamic host configuration protocol* (DHCP) tends to be the chosen protocol for session management instead of PPP. DHCP is better suited for non-connection-oriented technologies, such as Ethernet and IP. In order to provide security, DHCP must implement the relay agent identification option (option 82) defined in RFC 3046. With this option, the DHCP relay (usually the DSLAM) can forward information about itself and about its connection to the subscriber. This information makes it possible for the DHCP servers to implement security policies based on the port where the subscriber is connected instead of its MAC address (see Figure 8-30).

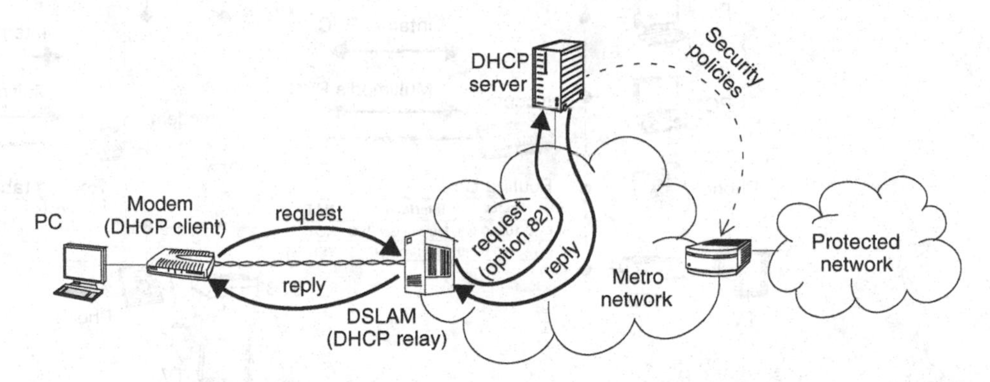

Figure 8-30. Usage example of DHCP with option 82 for providing connection parameters to users in a WAN. The DSLAM adds the option 82 to DHCP requests to inform the DHCP server on the port where the user is connected. This information is used by the DHCP server for providing the right information to that user and grant access to protected resources.

- ATM-based DSLAMs are being replaced by IP DSLAMs with Gigabit Ethernet interfaces, IEEE 802.1D bridging and IP routing features. IP DSLAMs are well suited for provision of IP multimedia services, and they are more cost-effective than ATM DSLAMs. Many of these new DSLAMs are being deployed out of the local exchange and closer to the subscribers as a part of deep-fibre roll-outs.
- ATM and FR switching are being replaced by Metro Ethernet bridging and *multi-protocol label switching* (MPLS). The new technology can provide point-to-point and multipoint-to-multipoint services of customize QoS.
- ATM encapsulation in the local loop is being replaced by Ethernet encapsulations like the one defined in the EFM standard.

Since the current development of the Internet is not QoS-aware, many operators are developing their own private backbones to deliver IPTV, VoD and other QoS-sensible services. These private backbones are carefully designed IP networks with support for QoS by means of IS, DS, MPLS or other mechanisms. Using private addressing for this private network segregated from the Internet helps to save scarce public IPv4 addresses. The network becomes unroutable from the Internet, but this is not a problem, because it does not need to be part of the Internet. Access to private backbones for multimedia service provisioning can be solved by using several PVCs or VLANs to deliver different services and allocating one of them for best-effort Internet access. It is even possible to use different mechanisms for IP address assignation in different PVCs. For example PPP can be used in the Internet PVC, and DHCP in PVCs for multimedia applications.

Another advantage of segregating Internet and multimedia PVCs is that it is possible to disable *network address translation* (NAT) only on the multimedia PVCs, thus improving transparency in service-provider applications while maintaining optimum usage of public

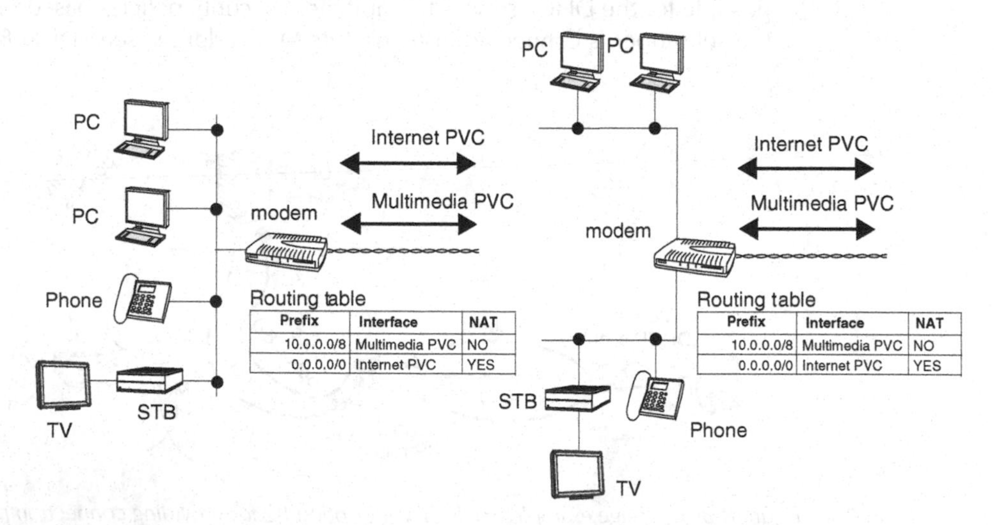

Figure 8-31. Two options for the subscriber LAN: (a) model based on a single network. Multimedia devices and PCs are connected to the network. (b) Model based on two subnetworks. One of them is for multimedia devices, the other for PCs.

addresses for the Internet access application. Disabling NAT is almost necessary in the multimedia PVCs, because some protocols that can be potentially used in this interface may have problems passing through the NAT filter. This happens, for example, with the *session initiation protocol* (SIP).

There are, however, certain issues if NAT is disabled in multimedia PVCs. The devices attached to the subscriber network become directly routable from the service provider private backbone. This means that the IP addresses of these devices must be unique, and they must be managed by the service provider rather than by the subscriber. When this approach is used, the service provider limits the maximum number of devices that can be connected to the subscriber LAN.

Some service providers deploy two different subnetworks for each subscriber. The first of them contains private addresses routable from the private backbone. This network is used for connecting multimedia devices. The second network has private addresses not routable from the private backbone, and it is used for Internet access (see Figure 8.31). With this second approach, the subscriber keeps full control over the Internet access subnetwork. The problem is that this solution makes it difficult to receive Internet-related contents in the multimedia subnetwork, or multimedia-related contents in the Internet access subnetwork.

Selected Bibliography

[1] Sargento S., Valadas R., Gonçalves J., Sousa H., IP-Based Access Networks for Broadband Multimedia Services, *IEEE Communications Magazine*, February 2003, pp. 146–154.

[2] Kerpez K., DSL Spectrum Management Standard, *IEEE Communications Magazine*, November 2002, pp. 116–123.

[3] Kerpez K., Waring D., Galli S., Dixon J., Madon P., Advanced DSL Management, *IEEE Communications Magazine*, September 2003, pp. 116–123.

[4] Kramer G., Pesavento G., Ethernet Passive Optical Network (EPON): Building a Next-Generation Optical Access Network, *IEEE Communications Magazine*, February 2002, pp. 66–73.

[5] Kramer G., Mukherjee B., Pesavento G., IPACT: A Dynamic Protocol for an Ethernet PON (EPON), *IEEE Communications Magazine*, February 2002, pp. 74–80.

[6] Effenberger F., Ichibangase H., Yamashita H., Advances in Broadband Passive Optical Networking Technology, *IEEE Communications Magazine*, December 2001, pp. 118–124.

[7] Maeda Y., Okada K., Faulkner D., FSAN OAN-WG and Future Issues for Broadband Optical Access Networks, *IEEE Communications Magazine*, December 2001, pp. 126–132.

[8] Ueda H., Okada K., Ford B., Mahony G., Homung S., Faulkner D., Abiven J., Durel S., Ballart R., Erikson J., Deployment Status and Common Technical Specifications for a B-PON System, *IEEE Communications Magazine*, December 2001, pp. 134–141.

[9] Pesavento G., Ethernet Passive Optical Network (EPON) architecture for broadband access, Optical Networks Magazine, January/February 2003.

[10] Eriksson P., Odenhammar B., VDSL2: Next important broadband technology, *Ericsson Review*, No. 1, 2006, pp. 36–47.

[11] ITU-T Recommendation G.992.3, Asymmetrical digital subscriber line transceivers 2 (ADSL2), January 2005.

[12] ITU-T Recommendation G.993.1, Very high speed digital subscriber line, June 2004.

[13] ITU-T Recommendation G.998.1, ATM-based multi-pair bonding, January 2005.

[14] ITU-T Recommendation G.998.2, Ethernet-based multi-pair bonding, January 2005.

Chapter 9: Quadruple Play

The expression 'Quadruple Play' refers to the commercial bundle of voice, video and data plus mobility. Quadruple Play makes sense because mobile operators are aiming to add to their offer advanced data and video services. The voice-centric business model of mobile operators is becoming much more rich and complex. The objective is to provide similar services to wireline operators but over mobile terminals. Multimedia conferencing or mobile TV are seen by mobile operators as strategic services. Online banking, shopping or gaming are popular for wireline subscribers but they are becoming popular for wireless users as well.

On the other hand, fixed operators are looking for mechanisms to add mobility to their services. The indoor home connection to the Internet based on wireless LAN (WLAN) rather than Ethernet or other wireline technology is now very common. Another example are mobile phones that become WLAN phones when a WLAN network is available.

Clearly, there have been two differentiated Quadruple Play strategies:

- *Migration of all services to mobile interfaces or fixed mobile substitution (FMS).* The basis of this strategy is that users prefer mobile services if price and performance are at the same level as wireline. It is assumed that mobile technology evolution will enable high service level and specifically high bandwidth interfaces. Mobile access networks are often considered easier to deploy than wireline networks. However, price for operation over licensed radio bands and user equipment price could influence the price of mobile

Triple Play: Building the Converged Network for IP, VoIP and IPTV Francisco J. Hens and José M. Caballero

services. The all-over-mobile strategy is easier to follow by pure mobile operators without fixed facilities.

- *Integration of fixed and mobile business divisions or fixed mobile convergence (FMC).* Integrated operators are able to provide services both over wireline and wireless interfaces. Ideally, the services would be the same without regard to the particular access network. Of course, different access networks may provide different performance levels or may be better suited to certain services. The control plane will provide the mechanisms needed to adapt services to the particularities of every access network. This strategy is good for companies that already have mobile and fixed business divisions. Many incumbent operators are examples of this.

Both the FMS and the FMC strategies require important shifts in business models and technology involving service definition, charging, roaming between operators, connection with service platforms and many others. The telecommunications industry is looking for the mechanisms that will allow interconnection of services and subscribers through a heterogeneous network made up of several wireline and wireless access technologies. Some of the most important initiatives are the standardization of an IP Multimedia Subsystem (IMS) by the 3rd Generation Partnership Project (3GPP) and the work on the Next Generation Network (NGN) led by the ITU-T and ETSI.

9.1 Cellular Communications Overview

Modern mobile communication networks are cellular. In such networks, mobile users are served by fixed radio transceivers called base stations. Base stations cover areas that are geographically limited, but if all base stations are considered together, they are able to serve a geographically wide area. As users move, communications are transferred between different base stations in a process called handoff. Therefore, the handoff mechanisms constitute the basis of mobility in cellular networks.

The first commercial cellular communications systems date to the early 1980s. These 1G systems were based on an analogue user plane and used different frequency bands to multiplex voice calls and allow multiple access. Many 1G mobile telephone systems were created in different countries, usually with very limited interoperability. Some examples of 1G mobile telephone systems are the Advanced Mobile Phone System (AMPS) for the American market, Total Access Communication System (TACS) for UK or the Nordic Mobile Telephone (NMT) for Northern Europe.

Analogue 1G mobile telephone systems were replaced by 2G digital systems by the end of 1980s. The new generation brought more quality, capacity and security at lower cost. In the USA, Canada, and other countries, AMPS evolved to the system known as time division multiple access (TDMA[1]) or digital AMPS (DAMPS). The original standard for TDMA/DAMPS is the Interim Standard-54 (IS-54). The IS-54 was later improved by the IS-136. Mobile communications based on code division multiple access (CDMA) were introduced

[1]The TDMA/DAMPS takes its name from the TDMA multiple access technology but the two concepts must not be confused.

by Qualcomm more or less at the same time as TDMA/DAMPS, and the two technologies became competitors. The CDMA standard IS-95 was approved in July 1993 by the Telecommunications Industry Association (TIA) and has been marketed with the branch name of cdmaOne in the USA and a number of other countries in North America, Latin America, and Asia. CDMA enables multiple access by associating different users, different and orthogonal code sequences rather than using time sequences like in TDMA or different slices of the frequency spectrum as is done with frequency division multiple access (FDMA). Traffic from every user in a CDMA system can be recovered thanks to the particular properties of orthogonal sequences. Since its introduction for commercial mobile communications in the 1990s, CDMA has represented a success due to the better performance exhibited when compared with its other 2G competitors. Particularly, cdmaOne systems brought broader coverage, better spectrum usage and lower power emission. Thanks to its features, CDMA technology is used in 3G mobile networks as well.

The Japanese 2G mobile telephony standard is the Personal Digital Cellular (PDC) system. The PDC standard was released in 1991 and NTT DoCoMo launched the first network based in PDC technology in 1993. PDC operates at the 800 and 1500 MHz bands and uses TDMA multiple access technology. PDC was successful only in Japan. After peaking at 80 million lines, it is currently being made obsolete by 3G technologies.

However, the most successful 2G mobile communications system was developed in Europe. In 1982, the *Conférence Européenne des aministrations des Postes et des Télécommunicacions*, created the *Groupe Spécial Mobile* (GSM) with the objective of developing a European standard for mobile telephony. Later, in 1989, this responsibility was transferred to the recently created European Telecommunications Standards Institute (ETSI). In 1990 the ETSI published phase I of the GSM specifications and the first commercial GSM service saw the light in 1991 in Finland. Since then, the number of GSM subscribers and networks has been growing steadily. Today, GSM has become almost the *de facto* standard for mobile communications and there are GSM networks in most countries. The number of GSM connections is approaching 2500 million (see Figure 9-1) and four out

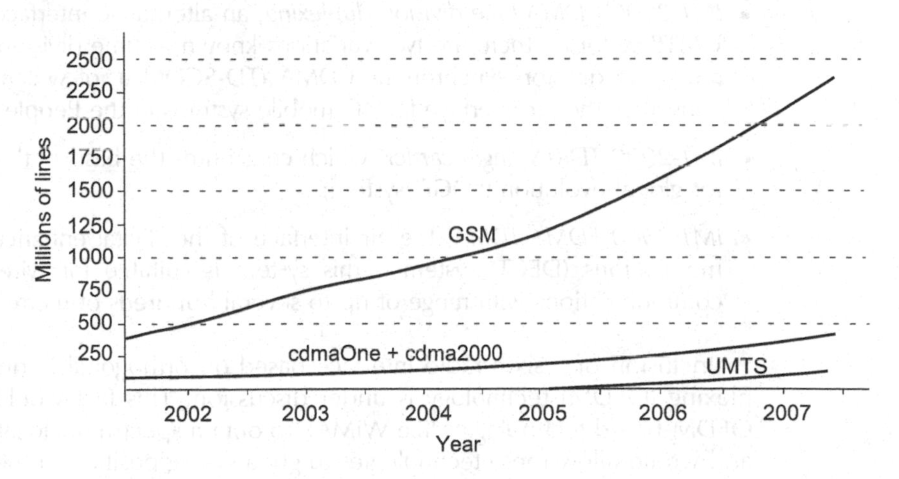

Figure 9-1. Evolution of the worldwide cellular telephony market.

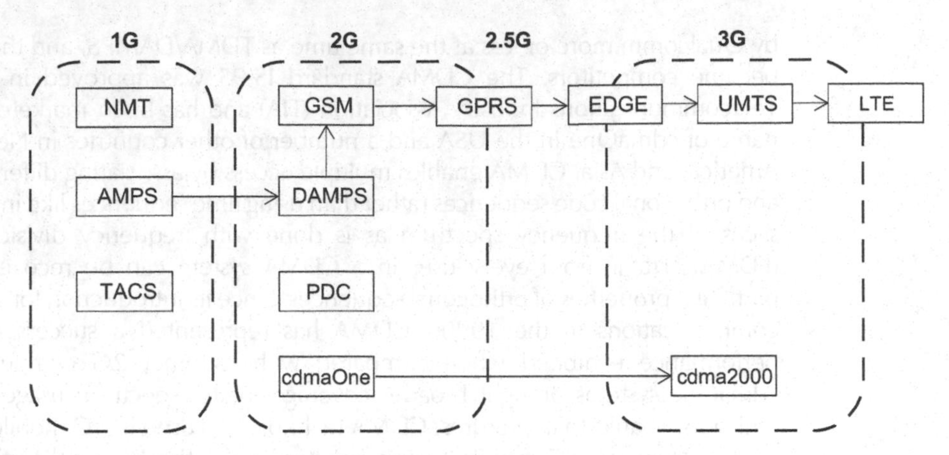

Figure 9-2. Evolution of mobile communications standards.

of every five mobile phones have GSM connectivity. These are the reasons why GSM has been renamed the *Global System for Mobile Communications*.

The driver for bringing onto the market 3G mobile systems was the demand for multimedia and data services for mobile devices. A global standard for 3G mobile systems was unsuccessfully sought in the late 1990s. Finally, the ITU-T standard for 3G mobile communications, known as International Mobile Communications 2000 (IMT-2000), accepted five different radio interfaces:

- *IMT-2000 CDMA Direct Spread*, also known as wideband CDMA (WCDMA) – this radio interface constitutes the basis of the 3G Universal Mobile Terrestrial System (UMTS);
- *IMT-2000 CDMA Multi-Carrier*, the radio interface for the cdma2000 system, successor of Qualcomm cdmaOne;
- *IMT-2000 CDMA time division duplexing*, an alternative interface to WCDMA for the UMTS system – there are two variations known as time division CDMA (TD-CDMA) and time division synchronous CDMA (TD-SCDMA), of which the latter has been chosen as the air interface for 3G mobile systems in the People's Republic of China;
- *IMT-2000 TDMA single-carrier*, which constitutes the basis of the enhanced data rate for global evolution (EDGE) system;
- *IMT-2000 FDMA/TDMA*, the air interface of the digital enhanced cordless telecommunications (DECT) system – this system is suitable for wireless voice and data communications with range of up to several hundreds of metres.

Inclusion of a sixth radio interface based on orthogonal frequency division multiplexing (OFDM) technology is under discussion. This fact would open the door for OFDM-based technologies like WiMAX to obtain spectrum allocations from the ITU-R and would allow these technologies to get a strong position in the definition of future mobile systems.

In practice, two important standards for 3G mobile communications systems arose from the IMT-2000 initiative: the 3GPP UMTS and the cdma2000 defined by the 3GPP2. UMTS takes GSM as a starting point. Some of the important features of UMTS are a new radio access network for increased bit rates in the air interface and a multiservice packet switched core. On the other hand, cdma2000 is the 3G successor of cdmaOne and it keeps the IS-95 1.25 MHz radio channel structure in all its variations:

- *cdma2000 1x* is in fact not a true 3G technology. It is rather a 2.5G mobile system. Performance was limited to symmetrical peak rates of 144 kbit/s in the first release but it was later improved to 307 kbit/s.
- *cdma2000 1xEV-DO*, where EV-DO means 'evolution – data only' or 'evolution – data optimized' and is a true 3G technology. Initially it provided peak rates of 2.4 Mbit/s in the downlink and 153 kbit/s in the uplink (later it improved to a maximum of 3.1 Mbit/s in the downlink and 1.8 Mbit/s in the uplink). The newest EV-DO release (Revision B) improves performance by aggregation of multiple radio channels in a single traffic channel.
- *cdma2000 1xEV-DV*, where EV-DV means 'evolution – data and voice' was designed to replace EV-DO, but has not got enough support from vendors and carriers so far.

9.1.1 The Global System for Mobile Communications

The advantages of the GSM telephony system over 1G technologies and other 2G technologies are a good quality voice service, attractive prices, the short message service (SMS) text-based application and extensive roaming services with many operators around the world.

9.1.1.1 Overview of the Air Interface

The GSM air interface, known as U_m, is a radio link between mobile stations and base stations. In most GSM networks, these radio links operate either in the 900 MHz or in the 1800 MHz bands. However, in some American countries, including the USA, 850 and 1900 MHz are used instead. There are also some GSM networks operating in the 450 MHz band working in Scandinavia, Eastern Europe and Russia (see Figure 9-3).

The basic GSM system uses two 25 MHz bands placed between 890.0 and 915.0 MHz for the uplink (link between the mobile station and the base station) and 935.0 and 960.0 MHz for the downlink (link between the base station and the mobile station). Each 25 MHz band contains 124 radio channels with 200 kHz carrier spacing. A full duplex radio channel is made up of an uplink and a downlink carrier pair. The GSM systems operating at 1800 MHz have 75 MHz for the uplink and 75 MHz for the downlink with 374 pairs of radio channels. However, not all carriers can be used simultaneously by every base station in the system. To avoid interferences only subsets of carriers can be assigned at every base station (see Figure 9-4).

Every radio channel in the GSM system transports an eight-slot time division multiplexing (TDM) frame. Each time interval in this frame is accessed by users with TDMA techniques and each radio channel in the GSM spectrum is accessed using FDMA. Multiple access in GSM is therefore a combination of FDMA and TDMA.

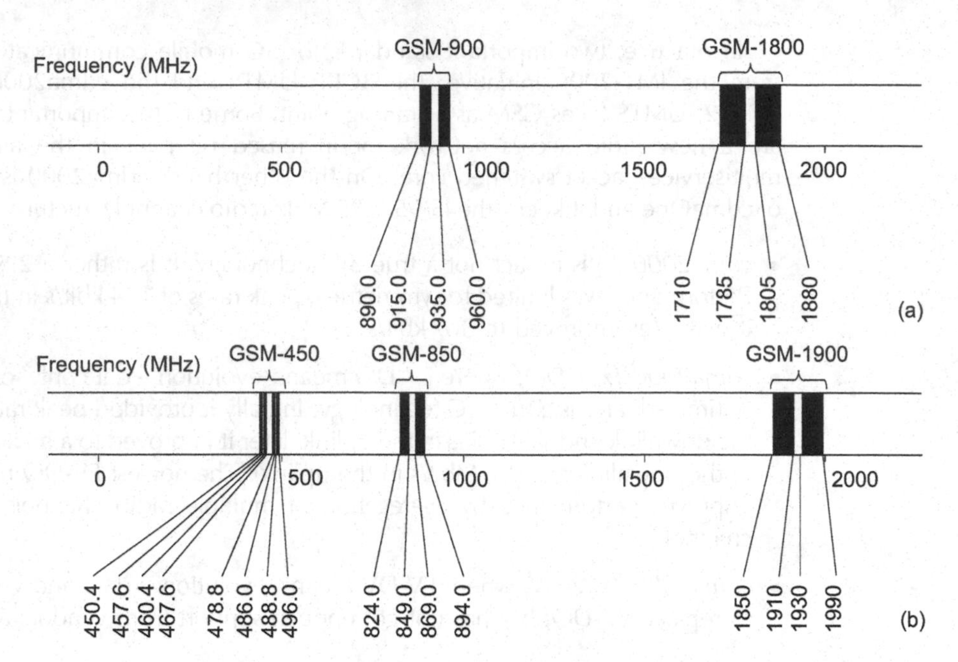

Figure 9-3. Frequency allocations of GSM cellular telephony system. (a) Common GSM-900 and GSM-1800 systems. (b) Less common GSM-450, GSM-850 and GSM-1900.

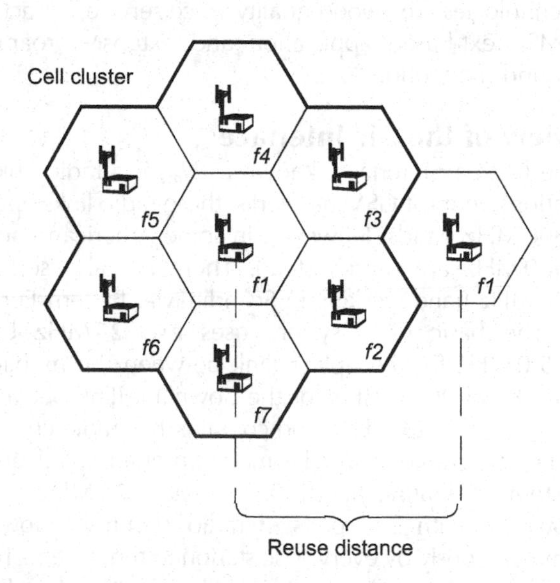

Figure 9-4. GSM cellular radio network structure. To minimize interferences not all the available carriers are used simultaneously. In the example, 124 available pairs of carriers are distributed in seven sets of 17 frequencies each. Carrier sets are assigned to the base stations of a seven-cell cluster. The structure of the cluster is then extended over a plane region of arbitrary dimensions. The maximum simultaneous number of calls a cluster can carry is 136 (17 × 8).

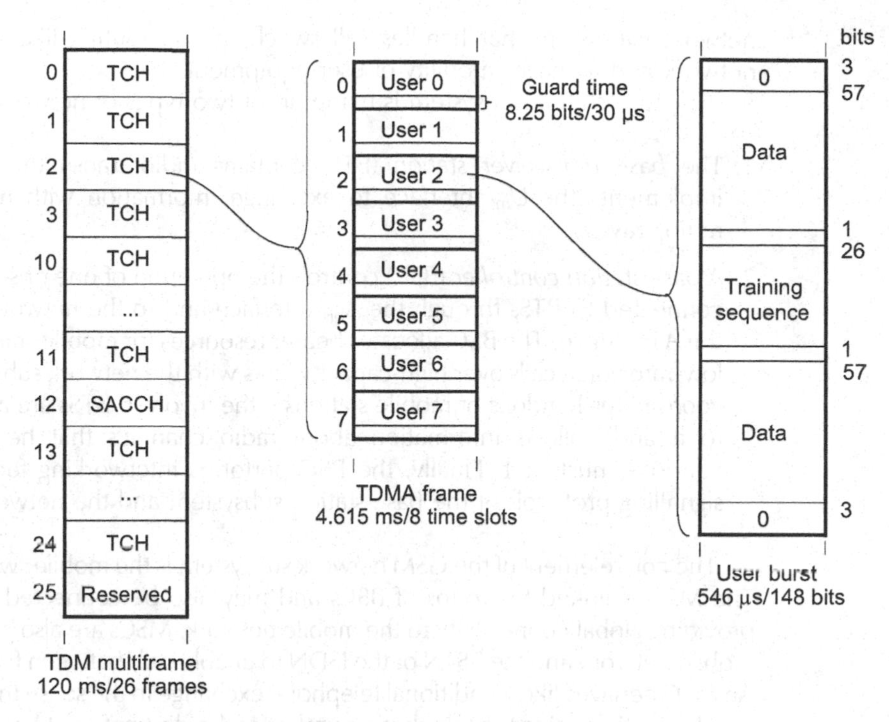

Figure 9-5. Simplified GSM framing structure.

The eight-slot TDM frame of the GSM system transports traffic channels (TCH), slow associated control channels (SACCH) and sometimes fast associated control channels (FACCH) in a multiframe structure. The user bits contained in one TDMA time slot are called a burst. A burst contains 148 bits but only two groups of 57 bits are usable. TDMA bursts are separated by guard times equivalent to 8.25 bits to prevent collisions (Figure 9-5). A TCH can place up to 22.8 kbit/s of user data but there also exists a half-rate TCH operating at 11.4 kbit/s. Even working at 22.8 kbit/s, a TCH cannot transport directly an ISDN 64 kbit/s speech signal. This problem is solved with the help of voice compression algorithms that reduce the speech bit rate to a maximum of 13 kbit/s and provide reasonable quality. Before being transmitted, the encoded voice is protected against bit errors with redundancy bits and the bitrate is increased from 13 to 22.8 kbit/s.

The GSM modulation is the Gaussian minimum shift keying (GMSK). The GMSK is a single carrier, phase digital modulation. In GSMK, the modulating signal is pre-filtered with a filter with gaussian impulse response. The GSMK provides good spectrum efficiency (about 1 bit/Hz), a compact spectrum with low radiation in adjacent radio bands, and a constant envelope signal in the time domain that allows the use of simple and efficient power amplifiers.

9.1.1.2 The GSM Network

A GSM network is made up of the mobile stations carried by users, the base station subsystem that controls the radio links between the network and mobile stations, and the network subsystem that handles call switching and routing like any other telephone network and manages mobility of user equipment.

The base station subsystem is made up of two types of network elements:

1. The *base transceiver station* (BTS) contains radio transmitters and receivers and implements the U_m interface to exchange information with mobile stations using radio waves.
2. A *base station controller* (BSC) controls the operation of one or several BTSs. BSCs are connected to BTSs through the A_{bis} interface and to the network subsystem through the *A* interface. The BSC allocates bearer resources for mobile stations and multiplexes low-rate voice calls over high-capacity links with the network subsystem. The BSC also coordinates handoffs of mobile stations if the involved BTSs are all directly connected to it and collects information about radio channels that helps determine when handoffs must start. Finally, the BSC performs interworking functions between the signalling protocols of the base station subsystem and the network subsystem.

The core element of the GSM network subsystem is the mobile switching centre (MSC). The MSC is linked to groups of BSCs and may also be connected to other MSCs, thus providing global connectivity to the mobile network. MSCs are also gateways between the mobile network and the PSTN or the ISDN to enable calls between fixed and mobile users. An MSC behaves like a traditional telephone exchange in the sense that it can route calls to the destination using the telephone number of the destination. However, MSCs are more than PSTN/ISDN exchanges because they are in charge of managing the mobility of user terminals. To do that, MSCs receive the support of different databases:

- The *home location register* (HLR) maintains and updates mobile subscribers' profiles and current locations. Subscribers' location in GSM networks are maintained in real time by means of paging procedures specifically defined for this purpose.
- *The visiting location register* (VLR) maintains profiles and locations of roaming subscribers in the visited network. Usually, the VLR is an integral part of the MSC, but this does not need to be the case. A mobile user is said to be roaming if it is being served by a network other than their home network. Roaming GSM subscribers can use other operator's network if there exists an agreement with the home operator.
- The *equipment identity register* (EIR) contains subscribers' equipment identities. It is useful to grant, limit or deny access to unauthorized mobile terminals.
- The *authentication centre* (AC) provides keys and algorithms to maintain security in the network.

When the MSC needs to route a call, it queries an HLR with the destination telephone number. The HLR then provides to the MSC the appropriate re-routing information. For roaming users, the HLR responds with the mobile station roaming number (MSRN) allocated for the user by the visited network. The MSRN contains the routing information necessary to extend the call from the home to the visited MSC. The visited MSC then contacts the destination mobile station.

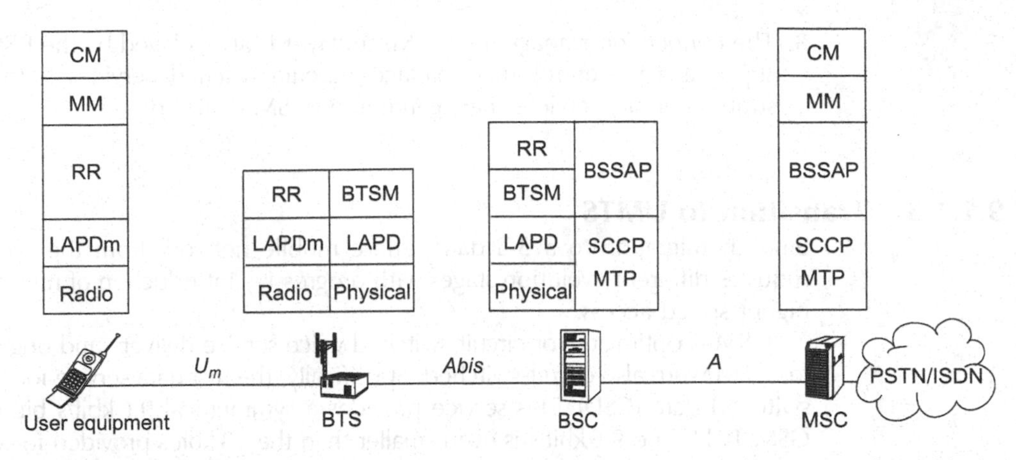

Figure 9-6. The GSM signalling protocol architecture is based on ISDN and SS7 protocols modified or extended when necessary.

GSM was defined as a mobile extension of the ISDN network and therefore many service definitions, interfaces and protocols are taken from ISDN. Signalling in GSM networks is strongly based on the ISDN and Signalling System 7 (SS7) (see Figure 9-6). The GSM signalling architecture is structured in three layers. Layers 1 and 2 are equivalent to the open systems interconnection (OSI) physical and data link layers. Layer 3 transports packetized signalling messages. The physical layer in the link between the user equipment and the BTS is the GSM FDMA/TDMA air interface. Other physical interfaces in the network are usually implemented with E1 trunks or other TDM interfaces. The data link layer in the radio interface is based on an adapted version of the ISDN link access procedure – channel D (LAPD), known as LAPD modified (LAPDm). In the A_{bis} reference point (link between the BTS and the BSC), LAPDm frames are translated to normal LAPD frames. Physical connectivity, error detection and retransmission, message segmentation and reassembly, addressing, and message multiplexing in the *A* reference point (link between the BSC and the MSC) are provided by the message transfer part (MTP) and the signalling connection control part (SCCP) SS7 protocols. The message layer is made up of three sublayers:

1. The *radio resource* (RR) management sublayer establishes and maintains a communications path between the user equipment and the MSC over which signalling and user data are delivered. Handoffs between cells are coordinated using the RR sublayer. The RR sublayer is terminated at the BSC, but some functions may be delegated to the BTS. Once interpreted and processed by the BSC, RR information can be delivered to the MSC using the base station subsystem management application part (BSSMAP) protocol
2. The *mobility management* (MM) sublayer is relayed transparently by the BSC and terminated at the MSC. This sublayer provides miscellaneous services related to mobility management, including mobile station location procedures along with authentication and ciphering.

3. The *connection management* (CM) sublayer is also relayed by the BSC to the MSC. It supports call control that manages circuit-switched services such as voice calls, supplementary services management, and SMS delivery.

9.1.1.3 Transition to UMTS

Smooth migration to a 3G data-centric mobile network from the voice-centric GSM requires different evolution stages with progressive introduction of new applications and higher speed access.

GSM is optimized for circuit-switched voice service delivery and original data services for GSM were also circuit-switched. Specifically, the first data service for GSM was circuit-switched data (CSD). This service provides a symmetrical 9.6 kbit/s bitrate over a single GSM TCH. The 9.6 kbit/s is even smaller than the 13 kbit/s provided for voice due to the stronger error protection required for data services. The high speed CSD (HSCSD), approved by ETSI in 1997, was an improvement to the CSD that allowed bitrates of several tens of kilobits per second by means of two new features:

1. A *flexible time slot allocation* scheme that allows multislot configurations. Some popular configurations are the symmetric service, made up of two slots for the downlink and two for the uplink, and the asymmetric service with three slots for the downlink and one for the uplink.
2. A new channel encoding that allows 14.4 kbit/s per slot rather than 9.6 kbit/s per slot. However the 14.4 kbit/s encoding is not as robust as the 9.6 kbit/s and some mobile units may need to switch to the 9.6 kbit/s mode under heavy interferences or noise.

Both CSD and HSCSD are circuit-switched and therefore they keep network resources busy while the communication lasts, even if transmission is discontinuous. This situation led the telecommunication industry to look for packet switching technologies, much better suited for data services, for cellular networks. The packet-switched evolution of CSD/HSCSD is the general packet radio service (GPRS). GPRS was specifically designed by ETSI for GSM networks but it has been integrated to DAMPS as well. GPRS greatly improves and simplifies wireless access to packet data networks like the Internet. Packet switching capabilities makes GPRS resemble a 3G technology, but speed limitations in the radio access network make GPRS pre-3G. For this reasons, GPRS is sometimes referred to as a 2.5G technology. The advantages of GPRS can be summarized in three points:

1. *Better usage* of radio resources in data applications thanks to statistical multiplexing.
2. *Shorter access times* to the data service from the mobile stations, usually 1 or less. With CSD/HSCSD the access time used to be several seconds.
3. *Flexible billing* based on the amount of transmitted data rather than connection duration, like in CSD/HSCSD.

Like HSCSD, GPRS uses flexible time slot allocation but GPRS allows statistic multiplexing of data from various users over the allocated space. One to eight slots can be allocated for GPRS. The uplink and the downlink are allocated separately and asymmetric configurations are allowed. Within the same radio channel, transmission capacity is dynamically shared between GPRS and circuit-switched services. Current traffic load and service priorities may be used to reallocate capacity periodically. Multiple access in the uplink is accomplished by individual grant messages delivered from the network to the mobile stations.

GPRS has its own channel coding mechanisms. In fact there are four different channel encoding schemes that provide different protection levels. CS-1 provides 9.02 kbit/s per time slot, CS-2 provides 13.4 kbit/s, CS-3 provides 15.6 kbit/s and CS-4 provides 21.4 kbit/s. CS-4 is the fastest but it can only be used under good transmission conditions. Under heavy interference or noise, it could be necessary to switch to CS-3, CS-2 or CS-1. The maximum theoretical bitrate provided by GPRS in one radio channel is 172.2 kbit/s (8 times 21.4 kb/s) to be shared among all mobile stations using the same radio channel. In practical situations, however, performance is limited to a few tens of kbits/s per user.

GPRS can be easily integrated in the GSM network. Requirements are fulfilled mainly by software upgrades of existing GSM network nodes and deployment of two new types of network elements, usually referred to as GPRS Support Nodes (GSN) (see Figure 9-7):

1. The *serving GPRS support node* (SGSN) is responsible for the delivery of data packets from and to the mobile stations within its service area. Its tasks include packet routing and transfer, mobility management, logical link management, and authentication, authorization and charging of mobile users.
2. The *gateway GPRS support node* (GGSN) is an interface between the GPRS backbone and external (IP, X.25, etc.) packet data networks. It translates between GPRS packet formats and the packet data protocol (PDP) format corresponding to the external network. In the GGSN, PDP addresses are converted to GSM identifiers.

The SGSN and the GGSN are connected via a GPRS backbone that is IP-based. Packets traversing the GPRS backbone are encapsulated and tunnelled using the GPRS tunnelling protocol (GTP).

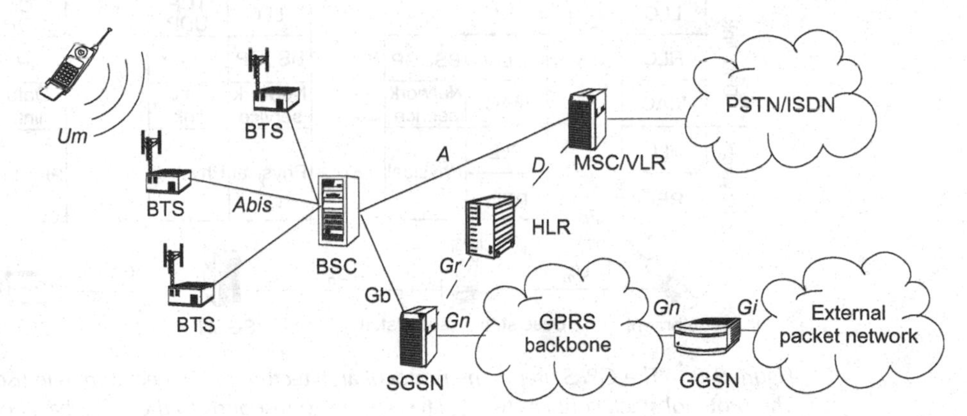

Figure 9-7. Simplified view of the 2.5G GPRS mobile network.

Mobile stations may have three different levels of support of GPRS. Class A devices support simultaneous operation of GPRS and conventional GSM services. Class B devices support GPRS and conventional GSM services but can only use one of them at a given time. Finally, class C devices cannot register simultaneously to GPRS and GSM networks.

GPRS users are registered in an SGSN before accessing any service provided by an external network. The SGSN authenticates users and assigns them a packet temporary mobile subscriber identity (P-TMSI). To exchange information with external packet networks, a GPRS mobile station must also be attached to one or more addresses from these networks. For example, if the mobile user wants to communicate with an IP network, it will need an IP address. GPRS users get the data they need to operate with external networks when they start a session. This procedure is known as the PDP context activation procedure. During the PDP context activation procedure, the network assigns a PDP address to the user but also other parameters like a QoS profile and the address of a GGSN to route packets to the external network.

One of the most important protocols within the GPRS network architecture (see Figure 9-8) is the GTP, used in the path between the SGSN and the GGSN (G_n reference point). It employs a tunnel mechanism to transfer user data packets. In the signalling plane, the GTP implements tunnel management and control mechanisms to create, modify and delete tunnels. The GTP can also transport charging information within the GPRS backbone. The GTP is a conventional IP protocol that transports IPv4, IPv6, PPP, X.25 or other packetized user data over a classical TCP/IP network stack. In the lower layers of the protocol stack, technologies like Asynchronous transfer mode (ATM), Ethernet, plesiochronous digital hierarchy (PDH) and synchronous digital hierarchy (SDH) are used. The GTP exists only between the SGSN and the GGSN. For connections between the user equipment and the SGSN, the protocol to be used is known as the subnetwork dependent convergence protocol (SNDCP). This protocol is compatible with

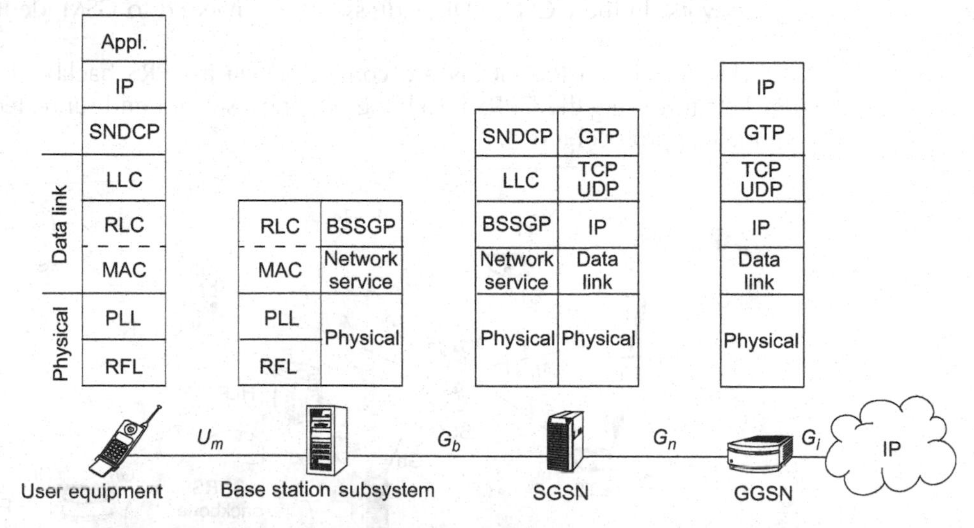

Figure 9-8. The GPRS user plane protocol architecture to connect a mobile user to an IP network. The protocol stack contains two IP layers. One corresponds to the GPRS backbone, the second one corresponds to an IP service network like the Internet.

the GSM protocol architecture and provides connection multiplexing and compression of redundant information.

In the radio interface, the protocol architecture has been adapted to support packetized data delivery. The physical layer is made up of two sublayers. The radio frequency layer (RFL) is in charge of modulation/demodulation and the physical link layer (PLL) provides digital transmission channels between the mobile station and the base station subsystem. The data link layer has three stacked sublayers. The media access control (MAC) sublayer provides multiple access to mobile stations, the radio link control (RLC) sublayer provides reliable links over the radio interface by means of error recovery mechanisms, and the logical link control (LLC) sublayer, based on a modified version of the LAPDm, provides flow control, packet reordering and additional error protection.

The interface between the base station subsystem and the SGSN can be built over frame relay (FR), ATM or Ethernet. Here, the base station subsystem GPRS application protocol (BSSGP) is used to deliver QoS and routing information.

Thanks to GPRS, the GSM network is able to efficiently deliver packet services. However, bit rates are still limited to a few tens of kbit/s. One of the technologies that attempts to solve this problem is the enhanced data rate for global evolution (EDGE) or enhanced GPRS (EGPRS). EDGE uses a new modulation and coding in the radio interface to provide increased data rates. However, the existing core infrastructure is conserved. The modified radio interface can be used to deliver both GPRS and circuit-switched services. Although this technology was first proposed by ETSI as a means to evolving GSM, EDGE is also compatible with IS-136 TDMA systems. This is the reason why some mobile operators, mainly in America, have adopted this technology as the path to 3G for their IS-136 networks.

The basis of the EDGE improvement comes from the 8 phase shift keying (8-PSK) modulation for the radio interface. This modulation offers better performance than the traditional GMSK in environments with low noise and interferences while keeping all other benefits from GMSK (see Table 9-1). EDGE uses 8-PSK in clean environments but it can switch to GMSK if necessary. Both GMSK and 8-PSK can be combined with different channel coding schemes referred to as modulation and coding schemes (MCSs). The different MCSs offer specific performance in terms of speed and protection against noise and interference. In multi-slot configurations EDGE supports peak rates over 384 kbit/s.

Table 9.1. EDGE modulation and coding schemes with their performance.

Scheme	Modulation	Speed (kbit/s/slot)
MCS-1	GSMK	8.8
MCS-2	GSMK	11.2
MCS-3	GSMK	14.8
MCS-4	GSMK	17.6
MCS-5	8-PSK	22.4
MCS-6	8-PSK	29.6
MCS-7	8-PSK	44.8
MCS-8	8-PSK	54.4
MCS-9	8-PSK	59.2

9.1.2 The Universal Mobile Telephone System

UMTS is envisioned as the 3G replacement for GSM. The new system was initially designed by ETSI but in December 1998 the 3GPP was formed to continue with the technical specification work. Currently, the 3GPP is also responsible for generating standards for the GSM and EDGE systems. Unlike ETSI, The 3GPP is a multiparty organization that involves collaboration of various telecommunications standard bodies known as 3GPP organizational partners: Association of Radio Industries and Businesses (ARIB) from Japan, China Communications Standards Association (CCSA) from China, European Telecommunications Standards Institute (ETSI) from the European Union, Alliance for Telecommunications Industry Solutions (ATIS) from USA, Telecommunications Technology Association (TTA) from South Korea, and Telecommunication Technology Committee (TTC) from Japan.

The 3GPP follows a release-oriented standardization process (see Table 9-2). New releases of the 3GPP specifications contain new features that improve previous standards or add new functionalities to the UMTS, GSM and EDGE systems.

UMTS is designed to support both high speed circuit- and packet-switched services. Theoretical maximum rates supported by UMTS are about 2 Mbit/s per mobile user, but in real implementations average performance is limited to a few hundreds of kilobits per second.

9.1.2.1 The WCDMA Air Interface Overview

Unlike GSM users, UMTS users are allowed to share the same frequency even when they are connected to adjacent cells. The reason is that UMTS uses a CDMA-based air interface rather than an FDMA/TDMA interface like GSM. Specifically, UMTS is based on direct sequence CDMA (DS-CDMA) technology, that obtains the spread spectrum signal by multiplying the original bit stream by a fast pseudo-random binary sequence

Table 9-2. Some features added by the 3GPP releases.

Release '99	2000	UTRAN and CDMA air interface ATM based transport within the UTRAN Enhanced data rates for GSM evolution (EDGE)
Release 4	2001	Low chip rate TDD air interface (TD-SCDMA) Split of the MSC into two functions: MSC server (decision point) and Media Gateway (enforcement point)
Release 5	2002	IP multimedia subsystem (IMS) IP-based transport within the UTRAN High-speed downlink packet access (HSDPA)
Release 6	2005	High-speed uplink packet access (HSUPA) Generic access network (GAN) WLAN interworking (I-WLAN)
Release 7	2007	High-speed packet access evolution (HSPA+) Voice call continuity (VCC) Multiple input and multiple output (MIMO) antennas

(chipping sequence). The original bit stream is recovered in the receiving side by multiplying again by the chipping sequence. In a DS-CDMA system, multiple access is accomplished by assigning chipping codes with low cross-correlation to different users. The orthogonality properties of chipping codes are therefore the basis of multiple access in a CDMA system, like the orthogonality of carrier frequencies in FDMA or time slot orthogonality in TDMA.

There are important reasons for adopting CDMA. Probably the most important of them is the ability of this technology to deal with multipath propagation typical of microwave propagation in mobile applications. Rake receivers for CDMA systems can discriminate multipath arrivals and combine them to make a stronger signal. CDMA signals also exhibit beneficial immunity properties against narrowband interferences. CDMA, however, needs more complex and expensive receivers than other mobile technologies.

The most important UMTS air interface is the WCDMA, which uses different frequency channels for the uplink and the downlink. The WCDMA interface is sometimes referred to as UMTS frequency division duplexing (FDD) for this reason. The alternative to the WDCMA FDD interface is the TD-CDMA that duplexes the uplink and downlink in different time slots within the same radio channel. For this reason, the TD-CDMA interface is known as UMTS time division duplexing (TDD). In fact, TD-SCDMA is also a TDD interface compatible with UMTS and it has been also accepted in the IMT-2000 initiative. The TD-SCDMA technology is a variation of the TD-CDMA developed by the China Academy of Telecommunication Technology (CATT) and Siemens operating with slower chipping sequences (1.28 Mcps instead 3.84 Mcps) than TD-CDMA. Generally speaking, TDD is better suited than FDD to data communications due to the increased capability to set asymmetric transmission channels made up of dynamically assigned time slot groups. On the other hand, FDD offers better performance in communications between users moving at high speeds and has less demanding synchronization requirements than TDD.

The ITU-R has allocated spectrum for IMT-2000/UMTS that has been adapted by national governments in accordance with particular needs and availability, leading to a complex scenario. Most of the spectrum used by current rollouts is in the 2 GHz band. Space has been specifically allocated both for FDD and TDD modes, but FDD technology is dominant (see Figure 9-9).

The bandwidth of WCDMA radio channels is about 5 MHz. This is larger than for cdma2000 (1.25 MHz) and much larger than for GSM (200 kHz). WCDMA network operators can build their networks with a single carrier frequency for all cells or they can build various overlapped macrocell and microcell structures operating at different frequencies. The standard provides many handoff possibilities:

- *Intra-frequency handoffs* enable mobility between cells operating at the same frequency. This type of handoff is referred to as hard handoff if the mobile station is disconnected from the origin base station as soon as it is connected to the destination base station, or soft handoff when the mobile station remains connected to two or more stations at the same time. A special case of soft handoff, known as softer handoff, happens when a mobile station is connected to different sectors of the same base station in a sectorial cell made up of directive antennae. Soft and softer handoffs do not exist in GSM.

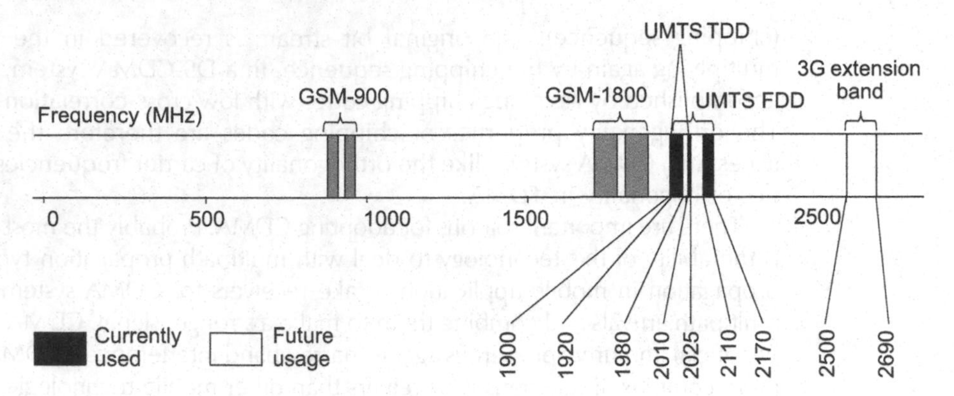

Figure 9-9. Frequency allocations of UMTS in Europe based on ITU allocations. Other 3G systems may operate at slightly different bands.

- *Inter-frequency handoffs* enable mobility between stations operating at different frequencies. This kind of handoff is useful to switch between macrocell and microcell layers and makes possible mobility between different WCDMA operators. Inter-frequency handoffs are always hard.

Handoffs between different systems are also considered by the UMTS standards. Examples are handoffs between WCDMA and TD-CDMA, and handoffs between WCDMA and GSM. Handoffs between WCDMA and non-3GPP networks like Wi-Fi or WiMAX are an active research and standardization area.

WCDMA signals contain 10 ms TDM frames divided into 15 slots. These frames are spread with 3.84 Mcps chipping sequences. There are always 2560 chips in a single time slot, but the number of bits per time slot may change. The result is a flexible interface that supports signals with multiple rates. The ratio between the chips and bits in a time slot is referred to as the spreading factor. In WDCMA, the spreading factor ranges between 256 and 4 in the uplink, and between 512 and 4 in the downlink.

The WCDMA standard define different types of chipping sequences. Some of them are orthogonal and some of them are not:

- *Spreading sequences* are used to identify channels. They must be mutually orthogonal if they are transmitted by the same device, otherwise the receiver will be unable to recover the original information.
- *Scrambling sequences* are used to reduce interference between mobile stations within the same cell and between base stations corresponding to different cells. They are not required to be orthogonal. However, they must have low mutual cross-correlation.

Time slots in WCDMA TDM frames contain traffic and control channels. These channels can be dedicated or shared between a group of users or broadcast channels. The WCDMA standards define a large number of channels with unique features and specific

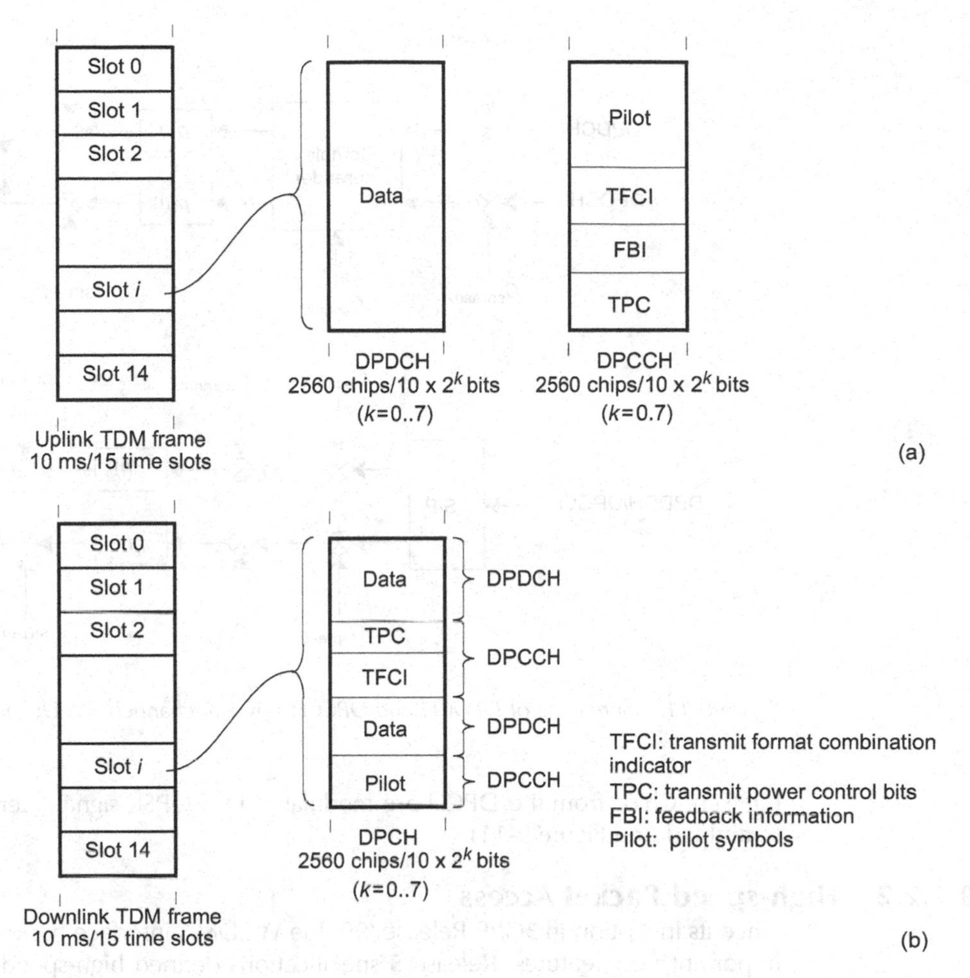

Figure 9-10. Framing of DPDCH, DPCCH and DPCH channels in WCDMA. (a) Uplink frame structure. (b) Downlink frame structure.

purposes. Depending on the protocol layer being considered by the standard, the channel structure is different and channels are identified by different names. As presented by the physical layer, channels are referred to as physical channels. Two important physical channels are the dedicated physical data channels (DPDCHs) and the dedicated physical control channels (DPCCHs). The former contain bearer data, and the latter contain control and signalling information (see Figure 9-10).

In the uplink, a DPDCH and a DPCCH (that may have different bit rates) are transmitted in the I and Q phases of a quaternary phase shift keying (QPSK) modulation after being spread and scrambled. Multiple DPDCHs, spread by unique channelization codes, may be allocated in the same physical layer connection. There is only one DPCCH per connection, however. In the downlink, the information corresponding to the DPDCH and the DPCCH is time multiplexed in a single physical channel known as the dedicated physical channel

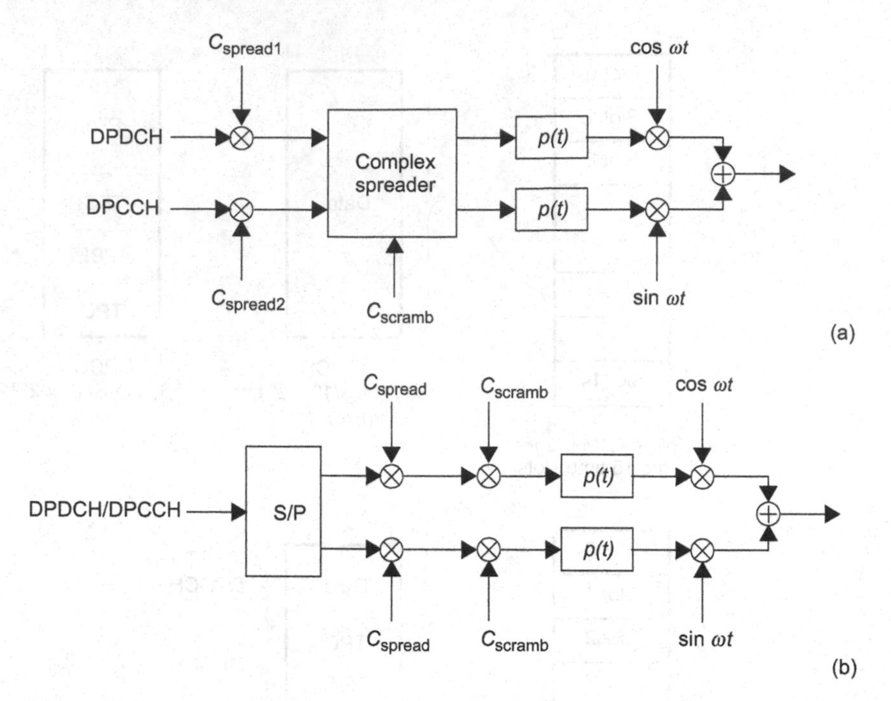

Figure 9-11. Spreading of DPDCH and DPCCH physical channels. (a) Uplink. (b) Downlink.

(DPCH). Chips from the DPCH are modulated in a QPSK signal after being spread and scrambled (see Figure 9-11).

9.1.2.2 High-speed Packet Access

Since its inception in 3GPP Release '99, the WCDMA interface has been improved with important new features. Release 5 specifications defined high-speed download packet access (HSDPA), and Release 6 introduced high-speed upload packet access (HSUPA). Evolved versions of WCDMA are often referred to as 3.5G technologies

HSDPA is a downlink technology that offers increased download bitrate and reduced latency to 3G mobile stations. Specifically, HSDPA increases the UMTS peak bitrate from 2 to 14 Mbit/s.

The HSDPA technology relies on a new type of CDMA channel, the high-speed downlink shared channel (HS-DSCH). Up to 15 of these can operate in a 5 MHz radio channel and user transmissions are dynamically assigned to one or more HS-DSCH channels. The new HS-DSCH is an improvement of the DSCH defined in 3GPP Release '99 that uses various new transmission mechanisms that rely on rapid adaptation of transmission parameters to instantaneous radio conditions. The most important of them are:

- *Fast link adaptation*: unlike other CDMA channels, the HS-DSCH lacks rapid power control for compensating varying radio conditions. It instead uses rate adaptation. Coding and modulation changes dynamically with time to compensate for the time varying radio channel. Spectrum efficiency is improved with the 16-QAM modulation if transmission conditions are good enough. Otherwise 16-QAM is replaced by the QPSK.

- *Hybrid automatic repeat request (HARQ)*: with this mechanism the system controls transmission errors and requests retransmissions when necessary. This technique combines the advantages of the forward error correction (FEC) and automatic repeat request (ARQ) techniques in a single algorithm. HARQ increases system robustness.
- *Channel dependent scheduling*: the base station scheduler considers instantaneous radio channel feedback to determine which is the mobile station best suited to receive data at every moment. This feature improves usage of available transmission resources.

HSUPA uses the new enhanced dedicated channel (E-DCH) to increase uplink performance of WCDMA networks. Many of the improvements of HSUPA were already known from HSDPA, including fast link adaptation and HARQ. The peak upload bitrate allowed by HSUPA is 5.76 Mbit/s. Average performance with realistic transmission conditions is above 1 Mbit/s.

HSPA technology is not yet exhausted. MIMO antenna systems and other improvements will enable HSPA+ to reach 40 Mbit/s in the downlink and up to 10 Mbit/s in the uplink.

9.1.2.3 The UMTS Network

The UMTS network is an evolution of the GSM/GPRS network and therefore both share many common points. The basis of UMTS is a new radio technology and, as a result, the main differences with GSM/GPRS are in the radio access network. Two new network elements are introduced: the Node-B is a replacement for the BTS and the radio network controller (RNC) replaces the base station controller (BSC). Together, the Node-B and the RNC make up the UMTS terrestrial radio access network (UTRAN), which replaces the GSM/GPRS/EDGE base station subsystem. Other GSM/GPRS network elements like the MSC, HLR, SGSN and GGSN do not need to be replaced. However, they have to be extended to adopt the UMTS requirements (see Figure 9-12):

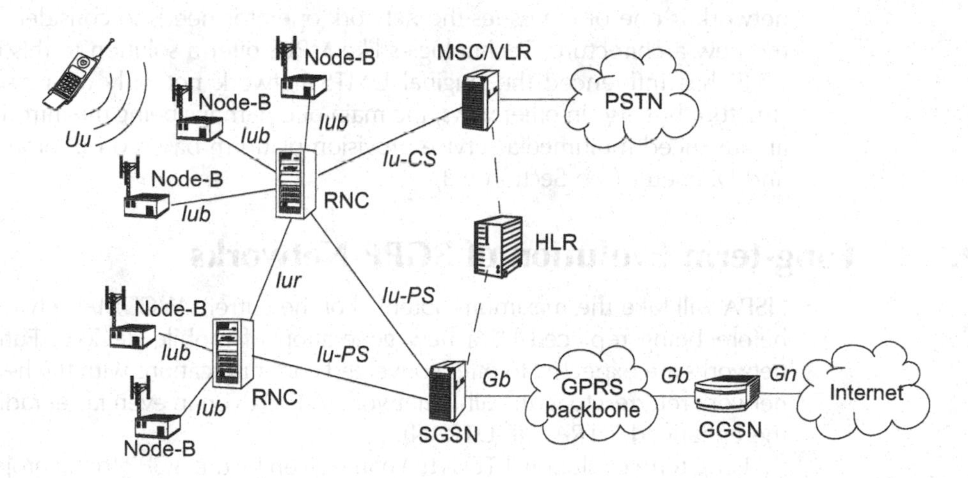

Figure 9-12. UMTS Release '99 network architecture.

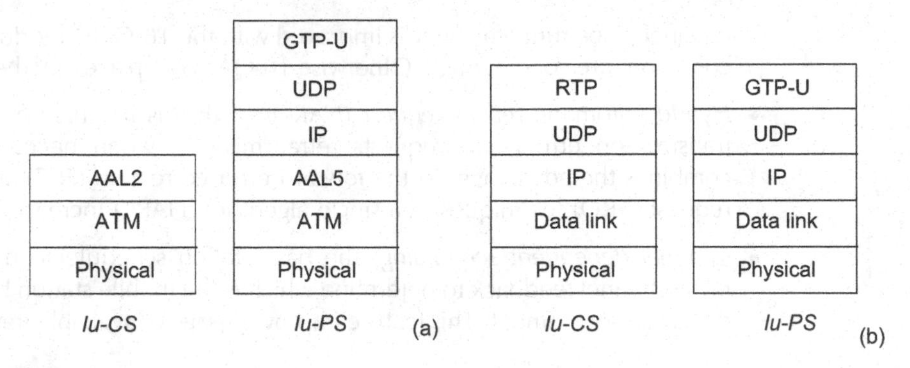

Figure 9-13. Protocol stacks for the different transport alternatives in the Iu interfaces (user plane): (a) ATM transport alternative; (b) IP transport alternative.

Release '99 of 3GPP specifications reuses GSM interfaces and defines new ones when necessary. The WCDMA interface that connects the user equipment and the Node-B is referred to as the U_u interface. The UTRAN nodes are connected through the I_{ub} and the I_{ur} reference points. The I_{ub} connects a Node-B with an RNC and it is similar to the GSM A_{bis} interface. The I_{ur} connects two RNCs and it has no parallel with GSM. The UTRAN is connected to the core network through the I_u interface. Packet-switched services are connected to the SGSN using the I_u-PS interface and circuit-switched services use the I_u-CS to access to the MSC. Both the SGSN and the MSC need to be upgraded to support the new interfaces.

Evolution of the UMTS network architecture is influenced by the development of IP as a new packet-centric transport technology. Releases '99 and 4 define an ATM network for transporting UMTS services but this approach was complemented by an IP based transport alternative defined in Release 5 of 3GPP specifications (see Figure 9-13). The IP transport alternative for the UTRAN is much more flexible than ATM transport. Thanks to the IP transport alternative, mobile network operators can use any third-party IP network to interconnect the UTRAN network nodes. QoS control within the IP transport network is one of the issues the network operator needs to consider when migrating to the new architecture. Technologies like MPLS offer a solution to this issue.

IP has influenced the original UMTS network not only as a new transport infrastructure but also in other ways, the main one perhaps being the introduction of the IMS, an advanced multimedia service provision platform based on IP protocols like SIP, SDP and Diameter (see Section 9.3).

9.1.3 Long-term Evolution of 3GPP Networks

HSPA will take the maximum potential of the current WCDMA networks for a few years before being replaced by a new generation of mobile devices. Future 3GPP mobile networks are expected to offer converged communications with the help of single IP core network referred to as the all-IP network (AIPN) over an even faster radio access network, the enhanced UTRAN (E-UTRAN).

Long-term evolution (LTE) is the name given by the 3GPP to the project that will evolve UMTS to an improved network. LTE will result in Release 8 of the 3GPP specifications.

Initial LTE deployments are targeted for 2009. Some important goals of the project are the following:

- Instantaneous peak data rate of 100 Mbit/s (downlink) and 50 Mbit/s (uplink) over 20 MHz radio channels.
- Performance targets should be met for 5 km cells, with slight degradation for 30 km cells and acceptable performance over 100 km cells.
- Flexible spectrum allocation with radio channels of different bandwidths, including 1.25, 1.6, 2.5, 5 and 20 MHz. FDD and TDD operation will be supported.
- Packet-switched architecture with end-to-end QoS support.
- Coexistence and interworking with legacy 3GPP systems including UMTS, EDGE and GSM. Handoff between LTE and these architectures should be supported.

The proposed radio interface for the LTE E-UTRAN is referred to as high-speed OFDM packet access (HSOPA). HSOPA uses single carrier FDMA (SC-FDMA) for the downlink and OFDM for the uplink. OFDM was adopted with success by the IEEE WiMAX standard (see Section 9.2.2) but, unlike HSOPA, WiMAX is based on OFDM both for the uplink and the downlink.

OFDM is not compatible with CDMA technologies that dominate 3G, including WCDMA. HSOPA has the same advantages as WCDMA with a simple implementation. Furthermore it offers a more flexible spectrum allocation scheme (spectrum does not need to be allocated in big blocks as happens with CDMA), and it is well suited for operation with smart antenna technologies, e.g.:

- *Beamforming* applies vectorial signal processing techniques to arrays of transmitting or receiving antennas to maximize the transmitted power or sensitivity of the array in chosen directions.
- *Receive diversity* relies in optimal combining of signals received by separate receiving antennas. Antenna spacing yields signals that have somewhat independent fading characteristics and as a result the combined signal may potentially be more effectively decoded than any of the components.
- *Space time coding* relies on multiple and redundant copies of a data stream transmitted by multiple antennas. Each copy is encoded using space time codes (STCs). Independent fading in the multiple antenna links makes it possible to decode the received signal.
- *Spatial multiplexing* requires multiple transmitting and receiving antennas in a configuration known as MIMO. This technique splits a high-rate signal into multiple lower rate streams. Each stream is transmitted from a different antenna in the same radio channel. Signals travelling across different communication paths subject to independent fading can be efficiently recovered.

Although many smart antenna techniques have been specified for WCDMA systems as well, they will become an essential part of the HSOPA interface. In fact, most of the

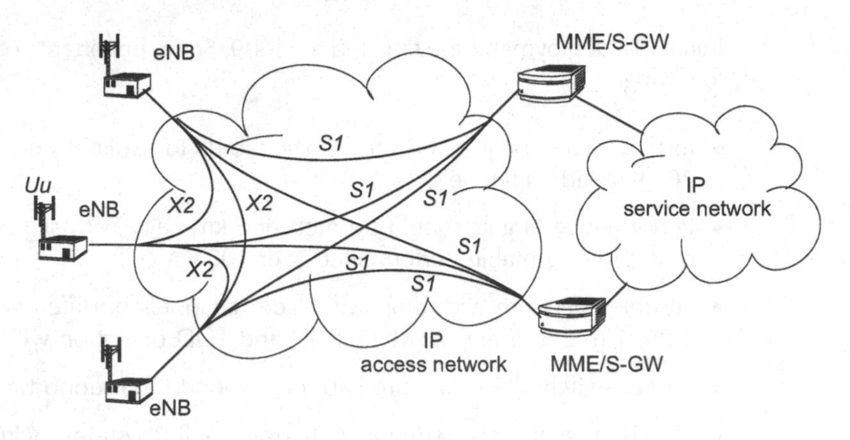

Figure 9-14. E-UTRAN network architecture.

improvement over WCDMA will come from the availability of radio channels with increased bandwidth and smart antenna techniques. HSOPA will possibly use up to four transmitting and up to four receiving antennas.

The access network for the LTE system is an IP network known as E-UTRAN. The E-UTRAN consists in a collection of enhanced Nodes-B (eNBs) connected to an evolved data core. Specifically, eNBs are connected to mobility management entities (MMEs) and serving gateways (SGWs) through the *S1* interface. The eNBs can be interconnected as well. The *X2* interface is used for this purpose (see Figure 9-14).

The E-UTRAN has a simpler and more efficient architecture than the UTRAN. The new access infrastructure does not have a circuit-switched domain. All services are provisioned over a single packet-switched access architecture connected to a multiservice IP core network: the AIPN. It will be possible for users to access the AIPN not only by means of the E-UTRAN but also by using other technologies like Wi-Fi, WiMAX and wireline.

9.2 Wireless Communications Overview

All communications based on a transmission media other than copper wire, optical fibre, or any other kind of cable, are wireless. This definition includes communications based on satellite microwave links, terrestrial radio links and free space optical communications. Mobile cellular communications are also included. However, here, the term 'wireless' applies only to a few microwave technologies designed for indoor or outdoor access to data networks. This kind of wireless communication is currently dominated by some technologies defined by the IEEE. The most important are the IEEE 802.11 family (Wi-Fi) and the IEEE 802.16 family (WiMAX).

Cellular communications technologies were designed as an extension of the fixed telephone network, but evolution has made cellular communications data friendly and suitable for multimedia service delivery. Evolution of IEEE wireless standards has followed the opposite path, but the destination appears to be the same. The newest IEEE wireless standards are adding sophisticated handoff mechanisms to support mobility, strong QoS for multimedia service delivery and other features that make these technologies suitable for the same applications as WCDMA, cdma2000 or any other mobile standard.

9.2.1 Wireless Local Area Networks

Wireless local area networks (WLAN) provide short-range wideband connectivity to computers and other electronic devices. WLAN technologies are designed to operate in indoor environments but they have been used for short-range outdoor applications as well. WLAN networks can be either *ad-hoc* or centralized. In *ad-hoc* networks, stations are connected through point-to-point wireless links. In centralized networks stations are connected through a central access point (AP) rather than by means of point-to-point connections.

The applications envisaged for WLANs include:

- *Replacement or complement of wireline Ethernet LANs*. Many businesses have deployed WLAN connectivity in their installations to allow staff to access corporate applications or the Internet. WLANs are also used in home networks to connect home computers to the Internet access gateway.
- *Internet access in hotspots*. Hotspots are public wireless networks placed in selected locations like libraries, airports, campuses and train stations. They already offer Internet access but other services are also possible. Hotspots usually have mechanisms to authenticate, authorize and charge users.
- *Complementary infrastructure for mobile operators*. Current WLAN technologies offer more bandwidth than any mobile access network. Owing to their short range, WLANs cannot replace mobile access networks but they can provide high-speed access in selected locations like airports, train stations or the subscriber's home. For this application, mobile operators need dual-mode user equipment with connectivity to mobile and WLAN networks. They must also define the mechanisms to authenticate, authorize and charge subscribers in WLAN mode.
- *Connectivity through mesh networks*. Mesh networks are *ad-hoc* networks with the special feature that connection with remote nodes can be achieved using intermediate user equipment. Nodes in a mesh network are at the same time terminals and routers. In mesh networks, connection with remote stations is possible, even if there are no direct links with them. WLAN technologies can be used to build these networks. Mesh networks may become an alternative wireless access technology to traditional mobile networks.

Currently, the WLAN market is dominated by the IEEE 802.11 family, also known as Wi-Fi. The alternatives to Wi-Fi are HiperLAN, a European ETSI standard, and HiSWAN, a Japanese standard developed by the ARIB. Both have a small market share compared with Wi-Fi.

9.2.1.1 Wi-Fi

Wi-Fi is the branch name of a family of WLAN technologies specified in the IEEE 802.11 standards. One of the advantages of Wi-Fi is that it operates over unlicensed bands of the radio spectrum. This fact eases deployment of Wi-Fi networks. However, interference from external devices is more likely to happen in unlicensed frequencies than in the licensed spectrum. Specifically, some Wi-Fi devices operate at the same frequencies as microwave ovens, bluetooth devices and cordless phones.

The first IEEE 802.11 standard was presented in 1997 and today is referred to as IEEE 802.11-1997. This standard allows 1 or 2 Mbit/s over both infrared and microwave channels. However, infrared operation was never widely accepted by the market. Microwave IEEE 802.11-1997 operates in the 2.4 GHz industrial, scientific and medical (ISM) radio band. Two spread spectrum modulations were defined for this technology:

- *Frequency hopping spread spectrum* (FHSS): the spread spectrum signal is built by modulating the original bit stream with a narrowband carrier of varying frequency. The carrier frequency must change in accordance with a fast pseudo-random sequence known as the frequency hopping sequence. To recover the original signal the receiver must know the frequency hopping sequence and be synchronized with it.
- *Direct Sequence Spread Spectrum* (DSSS): the spread spectrum signal is obtained by multiplying the original bit stream by a fast pseudo-random sequence (chipping sequence). Binary values '1' and '0' are mapped to '+1' and '−1' during the spreading operation. The spreading operation is combined with a PSK digital modulation to obtain a pass-band signal. The original signal is recovered by demodulating and multiplying the spread signal by the same chipping sequence again. Pseudorandom sequences must be synchronized in the transmitter and the receiver.

FHSS and DSSS signals have a broader spectrum than the original bit streams. Thanks to this property, signals become protected against narrowband interference. Despite having similar beneficial properties against interferences, it soon became clear that DSSS would offer more possibilities for scaling to higher speeds than the FHSS, and it therefore became the preferred modulation.

Wi-Fi did not became really popular until the release of the IEEE 802.11b standard in 1999. IEEE 802.11b devices operate at a maximum speed of 11 Mbit/s over the 2.4 GHz band. However, they can switch to 5.5, 2 and 1 Mbit/s when noise and interferences are high enough to degrade transmission. For 2 and 1 Mbit/s operation modes, devices are backwards compatible with IEEE 802.11-1997. For 11 and 5.5 Mbit/s, DSSS is replaced by complementary code keying (CCK). CCK is in fact a variation of DSSS that uses series of special 8-symbol complex sequences called complementary sequences to spread the signal rather than the binary 11-bit Barker sequence used by DSSS. Once modulated, the bandwidth of an IEEE 802.11b signal is about 22 MHz. This signal is placed in one of the 14 defined radio channels within the 2.4 GHz band.[2] Radio channels are spaced only by an amount of 5 MHz and therefore signal spectra are partially overlapped and interference between neighbouring systems is likely to occur.

The IEEE 802.11a standard was ratified almost at the same time as IEEE 802.11b. Unlike IEEE 802.11b, the new technology was designed to provide up to 54 Mbit/s over the 5 GHz

[2]In some countries only a subset of the 14 defined radio channels is allowed for 802.11b signals.

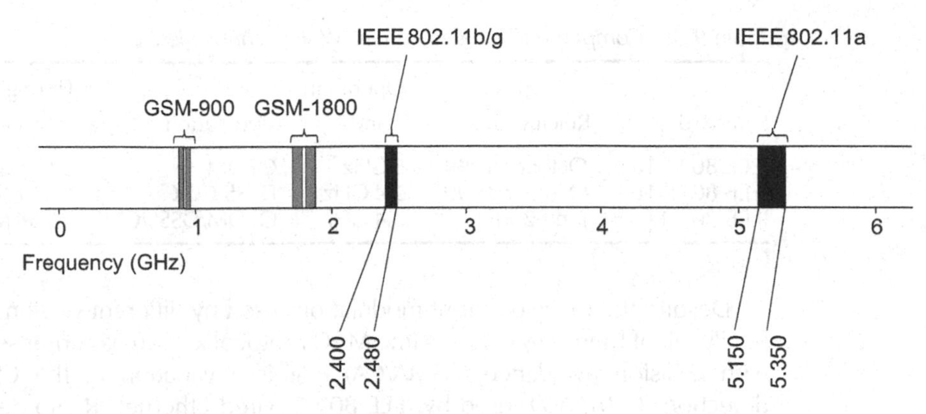

Figure 9-15. Frequency allocations for Wi-Fi.

Unlicensed National Information Infrastructure (U-NII) band (see Figure 9-15). However, EEE 802.11a devices are able to modify the transmission rate to 54, 48, 36, 24, 18, 12, 9 and 6 Mbit/s depending on the noise and interference level they register. Interferences in the 5 GHz band are less common than in the 2.4 GHz band and thus IEEE 802.11a may potentially provide better performance than IEEE 802.11b.

The IEEE 802.11a modulation is the orthogonal frequency division multiplexing (OFDM; see Section 9.2.2.2) that offers excellent protection against narrowband interferences and frequency selective fading caused by multipath propagation. Specifically, IEEE 802.11a OFDM uses 48 data sub-carriers plus four pilot sub-carriers. Individual data sub-carriers transport user information using binary phase shift keying (BPSK), quaternary phase shift keying (QPSK), 16-quadrature amplitude modulation (16-QAM) or 64-QAM. IEEE 802.11a OFDM signals have a bandwidth of 20 MHz. The standard defines 12 non-overlapped channels to carry this signal but only eight of them placed within the lower and mid U-NII band (5.150–5.350 GHz) are currently used.

Although IEEE 802.11a devices provide better performance than IEEE 802.11b, they arrived later to the market at a higher price and they were not widely adopted in the consumer space. The primary applications for IEEE 802.11a are found in enterprise network environments. The arrival of IEEE 802.11g in 2003, backwards-compatible with IEEE 802.11b and providing similar performance to IEEE 802.11a, reduced the chances of mass adoption of the 5 GHz standard. In fact, thanks to its advantages, IEEE 802.11g is currently the dominant WLAN technology in the consumer market.

IEEE 802.11g brings the advantages of OFDM modulation to the 2.4 GHz band. Specifically, OFDM is used for 54, 48, 36, 24, 18, 12, 9 and 6 Mbit/s operation modes, CCK for 11 and 5.5 Mbit/s, and DSSS for 2 and 1 Mbit/s (see Table 9-3). Diverse modulations for different bit rates make IEEE 802.11g compatible with older devices. Specifically, this feature makes IEEE 802.11b and 802.11g interoperable, although the presence of a IEEE 802.11b participant reduces the speed of the overall IEEE 802.11g network.

Compatibility between IEEE 802.11a and 802.11g devices is more difficult as they operate in different bands. However, there exist Wi-Fi access points with dual band capabilities. These devices can operate both at 2.4 GHz and at 5 GHz bands.

Table 9-3. Comparison between current Wi-Fi technologies.

Standard	Release date	Operation band	Modulation	Throughput (typical/maximum)	Range (indoor)
IEEE 802.11a	October 1999	5 GHz	OFDM	23/54 Mbit/s	30 m
IEEE 802.11b	October 1999	2.4 GHz	DSSS/CCK	5/11 Mbit/s	35 m
IEEE 802.11g	June 2003	2.4 GHz	OFDM/DSSS/CKK	19/54 Mbit/s	35 m

Despite the many different modulations used by different versions of the IEEE 802.11 family, all of them rely on the same MAC protocol known as carrier sense multiple access with collision avoidance (CSMA/CA) that is a variation of the CSMA with collision detection (CSMA/CD) used by IEEE 802.3 wired Ethernet. Radio devices are unable to detect collisions; therefore, they use collision avoidance based on a listen before talk and random backoff deferral mechanism. The contention-based multiple access mechanism of Wi-Fi is simple and does not need centralized control. Performance is good enough for most data applications and ideal for easy deployment of mesh networks. However, applications like wireless VoIP need a more predictable MAC protocol.

9.2.1.2 The Future of Wi-Fi

The market is waiting for a new generation of Wi-Fi devices driven by new applications like wireless VoIP, streaming video and music, gaming and network storage. The next release of IEEE 802.11n is expected to increase the average bitrate in the radio interface to at least 100 Mbit/s. IEEE 802.11n is still waiting for a final approval, but some vendors have started manufacturing using draft version specifications.

IEEE 802.1n is expected to be compatible with older Wi-Fi standards and at the same time it will offer important improvements. Perhaps the most important of them will be support of MIMO antenna systems. These systems use multiple transmitting and receiving antennas to improve transmission performance. Other improvement will probably come from usage of 40 MHz ODFM signals rather than operation limited to 20 MHz signals as in IEEE 802.11g.

9.2.1.3 WLAN Integration in Mobile Networks

The telecommunications industry has been searching for solutions to merge the high availability of cellular networks with the high speed and low cost of WLANs. Dual-mode user equipment remains connected to the cellular network like other mobile devices, but it can switch to WLAN operation when there is a WLAN available for access. 2G/3G cellular voice services are provisioned through the circuit-switched domain of the mobile core network while WLAN services are commonly offered through a packet-switched, fixed, IP-centric infrastructure. This is the reason why convergence between cellular and WLAN communications is considered a key step towards fixed mobile convergence (FMC). The requirements for any solution aiming to bring WLAN access to mobile services are:

- *Service transparency*: the services available to dual-mode terminals must be the same in cellular and in WLAN operation or, at least, the common set of services available must be large enough to be attractive to potential subscribers. Uninterrupted service provision by means of seamless handoff between WLAN and mobile networks is important as well.

- *Seamless integration to current mobile network*: the convergence solution should cause minimum disruption in available network infrastructure. It should be possible to reuse most or all of the current network nodes and the solution should not disturb services already deployed.
- *Scalability to support mass market deployments*: a successful WLAN/mobile convergence solution should be scalable enough to address the residential market. Fast rollout, availability of cheap terminals and simple service configuration and operation are vital.

So far, two main approaches for mobile/WLAN converged services have been considered. Both have some advantages and some inconveniences:

1. *Solutions based on specialized access gateways*: these approaches are based on a new network element that translates WLAN signals and protocols to the mobile network. Management of WLAN terminals connected to the mobile core network through this new gateway is done as though they are ordinary mobile terminals.
2. *Solutions based on IMS*: IMS is a platform that enables delivery of multimedia services through heterogeneous access networks (see Section 9.3). It is therefore normal to think that IMS should enable seamless handoff between different access networks for defined multimedia services.

The most important initiative based in access gateways is the generic access network (GAN), also known as unlicensed mobile access (UMA). The IMS-based solution is referred to as voice call continuity (VCC) (see Section 9.3.2). GAN is a 3GPP standard that connects WLAN users to the G_b and A interfaces of the GSM/GPRS/EDGE network with the help of a gateway known as the GAN controller (GANC; see Figure 9-16). The

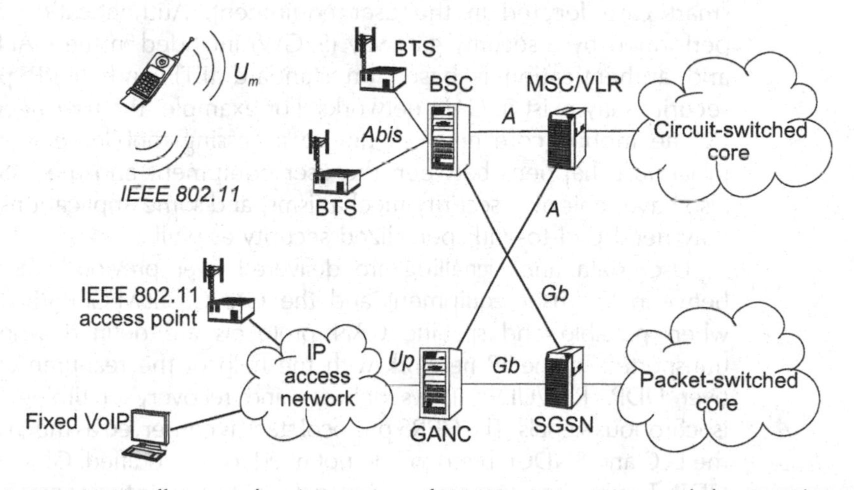

Figure 9-16. UMA offers seamless integration of WLAN access points and the GPRS/EDGE radio access network thanks to a new network element known as GANC.

current GAN standard only provides IP access to the GSM G_b and *A* reference points and, therefore, interworking with UMTS networks and services is limited. On the other hand, VCC, also being defined by the 3GPP, is a core network technology independent of the access network. Theoretically, it should work with Wi-Fi, WiMAX, WCDMA, wireline and any other access network, but it is limited to the voice service. It could be necessary to define more IMS continuity applications for other services in the future. In must be remarked that IMS and GAN may work together as well. Specifically, an operator may choose to deliver IMS multimedia services to WLAN and mobile access points through a GAN access infrastructure. In this case VCC is not needed.

An alternative to GAN is the architecture known as WLAN interworking (I-WLAN; see Section 9.2.2.3).

The basis of the GAN standard is the new U_p interface that maps GSM/GPRS/EDGE protocols to IP datagrams. The IP traffic is collected by the GANC and translated to G_b and *A* signals and protocols that can be interpreted by the GSM/GPRS/EDGE core network. For this reason, the GANC appears to the core network as a normal BSC and this enables mobile operators to use any IP access network for provision of mobile services. Usually, mobile operators attach to the IP access network's WLAN access points, but this does not need to be the case. For example, a DSL operator may use GAN to provide fixed VoIP services to their subscribers through a mobile network. This approach would need special VoIP phones with GAN support.

Any IP network, including the Internet, can theoretically be used by GAN as the access infrastructure. The QoS capabilities of this network are important for high-quality voice service provision. However, QoS is always limited in Wi-Fi access networks by the nature of the IEEE 802.11 interface. QoS may be further degraded if the access infrastructure is the Internet due to the lack of QoS mechanisms of this network.

Another issue with public IP access infrastructure is security. The GAN standard makes use of IPsec encryption security envelope (ESP) tunnels to provide encryption and data integrity to voice, data and signalling traffic (see Figure 9-17). IPsec tunnels are established as a result of authentication and authorization procedures carried by a smart card located in the user equipment. Authentication and authorization are performed by a security gateway (SEGW) included in the GANC. GAN authorization and authentication is based on standard IETF and 3GPP procedures. Additional security may exist in GAN networks. For example, the user needs to be authenticated by the mobile core network before accessing mobile services. This authentication procedure happens between the user equipment and the MSC/VLR. GPRS services also have their own security mechanisms, and some applications like the web or e-mail may need end-to-end specialized security as well.

User data and signalling are delivered over previously established IPsec tunnels between the user equipment and the GANC. GSM/GPRS/EDGE protocols are used when possible and specific GAN protocols are defined when necessary. Voice is transported in the IP network with the help of the real-time transport protocol (RTP) over UDP. RTP/UDP allows delivery and recovery of timing information carried by isochronous signals. The GPRS protocol stack is conserved as much as possible. Specifically, the LLC and SNDCP protocols do not need to be modified. GPRS data is transported over UDP. Timing recovery is not necessary for data applications and therefore RTP is not used. Control plane signalling is sent over reliable TCP connections.

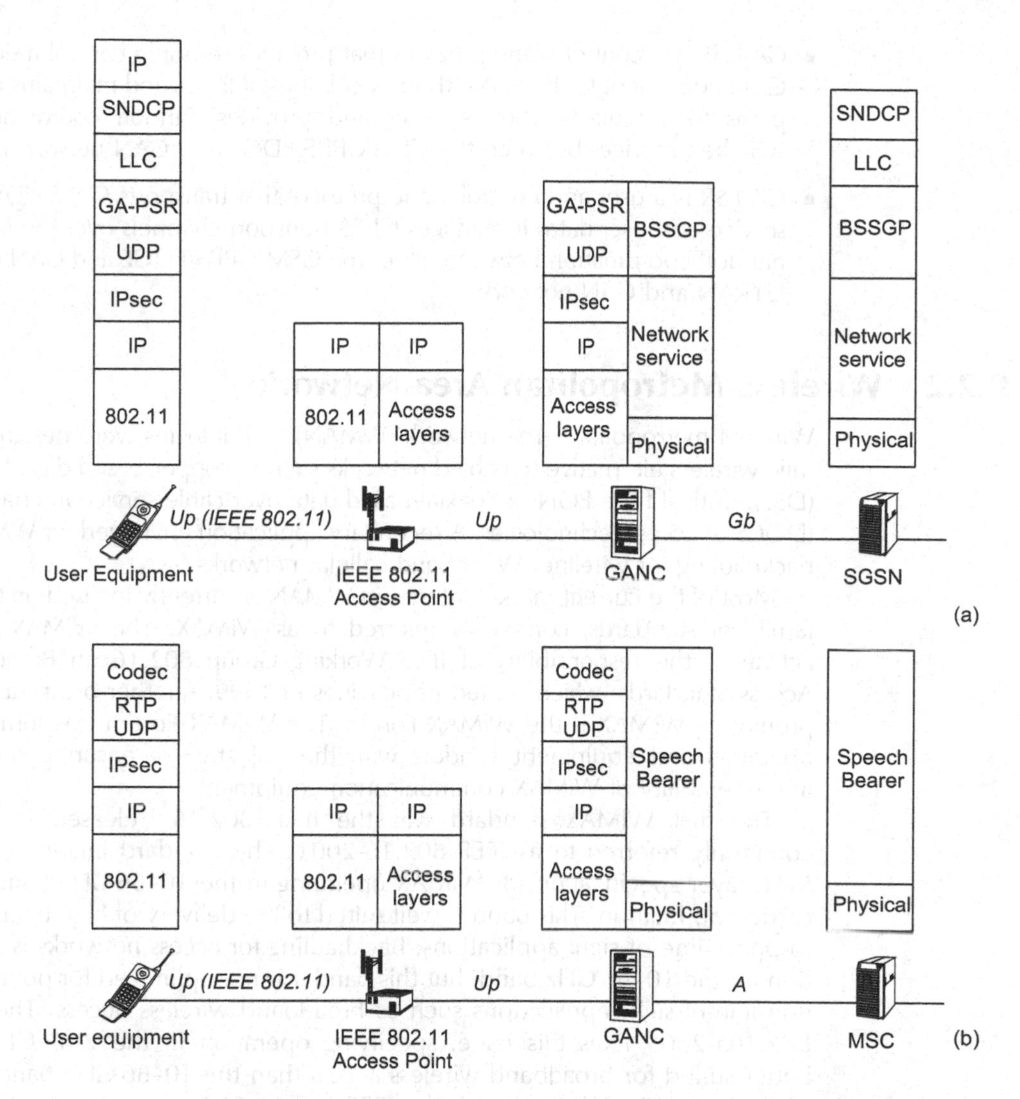

Figure 9-17. GAN user plane protocol architecture. (a) IEEE 802.11 access to the GPRS packet-switched domain. (b) IEEE 802.11 access to the GSM circuit-switched domain. Voice is encoded, packetized and delivered through the IP access network. Timing information is transported with the help of the RTP protocol.

There are three GAN-specific protocols for the U_p interface. The generic access – circuit-switched resources (GA-CSR) are used in circuit switched services, the generic access – packet-switched resources (GA-PSR) are used in packed-switched services and the generic access – resource control (GA-RC) is used for both circuit- and packet-switched services.

- *GA-RC* is a control plane protocol that connects users to the GANC. It provides discovery and registration mechanisms for mobile stations.

- *GA-CSR* is a control plane protocol that provides resource control mechanisms to the GAN equivalent to the GSM RR protocol. It establishes and maintains communication paths for circuit-switched services and provides handoff coordination of circuit-switched services between the GSM/GPRS/EDGE and GAN networks.
- *GA-PSR* is a user and control plane protocol that transports GPRS signalling, the SMS service and user data. It manages GPRS transport channels over the U_p and provides handoff coordination between either the GSM/GPRS/EDGE and GAN networks or the UTRAN and GAN networks.

9.2.2 Wireless Metropolitan Area Networks

Wireless metropolitan area network (WMAN) technologies were designed to be a first mile wireless alternative to cabled networks such as copper-based digital subscriber loop (DSL), optical fibre PON or coaxial-based data over cable service interface specification (DOCSIS) access technologies. A secondary application envisaged for WMAN is low-cost backhauling for wireline, WLAN and cellular networks.

Most of the current market interest in WMAN is currently focused on the IEEE 802.16 family of standards, commonly referred to as WiMAX. The WiMAX standardization activity is the responsibility of IEEE Working Group 802.16 on Broadband Wireless Access Standards, which started its activities in 1999. Another organization involved in promoting WiMAX is the WiMAX Forum. The WiMAX Forum was formed in 2001 by operators and equipment vendors with the objective of ensuring compatibility and interoperability of WiMAX communication equipment.

The first WiMAX standard was the IEEE 802.16, released in January 2001, commonly referred to as IEEE 802.16-2001. This standard included a physical and MAC layer specification for WiMAX operating in the 10–66 GHz band using a single carrier modulation. This band is well suited to the delivery of high data rates for point-to-point, line-of-sight applications. Backhauling for access networks is a good application for the 10–66 GHz band, but this band it is not optimized for point-to-multipoint, non-line-of-sight applications such as broadband wireless access. The standard IEEE 802.16a-2003 fixes this issue, specifying operation in the 2–11 GHz band, much better suited for broadband wireless access than the 10–66 GHz band. The currently active base WiMAX standard is the IEEE 802.16-2004, sometimes referred to as IEEE 802.16d. However, the latest important improvement in WiMAX is included in the standard IEEE 802.16e-2005, which includes handoff procedures for IEEE 802.16 networks, thus enabling terminal mobility. Thanks to IEEE 802.16e-2005, WiMAX becomes an alternative to 3GPP cellular networks. Mobile WiMAX will probably be an important ingredient of the future mobile network architecture.

Despite being both IEEE standards, WiMAX and Wi-Fi are different technologies. That means that WiMAX and Wi-Fi user and network equipment are not interoperable. The reason is that Wi-Fi and WiMAX were designed for different applications and therefore they have different technological requirements.

The performance of WiMAX technology in terms of range and bitrate is difficult to state because it depends largely on system configuration. Specifically, performance depends on the antennas, radio power, operation band, radio channel bandwidth and many other parameters. It has been reported that WiMAX operating in the 3.5 GHz band may provide a 10 Mbit/s downlink over 10 km line-of-sight paths in a pair of 3.5 MHz radio channels. However, more realistic results for non-line-of-sight paths point to 10 Mbit/s downlinks in 2 km paths. Performance tends to be better in rural and suburban areas than in urban environments. The given 10 Mbit/s bitrates are aggregated capacity that needs to be shared among all users of the same cell and radio channel.

The main alternatives to WiMAX are WiBro, a Korean standard, and HiperMAN, developed by ETSI. Both are very close to WiMAX and they are interoperable with this technology, or at least with some of its profiles.

9.2.2.1 Allocated Radio Bands for WiMAX

Allocation of frequency bands for WiMAX is more difficult than for Wi-Fi. Wi-Fi exists in unlicensed spectrum, and vendors can manufacture and sell products without heavy constraints. However, broadband wireless access business models demand licensed frequencies subject to close control by governments. Reaching international agreement on spectrum issues is complicated and this is an unsolved issue for WiMAX. This may delay or limit the adoption of WiMAX and affect the interoperability of network equipment.

There are, however, various licensed and unlicensed frequency bands that are the most likely to be used for broadband wireless access based on WiMAX.

- The 3.5 GHz band is the most widely available band outside the USA, including many European and Asian countries. This band contains licensed spectrum and it represents the largest potential market for WiMAX.
- The 5 GHz Unlicensed National Information Infrastructure (U-NII) and World Radio Conference (WRC) bands represent other major spectrum slices available for WiMAX. In this case they contain unlicensed frequencies. The low and mid U-NII are already being used by IEEE 802.11a Wi-Fi systems. Most WiMAX activities are in the upper U-NII band because it contains fewer interferences.
- The Wireless Communications Service (WCS) bands are two narrow spectrum bands around 2.3 GHz that can be used in the USA for licensed applications.
- The Multichannel Multipoint Distribution Service (MMDS) band is available in the USA for licensed applications. This band contains 31 channels with 6 MHz spacing originally allocated for the Instructional Television Fixed Service (ITFS) and, due to its under-utilization, has been allocated for broadband wireless access applications.
- The 2.4 GHz ISM radio band has been used for IEEE 802.11b/g Wi-Fi applications. It is also available for WiMAX broadband wireless access.

9.2.2.2 WiMAX Air Interface Specification Summary

For the physical layer, WiMAX uses different versions of orthogonal frequency division multiplexing (OFDM). A single carrier modulation from the first IEEE 802.16 release is still included in the standard but not implemented in commercial devices. None of the physical

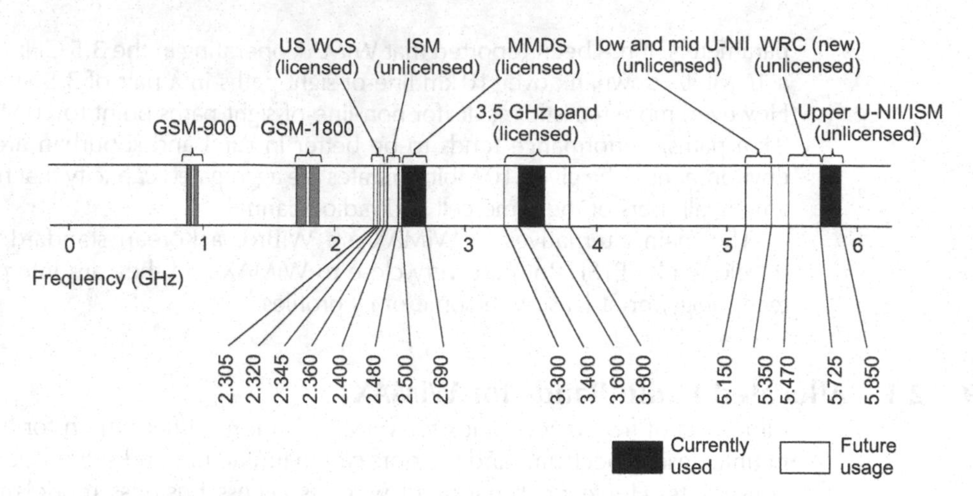

Figure 9-18. Frequency allocations for WiMAX.

layer specifications for WiMAX support CDMA. The OFDM waveform is composed of many narrowband orthogonal carriers. In an OFDM system, the input data stream is divided into several parallel sub-streams of reduced data rate and increased symbol duration. Each sub-stream is then transmitted on a separate orthogonal sub-carrier using single carrier digital modulations such as the BPSK, QPSK, 16-QAM or 64-QAM.

Increased symbol duration of OFDM provides good protection against inter-symbol interference (ISI) and delay spread typical of non-line-of-sight radio communications. The multi-carrier modulation is beneficial against frequency selective fading caused by multipath propagation because attenuation of some sub-carriers does not affect the others. OFDM radio systems have been demonstrated to be a good way to fight against multipath propagation with less complexity than CDMA systems. Moreover, OFDM systems can achieve higher spectral efficiency than CDMA systems.

Strictly speaking, OFDM does not provide multiple access in the same sense that CDMA does. To accommodate multiple users, OFDM relies either on TDMA or FDMA mechanisms. Specifically, in an OFDM system, transmission resources are available in the time domain by means of OFDM symbols and in the frequency domain by means of sub-carriers. Both sub-carriers and OFDM symbols are distributed among users in order to provide multiple access.

OFDM Access (OFDMA) is a variation of the OFDM modulation that achieves multiple access by assigning subsets of sub-carriers, known as sub-channels, to individual receivers (see Figure 9-19). Sub-channels can be resized or reassigned to different sub-carriers with the help of an adaptive algorithm and using the feedback information from the channel conditions. Differentiated service levels can be achieved by assigning different number of sub-carriers to different users. The version of OFDMA for WiMAX is referred to as scalable OFDMA (S-OFMDA) because it fixes frequency spacing to 10.94 kHz for radio channels of different bandwidth. This can be done by modifying the total number of sub-carriers in radio channels of different sizes (see Table 9-4). Thanks to this feature, S-OFDMA is able to support a wide range of bandwidths to address the need for various spectrum allocation and usage model requirements.

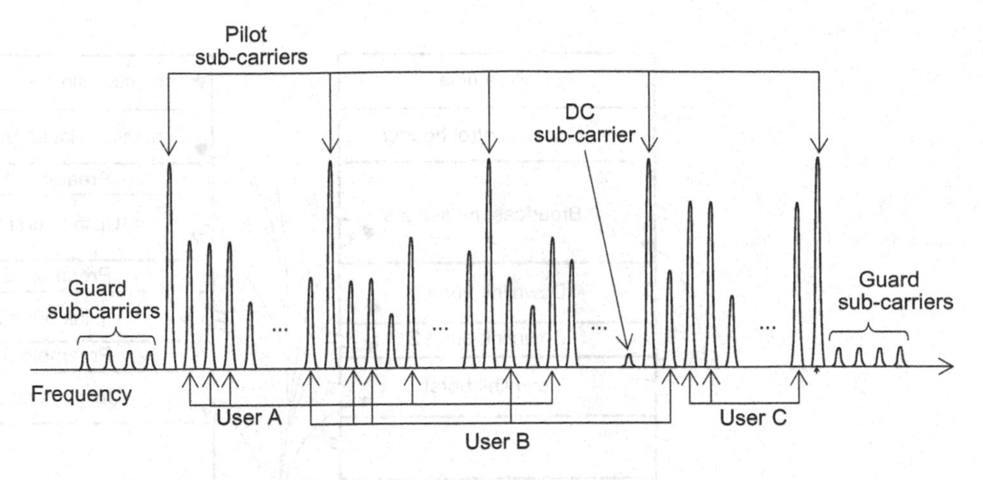

Figure 9-19. OFDMA operation principle.

S-OFDMA was introduced in the IEEE 802.16e-2005 standard for mobile WiMAX. Before the release of this standard, the WiMAX Forum established a transmission profile based on OFDM modulation with 256 sub-carriers for fixed services. The profiles that are most likely to be adopted by the market for mobile WiMAX are the S-OFDMA with 512 and 1024 sub-carriers. The latter matches Korea's WiBro specification.

WiMAX supports both TDD and FDD duplexing of the uplink and downlink. In TDD mode, uplink and downlink bit streams share a single frequency but are transmitted at different times. In FDD mode, two different frequencies are allocated for the uplink and the downlink. There is also a half duplex FDD mode that only needs one single frequency.

The WiMAX MAC is designed to support point-to-multipoint communications with uplink and downlink QoS handling that enable concurrent and reliable transmission of bandwidth demanding applications such as encoded video, bursty traffic such as data and delay-sensitive traffic such as voice. To meet these requirements WiMAX uses a connection-oriented MAC layer based on flexible TDMA to allocate bandwidth for different user equipment.

The WiMAX MAC layer has service-specific and common part sublayers. The service-specific sublayer defines encapsulations for cell services (ATM) and packet services (IP, Ethernet). The common part sublayer addresses management of multiple access over a shared radio channel.

Table 9-4. OFDMA sub-carriers and frequency spacing for different radio channels in mobile WiMAX.

Parameter	Transmission profiles			
Radio channel bandwidth (MHz)	1.25	5	10	20
Number of sub-carriers	128	512	1024	2048
Sub-carrier frequency spacing (kHz)	10.94	10.94	10.94	10.94

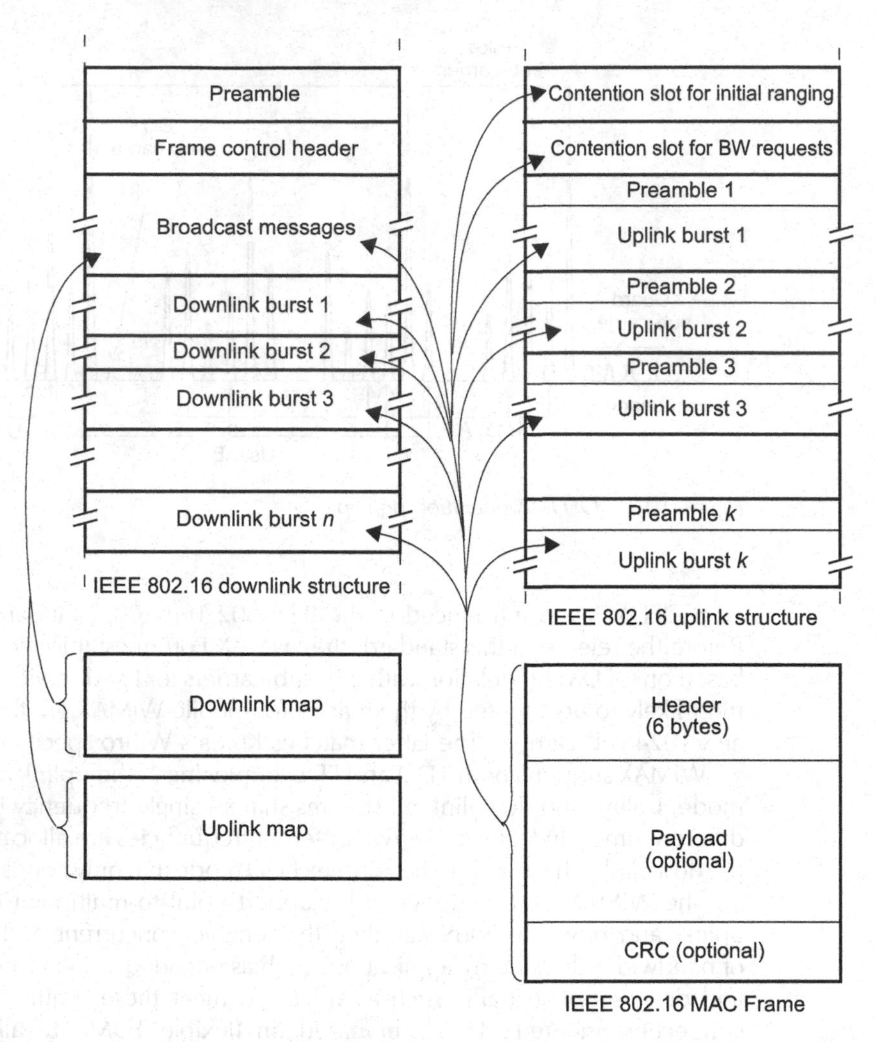

Figure 9-20. Simplified structure of the uplink and the downlink in IEEE 802.16 WiMAX. The downstream uses TDM and mobile stations share the radio channel using TDMA. MAC frames from different users are inserted in TDM bursts associated with different mobile stations. Uplink TDMA bursts are preceded by preambles. The uplink and downlink maps provide information about how time is distributed between users. Some uplink bursts are shared between some or all mobile stations. In these bursts collisions may happen.

All traffic between base stations and mobile stations is mapped into connections identified with a connection identifier (CID). Base and mobile stations use the CID to short and schedule MAC frames. Connections are therefore the basis of the QoS provision in the air interface. Base stations deliver traffic to mobile stations attached to them using TDM frames. The uplink is based on TDMA (see Figure 9-20). Mobile stations request uplink bandwidth to the base station and the base station grants bandwidth to mobile stations. The base station broadcasts information about how downlink and uplink transmission time is distributed between mobile stations in the downlink map (DL-map) and the uplink map (UL-map). The process by which base stations allocate bandwidth to the mobile stations for the specific purpose of making bandwidth requests is called polling. There are three different polling modes. QoS in the air interface strongly depends on the selected polling mode:

1. *Unicast polling* happens when mobile stations send bandwidth request messages in uplink bursts allocated for their exclusive use. No explicit polling message from the base station is necessary. The base station simply allocates bandwidth in the UL-map to allow the mobile station to send bandwidth request messages in its time slot.
2. *Multicast/broadcast* polling is possible if the bandwidth allocated in the UL-MAP is associated to a multicast or a broadcast CID rather than to a unicast CID. The mobile station is then allowed to issue bandwidth request messages in the contention slot allocated for that CID. Collisions in multicast/broadcast bandwidth requests may happen in contention slots. This means that bandwidth grants are subject to unpredictable delay when multicast/broadcast polling is used and therefore this operation mode is not recommended for real-time applications.
3. *Unsolicited grant* does not need explicit bandwidth request messages. A fixed amount of bandwidth is reserved in periodic time intervals known for the mobile station.

All uplink connections are mapped to one of the five possible uplink scheduling services. Every uplink scheduling service is suited to a specific family of applications. The defined scheduling services are as follows:

- *Unsolicited grant service* (UGS) is best suited to real-time services like PCM voice or emulation of circuit-switched services. UGS services never uses contention-based bandwidth request mechanisms to ensure maximum predictablity.
- *Real-time polling service* (rtPS) is designed to carry real-time service flows that generate variable size data packets on a periodic basis such as MPEG video. These services use unicast polling.
- *Extended rtPS* (ErtPS) is a mixture between the UGS and the rtPS scheduling services. ErtPS relies on unsolicited unicast grants like the UGS but can carry variable size data packets like the rtPS.
- *Non-real-time polling service* (nrtPS) is suited to carrying service flows that generate variable size data packets. Unlike rtPS, real-time operation is not required for these services. A typical example of nrtPS is FTP.
- *Best effort* (BE) service offers efficient transport of best effort traffic such as Internet data traffic. Contention-based bandwidth requests are allowed. Unicast polling and unsolicited grant operation is possible as well.

Other features supported by the WiMAX physical layer include different handoff modes, traffic encryption and key management. However, the most important WiMAX advanced feature is the availability of a full range of smart antenna technologies which enormously increase system range and capacity.

9.2.2.3 WiMAX and 3GPP Network Integration

WiMAX will play an important role in future mobile infrastructure. This is the reason why interworking between WiMAX and 3GPP networks is becoming an important topic. The same requirements and the same solutions apply to both WLAN and WMAN integration with 3GPP networks (see Section 9.2.1.3). That means that GAN, VCC and IMS are suitable solutions for different levels of integration between 3GPP networks.

There is a critical difference, however, between Wi-Fi/3GPP integration and WiMAX/3GPP integration. Unlike Wi-Fi, WiMAX aims to become a full alternative to 3GPP networks. Pointing in this direction, the WiMAX Forum has specified a reference model for mobile networks based on the IEEE 802.16 air interface that includes all the network elements, interfaces and procedures to operate as a standalone network. This does not happen with Wi-Fi, with reach usually limited to a few tens of metres. While integration between Wi-Fi and 3GPP networks is a problem of extending the original 3GPP infrastructure to a new air interface, integration between WiMAX and 3GPP resembles more a problem of interworking between two heterogeneous networks.

The WiMAX Forum network reference model is based on IETF and IEEE signals and protocols. This reference model is very flexible and enables network operators to choose between different physical configurations. The WiMAX network is made up of an access service network (ASN) and a connectivity service network (CSN) connected through an ASN gateway (ASN-GW; see Figure 9-21). The WiMAX Forum reference model defines several reference points (*R1*, *R2*, etc.) that enable communication between the different entities that made up the network. Every reference point has its own associated user plane and control plane protocols and procedures. IETF and IEEE protocols are reused when possible. The WiMAX Forum network includes procedures for network discovery, IP

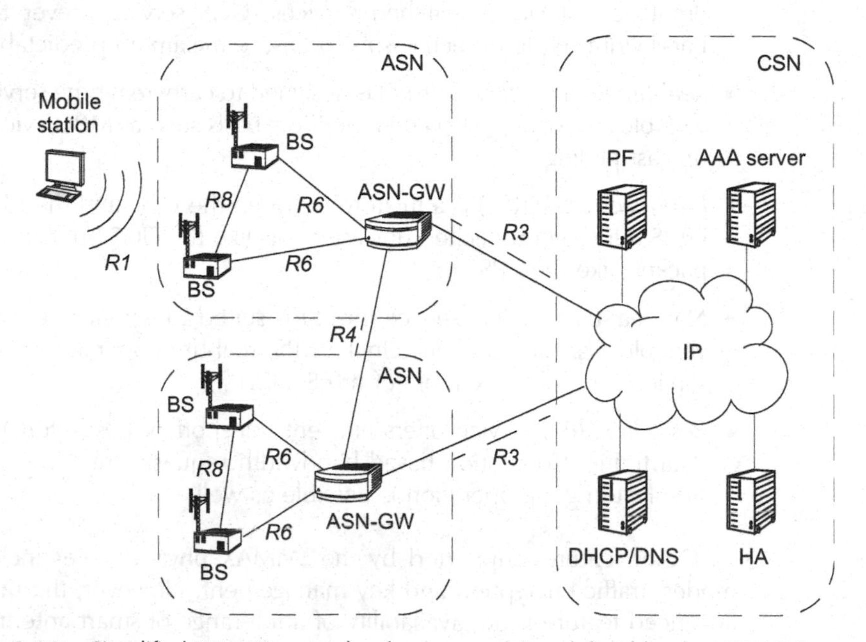

Figure 9-21. Simplified WiMAX network reference model as defined by the WiMAX Forum.

address management, authentication, authorization and accounting (AAA), QoS provision, CSN and ASN anchored handoffs, etc.

The WiMAX Forum has defined an interworking model between WiMAX and 3GPP networks based on the 3GPP I-WLAN standard. I-WLAN is a 3GPP standard defined in TS23.234 and TS33.234 that provides several interworking levels between WLAN and 3GPP mobile networks. In the simplest scenario, I-WLAN allows authentication and billing of wireless user equipment through the 3GPP mobile network. This has been also used for Internet access in public Wi-Fi hotspots managed by mobile operators. A more complex deployment scenario for I-WLAN also allows access to packet-switched services provided by the mobile network.

I-WLAN allows direct interconnection of the WiMAX and 3GPP core network without defining new interfaces or protocols. In fact, the WiMAX network is more complex than WLAN networks considered by the I-WLAN standard and therefore this standard needs to be customized for this particular application. The WiMAX Forum builds the WiMAX/3GPP interworking solution on top of the I-WLAN specification without modifying it.

Most interconnection issues are related with AAA. The interworking solution adds an AAA server/proxy in the AAA signalling path between the user equipment and the I-WLAN AAA server (see Figure 9-22). The I-WLAN AAA server is the entity that authenticates users and helps with any other AAA procedure over the WiMAX network and users connected to it. However AAA signalling messages always traverse the AAA server/proxy placed in the WiMAX CSN. The I-WLAN AAA server may also need to retrieve user profile information from the HLR or other 3GPP databases. For this reason, the I-WLAN standard defines interfaces with these databases.

Integration between WiMAX and the 3GPP network may be limited to unified registration and billing, but may also allow access to packet-switched services through

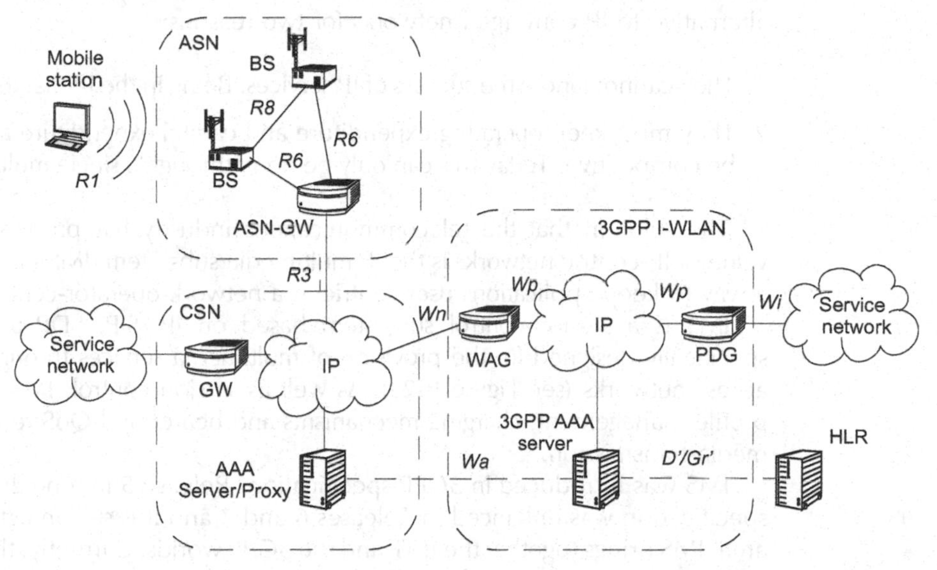

Figure 9-22. Simplified WiMAX and 3GPP interworking architecture based on the I-WLAN standard.

the 3GPP network. If this is the case, the WiMAX CSN is connected to the I-WLAN WLAN access gateway (WAG) through the W_n reference point. Information traversing the WAG is tunnelled to an I-WLAN packet data gateway (PDG) using the W_p interface before it enters the network that implements the services required by WiMAX users.

Work on integration of WiMAX and other networks is just starting. While the I-WLAN architecture is a good starting point, more integrated interworking between WiMAX, 3GPP and other networks is expected to appear in the future.

9.3 The IP Multimedia Subsystem

The public switched telephone network (PSTN) plays a central role in the business value chain associated to the POTS. However, things are less clear if the PSTN is replaced by the Internet and the POTS is replaced by IP telephony. In fact, many companies have succeeded in building viable business models around the Internet without the consent of any network operator. IP telephony service providers are just one of the many possible examples of a successful business model built over the Internet. Others are search engines, online shops, network games and document repositories. Moreover, peer-to-peer (P2P) applications are becoming increasingly important on the Internet, but network operators fail to control or manage these applications with the currently available mechanisms at their disposal.

It can be said that IP and the Internet have the effect of moving the value of telecommunications services from the network to the end equipment and, thus, IP makes it difficult for network operators to obtain a high return to their investments. Network operators are facing the risk of becoming mere 'bit pipe' administrators. Today, telcos are seeing how their network becomes a commodity but they are unable to find an alternative to IP converged networks for two reasons:

1. They cannot ignore the success of IP services. Being in the IP market is a must for them.
2. They must keep operating expenditure and capital expenditure as low as possible to be competitive. Today this can only be done through a single multiservice IP network.

The solution that the telecommunications industry has proposed to maintain the value of IP-centric networks is the IP multimedia subsystem (IMS). In other words, IMS is a way to keep applications user-centric in a network-operator-centric business model.

IMS is a session-control subsystem based on IP, SIP, SDP and other protocols, specifically designed for the provision of multimedia services through a wide variety of access networks (see Figure 9-23). As well as session control, IMS provides subscriber profile management, charging mechanisms and bearer and QoS resource allocation for media transmission.

IMS was introduced in 3GPP specifications Release 5 in June 2002. The initial IMS specification was enhanced in Releases 6 and 7 and it is still an active standardization area. IMS brings together the IETF and the 3GPP worlds. Currently, the organization that releases IMS standards is the 3GPP but most of the IMS protocols are largely based on IETF RFCs.

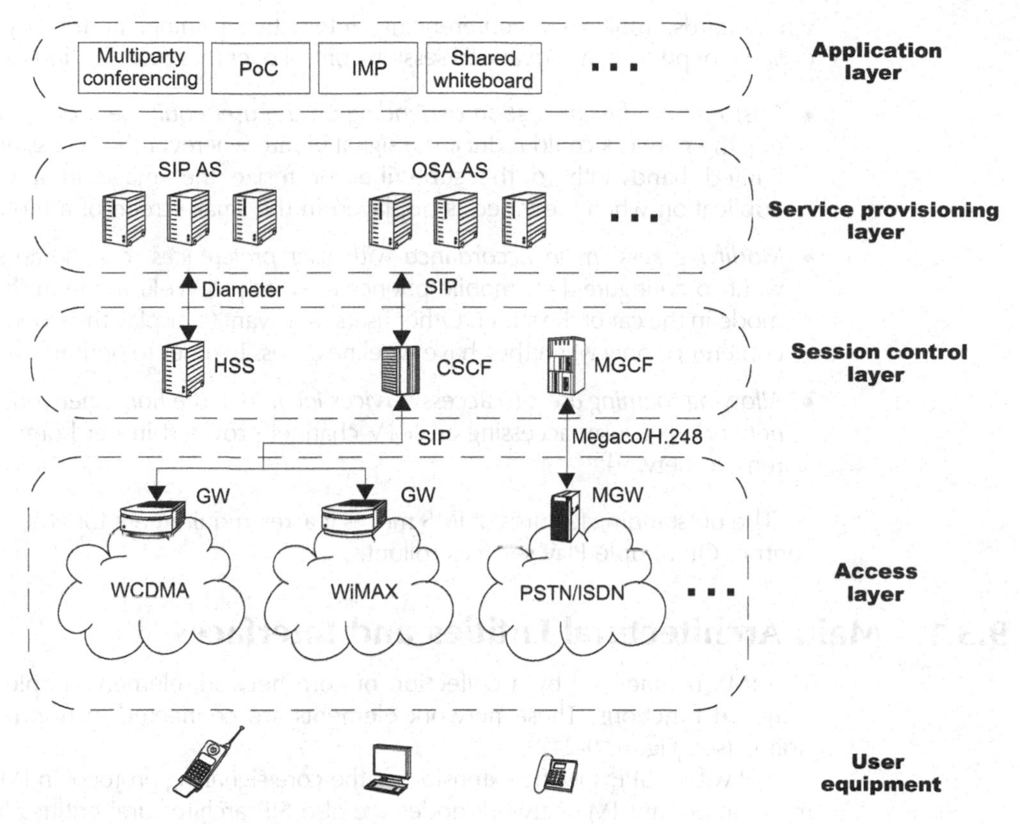

Figure 9-23. IMS layered architecture.

The key components of the IMS architecture are the call session control function (CSCF) and the home subscriber server (HSS). Together, these components control how users access the services in a highly versatile and customizable environment. The power of IMS is manifested through the great variety of user terminals that potentially could log on in the IMS capable network through many types of access mechanisms. Supported user equipment includes mobile phones but also PDAs, computers, VoIP phones, gaming consoles and many others. These devices must support the IPv6 protocol stack and run SIP user agents. Supported access networks are:

- *Wireline networks* based on DSL, PON, data over cable service interface specifications (DOCSIS) or Ethernet;
- *Cellular mobile networks*, including WCDMA, cdma2000, GSM and GPRS;
- *IEEE wireless networks*, including Wi-Fi and WiMAX.

There are also many applications potentially available for IMS subscribers, e.g. video telephony, push to talk over cellular (PoC), instant messaging and presence (IMP), shared whiteboards, multiparty conferencing, interactive gaming, and many others. IMS is capable of performing advanced session control over these applications in different ways:

- *Customizing the application depending on the user equipment or the access network*, e.g. the network could reduce the signal bitrate wherever the access network provides limited bandwidth to the subscriber or resize the image in a video telephony application when the video is displayed in the small screen of a mobile phone.
- *Modifying sessions in accordance with user preferences*, e.g., some subscribers may want to configure their mobile phones to switch to Wi-Fi mode in the office and 3G mode in the car or the street. Other users may want to display the video signal in a video conference only when they have wireline access in order to optimize bandwidth usage.
- *Allowing roaming users to access services located in the home network from the visited network*, e.g. user accessing an IPTV channel provided in her home network from a remote network.

The outstanding features of IMS makes it a key requirement for FMC and effective IP-centric, Quadruple Play services rollout.

9.3.1 Main Architectural Entities and Interfaces

The IMS is made up by a collection of core network elements implementing sets of standard functions. These network elements are connected in normalized reference points (see Figure 9-24).

SIP with 3GPP specific extensions is the core signalling protocol in IMS and therefore many important IMS network nodes are also SIP architectural entities like user agents, proxies or registrars.

One of the most important functions within the IMS is the CSCF. This function is implemented by SIP servers. The CSCF interconnects subscribers with services and performs session handling for them. The most important CSCF types are the following:

- *The proxy CSCF (P-CSCF)* is the first contact point for user equipment within the IMS. The P-CSCF is a SIP proxy that forwards SIP messages between the user equipment and

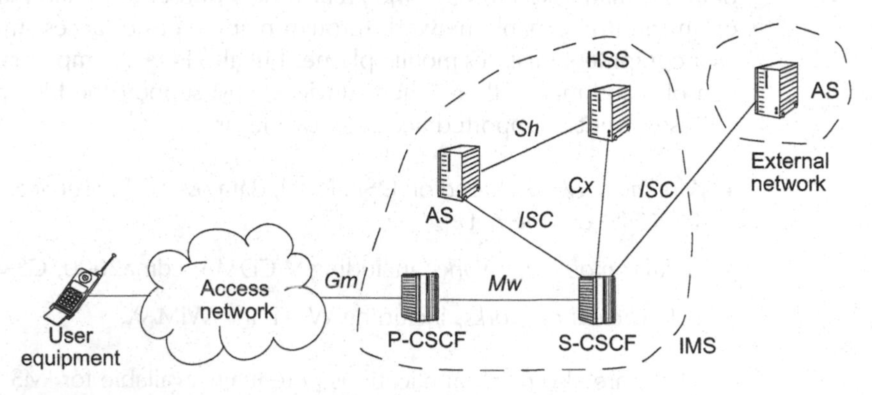

Figure 9-24. Simplified view of the IMS architecture.

the right nodes in the IMS. The P-CSCF is also involved in authorization of bearer resources, QoS management and securing communications between the user equipment and itself. To perform these tasks the P-CSCF needs to be connected to the user equipment (G_m reference point). This connection must support the SIP protocol.

- *The serving CSCF (S-CSCF)* is the function that provides services to subscribers. The S-CSCF behaves like a SIP registrar. It accepts or rejects registration requests from the user equipment. Once authenticated, subscribers are authorized to access network resources depending on their particular profiles. The S-CSCF may also behave as an ordinary SIP proxy forwarding SIP requests or replies towards the right destination endpoint. The S-CSCF is connected to the P-CSCF in the M_w reference point. This point supports the SIP protocol.

A P-CSCF and a S-CSCF is designated for every subscriber during registration in the IMS enabled network. Splitting the CSCF among different network nodes enables support of roaming users (see Figure 9-25). Services are always provided from the subscriber's home network. If the subscriber is in a roaming scenario, the visited network selects a P-CSCF for that user that acts as the contact point of the user with the IMS. If the user is in a non-roaming scenario, then the P-CSCF is still necessary but in this case it can be selected by her own home network.

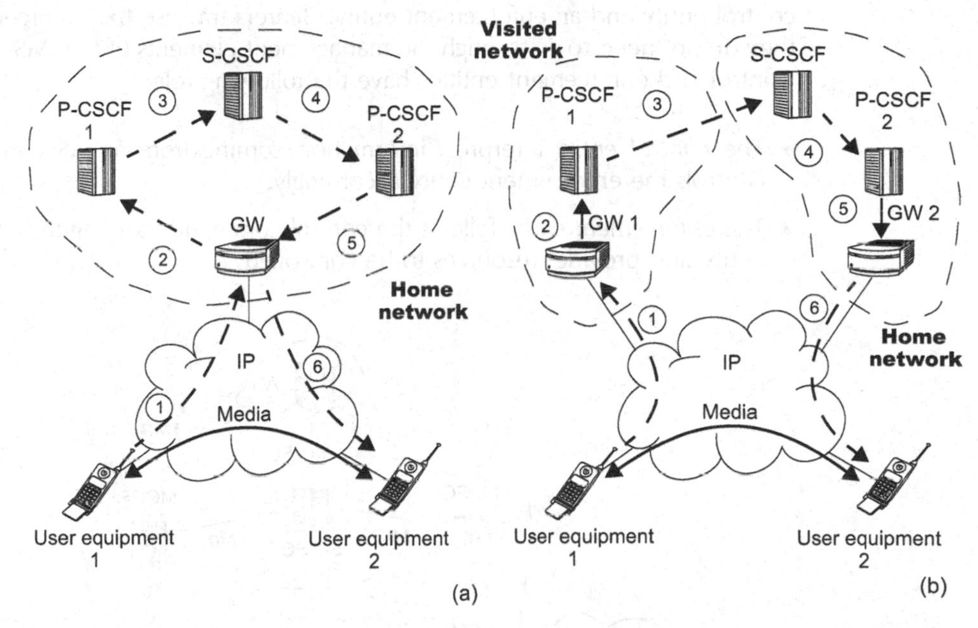

Figure 9-25. Simplified diagram of the network elements and information flows implied in session establishment for roaming and non-roaming users in IMS. (a) Conference in a non-roaming scenario. The media goes from origin to destination through the gateway GW. The SIP signalling traverses the P-CSCFs designated for source and destination subscribers. The common S-CSCF routes signalling between P-CSCFs. (b) Conference in a roaming scenario. The media goes from origin to destination through the gateways GW 1 and GW 2. The signalling traverses the P-CSCF 1 designated by the visited network for user equipment 1 and the P-CSCF 2 designated by the home network for user equipment 2. The common S-CSCF routes signalling between P-CSCFs.

The IMS can provide customized services to every subscriber depending on his profile. To do that, the IMS must store users' identities along with their subscription information, including services that they are allowed to use and servers and networks they are allowed to access. All this data is held in the HSS. The HSS is also useful to keep centralized information about which particular user is connected to the network, who is registered at a given moment, and which is his physical location. In order to allow the S-CSCF to authenticate users, authorize usage of network resources, and route signalling messages to the right destination, it needs to access the HSS. This connection is done at the C_x reference point. The C_x interface supports the Diameter protocol, specifically designed for AAA tasks. The HLR database is closely related to the HSS. In fact, the HLR provides a subset of the functions of the HSS.

Large networks may contain several HSSs. In such cases it may be necessary to implement a subscription locator function (SLF) in the IMS. The SLF is in charge of locating the HSS containing the data wanted by other network elements like an S-CSCF. The S-CSCF and the SLF are connected at the D_x reference point that supports the Diameter protocol.

The IMS hosts services in application servers (ASs). Examples of services hosted by ASs are presence, messaging, shared whiteboard and many others (see Section 9.3.2)

The IMS is designed to tightly control the operation of networks that carry user information but IMS does not carry user data itself. IMS gateways are divided into a control entity and an enforcement entity. Bearers traverse the enforcement entities but they do not need to go through the management elements of the IMS (see Figure 9-26). Control and enforcement entities have the following role:

- The *control entity* interprets information coming from an AS and an S-CSCF and controls the enforcement entity accordingly.
- The *enforcement entity* follows the control commands coming from the management entity and provides resources to be controlled.

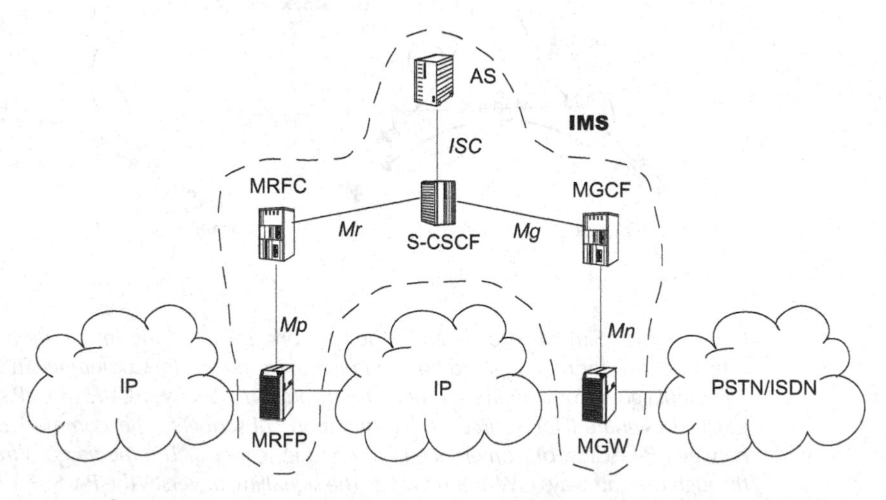

Figure 9-26. The IMS controls user data flows but it does not transport user data itself. Gateways are divided into management entities (MRFC, MGCF) and enforcement entities (MRFC, MGCF).

The multimedia resource function processor (MRFP), is an enforcement entity that performs operations on media streams in an IP network such as stream mixing and audio or video transcoding. The MRFP is controlled by the multimedia resource function controller (MRFC) following instructions from ASs and S-CSCFs. The interface between the management and the enforcement entities (M_p reference point) is compliant with Megaco/H.248.

Interworking with PSTN/ISDN is based in the same concept. In this case, the enforcement entity is a media gateway (MGW). The MGW terminates bearer channels in the PSTN/ISDN network and media streams in the IP network. It may need to perform transcoding functions on user data as well. Call control is performed by a media gateway controller function (MGCF). The MGFC controls the media translations performed by the MGW by means of the M_n interface based on Megaco/ H.248. The S-CSCF must interact with an MGCF when a user desires to communicate with an entity connected to the PSTN/ISDN; to do that, SIP is used. The MGCF translates SIP messages to SS7 signalling messages used by the circuit-switched network. To do this task, the S-CSCF is assisted by the breakout gateway control function (BGCF). The BGCF finds the network in which breakout is to occur and chooses the right MGCF there.

9.3.2 Services

Services in the IMS are hosted by entities known as application servers (ASs). ASs may be provisioned by the network operator that owns the IMS-enabled network or by an external service provider. In fact, ASs are envisioned as the way service and content providers will interact with subscribers in the IMS environment. With the help of this new approach, service providers and subscribers will have at their disposal IMS advantages like guaranteed QoS. Trust relationships between IMS network elements owned by the network operator and ASs owned by service providers greatly simplify charging to subscribers for the services. Users do not need to log every time they need to access a new service; only a single registration procedure with the IMS is necessary. These features, known as unified billing and single sign-on are two key advantages of IMS for service providers.

On the other hand, the IMS operator gets closer control of network resources and shares revenues with service providers. IMS and non-IMS provision models can coexist. However, IMS service providers have a competitive advantage against non-IMS service providers.

Within the IMS, the S-CSCF is connected to the ASs by means of the IMS service control (ISC) interface, which is SIP-based. ASs not based on SIP can also be connected to the IMS with the help of dedicated gateways (see Figure 9-27). Interworking with customized applications for mobile networks enhanced logic (CAMEL) and open services architecture (OSA) is currently supported by the 3GPP standards.

- *CAMEL* allow operators to define value-added services equivalent to the intelligent network (IN) services available for fixed networks but for GSM and UMTS mobile networks. CAMEL is designed to enable service provision even for roaming users. CAMEL standards were released in 1997 for first time and have been improved several times. CAMEL services are accessible to IMS subscribers through a gateway implementing the service switching function (SSF).

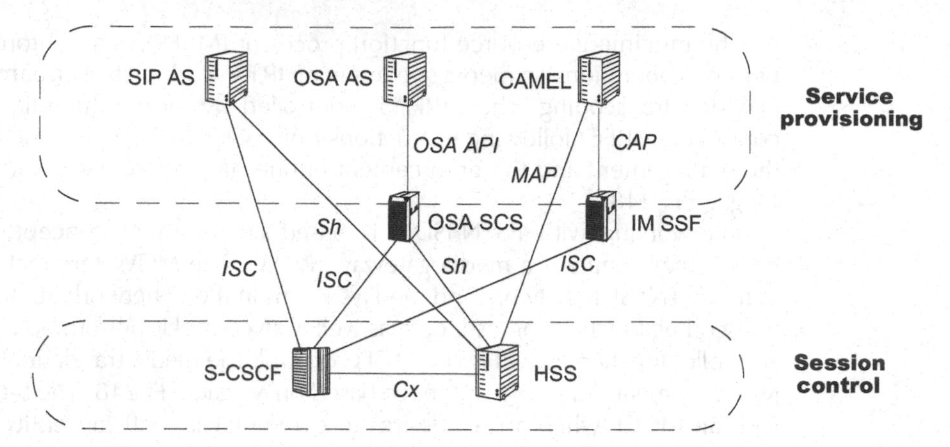

Figure 9-27. In IMS, services are provided by means of ASs connected to the S-CSCF. SIP ASs exchange information directly with the S-CSCF in the ISC interface. Other ASs rely on special gateways that translate information into ISC standards. AS can also directly access to the information stored in the HSS by means of the S_h interface.

- *The OSA* is a joint initiative of the 3GPP, ETSI, telecoms and Internet converged services and protocols for advanced network (TISPAN), and a multivendor consortium known as Parlay Group. The OSA standard defines a framework that makes the functionality of 3G mobile networks accessible to service application developers in a standard, flexible and scalable way. OSA is an advantage to service developers, since it abstracts them from the complexity of the underlying network infrastructure and low-level telecommunication protocols, allowing them to focus on the creation of the service logic. OSA applications are accessible to IMS users through an OSA service capability server (SCS).

Currently, definition of interoperable services for IMS depends on an organization that groups equipment vendors, mobile operators and software companies known as the Open Mobile Alliance (OMA). The OMA was created in June 2002 and its activity is essential for global adoption of mobile services. Specifications maintained by the OMA define requirements for the multimedia messaging service (MMS), digital rights management (DRM), instant messaging and presence (IMP), push to talk over cellular (PoC), gaming services, broadcast services, and many others.

An important example of IMS service is voice call continuity (VCC), introduced in the Release 7 of the 3GPP specifications. VCC enables transparent handoff of the voice service for any access network. VCC works for mobile or wireline access networks and for circuit-switched or packet-switched modes. However handoff of multimedia streams is beyond the scope of the current specification.

The VCC service is provided by a special IMS AS. On the request of dual user equipment, the VCC AS generates signalling to redirect voice traffic from the origin to the destination access networks (see Figure 9-28). Like other IMS ASs, the VCC AS is a SIP server. That means that it requires help of signalling gateways to interact with non-SIP devices. To operate with the GSM/UMTS circuit-switched domain, the IMS uses the MGCF to translate SIP messages to SS7 commands that can be interpreted by the MSC.

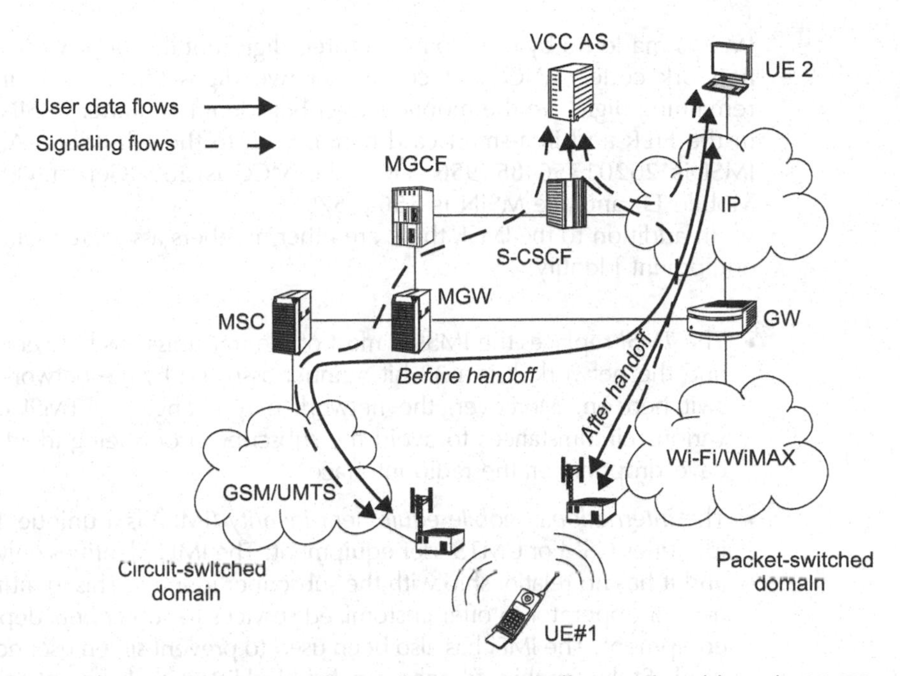

Figure 9-28. Simplified operation of VCC. The UE 1 voice call is transferred from the UMTS or GSM radio access network to a Wi-Fi or WiMAX access network. In the example, the receiver, identified as UE 2 is operating in packet-switched mode and is therefore receiving VoIP data.

9.3.3 User Identity

In IMS as well as in GSM and UMTS networks, user identification is independent of identification of user terminals. When compared with the PSTN, the concept of user identification is new because the PSTN identifies telephones rather than users.

User identification adds personal mobility to terminal mobility. Users can theoretically attach their smart cards to any mobile phone. The network stores the user identities in the HLR or HSS along with their profiles and many service preferences. During registration, the subscriber information from the smart card is authenticated rather than the user equipment itself. This allows subscribers to log into the network from different devices and enjoy a static environment (including the telephone number).

IMS Subscriber identities are very different from GSM and UMTS identities. Networks that are IMS and GSM/UMTS at the same time should provide their subscribers with GSM/UMTS identities as well as IMS identities.

9.3.3.1 User Identity in GSM and UMTS

Users in GSM and UMTS networks are identified by their international mobile subscriber identity (IMSI), a unique, non-dialable number of up to 15 digits. The

IMSI is made up by three parts: a three-digit mobile country code (MCC), the mobile network code (MNC) that consists of two digits (three digits in the USA) and the remaining digits are the mobile subscriber identity number (MSIN). The IMSI is stored in the HLR and in a smart card handed out to the subscriber. An example of a valid IMSI is 262013564857956. Here the MCC is 262 (Germany), the MNC is 01 (T-Mobile D) and the MSIN is 3564857956.

In addition to the IMSI, there are other numbers associated with subscriber and user equipment identity:

- The *TMSI* replaces the IMSI in most of the transmissions between the user equipment and the network. It is a 32 bit number assigned by the network when the mobile is switched on. Moreover, the network may change the TMSI of the mobile under various circumstances to avoid the subscriber from being identified and tracked by eavesdroppers on the radio interface.
- The *international mobile equipment identity* (IMEI) is a unique 15 digit number that identifies GSM or UMTS user equipment. The IMEI identifies only the user equipment and it has no relationship with the subscriber identity. This identifier can be used by a network operator to offer customized services to subscribers depending on their user equipment. The IMEI has also been used to prevent stolen user equipment from being used. Stolen mobile phones can be blacklisted and theoretically left unusable. An example of a valid IMEI number is 356057-00-481557-7.
- The *mobile station integrated services digital network* (MSISDN) is the name given by the 3GPP to the telephone number associated with a subscriber. The MSISDN conforms to ITU-T E.164 and should comply with each country's ISDN numbering plan. MSISDN numbers consists of a country code (CC), a national destination code (NDC) and a subscriber number (SN). An example of a valid MSISDN is 34659057404.

9.3.3.2 User Identity in IMS Networks

IMS subscribers are identified by means of one or various private user identities and one or various public user identities:

- A *private user identity* can be thought of as a user subscription identifier. In other words, there are as many private user identities as subscriptions to the IMS network the user has. The private user identifier uses a username@realm syntax as defined in RFC 2486. Storage of the private user identity in a subscriber's smart card and in the HSS is necessary to the user's subscription authentication during the registration process. The private user identity is used for transactions that may have security concerns (for example, registration, re-registration and de-registration), but it is never used for routing of SIP messages. This means that one user does not need to know other users' private identities to communicate with them.
- A *public user identity* identifies the user rather than the user subscription. One user may have several public identities for different purposes (see Figure 9-29). It is also possible to share one or more public user identities among various private user identities. SIP

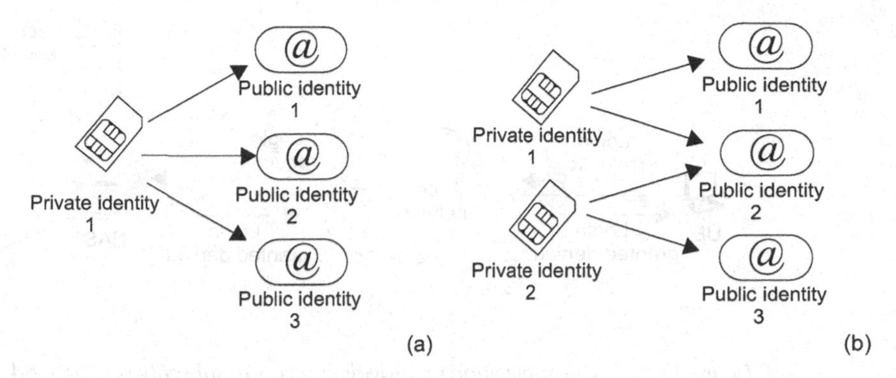

Figure 9-29. Relationship between private and public user identities in the IMS. (a) A single private user identity is associated with three public user identities; (b) two private user identities share the public user identity 2.

URIs like 'sip:alice@trend.com' or tel URIs like 'tel:+44-1628-503500' are both accepted for public user identity specification. Public user identities do not need to be authenticated during the registration process and their main purpose is to allow other users or services to reach the subscriber.

Public and private user identities are not changeable by the user. When a public user identity is shared among various user equipment the subscriber can be reached at the same time in different devices, e.g. a PDA and a computer. Some user equipment is able to select which user equipment it wants to communicate with, by means of globally routable user agent URIs (GRUUs). A GRUU is an entity that is made up of a unique combination of a public user identity and a code that identifies the user agent. A GRUU can be public or temporary. Public GRUUs expose the identity of the user and temporary GRUUs hide the user identity for privacy purposes. GRUUs are SIP URIs with the special *gr* parameter. A public GRUU may look like 'sip:alice@trend.com;gr=kjh29x97us97d' while a temporary GRUU looks like 'sip:8ffkas08af7fasklzi9@trend.com;gr'.

9.3.4 AAA with Diameter

Authentication, authorization and accounting (AAA) are three related terms that can be defined as follows:

- *Authentication* is the act of verifying the identity of an entity, e.g. the subscriber of a service provided by a network or a remote server that requests a specific network resource.
- *Authorization* is the act of determining whether a service subscriber or any other requesting entity will be allowed to access to a resource, a network or a server.
- *Accounting* is the act of collecting information on resource usage for the purpose of capacity planning, auditing, billing or cost allocation.

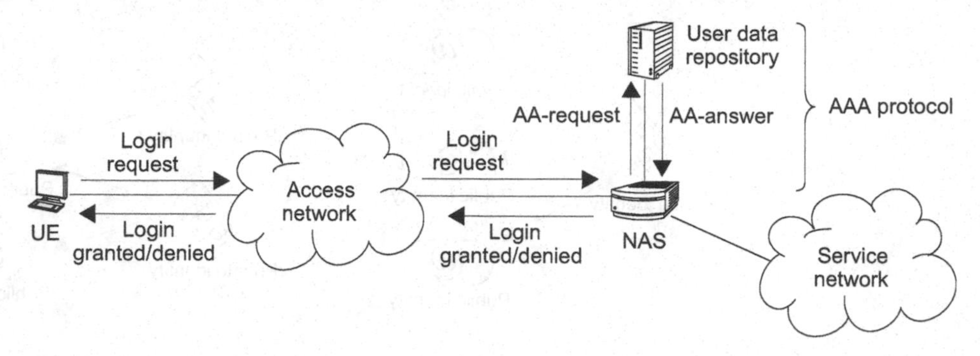

Figure 9-30. NAS application for providing access to subscribers to network resources. The NAS is an IP router that authenticates users and authorizes access to the network resources. However, user subscription data is stored in the user data repository and the NAS needs to check this information. Communication between the NAS and the user data repository is done with an AAA protocol like RADIUS or Diameter.

To perform AAA, the system needs to know the resources that every user is allowed to access. This information is part of the user subscription data and in small networks it can be stored in the server that provides access to the network. However, this solution becomes impractical for medium and large networks. When this happens, it is usually better to store user subscription information in a dedicated database and allow the access servers to read this information from the database. This solution provides enhanced security, allows users to login from different access servers without replicating user credentials and keeps track of user information records from a central location. The protocol that communicates the access servers and the user subscription database is an AAA protocol like the Remote Authentication Dial-In User Service (RADIUS) protocol (see Figure 9-30).

Within the IMS, AAA information exchange is performed by the Diameter protocol. Specifically, Diameter is the basis of the C_x interface that allows information exchange between the HSS and different types of CSCFs. Diameter, as specified in RFC 3588, is a replacement for the classical RADIUS. RADIUS was initially used to provide PPP dial-up access to network resources by means of a network access server (NAS). However, it has been necessary to develop a new AAA protocol to take into account the requirements of new access technologies like wireless cellular networks.

Diameter has been derived from RADIUS but the former provides many advantages when compared with the latter. RADIUS operates over the unreliable UDP transport protocol but Diameter works either with TCP or the stream control transmission protocol (SCTP) on port 3868. Both TCP and SCTP provide reliable communications and they are more suited to exchanging sensitive data transported in AAA applications than UDP. Moreover, the SCTP has been specifically designed to transport signalling messages over IP networks and provides even more benefits to AAA applications than TCP, e.g. path selection and monitoring or protection against denial of service (DoS) attacks. In fact, use of SCTP in IMS is mandatory. Other advantages of Diameter over RADIUS are a new end-to-end security framework, automatic peer discovery capability, better support of vendor-specific messages, improved negotiation of capabilities during session startup, support of longer

Table 9-5. Comparison between the Diameter and RADIUS AAA protocols.

	RADIUS	Diameter
Transport protocol	UDP	TCP and SCTP
Security	Hop-to-hop	Hop-to-hop, end-to-end
Peer discovery	Not supported	Supported
Server-initiated transactions	Not supported	Supported
Maximum attribute size	255 bytes	16 777 215 bytes
Vendor-specific support	Vendor-specific attributes	Vendor-specific messages and attributes
Capabilities negotiation	Not supported	Supported

attributes in messages and the possibility to start transactions from the Diameter server and not only from the client (see Table 9-5).

Diameter has been conceived as a peer-to-peer protocol. That means that, with Diameter, the communicating parties may act as both clients and servers and they may change their role in different transactions. In addition to Diameter clients and servers, there are Diameter agents. There are different types of Diameter agents (see Figure 9-31):

- *Relay agents and proxies* forward Diameter messages to the appropriate destination. This is useful in large networks to aggregate messages from different access servers to a central node. The difference between relay and proxy agents is that proxies are allowed to

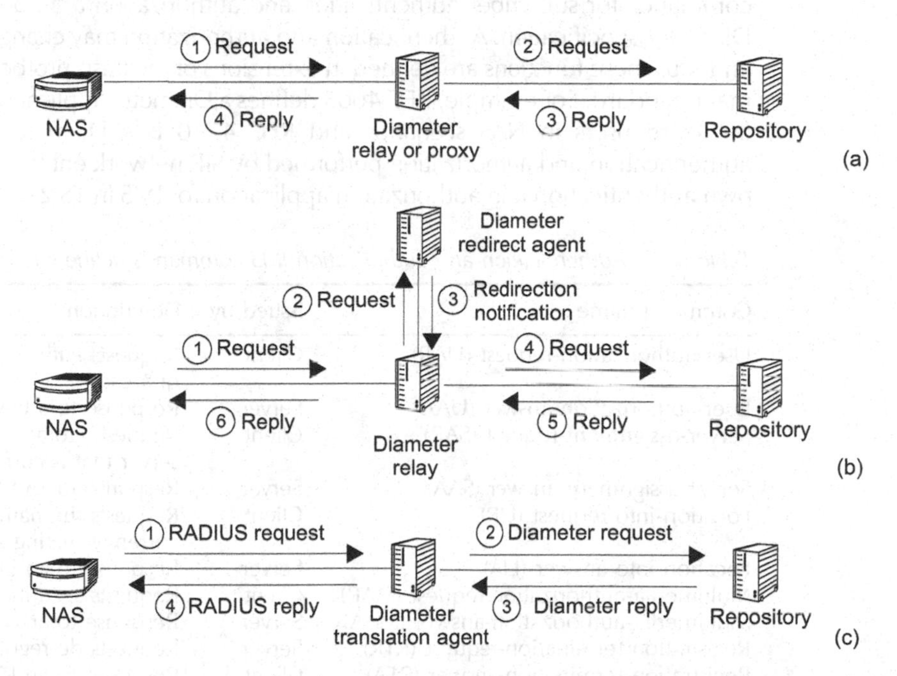

Figure 9-31. Diameter agents: (a) relays and proxies route messages to a remote location; (b) redirect agents notify how to reach a destination; (c) translation agents translate between different AAA protocols.

modify messages with different purposes but relay agents are only allowed to append or remove service layer routing information to Diameter messages but they do not modify other portions of them.

- *Redirect agents* do not relay Diameter messages; they rather notify other agents how to reach a destination.
- *Translation agents* translate messages between one AAA protocol and another. For example a translation agent may translate RADIUS messages into Diameter messages.

Diameter messages carry sensible information. For example an operator may charge its subscribers by means of an application based on Diameter. In order to provide proper confidentiality, authentication and integrity, Diameter relies on features provided by its own Diameter and other external protocols. End-to-end security is optional and provided by an extension of the base Diameter protocol. Hop-by-hop protection is based either on IP security (IPsec) or transport layer security (TLS). However, for IMS the use of IPsec in the network layer is mandatory. Specifically, IMS standards mandate that the IPSec encryption security envelope (ESP), as defined in RFC 2406, must be used for all inter-domain signalling traffic to achieve confidentiality, integrity and data source authentication. Peer authentication at session startup, negotiation of security associations and secure key management in IMS use the standard mechanisms defined by the Internet key exchange (IKE) as per RFC 2409.

Peer authentication and authorization in Diameter sessions must not be confused with subscriber authorization and authentication (see Section 9.3.6.1). In fact, the necessary commands for subscriber authentication and authorization are not included in the base Diameter specification. Authentication and authorization may change among applications and thus these functions are defined in extensions of the basis protocol not included in the main standard. For example, RFC 4005 defines a Diameter application to authenticate and authorize users in NAS scenarios and RFC 4740 is a Diameter application for user authentication and authorization performed by SIP network entities. The 3GPP defines its own authentication and authorization application for IMS in TS 29.229 (see Table 9-6).

Table 9-6. Authentication and authorization IMS commands defined in TS 29.229.

Command name	Issued by	Description
User-authorization-request (UAR)	Client	Requests authorization for registration of a user
User-authorization-answer (UAA)	Server	Response to a UAR command
Server-assignment-request (SAR)	Client	Requests storage of the name of the server that is currently serving a user
Server-assignment-answer (SAA)	Server	Response to an SAR command
Location-info-request (LIR)	Client	Requests the name of the server that is currently serving a user
Location-info-answer (LIA)	Server	Response to an LIR command
Multimedia-authorization-request (MAR)	Client	Requests security information
Multimedia-authorization-answer (MAA)	Server	Response to an MAR command
Registration-termination-request (RTR)	Server	Requests de-registration of a user
Registration-termination-answer (RTA)	Client	Response to an RTA command
Push-profile-request (PPR)	Server	Updates the subscription data of a user
Push-profile-answer (PPA)	Client	Response to a PPR command

9.3.5 Policy and Charging Control

Charging and QoS control are different but related topics: in a network, users pay to use transmission resources that can be specified by a set of QoS metrics. Policy control and charging control were separate topics in old 3GPP releases. However, they have been unified in 3GPP Release 7. The current policy and charging control (PCC) framework is a joint evolution of the session-based local policy (SBLP) and the flow-based control (FBC) framework.

- As defined in Release 5 of 3GPP specifications, SBLP allows the IMS to control and manage the IP network resources devoted to a specific application. Specifically, the SBLP framework allows coordination of the QoS advertised during session setup (for example using an SDP body in an SIP INVITE transaction) and the actual QoS reserved in the bearer domain. Without this kind of coordination and control, user equipment would be able to reserve and use more resources than advertised during session setup without being charged for it. The policy decision function (PDF) is a key entity of the SBLP framework. The PDF authorizes specific service levels for different applications and users taking as a basis subscriber profiles and SIP/SDP control plane information.
- The FBC is defined in Release 6 of 3GPP specifications. The FBC framework solves the problem of charging at the bearer layer with flow granularity rather than with session granularity. Thanks to the FBC it becomes feasible to charge individual SIP VoIP calls using SIP signalling generated by the communicating parties, to charge individual FTP data flows and web services between specific subscribers and servers, and many other alternatives. Furthermore, FBC enables the IMS operator to use different charging models like volume-based charging, time-based charging, event-based charging, and combinations of them. An important element in the FBC framework is the charging rules function (CRF). This entity generates charging rules for the appropriate network elements where charging occurs.

In the new PCC framework, there is a single entity that generates charging rules and allocates network resources. This unified entity, known as policy control and charging rules function (PCRF), generates PCC rules that are used at the same time for policy and charging purposes. PCC rules are applied to an enforcement point known as policy and charging enforcement function (PCEF), that is usually integrated in a media gateway such as a GGSN (see Figure 9-32). The interface between the PCEF and the PCRF is based on Diameter as specified in RFC 3588 with appropriate extensions. The PCEF has the ability to process and filter the user plane traffic to ensure that policy defined by the PCC rules is correctly applied. To do that, the PCEF may discard or delay user traffic and may also deny resource reservations carried out by user equipment. The PCEF also generates charging information when a so-called chargeable event occurs, that is, user activities that utilize or consume network resources. Charging information is then processed either by the offline charging system (OfCS) or the online charging system (OCS), depending on the agreement with the subscriber. Offline and online charging are defined as follows:

- *Offline charging* is a billing mechanism where charging information does not affect the service in real time. Charging information is delivered to the operator after the resource usage occurred and as a result interaction with the service is not possible.

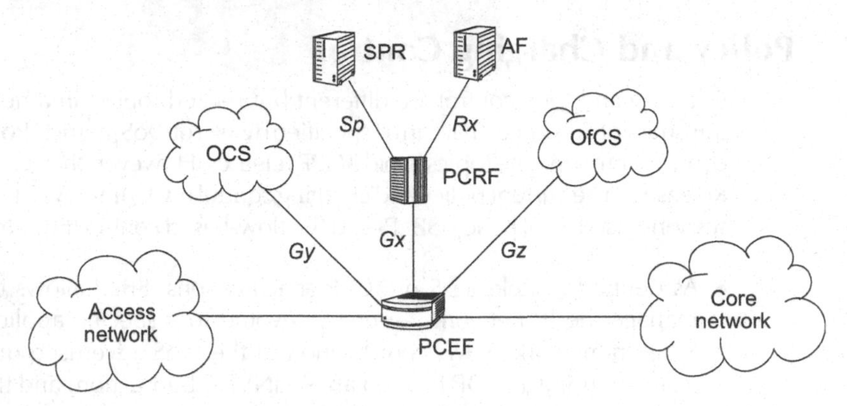

Figure 9-32. PCC architecture for IMS.

- *Online charging* involves real-time interaction between the charging framework and the service to accomplish credit-control or budget supervision. Authorization for resource usage must be obtained by the network prior to the actual resource usage occurring. Resource usage permissions may be limited in their scope. For example, permission could be granted for a limited amount of data or time.

The OfCS and OCS generate charging records that are delivered to a billing domain (BD), an operator's postprocessing system to process and store charging information. The PCEF and the OCS are connected through the G_y reference point that is Diameter-based and uses the Diameter credit-control application defined by RFC 4006. Connection between the PCEF and the OfCS is done by means of the G_z reference point that is Diameter-based as well and is defined in 3GPP TS 32.240.

To generate PCC rules, the PCRF receives dynamic session information from an application function (AF) that is usually represented by the P-CSCF designated for the subscriber. Examples of information that can be used for charging and policy control are registration in the IMS, a SIP call origination or release and many others. Generation of PCC rules may also depend on the user profiles. For this reason the PCRF is allowed to retrieve information from the database where profiles are stored. Within the PCC framework, this database is known as the subscription profile repository (SPR). The PCRF communicates with the AF by means of the R_x reference point and with the SPR by means of the S_p reference point. Both are based on the Diameter protocol.

9.3.6 Basic Procedures

IMS procedures include registration and de-registration, call setup and termination, messaging and many others. These procedures are based on dialogs between IMS network elements. Most of them use SIP and Diameter.

9.3.6.1 Registration

Before being served, users must register in the IMS network. However, there are at least three basic conditions that must be fulfilled before registration.

1. Establishment of a bearer channel in the access network is necessary to enable information exchange with core nodes. Sign on within these access networks may be necessary before IMS registration. For example, GSM and UMTS technologies need to authenticate users before allowing them to use network resources.
2. IMS registration is a SIP-based procedure that needs IP connectivity to work. Therefore, the user equipment must receive at least a valid IP address that will allow it to communicate with remote IP peers. Some networks have special procedures to assign IP addresses to users. For example, GPRS networks can do that in the PDP context activation procedure. The DHCP protocol constitutes a generic way to assign IP addresses to users in networks without specific procedures.
3. The user equipment needs the IP address of at least one P-CSCF to establish communication with the IMS core. The address can be retrieved during the IP connectivity establishment phase if the access network provides such means. Otherwise, a generic procedure based on the DNS and DHCP protocols can be used. In the generic procedure, the user equipment requests from a DHCP server the IP address of one or several DNS servers and the domain name of at least one P-CSCF. The user equipment can then resolve the domain names of the P-CSCFs by means of one or various DNS queries. The P-CSCF does not need to be in the home network. P-CSCFs placed in visited networks are used in roaming scenarios.

Once the user equipment has the IP address of a P-CSCF, registration in the IMS network can start. Registration involves mutual authentication of the subscriber and the network. The mechanism defined for this purpose is known as IMS authentication and key agreement (IMS AKA). IMS AKA is a challenge-based procedure derived from the UMTS authentication (UMTS AKA). IMS AKA dialogues are implemented with SIP and SIP 3GPP specific extensions. Furthermore, extra security is needed to protect signalling between the user equipment and the P-CSCF. Setup of security associations (SA) between these network elements also uses SIP and follows RFC 3329.

The following illustrates the registration sequence for a roaming user (see Figure 9-33):

1. The user equipment sends a SIP Register request to the P-CSCF. This message contains public and private IMS user identities, home network domain name and other information.
2. IMS services are always provided from the home network. For this reason, the P-CSCF must forward the register information to the user's home network. To do that, the P-CSCF examines the home network domain name included in the register request. The registration information flow reaches a special CSCF in the home network, known as interrogating CSCF (I-CSCF). This server acts as the entry point of signalling flows from external IMS networks. The IP address of the I-CSCF is published in DNS servers to enable roaming users to reach this server using the home server domain name.
3. The I-CSCF issues a Diameter query in the C_x interface to the HSS to ask whether the user is allowed to register in the network through this P-CSCF or not. The Diameter message user authorization request (UAR) defined in 3GPP TS 29.229 can be used for this purpose.

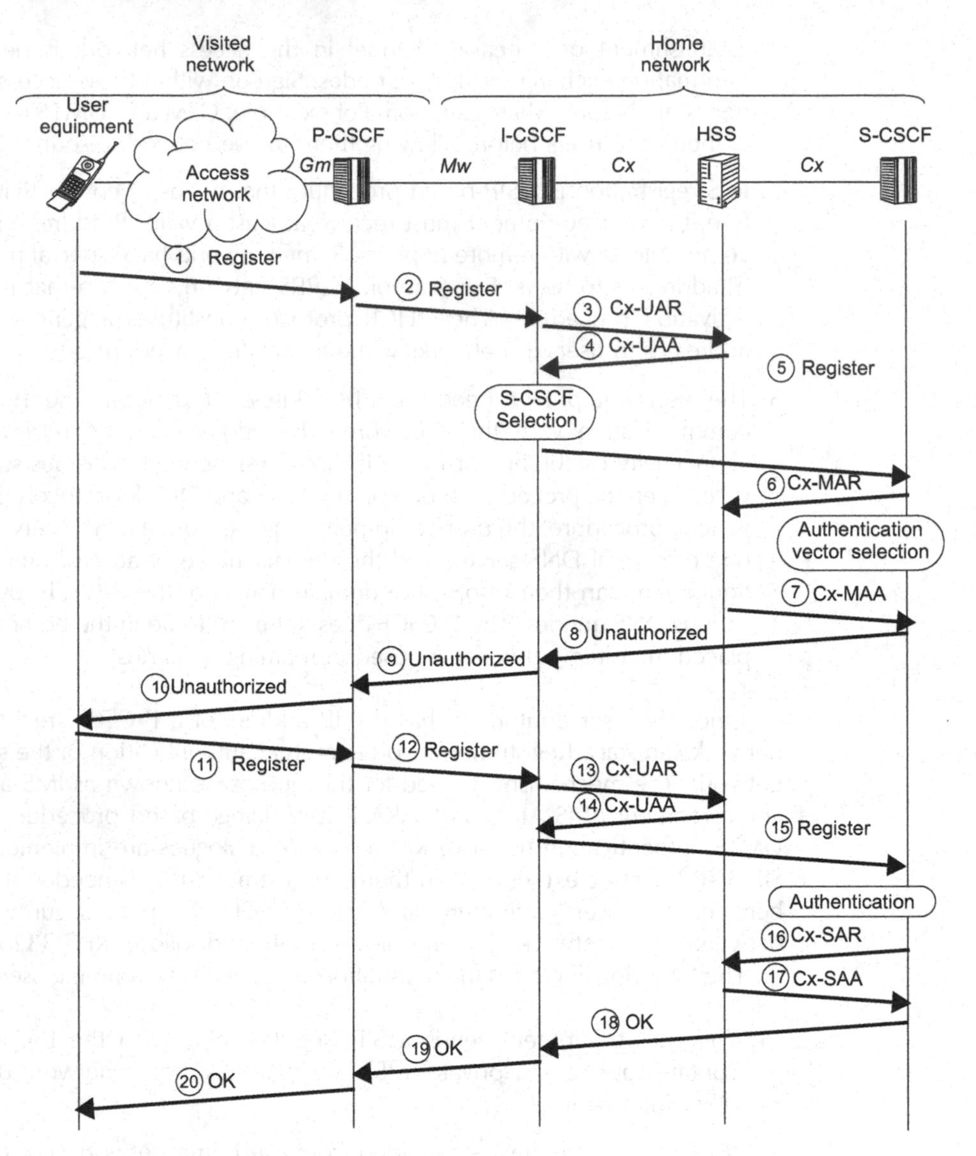

Figure 9-33. Typical registration information flow in an IMS network. The user equipment is considered to access to the home network through an external visited network.

4. The HSS determines whether the user is allowed to register and answers the query from the I-CSCF. The answer to the UAR query is the user authorization answer (UAA). Sometimes the answer from the HSS contains the domain name of the S-CSCF designated for the user, sometimes it only contains information about the S-CSCF that enables the I-CSCF to perform the right choice.

5. The I-CSCF determines the IP address of the S-CSCF designated for the user and resends the register information flow to that server.
6. The S-CSCF challenges the user before allowing registration. To do that, the S-CSCF may need to request IMS AKA authentication vectors from the user profile stored within the HSS. The Diameter multimedia authorization request (MAR) message is used for this purpose.
7. The HSS retrieve authorization information for the user to be authenticated and delivers it to the S-CSCF by using a Diameter multimedia authorization answer (MAA) message.
8. The S-CSCF uses the authentication information received from the HSS to compute a challenge for the user. Then it attaches the challenge within a 401 Unauthorized SIP response. The response is delivered to the I-CSCF.
9. The 401 Unauthorized response propagates through the signalling path and it arrives at the P-CSCF.
10. The P-CSCF receives the 401 Unauthorized SIP message and resends it to the user equipment.
11. Upon receiving the 401 Unauthorized response, the user equipment generates a new SIP Register request. This time it calculates the challenge response and attaches it in the new Register message before being sent.
12. The P-CSCF receives the new Register request and forwards it to the user's home network via the I-CSCF.
13. The I-CSCF queries again the HSS for information about the subscriber registration status. The Diameter UAR message is used for this purpose.
14. The HSS replies to the I-CSCF with the requested information by means of the Diameter UAA message.
15. The I-CSCF forwards the Register message with the challenge response to the S-CSCF.
16. The S-CSCF receives the Register request from the user and authenticates it. If the result of this operation is successful, it issues a query to the HSS so that the HSS identifies it as the service control server for that user. The Diameter server assignment request (SAR) is used for this purpose.
17. The HSS bounds the user identity to the S-CSCF included in the Diameter query and issues a response. The response message to a SAR is the server assignment answer (SAA). The answer may include names of platforms which can be used for service control while the user is registered and other information.
18. If the S-CSCF is able to accept the registration request it delivers a SIP OK response to the I-CSCF.
19. The I-CSCF resends the information flow to the P-CSCF in the visited network.
20. The information flow reaches the user equipment. Starting from this moment, the user is registered in the IMS network.

9.3.6.2 Call Origination

Like other IMS procedures, call origination is based on SIP with extensions specifically designed by 3GPP for call control (see Figure 9-34). A user who wants to establish a call with a remote user issues a SIP Invite to the recipient. The Invite request contains an SDP body with descriptions of the media formats and codecs to be used during the session. The SDP body may also contain descriptions of the required QoS for the call. The Invite request with the initial SDP description is propagated through the IMS network. Messages from roaming users are routed through an I-CSCF to the S-CSCF that is always in the home network. As the recipient is potentially a mobile user, the S-CSCF will need to request location information for that user from the HSS. The Diameter message location info request (LIR) is used for this purpose.

When the called user equipment receives the initial SIP Invite it starts a session negotiation stage using provisional SIP requests and responses and SDP. Specifically, the called user equipment responds with a 183 session progress to the initial Invite and the originator user equipment uses the provisional acknowledge (PRACK) procedure to send provisional information. During this stage two important things happen:

- Both communication ends agree the media flows to be used for the session and the media codecs to be used for every media flow. Media flow and codec negotiation may require several SDP offers before an agreement is reached.
- During the negotiation stage, the P-CSCFs authorize QoS resources necessary for the call.

Media flow and codec negotiation finishes with a 200 OK reply. After delivery of the last negotiation message, the communication ends start the resource reservation procedure. Actual reservation depends on the particular network infrastructure but it should agree with the QoS resources allocated by the P-CSCF. Otherwise, resource reservation may be denied. The call originator uses an SDP body sent within a SIP Update request to inform the remote end that QoS resources have been successfully allocated. The call recipient does the same by replying with a 200 OK reply to the Update request. This reply also uses the SDP protocol to communicate successful resource reservation.

When bearer resources are ready for communication, the call recipient may perform alerting by issuing a 180 Ringing provisional response to the previous Update request. This response is propagated through the signalling path like the previous SIP messages. When the call originator receives the 180 Ringing response it issues a new PRACK request that is answered by a new 200 OK reply.

If the call recipient decides to answer the call, a final OK reply is issued and delivered towards the call originator. The originator then opens the media flows for the call and generates an ACK procedure that is propagated through the signalling path towards the call recipient. The media session can then start.

9.3.7 The Next-generation Network

The concept of an integrated broadband network has developed over the last years and has been labelled the Next Generation Network (NGN). Requirements, features and implementation

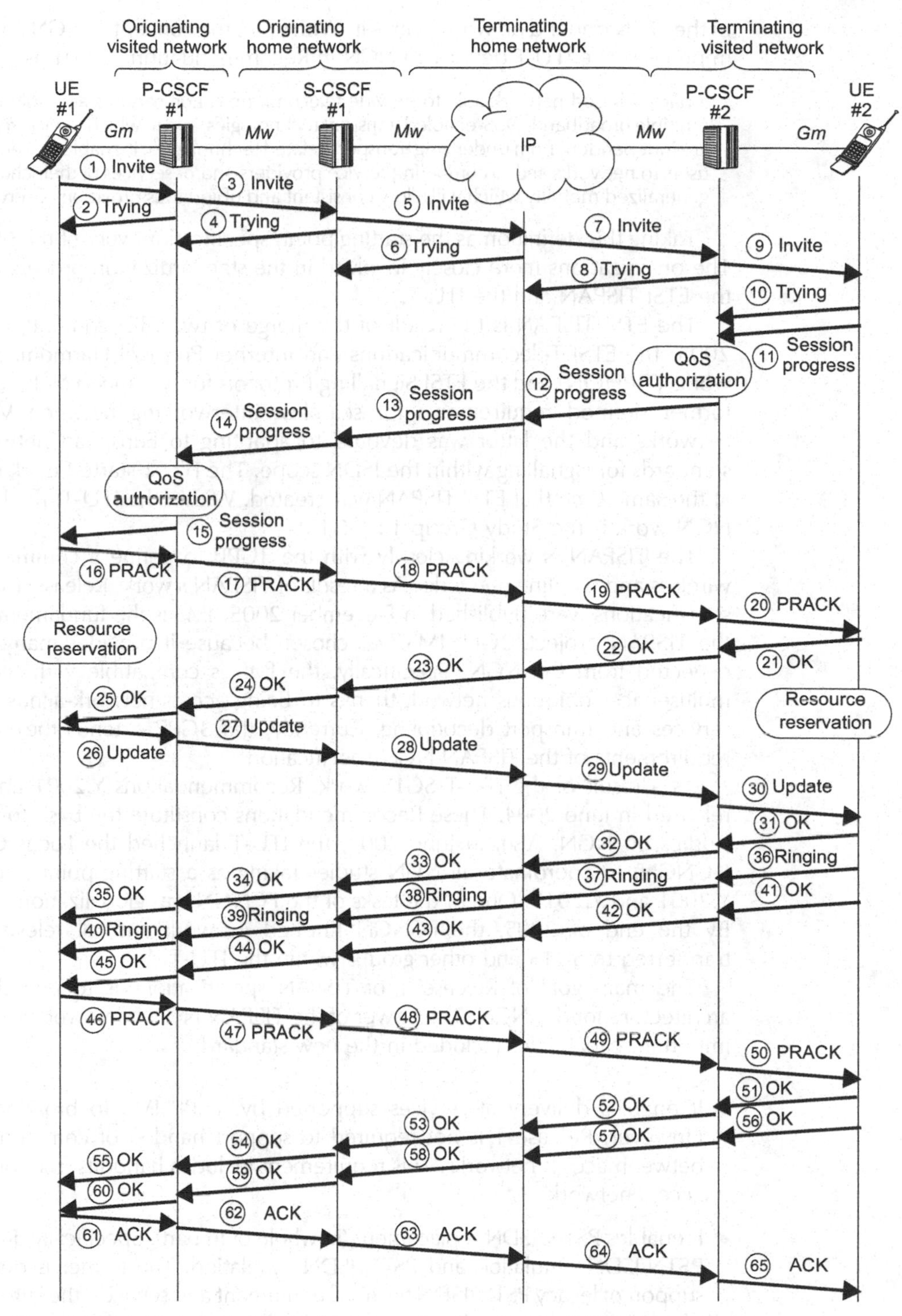

Figure 9-34. Typical session establishment signaling flow in IMS (roaming case).

of the NGN depend on who defines it. With time, the concept of NGN has become rather imprecise. The ITU-T defines an NGN in Recommendation Y.2001 as

> a packet-based network able to provide telecommunication services and able to make use of multiple broadband, QoS-enabled transport technologies, and in which service-related functions are independent from underlying transport-related technologies. It enables unfettered access for users to networks and to competing service providers and/or services of their choice. It supports generalized mobility which will allow consistent and ubiquitous provision of services to users.

Taking this definition as the starting point, specification work of the NGN has started. The organizations more closely involved in the standardization process of the NGN are the ETSI TISPAN and the ITU-T.

The ETSI TISPAN is the result of the merge of two ETSI committees in September 2003: the ETSI Telecommunications and Internet Protocol Harmonization Over Networks (TIPHON) and the ETSI Signalling Protocols for Advanced Networks (SPAN). The former defined requirements for seamless interworking between VoIP and PSTN networks and the latter was devoted to adapting to European networks the ITU-T standards for signalling within the ISDN scope. The ITU-T started work on NGN almost at the same time that ETSI TISPAN was created. Within the ITU-T, the leading group in NGN work is the Study Group 13 (SG13).

The TISPAN is working closely with the 3GPP to define a common core for both wireless and wireline networks. As a result of TISPAN's work, Release 1 of NGN TISPAN specifications were published in December 2005. IMS is the fundamental ingredient of the TISPAN project. 3GPP IMS was chosen because it provides many of the features expected from the NGN. Specifically, the IMS is compatible with the concept of a multiservice, ubiquous network thanks to being access-network-agnostic and allowing services and transport decoupling. Currently, the 3GPP extends the IMS to meet the requirements of the TISPAN NGN specifications.

As a result of the ITU-T SG13 work, Recommendations Y.2001 and Y.2011 were released in June 2004. These Recommendations constitute the basis for all other ITU-T studies on NGN. Also, in June 2004, the ITU-T launched the Focus Group on NGN (FGNGN) to coordinate all NGN studies taking as a starting point Recommendations Y.2001 and Y.2011. One of the tasks of the FGNGN was globalization of TISPAN work. By the end of 2005, the FGNGN finished its work and all relevant results were transferred to SG13 and other groups within the ITU-T.

The main goal of Release 1 of TISPAN specifications is to provide an extensive architecture for the NGN. The power of the TISPAN NGN framework is exhibited in two important capabilities included in the new standard:

- It enables delivery of services supported by 3GPP IMS to broadband fixed lines. However, Release 1 is not required to support handoff of communication sessions between access networks. This requirement includes handoffs from or to the wireline access network.
- It enables PSTN/ISDN replacement (in whole of in part). Specifically, Release 1 enables PSTN/ISDN emulation and PSTN/ISDN simulation. The former is defined to be the support of legacy PSTN/ISDN terminal equipment and services; the latter is the provision of services similar to the PSTN/ISDN to NGN voice or other multimedia terminal

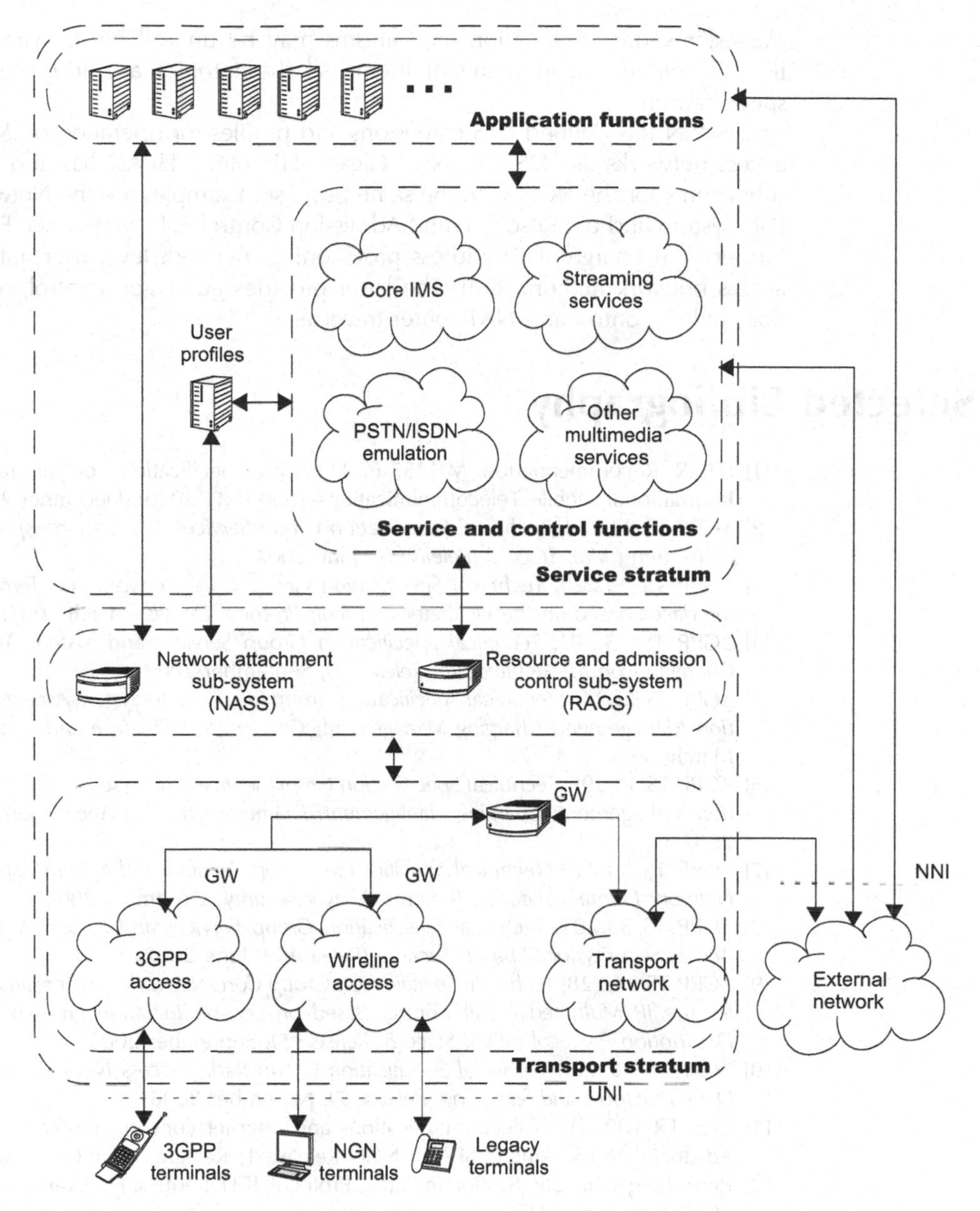

Figure 9-35. TISPAN NGN architecture.

equipment. The purpose of PSTN/ISDN emulation and simulation is to enable gradual migration from the legacy infrastructure to the NGN.

Wireline access networks exhibit significant differences with 3GPP mobile access networks. For example, location information is fundamentally different in both network classes, resource reservation mechanisms may be unavailable in wireline access, and IPv4 is still in use in many of them. All these topics are addressed by the NGN specification.

TISPAN has defined IMS extensions and profiles for operation of IMS over wireline access networks like DSL, PON or Gigabit Ethernet. TISPAN has also worked on new subsystems for the NGN with the same purpose. Examples are the Network Attachment Sub-System and the Resource and Admission Control Sub-system (see Figure 9-35). The former is in charge of IP address provisioning, network level user authentication and access network authorization. The latter provides admission control, resource reservation, policy control and NAT router traversal.

Selected Bibliography

[1] ITU-R Recommendation M.1457-6, Detailled specifications of the radio interfaces of International Mobile Telecommunications-2000 (IMT-2000), December 2006.

[2] 3GPP TS 23.228, *Technical Specification Group Services and System Aspects; IP Multimedia Subsystem (IMS); Stage 2 (Release 8)*, June 2007.

[3] 3GPP TS 29.229, *Technical Specification Group Core Network and Terminals; Cx and Dx Interfaces based on the Diameter Protocol; Protocol Details*, March 2007.

[4] 3GPP TS 23.203, *Technical Specification Group Services and System Aspects; Policy and Charging Control Architecture (Release 7)*, September 2007.

[5] 3GPP TS 32.240, *Technical Specification Group Services and System Aspects; Telecommunication Management; Charging Management; Charging Architecture and Principles (Release 7)*, March 2007.

[6] 3GPP TS 32.299, *Technical Specification Group Services and System Aspects; Telecommunication Management; Charging Management; Diameter Charging Applications (Release 7)*, June 2007.

[7] 3GPP TS 33.210, *Technical Specification Group Services and System Aspects; 3G Security; Network Domain Security, IP network layer security*, December 2006.

[8] 3GPP TS 33.203, *Technical Specification Group Services and System Aspects; 3G security; Access Security for IP-based Services (Release 7)*, June 2007.

[9] 3GPP TS 24.228, *Technical Specification Group Core Network and Terminals; Signalling flows for the IP Multimedia Call Control Based on Session Initiation Protocol (SIP) and Session Description Protocol (SDP); Stage 3 (Release 5)*, September 2006.

[10] 3GPP TS 25.414, *Technical Specification Group Radio Access Network; UTRAN Iu interface Data Transport and Signalling (Release 7)*. November 2006.

[11] ETSI TR 180.001, Telecommunications and Internet converged Services and Protocols for Advanced Networking (TISPAN), NGN Release 1; Release definition; March 2006.

[12] Rosenberg *et al.*, SIP: Session Initiation Protocol, IETF Request For Comments RFC 3261, June 2002. Obsoletes RFC 2543.

[13] Calhoun P., Loughney J., Guttman E., Zorn G., Arkko J., Diameter Base Protocol, IETF Request For Comments RFC 3588, September 2003.

[14] Eklund C., Marks R., Stanwood K., Wang S., IEEE Standard 802.16: A Technical Overview of the WirelessMAN Air Interface for Broadband Wireless Access, *IEEE Communications Magazine*, June 2002, pp. 98–107.

[15] Ghosh A., Worter D., Andrews J., Runhua C., Broadband Wireless Access with WiMaz/802.16: Current Performance Benchmarks and Future Potential, *IEEE Communications Magazine*, April 2005, pp. 129–136.

[16] WiMAX Forum, *Mobile WiMAX – Part I: A Technical Overview and Performance Evaluation*, August 2006.
[17] WiMAX Forum, *Mobile WiMAX – Part II: A Comparative Analysis*, May 2006.
[18] Rahnema M., Overview Of the GSM System and Protocol Architecture, *IEEE Communications Magazine*, April 1993, pp. 92–100.
[19] De Vriendt J., Lainé P., Lerouge C., Xu X., Mobile Network Evolution: A Revolution on the Move, *IEEE Communications Magazine*, April 2002, pp. 104–111.
[20] Camarillo G., Kauppinen T., Kuparinen M., Más I., Towards an Innovation Oriented IP Multimedia Subsystem, *IEEE Communications Magazine*, March 2007, pp. 130–136.
[21] Cuevas A., Moreno I., Vidales P., Einsiedler H., The IMS Service Platform: A Solution for Next-Generation Network Operators to Be More than Bit Pipes, *IEEE Communications Magazine*, February 2006, pp. 75–81.
[22] Kühne R., Görmer G., Schläger M., Carle G., Charging in the IP Multimedia Subsystem: A Tutorial, *IEEE Communications Magazine*, July 2007, pp. 92–98.
[23] Salsano S., Veltri L., QoS Control by Means of COPS to Support SIP-Based Applications, *IEEE Network*, March/April 2002, pp. 27–33.
[24] Bettstetter C., Jörg Vögel H., Eberspächer J., GSM Phase 2+ General Packet Radio Service GPRS: Architecture, Protocols, and Air Interface, Third Quarter 1999, Vol. 2, No. 3, pp. 2–14.

Chapter 10: Carrier-class Ethernet

Incumbent and competitive operators have started to provide telecommunications services based on Ethernet. This technology is arising as a real alternative to support both traditional data-based applications such as *virtual private networks* (VPN), and new ones such as Triple Play.

Ethernet has several benefits, namely:

- It improves the flexibility and granularity of legacy TDM-based technologies. Frequently, the same Ethernet interface can provide a wide range of bit rates without the need to upgrade network equipment.
- Ethernet is cheaper, more simple and more scalable than ATM and *frame relay* (FR). Today, Ethernet scales up to 10 Gbit/s, and soon it will arrive at 100 Gbit/s.

Furthermore, Ethernet is a well-known technology, and it has been dominant in enterprise networks for many years. However, Ethernet, based on the IEEE standards, has some important drawbacks that limit its rollout, especially when the extension, number of hosts and type of services grow. This is the reason why, in many cases, Ethernet must be upgraded to carrier-class, to match the basic requirements for a proper telecom service in terms of quality, resilience and OAM (see Figure 10-1).

Triple Play: Building the Converged Network for IP, VoIP and IPTV Francisco J. Hens and José M. Caballero

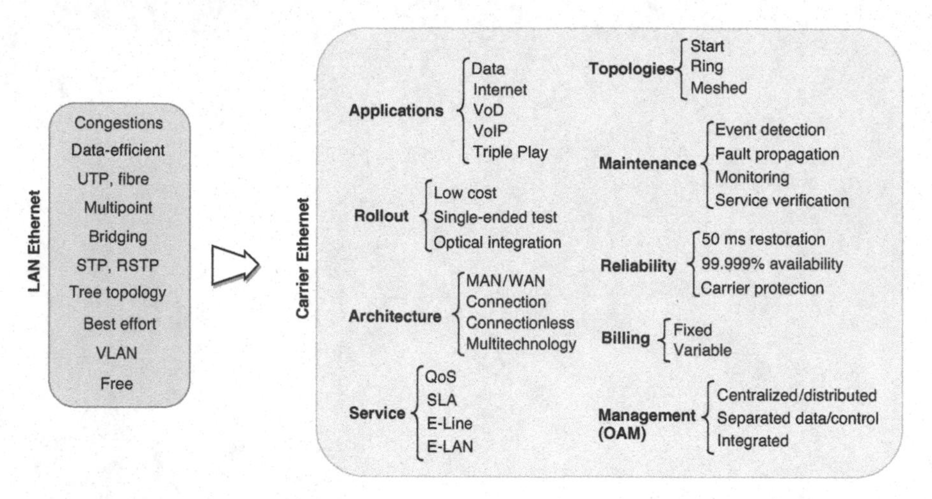

Figure 10-1. The path to carrier-class Ethernet.

10.1 Ethernet as a MAN/WAN Service

Ethernet has been used by companies for short-range and medium/high-bandwidth connections, typical of LANs. To connect hosts from remote LANs, up until now it has been necessary to provide either FR, ATM or leased lines. This means that the Ethernet data flow must be converted to a different protocol to be sent over the service-provider network and then converted back to Ethernet again. Using Ethernet in MAN and WAN environments would simplify the interface, and there would be no need for total or partial protocol conversions (see Figure 10-2).

Currently, the Metro Ethernet Forum (MEF), the IETF, the IEEE and the ITU are working to find solutions to enable the deployment of carrier-class Ethernet networks, also known as metro-Ethernet networks (MEN). This includes the definition of generic services, interfaces, deployment alternatives, interworking with current technologies, and much more. Carrier-class Ethernet is not only a low-cost solution to interfacing with the subscriber network and carrying its data across long distances, but it is also part of a converged network for any type of information, including voice, video and data.

10.1.1 Network Architecture

The ideal MEN makes use of pure Ethernet technology: Ethernet switches, interfaces and links. However, in reality, Ethernet is often used together with other technologies currently available in the metropolitan network environment. Most of these technologies can inter-network with Ethernet, thus extending the range of the network. Next-generation SDH (NG SDH) nodes can transport Ethernet frames transparently (see Section 10.2.3). Additionally, Ethernet can be transported by layer-2 networks, such as FR or ATM.

Today, many service providers are offering Ethernet to their customers simply as a service interface. The technology used to deliver the data is not an issue. In metropolitan

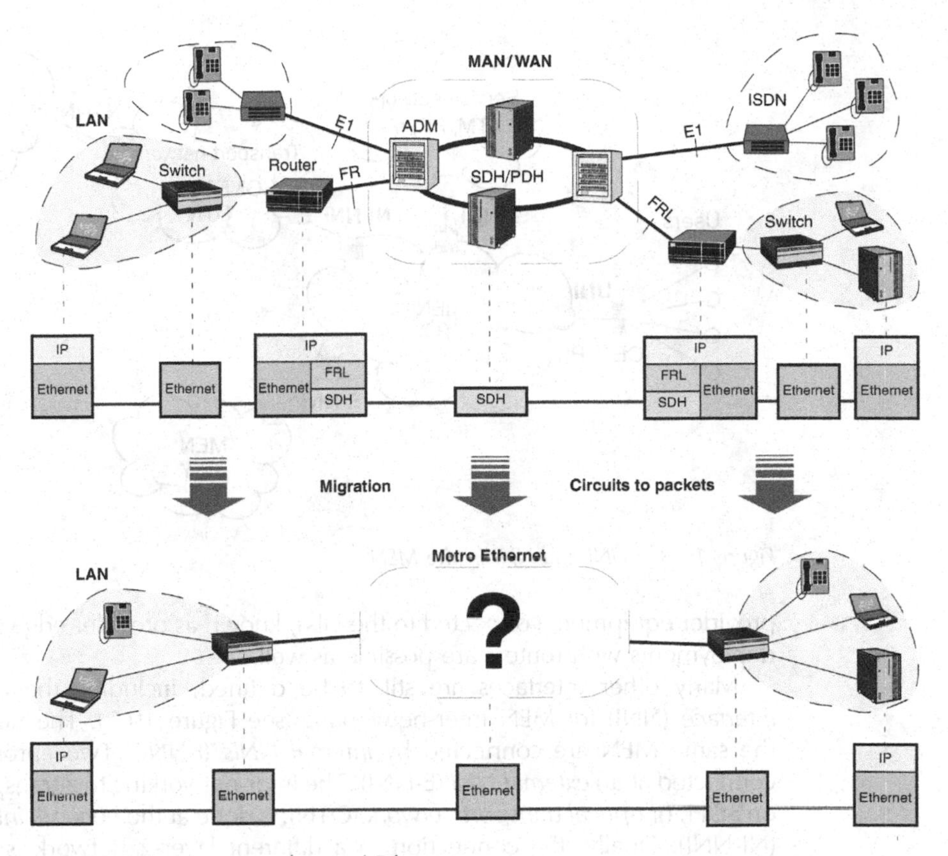

Figure 10-2. Migration to end-to-end Ethernet.

networks this technology can be Ethernet or SDH. Inter-city services are almost exclusively transported across SDH.

The interface between the customer premises equipment and the service-provider facilities is called the user-to-network interface (UNI). The fact that Ethernet is being offered as a service interface makes the definition of the Ethernet UNI very important. In fact, this is one of the main points addressed by standardization organizations. The deployment plans for the UNI include three phases:

1. UNI type 1 focuses on the Ethernet users of the existing IEEE Ethernet physical and MAC layers.
2. UNI type 2 requires static service discovery functionality with auto-discovery and OAM capabilities.
3. UNI type 3 requires a dynamic connection setup such that *Ethernet virtual connections* (EVC) can be set up and/or modified from the customer UNI equipment.

The customer premises equipment that enables access to the MEN can be a router or a switch. This equipment is usually called customer edge (CE) equipment. The service

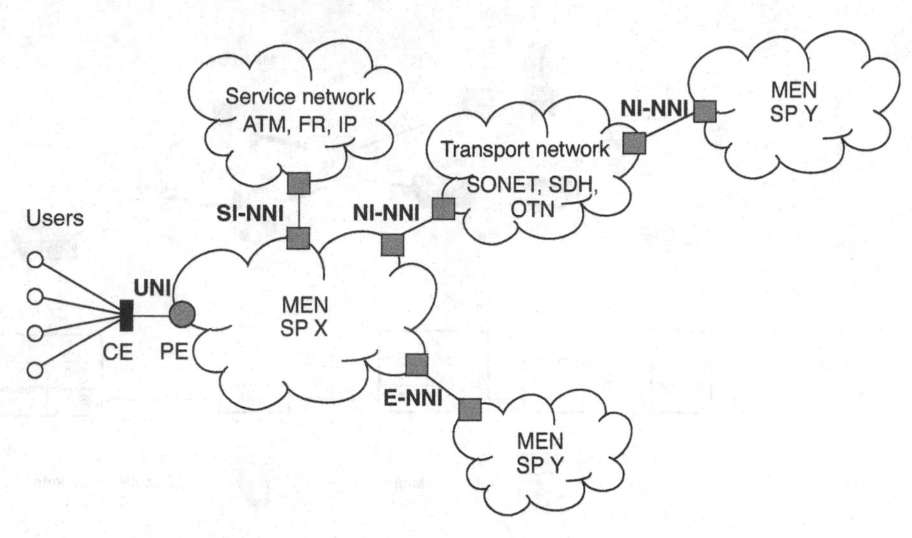

Figure 10-3. UNI and NNI in the MEN.

provider equipment connected to the UNI, known as provider edge (PE), is a switch but deployments with routers are possible as well.

Many other interfaces are still to be defined, including the *network-to-network interface* (NNI) for MEN inter-networking (see Figure 10-3). The network elements of the same MEN are connected by *internal NNIs* (I-NNI). Two autonomous MENs are connected at an *external NNI* (E-NNI). The inter-networking to a transport network based on SDH, or *optical transport network* (OTN), is done at the *network inter-networking NNI* (NI-NNI). Finally, the connection to a different layer-2 network is established at the *services inter-networking NNI* (SI-NNI).

10.1.2 Ethernet Virtual Connections

An *Ethernet virtual connection* (EVC) is defined as an association of two or more UNIs. A point-to-point EVC is limited to two UNIs, but a multipoint-to-multipoint EVC can have two or more UNIs that can be dynamically added or removed.

An EVC can be compared with the *virtual circuits* (VC) used by FR and ATM – however, the EVC has multipoint capabilities, whilst VCs are strictly point-to-point. This feature makes it possible to emulate the multicast nature of Ethernet. An EVC facilitates the transmission of frames between UNIs, but also prevents the transmission of information outside the EVC.

Origin and destination MAC addresses and frame contents remain unchanged in the EVC, which is a major difference compared with routed networks where MAC addresses are modified at each Ethernet segment.

10.1.3 Multiplexing and Bundling

An Ethernet port can support several EVCs simultaneously. This feature, called service multiplexing, improves port utilization by lowering the number of ports per switch. It also

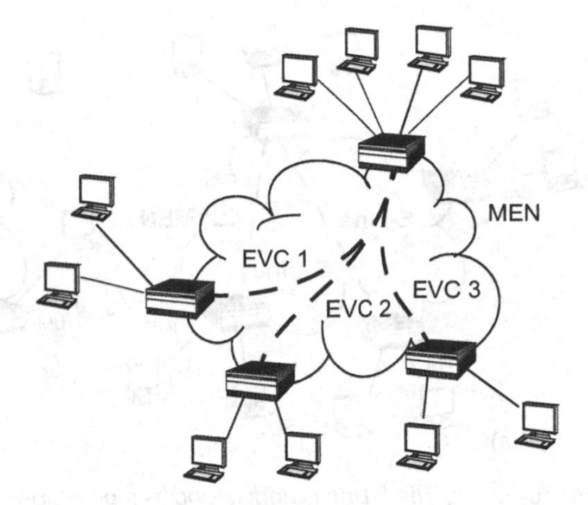

Figure 10-4. EVC Service multiplexing in a single port.

makes service activation simpler (see Figure 10-4). Service multiplexing is achieved using the IEEE 802.1Q *virtual LAN* (VLAN) ID as a connection identifier.

Service multiplexing makes it possible to provide new services without installing new cabling or nodes. This, consequently, reduces capital expenditure.

Bundling occurs when more than one subscriber's VLAN ID is mapped to the same EVC. Bundling is useful when the VLAN tagging scheme must be preserved across the MEN when remote branch offices are going to be connected. A special case of bundling occurs when every VLAN ID is mapped to a single EVC. This is called *all-to-one bundling*.

10.1.4 Generic Service Types

Currently, the MEF has defined two generic service types: *Ethernet line* (E-Line) and *Ethernet LAN* (E-LAN).

10.1.4.1 E-Line Service Type

The *E-Line service* is a point-to-point EVC with attributes such as QoS parameters, VLAN tag support and transparency to layer-2 protocols (see Figure 10-5). The E-Line service can be compared, in some ways, with *permanent VCs* (PVCs) of FR or ATM, but E-Line is more scalable and has more service options.

An E-Line service type can be a simple Ethernet point-to-point with best effort connection, but it can also be a sophisticated TDM private line emulation.

10.1.4.2 E-LAN Service Type

The *E-LAN service* is an important new feature of carrier-class Ethernet. It provides a multipoint-to-multipoint data connection (see Figure 10-5). UNIs are allowed to be connected or disconnected from the E-LAN dynamically. The data sent from one UNI is sent to all other UNIs of the same E-LAN in the same way as happens in a classical Ethernet LAN. The E-LAN service offers many advantages over FR and ATM hub-and-spoke

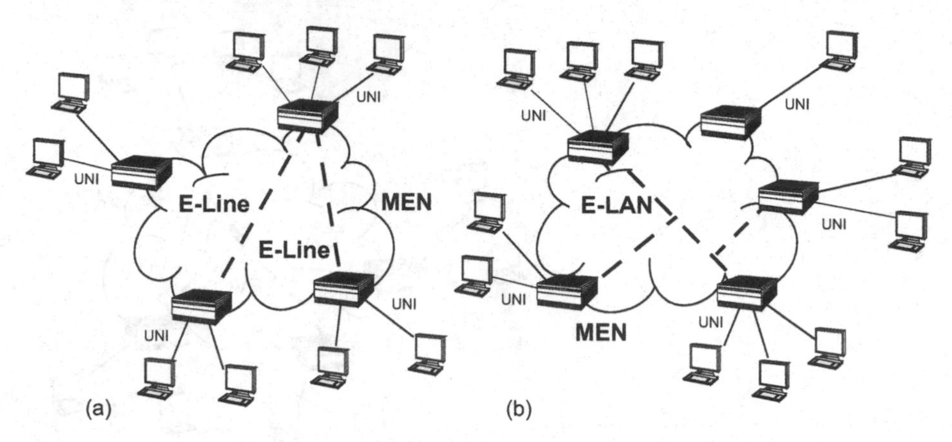

Figure 10-5. (a) The E-Line is understood as a point-to-point virtual circuit. (b) The E-LAN service is multipoint to multipoint.

architectures that depend on various point-to-point PVCs to implement multicast communications.

The E-LAN can be offered simply as a best-effort service type, but it can also provide a specific QoS. Every UNI is allowed to have its own bandwidth profile. This could be useful when several branch offices are connected to one central office. In this case the *committed information rate* (CIR) in the UNI for every branch office could be 10 Mbit/s, and 100 Mbit/s, for the central office.

10.1.5 Connectivity Services

An Ethernet service arises when a generic service type (E-Line or E-LAN) is offered with particular EVC and UNI features. When a service is port-based – that is, one single service per port is provided at the UNI, it is called an *Ethernet private line* (EPL) or *Ethernet private LAN* (EPLAN), depending on whether it is point-to-point or multipoint-to-multipoint. Multiplexed services are called virtual. An *Ethernet virtual private line* (EVPL) service and an *Ethernet virtual private LAN* (EVPLAN) service can be defined (see Table 10-1).

From the point of view of the customer, the main differences between virtual and non-virtual services are that EPLs and EPLANs provide better frame transparency, and they are subject to more demanding SLA margins than EVPLs and EVPLANs.

The meaning of multiplexed services in the case of EVPLs and EVPLANs needs to be further explained. For example, several E-Line service types may be multiplexed in different SDH timeslots and be still considered EPLs. This is because the *time division multiplexing* (TDM) resource-sharing technique of SDH makes it possible to divide the available bandwidth in such a way that congestion in some timeslots does not affect other timeslots. This way, it is possible to maintain the strong SLA margins typical of EPLs and EPLANs in those timeslots that are not affected by congestion.

EVPLs and EVPLANs are statistically multiplexed services. They make use of service multiplexing (see Section 10.1.3), and thus VLAN IDs are used as EVC identifiers at the UNI.

Table 10-1. Ethernet connectivity services

	Generic Etherservice type	
EVC to UNI relationship	E-Line Point-to-point Best-effort or guaranteed QoS Optional multiplexing and bundling	E-LAN Multipoint-to-multipoint Best-effort or guaranteed QoS Optional multiplexing and bundling
Port-based service No service multiplexing Dedicated bandwidth	Ethernet private line	Ethernet private LAN
VLAN-based service Service multiplexing Shared bandwidth	Ethernet virtual private line	Ethernet virtual private LAN

10.1.5.1 Ethernet Private Lines

The EPL service is a point-to-point Ethernet service that provides high frame transparency, and it is usually subject to strong SLAs. It can be considered as the Ethernet equivalent of a private line, but it offers the benefit of an Ethernet interface to the customer.

The EPLs make use of all-to-one bundling and subscriber VLAN tag transparency. This allows the customer to easily extend the VLAN architecture between sites at both ends of the MAN/WAN connection. Frame transparency enables typical layer-2 protocols, such as IEEE 802.1D Spanning Tree Protocol (STP), to be tunnelled through the MAN/WAN.

EPLs are sometimes delivered over dedicated lines, but they can be supplied by means of layer-1 (TDM or lambdas) or layer-2 (MPLS, ATM, FR) multiplexed circuits. Some service providers want to emphasize this, and they talk about dedicated EPLs, if dedicated lines or layer-1 multiplexed circuits are used to deliver the service, or shared EPLs if layer-2 multiplexing is used.

EPLs are the most extended metro Ethernet services today. They are best suited for critical, real-time applications.

10.1.5.2 Ethernet Virtual Private Lines

The EVPL is a point-to-point Ethernet service similar to the EPL, except that service multiplexing is allowed, and it can be opaque to certain types of frames. For example, STP frames can be dropped by the network-side UNI.

Shared resources make it difficult for the EVPL to meet SLAs as precise as those of EPLs. The EVPL is similar to the FR or ATM PVCs. The VLAN ID for EVPLs is the equivalent of the FR *data link connection identifier* (DLCI) or the ATM *virtual circuit identifier* (VCI)/*virtual path identifier* (VPI).

One application of EVPLs could be a high-performance ISP-to-customer connection.

10.1.5.3 Ethernet Private LANs

EPLANs are multipoint-to-multipoint dedicated carrier-class Ethernet services. The EPLAN service is similar to the classic LAN Ethernet service, but over a MAN or a WAN. It is a dedicated service in the sense that Ethernet traffic belonging to different customers is not mixed within the service-provider network.

Ethernet frames reach their destination thanks to the MAC switching supported by the service-provider network. Broadcasting, as well as multicasting are supported.

EPLAN services make use of all-to-one bundling and subscriber VLAN tag transparency to support the customer's VLAN architecture. Frame transparency is implemented in EPLANs to support LAN protocols across different sites.

10.1.5.4 Ethernet Virtual Private LANs

The EVPLAN is similar to the EPLAN, but EVPLANs are supported by a shared architecture instead of a dedicated one. The EVPLAN also has some common points with the EVPL. For example, the VLAN tag is used for service multiplexing, and EVPLANs could be opaque to some LAN protocols, such as the STP.

Both EPLAN and EVPLAN will probably be the most important carrier-class Ethernet services in the future. They have many attractive features. The same technology, Ethernet, is used in LAN, MAN and WAN environments. One connection to the service-provider network per site is enough, and EPLAN and EVPLAN offer an interesting alternative to today's layer-3 VPNs. EPLANs and EVPLANs enable the customers to deploy their own IP routers on top of the layer-2 Ethernet VPN.

10.2 End-to-End Ethernet

Today's installations use Ethernet on LANs to connect servers and workstations. Data applications and Internet services use WANs to get or to provide access to/from remote sites by means of leased lines, PDH/SDH TDM circuits and ATM/FR PVCs. Routers are the intermediate devices that use IP as a common language can also talk to the LAN and WAN protocols. This has been a very popular solution, but it is not a real end-to-end Ethernet service. This means that MAC frames 'die' as soon as IP packets enter the PDH/SDH domain, and they are created again when they reach the far end.

This option, which is now often considered as legacy, has been the most popular networking data solution. During the past couple of decades, routing technologies have formed flexible and distributed layer-3 networks. Since Ethernet is present in both LANs, why not use Ethernet across the WAN as well?

The first approaches for extending Ethernet over a WAN are based on mixing Ethernet with legacy technologies, for example Ethernet over ATM, as defined in the IETF standard RFC-2684, or Ethernet over SDH by means of the *Link Access Procedure - SDH* (LAPS) as per ITU-T Recommendation X.86 (see Figure 10-6).

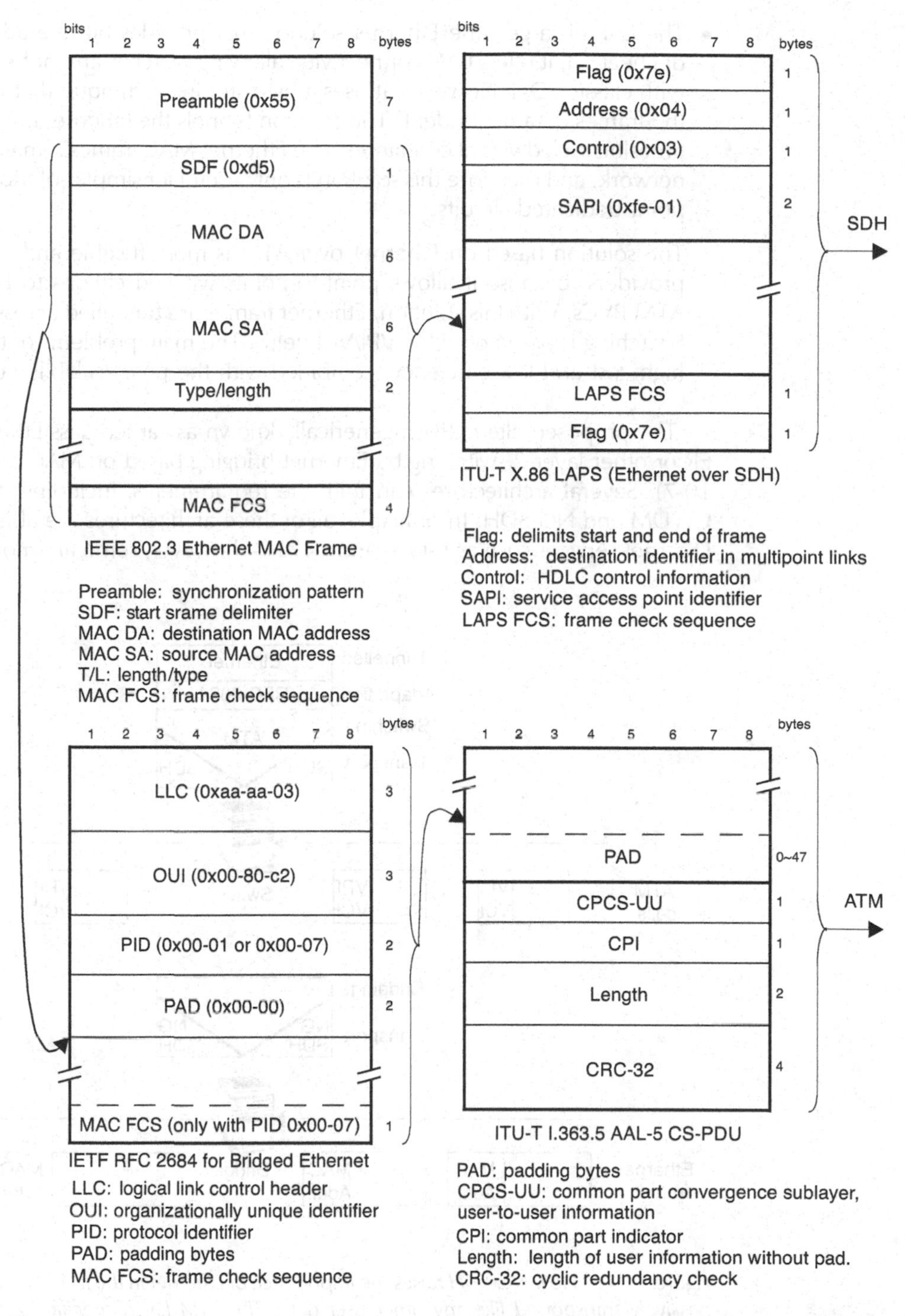

Figure 10-6. Legacy encapsulations for transporting Ethernet over SDH or ATM networks. The SDH solution makes use of the LAPS encapsulation. The ATM mapping uses RFC-2684 and AAL-5 encapsulations.

- The LAPS is a genuine Ethernet solution that provides bit rate adaptation and frame delineation. It offers LAN connectivity, allowing switches and hubs to interface directly with classic SDH. However, it uses a byte-stuffing technique that makes the length of the frames data-dependent. This solution tunnels the Ethernet frames over SDH TDM timeslots called virtual containers. The Ethernet MAC frames remain passive within the network, and therefore this solution is only useful for simple solutions, such as point-to point dedicated circuits.
- The solution based on Ethernet over ATM is more flexible and attractive for service providers, because it allows point-to-point switched circuits to be set up based on ATM PVCs. With this solution, Ethernet frames are tunnelled across the ATM network. Switching is based on ATM VPI/VCI fields. The main problems of this architecture are high cost and low efficiency, combined with the poor scalability of ATM.

The proposed alternatives, generically known as carrier-class Ethernet, replace ATM, FR or other layer-2 switching by Ethernet bridging based on MAC addresses (see Figure 10-7). Several architectures can fulfil the requirements, including dark fibre, DWDM/CWDM and NG-SDH. In principle, all of these architectures are able to support Carrier Ethernet services such as E-Line and E-LAN – however, some are more appropriate than others.

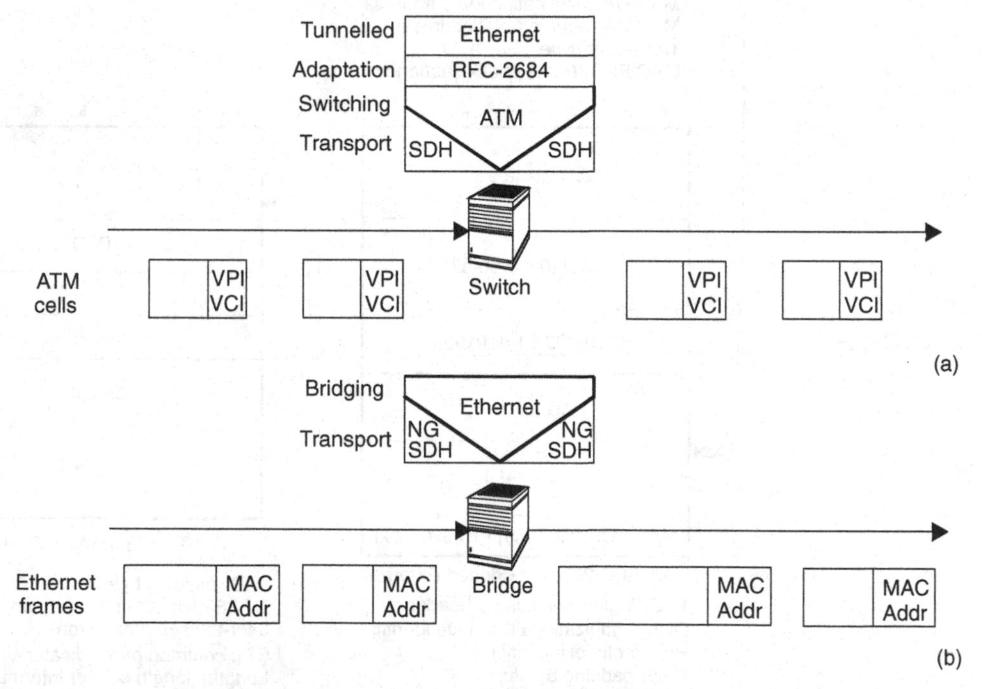

Figure 10-7. How NG-SDH raises the importance of Ethernet in the MAN/WAN. (a) Ethernet traffic is passively transported like any other user data. The ATM layer, specific for the WAN, is used for switching traffic. (b) The ATM layer disappears and the Ethernet layer becomes active. Traffic is now guided to its destination by means of Ethernet bridging.

10.2.1 Optical Ethernet

Ethernet can now be used in metropolitan networks due to the recent standardization of new long-range, high-bandwidth Ethernet interfaces. It can be said that Ethernet bandwidths and ranges are now of the same order as the bandwidths and ranges provided by classical WAN technologies.

MENs based on optical Ethernet are typical of early implementations. They are built by means of standard IEEE interfaces over dark optical fibre. They are therefore pure Ethernet networks. Multiple homing, link aggregation and VLAN tags can be used in order to increase resilience, bandwidth and traffic segregation. Interworking with the legacy SDH network can be achieved with the help of the WAN interface subsystem (WIS). The WIS is part of the WAN PHY specification for 10 Gbit Ethernet. It provides multigigabit connectivity across SDH and WDM networks as an alternative to the LAN PHY for native-format networks.

With this simple solution, a competitive operator can take advantage of packet switching, multipoint-to-multipoint applications and quick service rollout. This option is a cost-effective in those areas where spare dark fibre is available and tree topologies are likely. Despite its simplicity, pure Ethernet solutions for MEN have big scalability problems. Furthermore, they suffer from insufficient QoS, OAM and resilience mechanisms (see Section 10.3).

Optical Ethernet is often the architecture implemented by new operators to compete with the incumbent ones (see Figure 10-8). This kind of solution is still practical in metropolitan environments with a small number of connected subscribers or subsidiaries. Specifically, the

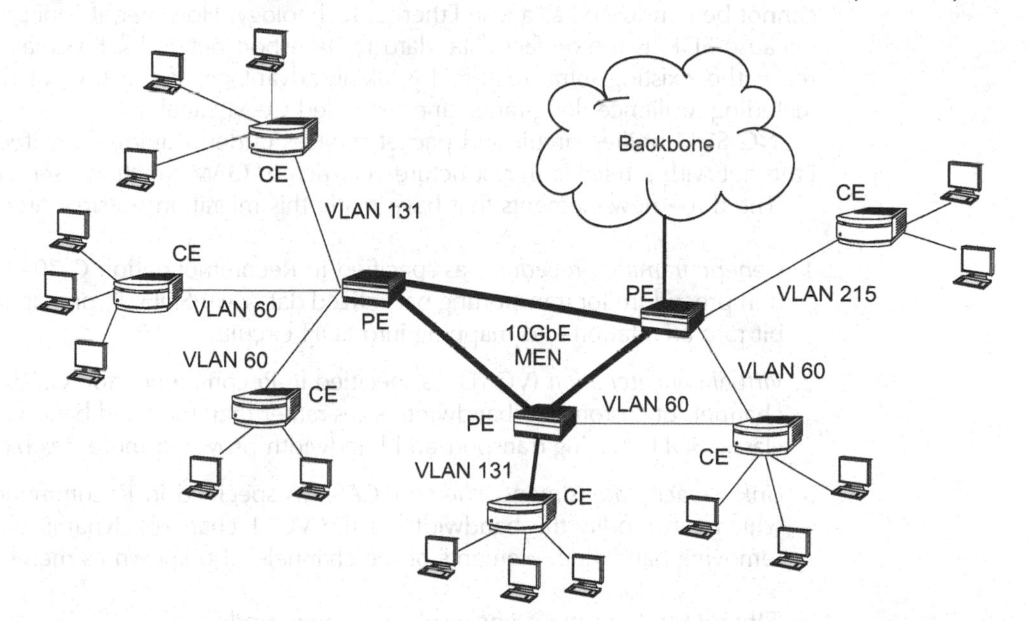

Figure 10-8. Optical Ethernet carrier network. The trunk MEN links are implemented with optical 10 Gbit Ethernet. Access links can be based on 1 Gbit Ethernet or EFM, depending on the bandwidth requirements of every placement. Segregation of traffic from different subscribers or work groups is done using VLAN tagging.

pure Ethernet over dark fibre approach is discouraged for operators who want to provide services to a large number of residential and small office/home Office (SOHO) customers.

10.2.2 Ethernet Over WDM

The transport capability of the existing fibre can be multiplied by 16 or more if *wavelength-division multiplexing* (WDM) is used. The resulting wavelengths are distributed to legacy and new technologies such Ethernet (see Figure 10-14) that will get individual lambdas while sharing fibre optics. WDM is a good option for core networks serving very high bandwidth demands from applications like Triple Play, remote backups or hard disk mirroring. However, cost can be a limiting factor.

One of the inconveniences of this approach is the need to keep track of different and probably incompatible management platforms: one for Ethernet, another one for lambdas carrying SDH or other TDM technologies, and finally a third one for WDM. That makes OAM, traffic engineering and maintenance difficult.

10.2.3 Ethernet Over SDH

Solutions for transporting Ethernet over SDH based on the *generic framing procedure* (GFP), virtual concatenation and the *link capacity adjustment scheme* (LCAS) are generically known as *Ethernet over SDH* (EoS). The idea behind EoS is to substitute the native Ethernet layer 1 by SDH. The Ethernet MAC layer remains untouched to guarantee as much compatibility as possible with the IEEE Ethernet. Owing to this, EoS cannot be considered as a true Ethernet technology. However, it is of great importance, because SDH is the *de facto* standard for transport networks. EoS makes it possible to reuse the existing infrastructure by taking advantage of the best of the SDH world, including resilience, long-range and extended OAM capabilities.

NG-SDH unifies circuit and packet services under a unique architecture, providing Ethernet with a reliable infrastructure very rich in OAM functions (see Figure 10-9).

The three new elements that have made this migration possible are:

1. *Generic framing procedure*, as specified in Recommendation G.7041, an encapsulation procedure for transporting packetized data over SDH. In principal, GFP performs bit rate adaptation and mapping into SDH circuits.
2. *Virtual concatenation* (VCAT), as specified in Recommendation G.707, which creates channels of customized bandwidth sizes rather than the fixed bandwidth provision of classic SDH, making transport and bandwidth provision more flexible and efficient.
3. *Link capacity adjustment scheme* (LCAS), as specified in Recommendation G.7042, which can modify the bandwidth of the VCAT channels dynamically, by adding or removing bandwidth elements of the channels, also known as members.

Ethernet traffic can be encapsulated in two modes:

1. *Transparent GFP* (GFP-T) is equivalent to a leased line with the bandwidth of the configured bit rate. No delays, but expensive.

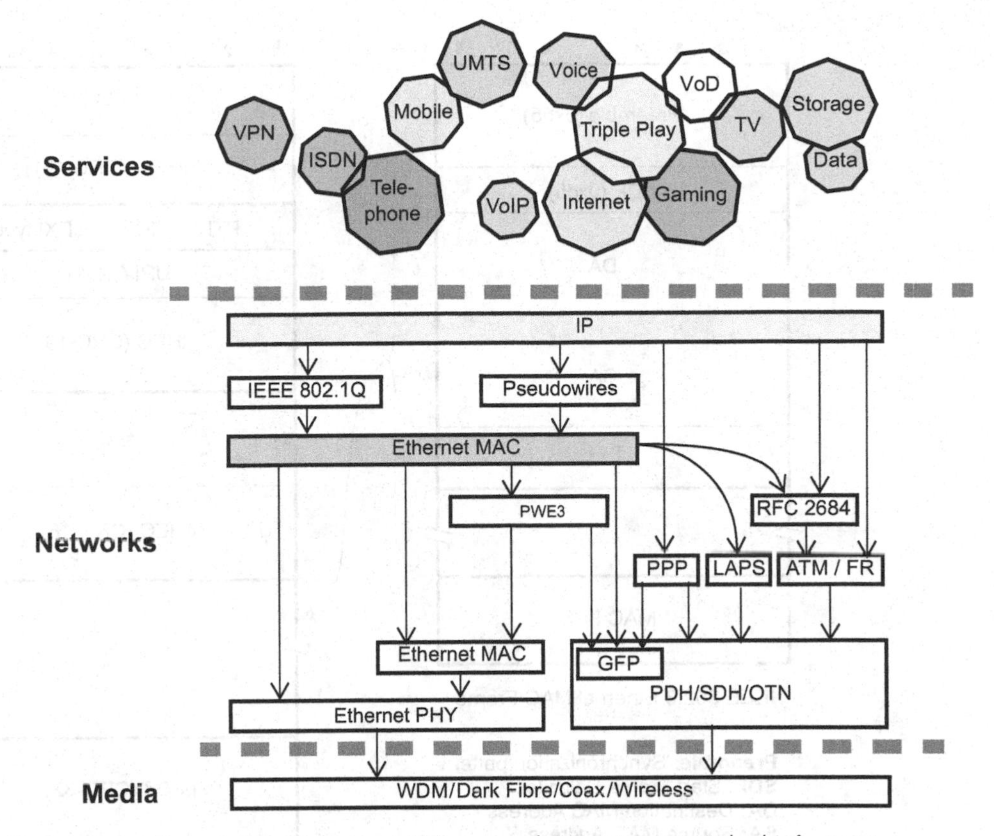

Figure 10-9. Transporting Ethernet and IP over packet- or circuit-switched infrastructures.

2. *Framed GFP* (GFP-F) is more efficient because it removes interframe gaps and unnecessary frame fields. It also allows bandwidth sharing among several traffic flows. With GFP-F, service providers benefit from the statistical multiplexing gain, although subscribers may receive reduced performance when compared with GFP-T. This is due to the use of queues that increase end-to-end delay. Differentiated traffic profiles can be offered to customer signals (see Figure 10-10).

Compared with ATM, the GFP-F encapsulation has at least three critical advantages:

1. It adds very little overhead to the traffic stream. ATM adds 5 overhead bytes for every 53 delivered bytes plus AAL overhead.
2. It carries payloads with variable length, as opposed to ATM that can only carry 48-byte payloads. This makes it necessary to split long packets into small pieces before they are mapped in ATM.
3. It has not been designed as a complete networking layer like ATM – it is just an encapsulation. Specifically, it does not contain VPI/VCI or other equivalent fields for switching traffic. Switching is left to the upper layer, usually Ethernet.

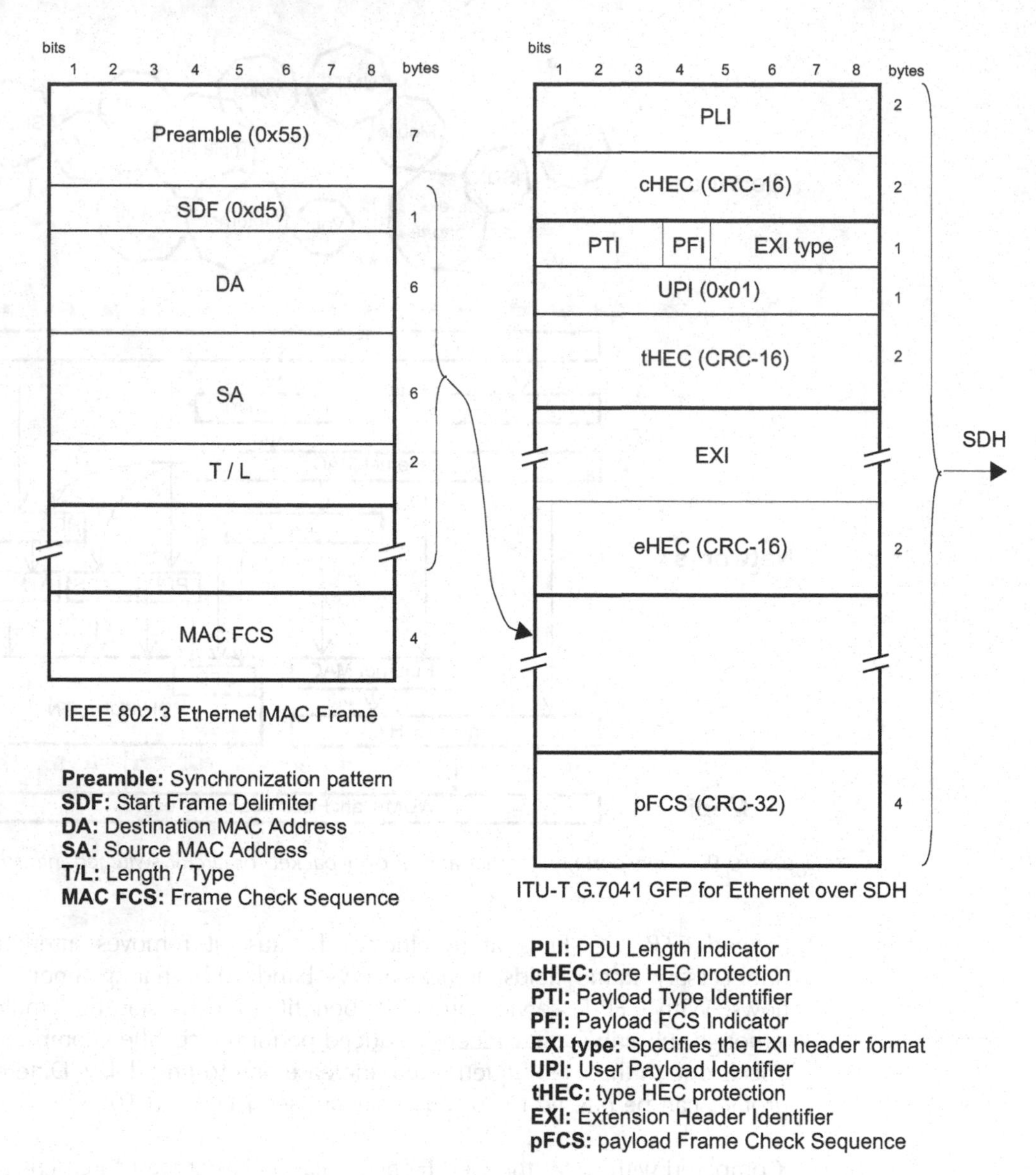

Figure 10-10. The GFP-F mapping for Ethernet makes ATM unnecessary. Now there is no VPI/VCI to switch the traffic, but the Ethernet MAC addresses can be used for similar purposes.

Like LAPS, GFP can be used for tunnelling of Ethernet traffic over an SDH path, but the importance of this new mapping is that it allows Ethernet traffic to be active within the WAN. With the help of GFP, SDH network elements are able to bridge MAC frames like any other switch based on the Ethernet physical layer. The features of SDH MAC switches include MAC address learning and flooding of frames with unknown destination MAC

(a) **NG-SDH – customer switching**

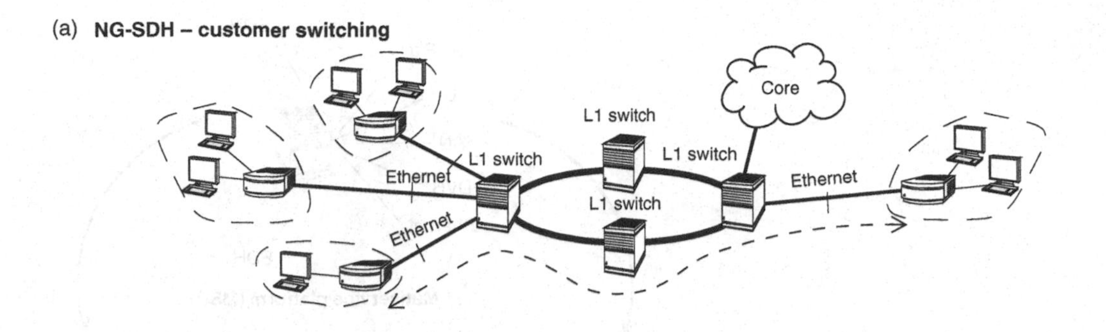

(b) **NG-SDH – network switching**

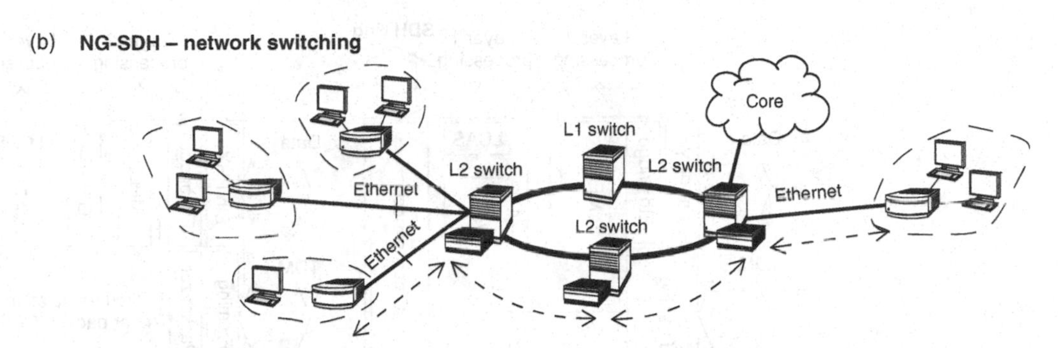

Figure 10-11. Ethernet over NG-SDH. Depending on the requirements, two approaches are possible: (a) customer switching – simple; the transport network is just a link between the customer switches. (b) Network switching – more flexible; one step forward towards a more sophisticated service based on MPLS.

address. In a few words, SDH MAC switches enable us to emulate an Ethernet LAN over an SDH network (see Figure 10-11).

Deploying Ethernet in MAN/WAN environments makes it necessary to develop new types SDH *add/drop multiplexers* (ADMs) and *digital cross-connects* (DXC) with layer-2 bridging capabilities (see Figure 10-12):

- Enhanced ADMs are like a traditional ADM, but they include Ethernet interfaces to enable access to new services, and TDM interfaces for legacy services. Many of these network elements add Ethernet bridging capabilities, and some support MPLS and *resilient packet ring* (RPR). New services benefit from the advantages of NG-SDH. New and legacy services are segregated in different SDH TDM timeslots.
- Packet ADMs have a configuration similar to enhanced ADMs: they include TDM and packet interfaces but packet ADM offers common packet-based management for both new and legacy services. The TDM tributaries are converted into packets before being forwarded to the network. *Circuit emulation over packet* (CEP) features are needed. MPLS is likely to be the technology in charge of multiplexing new and legacy services together in packet ADMs, due to the flexibility given by MPLS connections known as *label-switched paths* (LSP). Packet ADMs provide the same advantages as enhanced

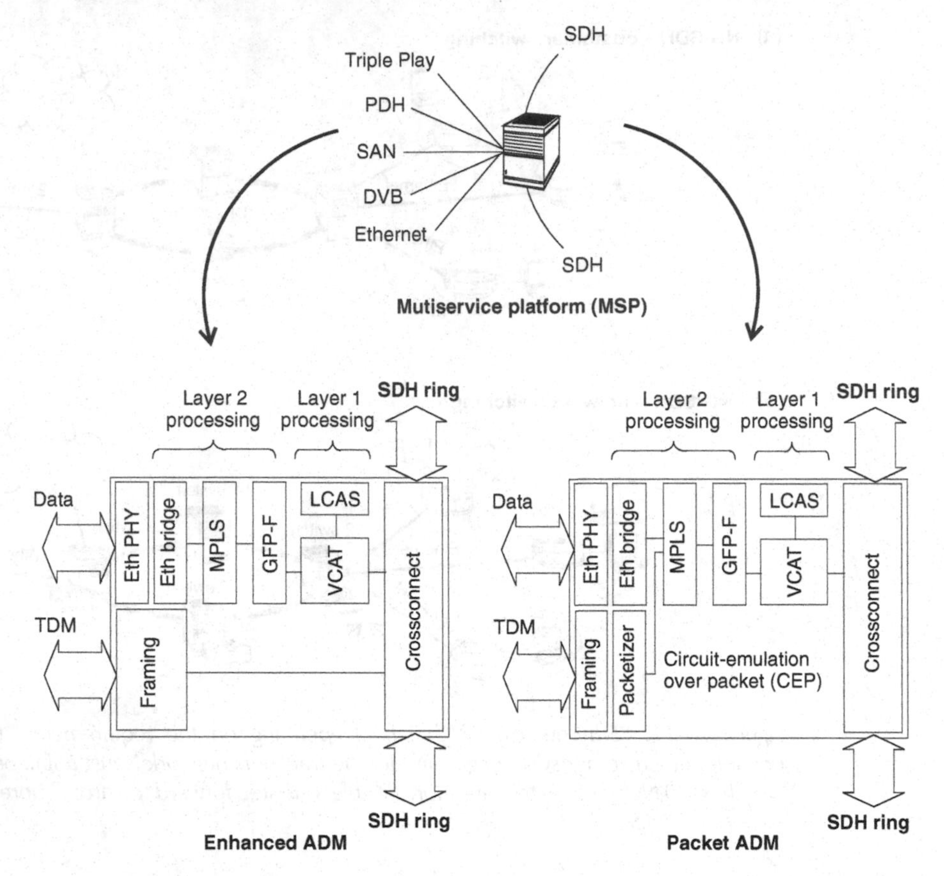

Figure 10-12. New SDH network elements. The enhanced ADM offers packet and TDM interfaces in the same network element. Packet ADMs offer the same, but over a unified packet-based switching paradigm for all tributaries.

ADMs, but additionally, the network operator can benefit from increased efficiency and simplified management due to a unified switching paradigm.

EoS is the technology preferred by incumbent operators, as they already have a large basis of SDH equipment in use. On the other hand, new operators generally prefer carrier Ethernet directly implemented over optical layers.

10.3 Limitations of Bridged Networks

Metro Ethernet architectures, based only on native Ethernet switches and any combination of dark fibre, SDH and WDM, are like a big LAN – they have the same advantages and similar disadvantages. We know that in low traffic conditions and with a limited number of

Table 10-2. *Alternatives for providing Carrier Ethernet services*

	Fibre	Fibre/WDM +MPLS	CWDM or DWDM	NG-SDH	NG-SDH + Pseudowires	Classic SDH (VC/LAPS)
	Low latency, low jitter, high resiliency, good OAM					
Service	E-Line E-LAN	E-Line E-LAN if VPLS	E-Line E-LAN if VPLS	E-Line	VPWS: E-Line VPLS: E-LAN	E-Line
Bandwidth	High	High	High	High	High	Inefficient
Cost	Medium-low	Medium-low	Very expensive	Medium, if SDH exits	Medium, if SDH exits	Low
QoS	Best effort	Best effort Carrier if MPLS	Carrier	Carrier	Carrier	Not differentiated
SLA	EIR, PIR	CIR, EIR	Guaranteed	GFP-F: EIR>CIR GFP-T: EIR=CIR	EIR, CIR	EIR = CIR
Legacy TDM	No	Yes	No	Yes	Yes	Reliable Predictable
Statistical	Yes	Yes	Yes	Yes	Yes	No
Latency	Medium	Medium	Very low	Low	Very low	Very low
Jitter	High	High	Low	Very low	Very low	Very low
Topology	Meshed	Meshed	Meshed	Ring Linear	Ring Linear	Ring Linear
Network	Access Metro	Access Metro	Core Data Centre	Metro Core	Metro Core	Access metro
Good for	VPN Triple Play if MPLS	Internet Triple Play if MPLS	Any including storage	Data, VoIP, video, connectivity	Time-sensitive applications	Superseded
Resilience	STP ~10 s RSTP ~1 s RPR 50 ms	STP ~10 s RSTP RSTP ~1 s RPR 50 ms	OTN 50 ms	APS 50 ms LCAS 50 ms	Path protection APS 50 ms LCAS 50 ms	APS 50 ms
MPLS	Yes	Yes	Yes	Yes	VPLS and VPWS	No
Management	Distributed	Distributed	Distributed	Hierarchical and distributed	Hierarchical and distributed	Hierarchical

hosts, this network works very well. However, as soon as the installation begins to grow, aspects like scalability, quality of service, topologies and protection tend to fall down. Let us examine the drawbacks in more detail.

10.3.1 Scalability

Ethernet switches use promiscuous broadcasting to learn addresses constantly (IEEE 802.1D). When a request to forward a frame to an unknown address arrives, the switch has to flood the frame to all the ports, waiting for a response to know where the address is. This way is not very efficient, nor secure.

MAC addresses are not hierarchical, and the switching table does not scale very well, slowing down the performance of switches when there is a large number of client hosts (this is known as the MAC switching table explosion problem). Furthermore, all the switches in the network have to constantly learn the MAC addresses of new stations connected to the network.

VLANs (IEEE 802.1Q) offer an easy solution to some of the problems mentioned before. A switch can be divided into smaller virtual switches, each of them belonging to a different VLAN. VLANs are used to split one big broadcast domain into several smaller domains, thus improving security and reducing broadcast traffic.

Another advantage of 802.1Q VLANs is the ability to offer QoS by means of the three 802.1p bits (VLAN CoS bits).

However, the number of VLAN identifiers is limited to 4096. This limits the number of subscribers a service provider can have in the same network. Furthermore, subscribers may have their own VLAN structure, and it is desirable to support the customer and the provider VLAN structure simultaneously.

A solution is to stack VLAN tags (Q-in-Q solution) to obtain a larger number of VLAN identifiers as is specified in the IEEE 802.1ad standard for provider Ethernet bridges. *MAC address stacking* (MAS) is a more powerful solution, designed to deal with MAC switching table explosions (see Figure 10-13). The customer equipment is not supposed to understand the Q-in-Q and MAS frame formats. This means that the service provider VLAN labels and MAC addresses are added by the first hop and removed by the last hop in the carrier network.

10.3.2 Protection

The *spanning tree protocol* (STP) and the *rapid spanning tree protocol* (RSTP) work well enough in LAN and also in data services. However, they do not meet the challenge when it comes to the mass rollout of IP services, or the 50 ms restoration time paradigm of carrier-class services.

10.3.3 Topologies

All topologies are possible, but some may become inefficient due to the operation of the STP or any of its variations. This is the case with rings which are used only partially; they are transformed into trees, and unused links are left just for protection. The IEEE 802.17 for *Resilient Packet Rings* (RPR) was designed to solve this problem, but it has not gained market acceptance, probably because it is rather expensive and not flexible enough, as it requires the same bit rate for all nodes.

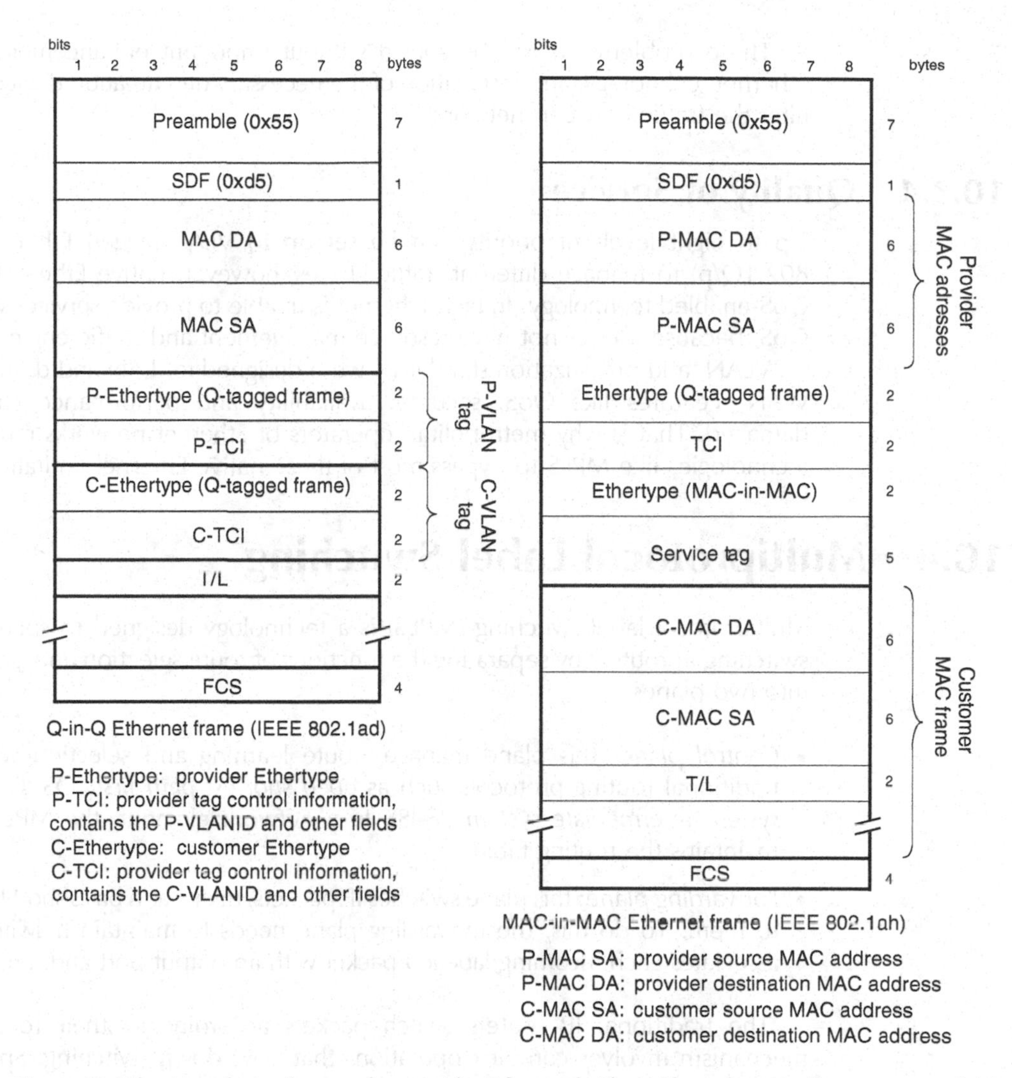

Figure 10-13. New frame formats for provider Ethernet bridges. These frames offer a more scalable Ethernet for carrier networks.

Another problem is network demarcation, when the same technology is everywhere without a clear border between customer and provider installations. It is necessary to find a solution to questions like:

- Can the typical LAN protocols like STP, deployed by subscribers in their networks, influence the overall service-provider network operation?
- How will uncontrolled traffic generated by some subscribes affect the global MAN/WAN operation?
- How will the continuous connection and disconnection of stations in the subscriber network affect the service provider network?

These problems cannot be solved without important enhancements in the classic Ethernet technology, and installation of the necessary demarcation devices to isolate and filter the traffic between networks.

10.3.4 Quality of Service

Up to eight levels of priority can be set up to VLAN-tagged Ethernet frames (IEEE 802.1Q/p) to manage different traffic classes–however, native Ethernet is not really a QoS-enabled technology. In fact, Ethernet is unable to provide services with guaranteed QoS, because it does not have resource management and traffic engineering tools.

VLAN, and prioritization standards[1] were designed for LAN and do not scale well on WAN. Features like QoS, security, availability and performance can be seriously damaged. That is why metropolitan operators of Ethernet networks must rely on other technologies like MPLS to bypass most of these native Ethernet limitations.

10.4 Multiprotocol Label Switching

Multiprotocol label switching (MPLS) is a technology designed to speed up IP packet switching in routers by separating the functions of route selection and packet forwarding into two planes:

- *Control plane*: this plane manages route learning and selection with the help of traditional routing protocols such as *open shortest path first* (OSPF) or *intermediate system–intermediate system* (IS-IS). In a network element, the MPLS control plane maintains the routing table.
- *Forwarding plane*: this plane switches IP packets, taking as a basis short labels appended to them. To do this, the forwarding plane needs to maintain a switching table that associates each incoming labelled packet with an output port and a new label.

The traditional IP routers switch packets according to their routing table. This mechanism involves complex operations that slow down switching. Specifically, traditional routers must find the longest network address prefix in the routing table that matches the destination of every IP datagram entering the router.

On the other hand, MPLS routers, also known as *label-switched routers* (LSR), use simple, fixed-length label forwarding instead of a variable-length IP network prefix for fast forwarding of packetized data (see Fig ure 10-15).

MPLS enables the establishment of a special type of virtual circuits called *label-switched paths* (LSP) in IP networks. Thanks to this feature, it is possible to implement resource management mechanisms for providing hard QoS on a per-LSP basis, or to deploy advanced traffic engineering tools that provide the operator with tight control over the path that follows every packet within the network. Both QoS provision and advanced traffic engineering are difficult, if not impossible to solve in traditional IP networks.

[1]IEEE 802.1Q for VLAN, IEEE 802.1Q/p for priority, IEEE 802.1D for bridging/switching.

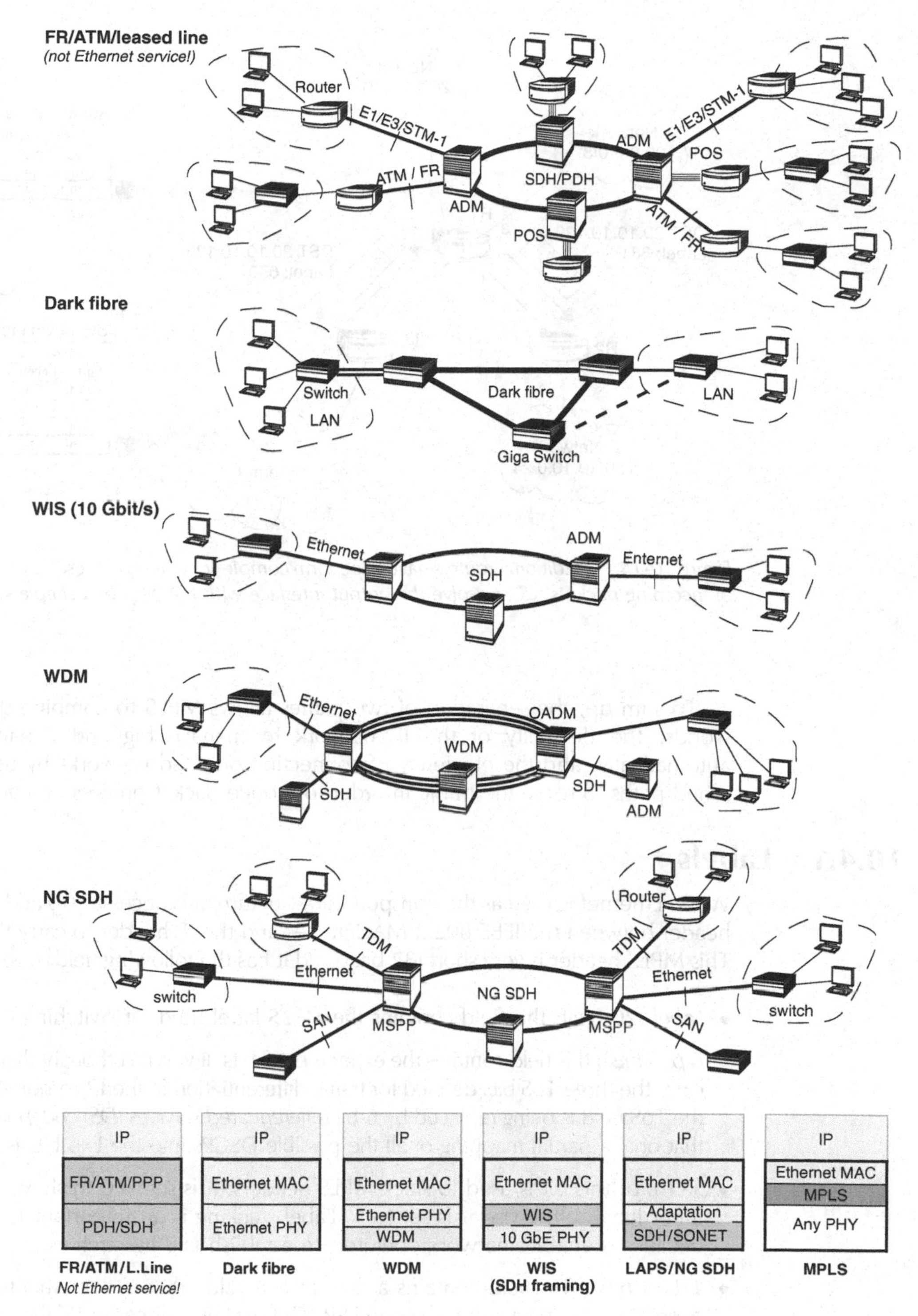

Figure 10-14. Alternatives for providing carrier Ethernet services.

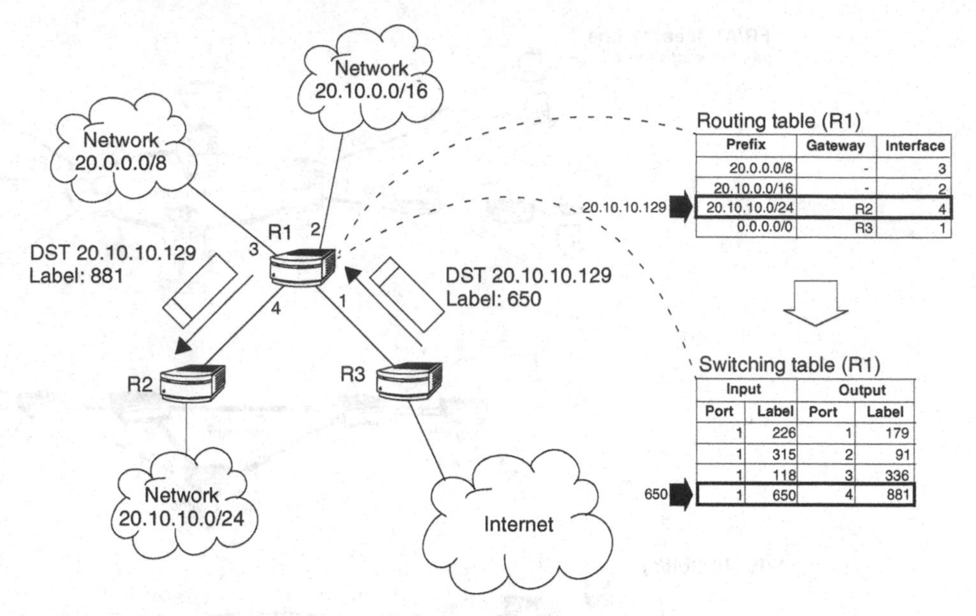

Figure 10-15. Traditional routers have to perform complex operations to resolve the output interface of incoming packets. LSRs resolve the output interface with the help of a simple switching table.

To sum up, the separation of two planes allows MPLS to combine the best of two worlds: the flexibility of the IP network to manage big and dynamic topologies automatically, and the efficiency of connection-oriented networks by using pre-established paths to route the traffic in order to reduce packet process on each node.

10.4.1 Labels

When Ethernet is used as the transport infrastructure, it is necessary to add an extra 'shim' header between the IEEE 802.3 MAC frames and the IP header to carry the MPLS label. This MPLS header is very short (32 bits), and it has the following fields (see Figure 10-16):

- *Label (20 bits)*: this field contains the MPLS label used for switching traffic.
- *Exp (3 bits)*: this field contains the experimental bits. It was first thought that this field could carry the three ToS bits defined for traffic differentiation in the IP version 4, but currently, the ToS field is being replaced by 6-bit *differentiated services (DS) codepoints*. This means that only a partial mapping of all the possible DSCPs into the Exp bits is possible.
- *S (1 bit)*: this bit is used to stack MPLS headers. It is set to 0 to show that there is an inner label, otherwise it is set to 1. Label stacking is an important feature of MPLS, because it enables network operators to establish LSP hierarchies.
- *TTL (8 bits)*: this field contains a *time to live* value that is decremented by one unit every time the packet traverses an LSR. The packet is discarded if the value reaches 0.

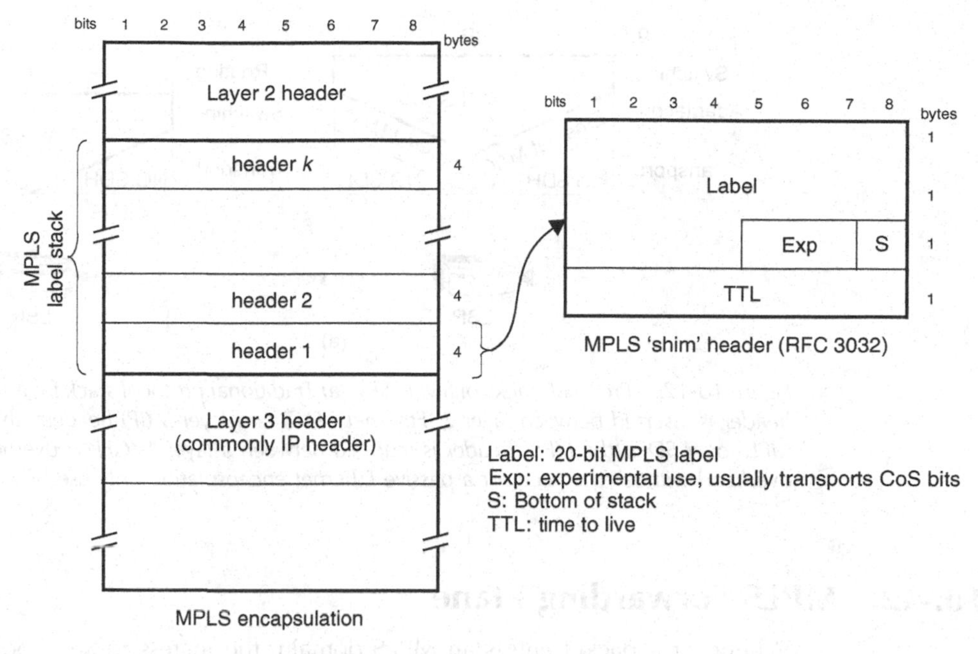

Figure 10-16. MPLS 'shim' header format. The label is usually inserted between the layer-2 and layer-3 headers.

MPLS can be used in SDH transport infrastructures as well. IP routers with SDH interfaces can benefit from the advantages of MPLS like any other IP router. Since the MPLS header must be inserted between layer-2 and layer-3 headers, it is necessary to encapsulate MPLS-labelled frames into Ethernet MAC frames before they are mapped to SDH. However, recent ITU-T recommendations allow direct mapping of MPLS-labelled packets to GFP-F for transport across NG-SDH circuits. This is an important exception of the common frame labelling, because in this case labels are inserted between a layer-1 header (GFP-F) and a layer-3 header (IP). This new mapping improves the efficiency of SDH LSRs by eliminating the need for a passive Ethernet encapsulation used only for adaptation (see Figure 10-17).

The MPLS label is sometimes included in the 'shim' header inserted between the layer-2 and layer-3 headers, but this is not always true. Almost any header field used for switching can be reinterpreted as an MPLS label. The FR 10-bit *data link connection identifier* (DLCI) field or the ATM *virtual path identifier* (VPI) and *virtual circuit identifier* (VCI) are two examples of this. The ATM VPI/VCI example is of special importance, because it allows a smooth transition from the ATM-based network core to an IP/MPLS core. An ATM switch can be used as an LSR with the help of relatively simple upgrade that will probably involve only new software.

MPLS is proving to be a technology with incredible flexibility. Timeslot numbers in TDM frames, or even wavelengths in WDM signals, can be re-interpreted as MPLS labels as well. This approach opens the door to a new way of managing TDM/WDM networks. The MPLS-based management plane for TDM/WDM networks uses distributed IP routing, and at the same time it benefits from the powerful traffic engineering features of MPLS. This, in fact, forms a new technology and an active investigation field called generalized MPLS (GMPLS).

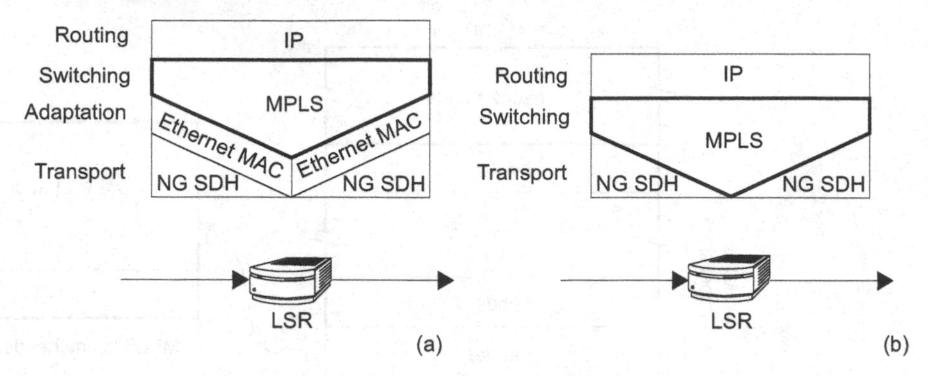

Figure 10-17. Protocol stacks of SDH LSRs. (a) Traditional protocol stack for an SDH LSR: the MPLS header is inserted between layer-2 (Ethernet MAC) and layer-3 (IP) headers. (b) Direct mapping of MPLS over SDH: the MPLS header is mapped between a layer-1 (GFP-F) overhead and layer-3 (IP) overhead without the need for a passive Ethernet encapsulation only used for adaptation.

10.4.2 MPLS Forwarding Plane

Whenever a packet enters an MPLS domain, the ingress router, known as the ingress *label edge router* (LER), inserts a header that contains a label that will be used by the LSR to route packets to their destination. When the packet reaches the edge where the egress router is, the label is dropped and the packet is delivered to its destination (see Figure 10-18). Only input labels are used for forwarding the packets within the network, while encapsulated addresses like IP or MAC are completely ignored.

In typical applications, labels are chosen to force the IP packets to follow the same paths they would follow if they were switched with routing tables. This means that the entries in the LSR routing tables must be taken into account when assigning labels to packets and building switching tables.

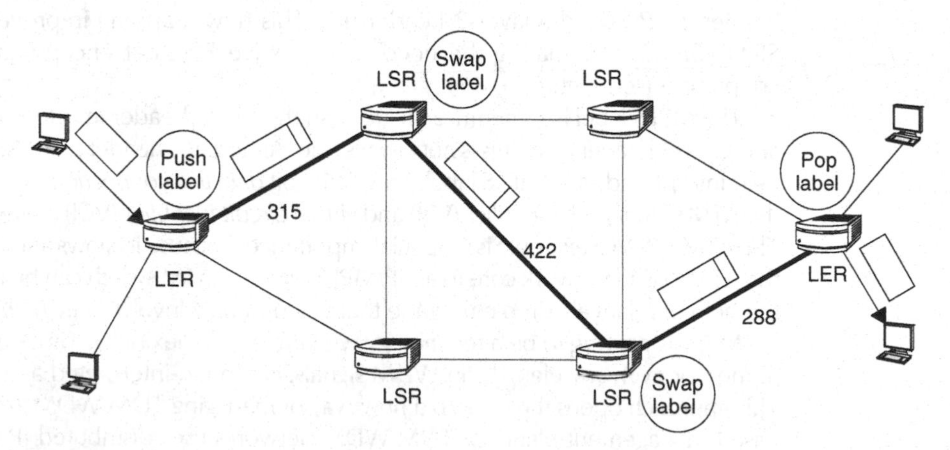

Figure 10-18. Label processing within an MPLS domain. A label is pushed by the ingressing LER, swapped by the intermediate LSR across the LSP, and popped by the egressing LER.

The set of packets that would receive the same treatment by an LSR (i.e. packets that will be forwarded to the same port towards the same destination network) is called a *forwarding equivalence class* (FEC). LSRs bind FECs with label/port pairs. For example, all packets that must be delivered to the network 20.10.10.0/24 constitute an FEC that might be bound to the pair (4, 881). All packets directed to that network will be switched to the port 4 with label 881. The treatment that packets will receive on the next hop depends on the selection of the outgoing label. In our example, a packet switched to the port 4 with label 882 will probably never arrive at network 20.10.10.0/24. An LSR may need to request the right label at the next hop to ensure that the packets receive the desired treatment and that they will be forwarded to the correct destination.

The most common FECs are defined by network address prefixes stored in the routing tables of LSRs. In the routing table, the network prefix determines the outgoing interface for the set of incoming packets matching this prefix. If we wish to emulate the behaviour of a traditional IP router, every network prefix must be bound with a label.

Within the MPLS domain, labels only have a local meaning, which is why the same label can be re-used by different LSRs. For the same packet, the value of the label can be different at every hop, but the path a packet follows in the network is totally determined by the label assigned by the ingressing LER. The sequence of labels [315, 422, 288] defines an LSP route; all packets following the LSP receive the same treatment in terms of bandwidth, delays or priority, enabling specific treatment for each traffic flow like voice, data or video. There are two LSP types (see Figure 10-19):

- *Hop-by-hop LSPs* are computed with routing protocols alone. MPLS networks with only hop-by-hop LSP route packets are like traditional IP networks but with enhanced forwarding performance provided by label switching.
- *Explicit LSPs* are computed by the network administrators for specific purposes, and configured either manually in the LSRs, or with the help of the management platform.

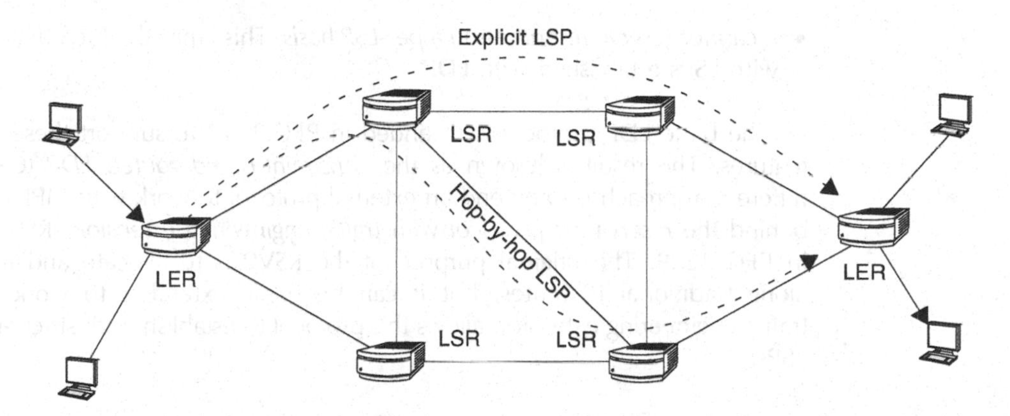

Figure 10-19. A hop-by-hop LSP and an explicit LSP between the same source and destination. The hop-by-hop LSP is computed by the routing protocols running in the LSRs. The explicit route is computed by an external network management system (NMS).

> The path followed by the packets forwarded across explicit LSPs may be different from the paths computed by routing protocols. They can be useful to improve network utilization or select custom paths for certain packets. The ability to provide explicit LSPs converts MPLS into a powerful traffic engineering tool.

An explicit LSP can be strict or loose, depending on how it is established:

- If all the hops that constitute the explicit LSP are specified one by one, the LSP is said to be strictly specified.
- If some but not all of the hops that constitute the explicit LSP are specified but some others are left to the decision of the distributed routing algorithms, the LSP is said to be loosely specified.

10.4.3 Label Distribution

The LSR needs to know which label to assign to outgoing packets to make sure they arrive at the correct destination. The obvious way to do this is to configure the switching tables manually in every LSR. Of course, this approach is not the best possible if there are many LSPs dynamically established and released. To deal with this situation a label distribution protocol is needed.

A label distribution protocol enables an LSR to tell other LSRs the meaning of the labels it is using, as well as the destination of the packets that contain certain labels. In other words, by using a label distribution protocol the LSR can assign labels to FECs.

The RFC 3036 defines the *label distribution protocol* (LDP) that was specifically designed for distributing labels. As MPLS technology evolved, this protocol showed its limitations:

- *It can only manage hop-by-hop LSPs.* It cannot establish explicit LSPs and therefore does not allow traffic engineering in the MPLS network.
- *It cannot reserve resources on a per-LSP basis.* This limits the QoS that can be obtained with LSPs established with LDP.

The basic LDP protocol is extended in RFC 3212 to support these and some other features. The result is known as the *constraint-based routed LDP* (CR-LDP). Another different approach is to extend an external protocol to work with MPLS. This is the idea behind the *reservation protocol with traffic engineering extension* (RSVP-TE) as defined in RFC 3209. The original purpose of the RSVP is to allocate and release resources along traditional IP routes, but it can be easily extended to work with LSPs. The traffic engineering extension allows this protocol to establish both strict and loose explicit LSPs.

10.4.3.1 The Label Distribution Protocol

The LDP enables LSRs to request and share MPLS labels. To do this it uses four different message types.

1. *Discovery messages* announce the presence of LSRs in the network. LSRs send 'Hello' messages periodically, to announce their presence to other LSRs. These 'Hello' messages are delivered to the 646 UDP port. They can be unicast to a specific LSR or multicast to all routers in the subnetwork.
2. *Session messages* establish, maintain and terminate sessions between LDP peers. To share label to FEC binding information, two LSRs need to establish an LDP session between them. Sessions are transported across the reliable TCP protocol and they directed to port 646.
3. *Advertisement messages* create, modify or delete label mappings for FECs. To exchange advertisement messages, the LSRs must first establish a session.
4. *Notification messages* are used to deliver advisory or error information.

The most important LDP messages are (see Figure 10-20):

- The *label request message*, used by the LSR to request a label to bind with an FEC that is attached to the message. The FEC is commonly specified as a network prefix address.
- The *label mapping message*, distributed by the LSR to inform a remote LSR on which label to use for a specific FEC.

The LSR can request a label for an FEC by using request messages, but it can also deliver labels to FEC bindings without explicit request from other LSRs. The former is an operation mode called *downstream on demand*, and the latter is known as *downstream unsolicited*. Both modes can be used simultaneously in the same network.

Regarding the behaviour of LSRs when they operate in the downstream on demand mode, receiving label request messages, there are two different options:

- *Independent label distribution control*: LSRs are allowed to reply to label requests with label mappings whenever they desire, for example immediately after the request arrives. This mode can be compared to the *address resolution protocol* (ARP) used in LANs to request mappings between destination IP addresses and MAC addresses.
- *Ordered label distribution control* (see Figure 10-21): LSRs are not allowed to reply to label requests until they know what to do with the packets belonging to the mapped FEC. In other words, LSRs cannot map an FEC with a label unless they have a label for the FEC, or if they are egress LERs themselves. When an LSR operates in this mode, it propagates the label requests downstream and waits for a reply before replying upstream.

10.4.4 Martini Encapsulation

In the MPLS network, only the ingress and egress LERs are directly attached to the end-user equipment. This makes them suitable for establishing edge-to-edge sessions to enable

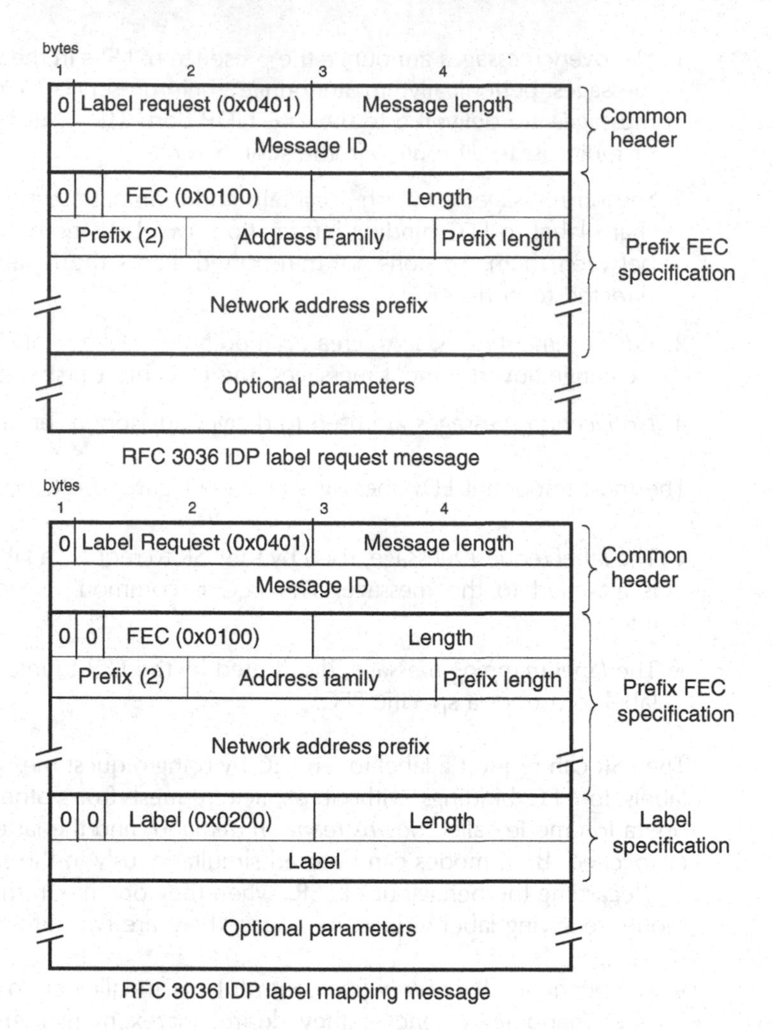

Figure 10-20. Two important LDP messages. The label request message requests a label from a remote LSR for binding with an FEC that is attached to the message. The label mapping message is used to inform a remote LSR on which label to use for a specific FEC.

communications between remote users. In this network model, the roles of LSRs and LERs would be:

- *LSRs* are in charge of guiding the frame through the MPLS network, using either IP routing protocols or paths that the network administrator has chosen by means of explicit LSPs.
- The *Ingress LER* is in charge of the same tasks as any other LSR, but it also establishes sessions with remote LERs to deliver traffic to the end-user equipment attached to them.

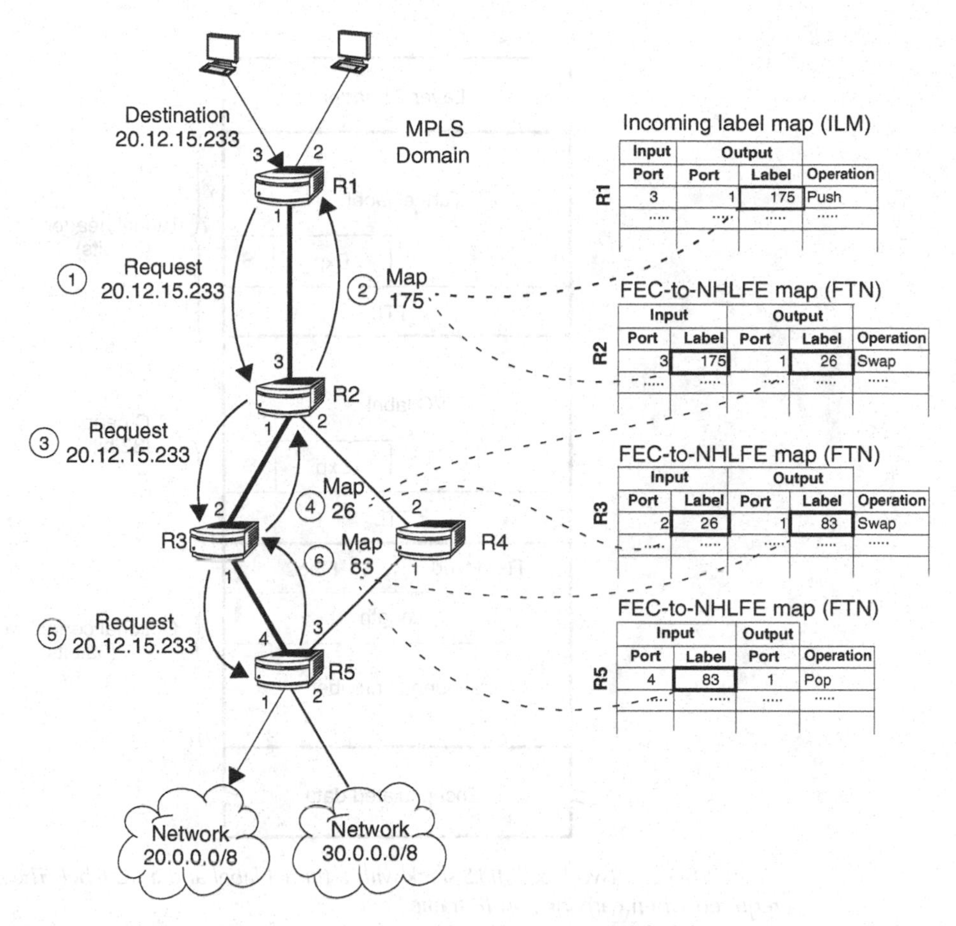

Figure 10-21. LDP in downstream-on-demand and independent label distribution mode. LSRs in the LSP generate label requests. The replies they receive from LSRs upstream are used to fill their switching tables.

- The *Egress LER* acts as the peer of the ingress LER in the edge-to-edge session, but it does not need to guide the traffic through the MPLS network, because the traffic leaves the network in this node and it is not routed back to it.

There is an elegant way to implement the discussed model without any new overhead or signalling: by using label stacking. This model needs an encapsulation with a two-label stack known as the *Martini encapsulation* (see Figure 10-22):

- The *tunnel label* is used to guide the frame through the MPLS network. This label is pushed by the ingress LER and popped by the egress LER, but it can also be popped by the penultimate hop in the path, because this LSR makes the last routing decision within the MPLS domain, thus making the tunnel label unnecessary for the last hop (the egress LER).

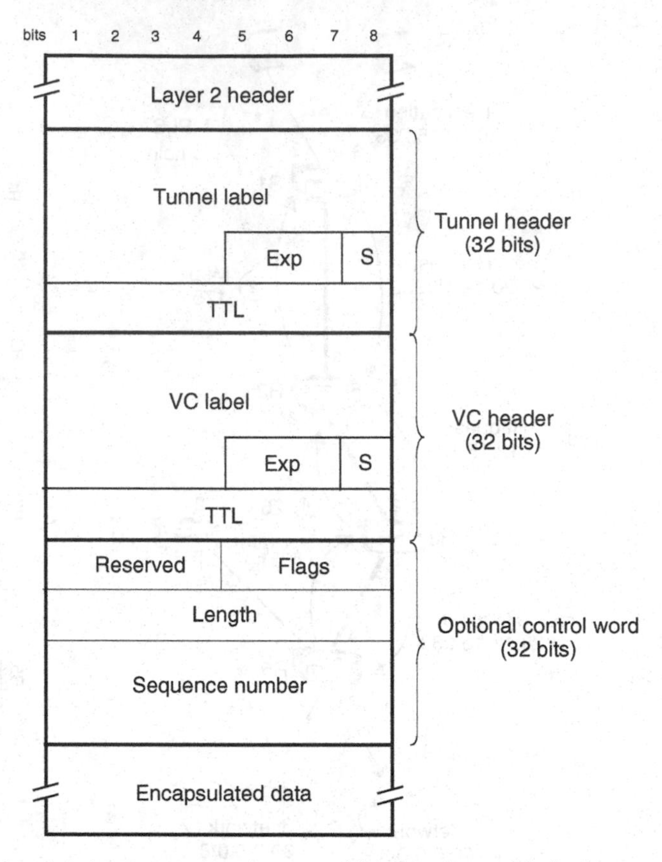

Figure 10-22. Two-label MPLS stack with a tunnel label and a VC label. The control word may be required when carrying non-IP traffic.

- The *VC label* is used by the egress LER to identify client traffic and forward the frames to their destination. The way the traffic reaches end users is a decision taken by the ingress and egress nodes, and it does not involve the internal LSRs. The VC label is therefore pushed by the ingress LSR and popped by the egress LSR.

In the non-hierarchical one-label model, all the routers in the LSP participate in establishing an edge-to-edge session, and all are involved in routing decisions as well. A two-label model involves two types of LSPs. The tunnel LSP may have many hops, but the VC LSP has only two nodes, the ingress and egress LERs. VC LSPs can be interpreted as edge-to-edge sessions that are classified into groups and delivered across the MPLS network within Tunnel LSPs (see Figure 10-23). Tunnel LSPs are established and released independently of the VC LSPs. For example, tunnel LSPs can be established or modified when new nodes are connected to the network, and VC LSPs could be set up when users wish to communicate between them.

The two-label model makes routing and session management independent of each other. It is not necessary to maintain status information about sessions in the internal LSRs. All these tasks are carried out by LERs. The signalling of the VC LSP is also different from that

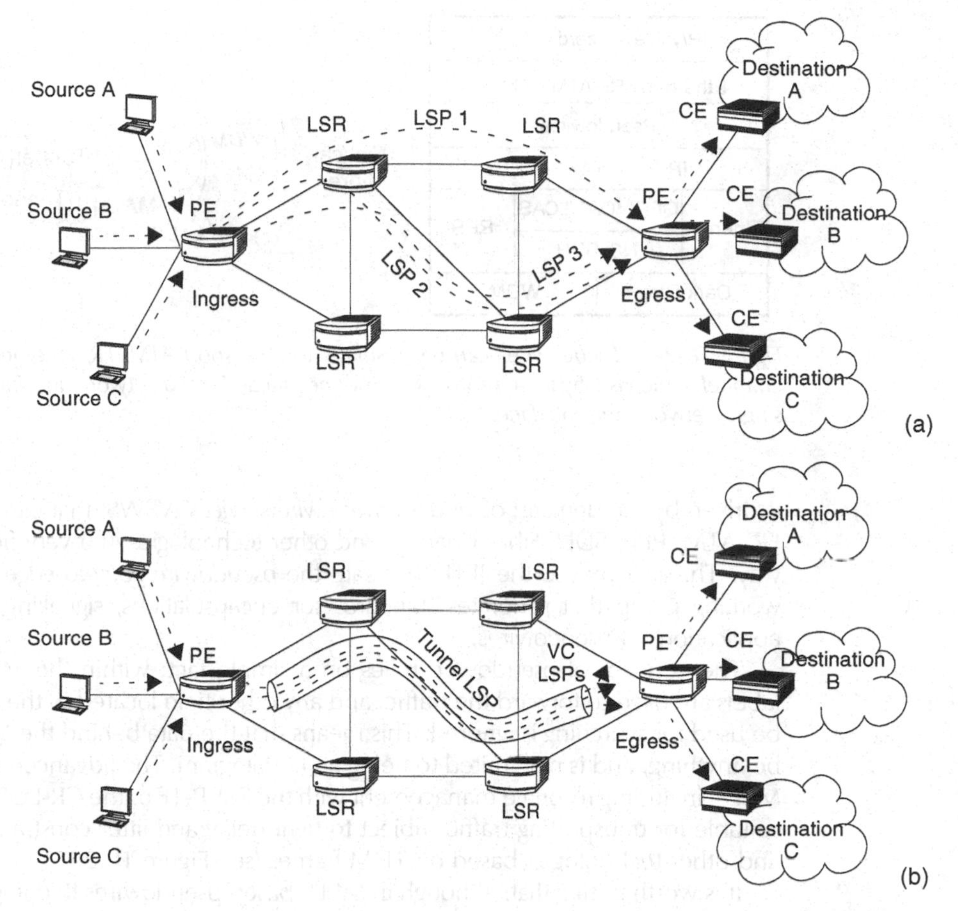

Figure 10-23. (a) One-label approach: the decision to establish routing and edge-to-edge sessions is shared between all the routers. (b) Two-label model: edge-to-edge sessions are tunnelled, and internal LSRs are unaware of them.

of the tunnel LSP. While establishing a tunnel LSP may require specific QoS or it may depend on administrative policies relying on traffic engineering, VC LSPs are much more simple. This is the reason why label distribution of tunnel LSPs is carried out with the CR-LDP or the RSVP-TE protocols, but VC LSPs can be managed with the simple LDP.

Although the two-label approach is valid for any MPLS implementation, it has been defined to be used with pseudowires (see Section 10.4.5).

10.4.5 Pseudowires

Pseudowires are entities that carry the essential elements of layer-2 frames or TDM circuits over a packet-switched network with the help of MPLS.[2] The standarization of pseudowires

[2]Although it is possible to implement pseudowires without MPLS, it is used in all the important solutions due to its better performance when compared with other options.

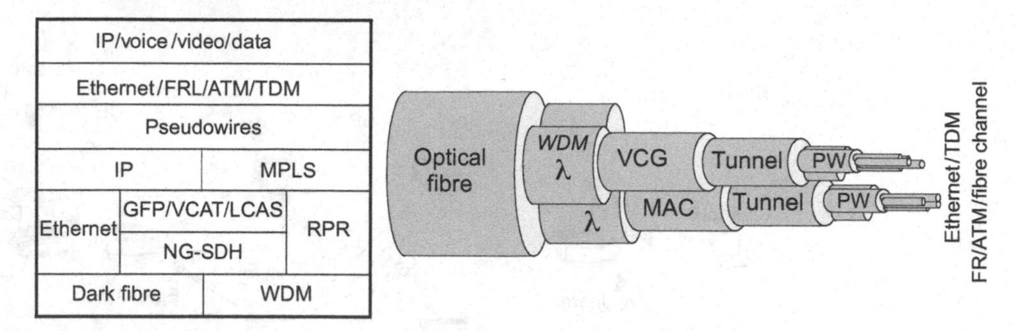

Figure 10-24. Pseudowires can encapsulate and transport ATM, FR, Ethernet, PPP, TDM or fibre channel, which is why these protocols do not need a dedicated network unifying the transport in one single network and interface.

is driven by the demand of *virtual private wire services* (VPWS) that can transport Ethernet, FR, ATM, PPP, SDH, Fibre Channel and other technologies in a very flexible and scalable way. This fact moved the IETF to create the *pseudowire edge-to-edge emulation* (PWE3) working group that generates standards for encapsulations, signalling, architectures and applications of pseudowires.

The concept of pseudowire relies on a simple fact: within the MPLS network, only labels are used to forward the traffic, and any other field located in the payload that could be used for switching is ignored. This means that the data behind the MPLS header could be anything, and is not limited to being an IP datagram. The advanced QoS capabilities of MPLS, including resource management with the RSVP-TE or the CR-LDP protocols, make it suitable for transporting traffic subject to tight delay and jitter constraints, including SDH and other technologies based on TDM frames (see Figure 10-24).

It is worth noting that, although in MPLS-based pseudowires IP datagrams are replaced by layer-2 or TDM data, IP routing is still an important part of the network. OSPF, IS-IS or other routing protocols are still necessary to find routes in the service provider network when they are not explicitly defined in the LSP setup process. This means that, in the MPLS network carrying pseudowires, IP numbering must be maintained in the network interfaces, because IP routing protocols need IP addresses to work.

Pseudowires are tunnelled across the packet-switched network (see Figure 10-25). Any network capable of providing tunnels can be used as a transport infrastructure. MPLS is by far the most common transport infrastructure for pseudowires, but pure IP networks can be used for the same purpose as well. The MPLS-based pseudowires use LSPs as tunnels, but IP networks need to use other tunnels, such as *generic routing encapsulation* (GRE) or *layer-2 tunnelling protocol* (L2TP) tunnels.

Many pseudowires are allowed to be multiplexed in the same tunnel, and therefore it is necessary to identify them. An MPLS label can be used for this purpose, although the IP transport infrastructure uses different alternatives again. MPLS architectures need two labels for carrying pseudowires: the first to identify the tunnel and the second to identify the pseudowire. The tunnel/VC double labelling (see Section 10.4.4) is applied to this case. Here, the VC label becomes the pseudowire identifier, and it is therefore known as the *pseudowire* (PW) label.

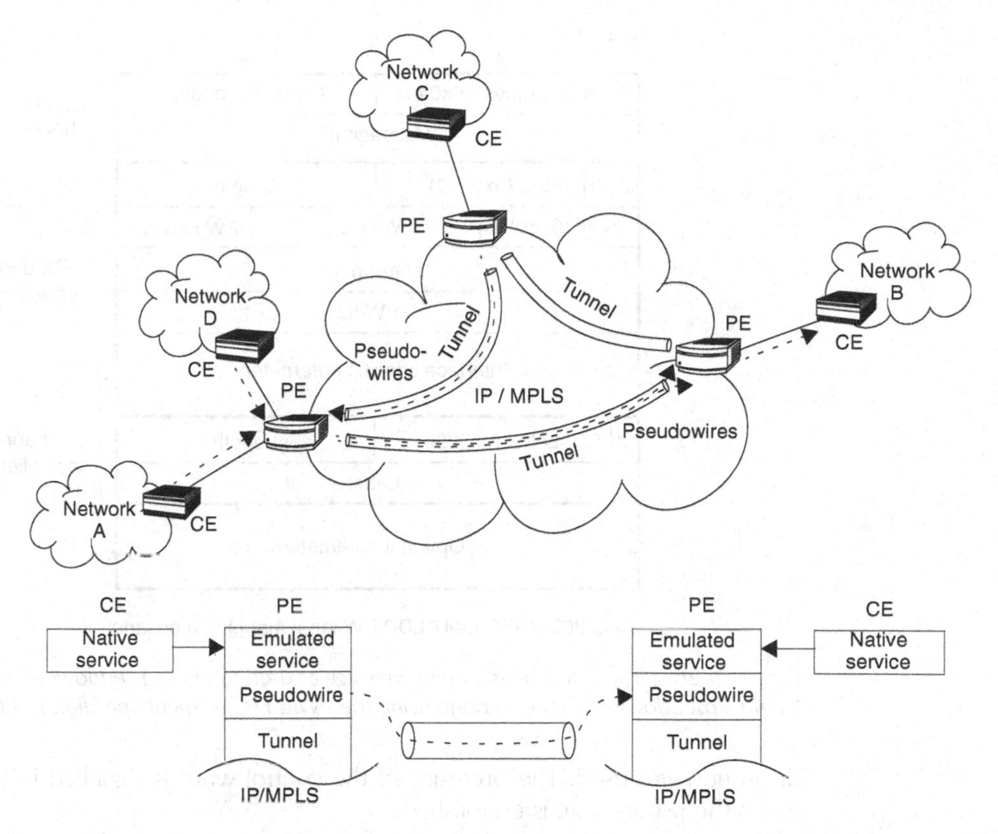

Figure 10-25. Emulation of connectivity services over pseudowires and tunneling across an IP/MPLS network.

In the traditional MPLS applications, FECs are specified by means of IP addresses or IP network prefixes. Once a label is bound with a network prefix, the network node automatically knows how to forward those packets that carry this label. However, this simple approach does not work with pseudowires, because they carry non-IP data. It is necessary to specify a new way to tell the pseudowire end points how to process the data carried by the pseudowire. This means that new ways of specifying FECs must be defined. Furthermore, each technology may need its own FEC specification. For example, forwarding Ethernet frames from or to pseudowires depends on the physical port and the VLAN tag, but this is not necessarily true for ATM or SDH pseudowires. This problem is addressed by extending the LDP protocol to work with pseudowires (see Figure 10-26).

The existing definitions are generalistic and have different interpretations for different types of pseudowire. This is the reason why the new FEC specifications include a 16-bit field for choosing the service emulated over the packet-switched network (see Table 10-3).

Sometimes, it must be ensured that packets are received in the correct order. Other times it is necessary to pad small packets with extra bits, or add technology- specific control bits. To deal with these issues, an extra 32-bit word is inserted between the PW label and the encapsulated data (see Figure 10-22). The presence of this control word is sometimes required, other times optional, and occasionally not required at all, depending on the type

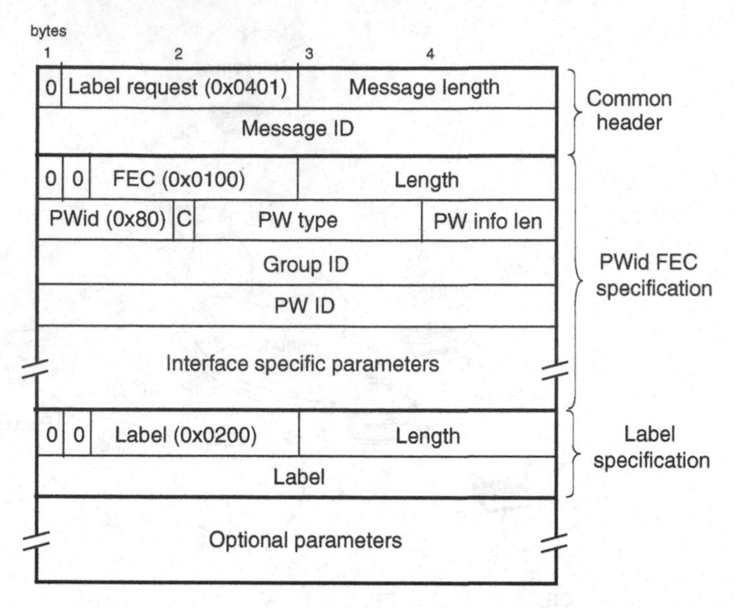

Figure 10-26. The LDP label-mapping message as used to map a pseudowire to an MPLS label. The label-to-pseudowire binding is done using the PWid FEC element specified in RFC 4447.

of pseudowire used. The presence of the control word is signalled in the LDP protocol when the pseudowire is established.

10.4.6 Ethernet Pseudowires

The aim of Ethernet pseudowires is to enable transport of Ethernet frames across a packet-switched network and to emulate the essential attributes of Ethernet LANs, such as MAC frame bridging or VLAN filtering across that network.

Standardization of pseudowires enables IP/MPLS networks to transport Ethernet efficiently. The Ethernet pseudowire is perhaps the most important type of pseudowire, because it can be used by network operators to fix some of the scalability, resilience, security and QoS problems of provider Ethernet bridges, thus making it possible to offer a wide range of carrier-grade, point-to-point and multipoint-to-multipoint Ethernet services, including EPL, EVPL, EPLAN and EVPLAN.

PE routers with Ethernet pseudowires can be understood as network elements with both physical and virtual ports. The physical ports are the attachment circuits where CEs are connected through standard Ethernet interfaces. The virtual ports are Ethernet pseudowires. Frames are forwarded to physical or virtual ports, depending on their incoming port and VLAN tags. These network elements may also include flooding and learning features to bridge frames to and from physical ports and Ethernet pseudowires, thus making it possible to offer emulated multipoint-to-multipoint LAN services. Many of these PE routers are also able to shape and police Ethernet traffic to limit traffic ingressing in the service provider network.

Table 10-3. The existing types of pseudowire

PW type	Description
0x0001	Frame relay DLCI (Martini mode)
0x0002	ATM AAL5 SDU VCC transport
0x0003	ATM transparent cell transport
0x0004	Ethernet tagged mode
0x0005	Ethernet
0x0006	HDLC
0x0007	PPP
0x0008	SONET/SDH circuit emulation service over MPLS
0x0009	ATM n-to-one VCC cell transport
0x000a	ATM n-to-one VPC cell transport
0x000b	IP layer 2 transport
0x000c	ATM one-to-one VCC cell mode
0x000d	ATM one-to-one VPC cell mode
0x000e	ATM AAL5 PDU VCC transport
0x000f	Frame relay port mode
0x0010	SONET/SDH circuit emulation over packet
0x0011	Structure-agnostic E1 over packet
0x0012	Structure-agnostic T1 (DS1) over packet
0x0013	Structure-agnostic E3 over packet
0x0014	Structure-agnostic T3 (DS3) over packet
0x0015	CESoPSN basic mode
0x0016	TDMoIP AAL1 mode
0x0017	CESoPSN TCM with CAS
0x0018	TDMoIP AAL2 mode
0x0019	Frame relay DLCI

When a new PE router is connected to the network, it must create tunnels to reach remote PE routers. The remote router addresses may be provided by the network administrators but many PE routers have autodiscovery features. Once the tunnels are established, it is possible to start the pseudowire setup with the help of LDP signalling. LDP mapping signals tell the remote PE routers to which physical port and to which VLANs frames with specified PW labels (see Figure 10-27) will be switched.

The physical attachment circuits of the PE router are standard Ethernet interfaces. Some of them may be trunk links with VLAN-tagged MAC frames, or even double VLAN-tagged Q-in-Q frames. Regarding how VLAN tags are processed, the PE routers have two operation modes:

- *Tagged mode*: the MAC frames contain at least one service-delimiting VLAN tag. Frames with different VLAN IDs may belong to different customers, or if they belong to the same customer, they may require different treatment in the service provider network. MAC frames with service-delimiting VLAN tags may be forwarded to different pseudowires or mapped to different Exp values for custom QoS treatment.
- *Raw mode*: the MAC frames may contain VLAN tags, but they are not service-delimiting. This means that any VLAN tag is part of the customer VLAN structure and must be transparently passed through the network without processing.

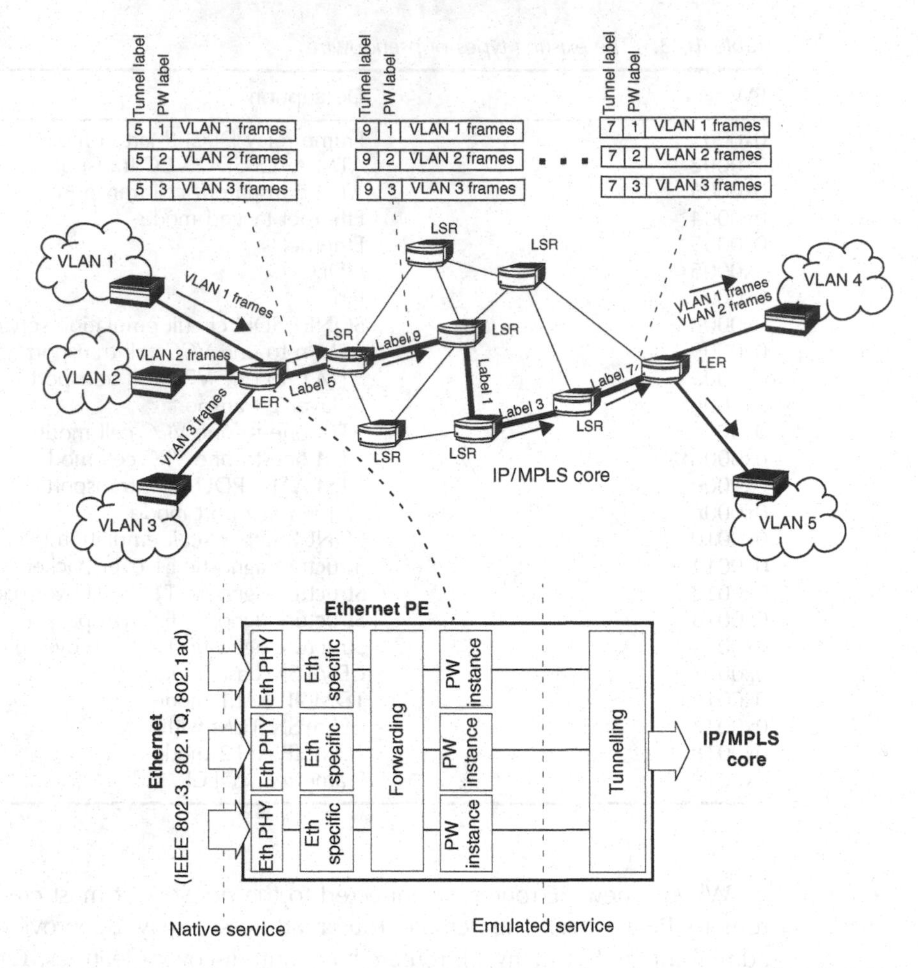

Figure 10-27. Operation of Ethernet pseudowires. The PE router becomes an Ethernet bridge with physical and virtual ports. Physical ports are connected to CEs with standard Ethernet interfaces. Virtual ports are Ethernet pseudowires tunnelled across the IP/MPLS core.

10.4.6.1 Virtual Private LAN Service

The VPLS is a multipoint-to-multipoint service that emulates a bridged LAN across the IP/MPLS core. VPLS is an important example of a layer-2 VPN service. Unlike more traditional layer-3 VPNs, based on network layer encapsulations and routing, layer-2 VPNs are based on bridging to connect two or more remote locations as if they were connected to the same LAN. Layer-2 VPNs are simple and well suited to business subscribers demanding Ethernet connectivity. VPLS also constitutes a key technology for metropolitan networks. This technology is currently available for network operators who want to provide broadband Triple Play services to a large number of residential customers.

When running VPLS, the service provider network behaves like a huge Ethernet switch that forwards MAC frames where necessary, learns new MAC addresses dynamically and performs flooding of MAC frames with unknown destination. In this architecture, PE routers behave like Ethernet bridges that can forward frames both to physical ports and to pseudowires.

As with physical wires, bridging loops may also occur in pseudowires. If fact, it is likely that this occurs if the pseudowire topology is not closely controlled, because pseudowires are no more than automatically established LDP sessions. A bridged network cannot work with loops. Fortunately, the STP or any of its variants can be used with pseudowires, as is done with physical wires to avoid them. However, there is another approach recommended by the standards. The most dangerous situation occurs when a PE router relays MAC frames from a pseudowire to a second pseudowire. To avoid pseudowire-to-pseudowire relaying, a direct pseudowire connection must be enabled between each PE router in the network. This implies a full-mesh pseudowire topology (see Figure 10-28). The full-mesh topology is completed with the *split-horizon rule*: it is forbidden to relay a MAC frame from a pseudowire to another one in the same VLPS mesh. Relaying would anyway be unnecessary because there is a direct connection with every possible destination.

To understand how VPLS works we can think of two end users, S and D, who want to communicate to each other (see Figure 10-29). User S wants to send a MAC frame to user D across a shared network running VPLS.

1. S sends the MAC frame towards D. LAN A is unable to find a local connection to D and finally the frame reaches bridge CE 1 that connects LAN A to a service provider network.
2. Bridge CE 1 forwards S's frame to PE 1 placed at the edge of a VPLS mesh. If PE 1 has not previously learnt S's MAC address, it binds it with the physical port where the frame came from.
3. The PE 1 bridge has not previously learnt the destination address of the MAC frame (D's MAC address), and therefore it floods the frame to all its physical attachment circuits. S's frame reaches LAN B, but D is not connected to it.

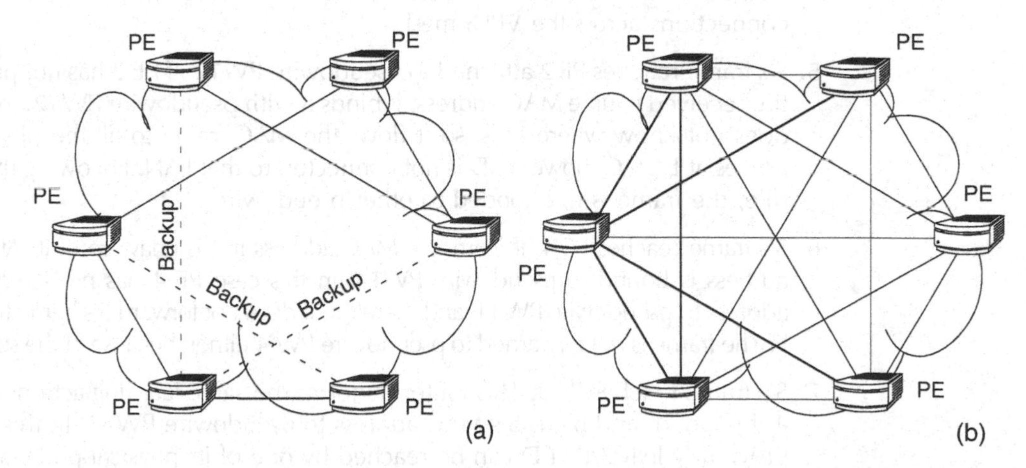

Figure 10-28. Pseudowire topologies in VPLS: (a) partial mesh with STP. Some of the pseudowires are disabled to avoid loops. (b) Full mesh of pseudowires. The split-horizon rule is applied to avoid bridging loops.

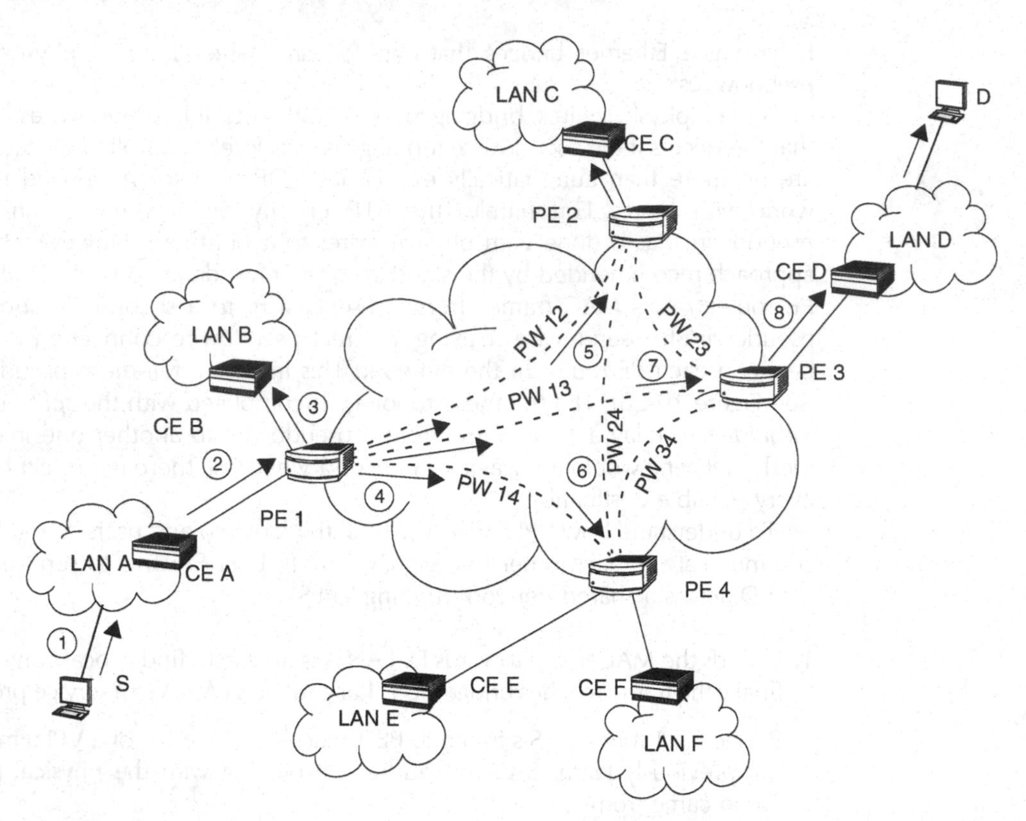

Figure 10-29. Flooding and learning in VPLS emulates a LAN broadcast domain.

4. PE 1 not only performs flooding on its physical ports, but also on the pseudowires. S's frame is thus forwarded to all other PEs in the network by means of direct pseudowire connections across the VPLS mesh.

5. S's frame reaches PE 2 attached to pseudowire PW12. If PE 2 has not previously learnt the received source MAC address, it binds it with pseudowire PW12. In this case, PE 2 does not know where D is, so it flows the MAC frame to all the physical ports and arrives at LAN C; however D is not connected to that LAN. Following the split-horizon rule, the frame is not flooded to other pseudowires.

6. S's frame reaches PE 4. It learns S's MAC address if it is unaware of it. After learning, S's address is bound to pseudowire PW14. In this case PE 4 has previously bounded D's address to pseudowire PW34, and therefore it does not forward S's frame to LAN E or LAN F. The frame is not forwarded to pseudowire PW 4 either, because of the split-horizon rule.

7. S's frame reaches PE 3. This router performs the same learning actions as PE 2 and PE 4 if needed, and binds S's MAC address to pseudowire PW13. In this case, PE 3 has previously learnt that D can be reached by one of its physical ports, and therefore it forwards S's frame to it.

8. S's frame reaches CE D, which forwards this frame to its final destination.

The previous example deals with a single broadcast domain that appears as a single distributed LAN. However, this may not be acceptable when providing services to many customers. Every customer will normally require its own broadcast domain. The natural way to solve this is by means of VLANs. Every subscriber is assigned a service-delimiting VLAN ID. Every VLAN is then mapped to a VPLS instance (i.e. a broadcast domain) with its own pseudowire mesh and learning tables. The link between CE and PE routers is multiplexed, and customers are identified by VLAN tags. This deployment is useful for offering EVPLAN services as defined by the MEF.

However, VLAN tags are not always meaningful for the service provider network. All VLAN tags can be mapped to a single VPLS instance and therefore all of them are part of the same broadcast domain within the service provider network. In this case VLAN-tagged frames are filtered by the subscriber network, but they are left unchanged in the service provider network. Different customers can still be assigned to different broadcast domains, but not on a per-VLAN-ID basis. Mapping customers to VPLS instances on a per-physical-port basis is the solution in this case. This second deployment option is compatible with the EPLAN connectivity service definition given by the MEF.

VPLS has been demonstrated to be flexible, reliable and efficient, but it still lacks scalability due to excessive packet replication and excessive LDP signalling. The origin of the problem is in the full meshed pseudowire topology. The total number of pseudowires needed for a network with n PE routers is $n(n-1)/2$. This limits the maximum number of PE routers to about 60 units with current technology.

Hierarchical VPLS (HVPLS) is an attempt to solve this problem by replacing the fully meshed topology with a more scalable one. To do this it uses a new type of network element, the *multitenant unit* (MTU). In HVPLS, the pseudowire topology is extended from the PE to the MTU. The MTU now performs some of the functions of the PE, such as interacting with the CE and bridging. The main function of the PE is still frame forwarding based on VLAN tags or labels. In some HVPLS architectures, the PE does not implement bridging. The result is a two-tier architecture with a full mesh of pseudowires in the core and non-redundant point-to-point links between the PE and the MTU (see Figure 10-30). A full mesh between the MTUs is not required, and this reduces the number of pseudowires. The core network still needs the full mesh, but now the number of PEs can be reduced, because some of their functions have been moved to the access network.

The MTUs behave like normal bridges. They have one (and only one) active pseudowire connection with the PE per VPLS instance. Flooding, as well as MAC address learning and aging, is performed in the pseudowire as if it were a physical wire. The PE operates the same way in an HVPLS as in a flat VPLS, but the PE-MTU pseudowire connection is considered as a physical wire. This means that the split-horizon rule does not apply to this interface.

In practical architectures, the MTUs are not always MPLS routers. Implementations based on IEEE 802.1ad service provider bridges are valid as well. These bridges make use of Q-in-Q encapsulation with two stacked VLAN tags. One of these tags is the service delimiting P-VLAN tag added by the MTU. The P-VLAN designates the customer, and is used by the PE for mapping the frames to the correct VPLS instance.

HVPLS can be used to extend the simple VPLS to a multioperator environment. In this case, the PE-MTU non-redundant links are replaced by PE-PE links where each PE in the link belongs to a different operator.

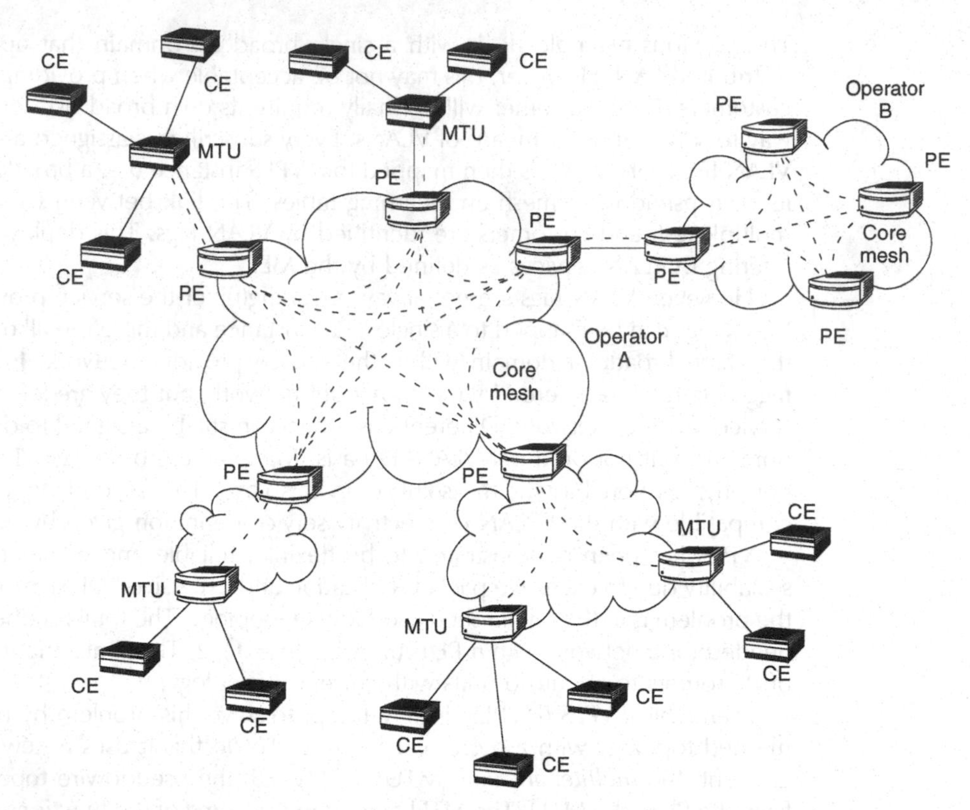

Figure 10-30. In HVPLS, the full mesh of pseudowires is replaced by a two-tier topology with full mesh only in the core and non-redundant point-to-point links in the access.

The main drawback of the HVPLS architecture is the need for non-redundant MTU-PE pseudowires. A more fault-tolerant approach would cause bridging loops. One solution is a multi-homed architecture with only one simultaneous MTU-PE pseudowire active. The STP can help in managing active and backup pseudowires in the multi-homed solution.

10.4.7 Pseudowires and NG-SDH

The VPLS is the enabling technology for QoS and protection, allowing some operators to build SDH-free backhaul networks, especially if their traffic is only IP. However, PWE3 is being implemented in NG-SDH networks as well as in those cases where there are high QoS, TDM or OAM requirements.

The use of pseudowires over NG-SDH results in an efficient network that combines the MPLS features with the carrier-class approach of SDH. Edge SDH nodes push labels to the entering traffic, or they pop them for the egressing traffic. Intermediate LER nodes check the labels at the input VCGs, swap the labels, and then switch the packets to the correct output SDH circuit.

10.4.8 Advantages of the MPLS

IP routers operate by matching the IP address with its internal tables to calculate the next hop before the packet reaches its destination. In IP networks, all packets with the same destination are treated in a very similar way. The MPLS adopted virtual circuit connections that help to set up paths that can provide QoS.

MPLS was created to optimize packet forwarding in IP networks to meet the required QoS, as routers had become a bottleneck. The original aim of the MPLS was to reduce processing time to make routing more efficient. Although the original intention is now irrelevant, the MPLS still has many important advantages that make it a key piece in the puzzle (see Figure 10-31):

1. *MPLS is protocol-agnostic*
    - SDH, NG-SDH, Ethernet, dark fibre and WDM can be used as the transport layer.
    - Most of the protocols can be encapsulated, like Ethernet, FR, ATM and PPP.
    - Interoperatibility between different vendors is easier and more direct
2. *MPLS is a traffic engineering tool*
    - MPLS paths can be pre-established to route traffic and avoid bottlenecks, or to give priorities.
    - Data flows can be separated to provide privacy for each customer's traffic.
    - A network with MPLS can provide different types of QoS.
    - MPLS can set up protection mechanisms.

10.5 Migration

For most of today's communications services, telecom operators, carriers and providers have SDH networks supplied with TDM circuits. The emerging demand for Ethernet-based services is pushing the migration to those services, as well as the reuse of the existing infrastructures. The path and the tempo of such migration depends on many factors, so each case should be studied separately. However, any migration route analysis should consider the following:

- market demand, to identify which connectivity services are required, at what price and what is the cost of supplying them (CAPEX);
- competition analysis, to define a realistic business case;
- existing network architecture, including topologies and bit rates;
- optical resources, spare fibre or wavelength may make it more attractive to use an overlapped network rather than a network integrated into SDH;
- existing and new traffic.

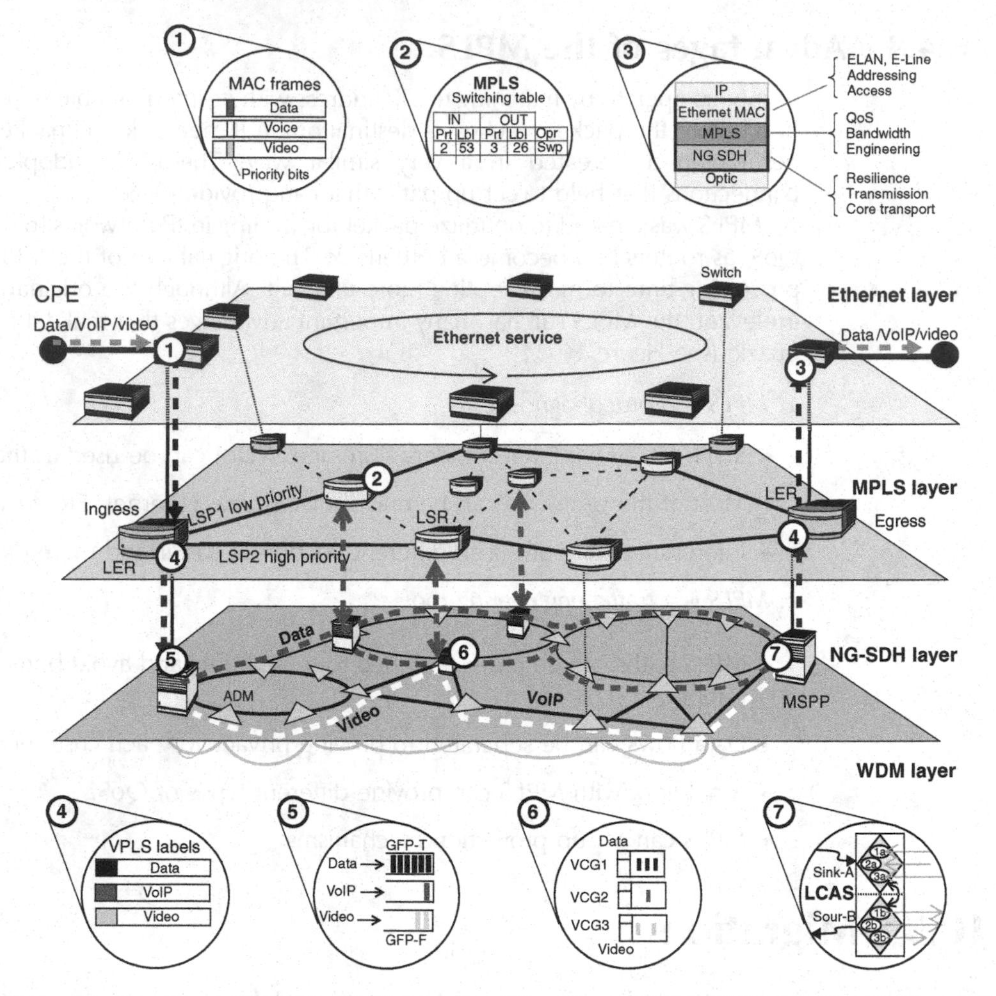

Figure 10-31. Ethernet customer traffic is sent to the service provider's MPLS-domain LER. The label attached to each packet defines an MPLS path with a certain QoS. Packets are then passed to one of the available VCGs according to the priority of the label. Immediately NG-SDH transmits the packet to the next LSR, which checks the label again, and forwards the packet accordingly. When the packet finally reaches the LER, the encapsulation with the label is removed and the MAC frame is delivered to its destination.

10.5.1 Migrating the Architecture

Most of today's service providers rely on PDH/SDH networks (see Figure 10-32).

1. The starting architecture is an SDH network providing only TDM services. Ethernet services end where the routers are. Some are also mapped on TDM.
2. NG-SDH is used to allow native Ethernet interfaces, and submultiplexing is possible to increase the efficiency of packet services.

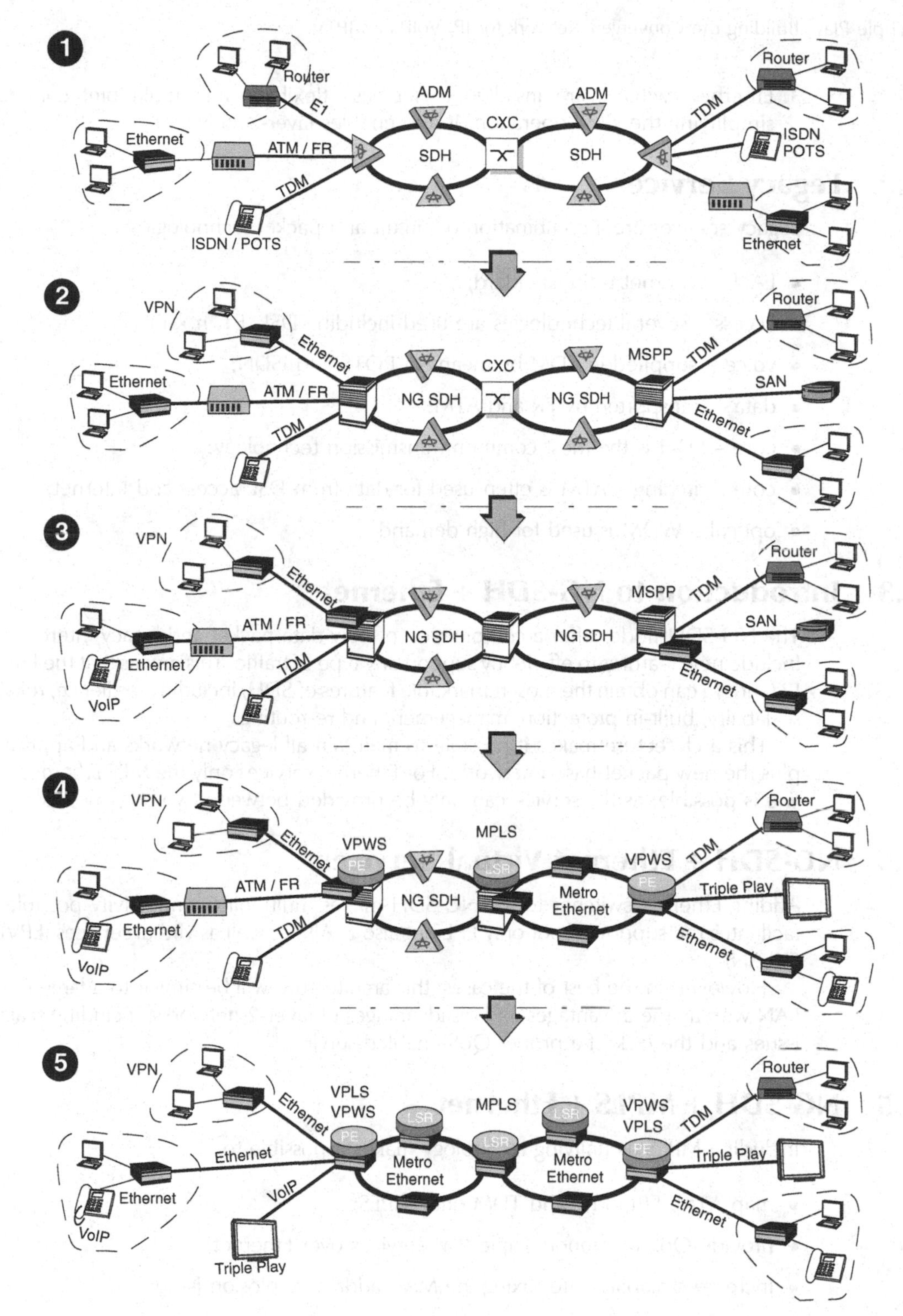

Figure 10-32. Migration strategies from a circuit-oriented (TDM) to a packet (STDM) network.

3. Ethernet switches are installed to increase flexibility and multipoint connections simplifying the NMS operation. It also enables layer-2 VPN.

10.5.2 Legacy Services

Legacy services are a combination of circuit and packet technologies:

- LAN – Ethernet is the standard;
- access – several technologies are used including DSL, E1, $n \times 64$;
- voice – supplied in TDM by means of POTS and ISDN;
- data – transported by FR and ATM;
- core – SDH is the most common transmission technology;
- core switching – ATM is often used for data from DSL access and Internet;
- optical – WDM is used for high demand.

10.5.3 Introduction to NG-SDH + Ethernet

The NG-SDH nodes offer a combination of new data-packet and legacy interfaces, and include new features to efficiently support any type of traffic. This means that the Ethernet/IP tandem can obtain the most remarkable features of SDH, including resilience, reliability, scalability, built-in protection, management and re-routing.

This architecture makes it possible to maintain all legacy networks and applications, plus the new packet-based networks. For Ethernet services only the MEF *Ethernet private line* is possible, as the service can only be provided between two edge points.

10.5.4 NG-SDH + Ethernet Virtual Services

Adding Ethernet switches to the NG-SDH makes multipoint connectivity possible. This facilitates the support of not only EPL but also ELAN, as well as virtual services (EPVL and EVLAN).

However, in the best of the cases, this architecture will be similar to a large Ethernet LAN with all the advantages and disadvantages of layer-2 networks, including scalability issues and the lack of a proper QoS-enabled service.

10.5.5 NG-SDH + MPLS + Ethernet

Including MPLS as unifying technology makes it possible to:

- map ATM, FRL, PPP and TDM onto MPLS;
- provide QoS to support Triple Play services over Ethernet;
- increase scalability after fixing the MAC address explosion issue;
- use MPLS-based protection architectures.

10.5.6 Service Interworking

The new architectures can provide a smooth migration from full TDM-oriented solutions to full packet-oriented solutions. However, that is a long journey, and it will depend on many circumstances, including legacy installations, type of customers, applications, and the model that has been adopted for the new architecture.

When the data network is based on TDM, connection-oriented technologies like FRL, ATM and leased lines, the use of layer-2 MPLS may mean a smooth migration to packet services. Large corporations can roll out Ethernet on selected premises that may interact with the existing TDM networks:

- *Ethernet- FRL* – RFC 2427 multiprotocol interconnect over frame relay;
- *Ethernet-ATM* – RFC 2684 Multiprotocol encapsulation over ATM adaptation layer 5.

10.5.7 Ethernet + MPLS – *urbi et orbe*?

In the short or middle term, will Ethernet be the only technology providing any type of service? We do not know, but consider this:

- Ethernet has always been a winner;
- historically, telecom services have had a long life.

Statistical multiplexing everywhere? Maybe. We do not know yet, but it is true that there is a constant tendency to substitute TDM with packet networking at layer 3, layer 2 and layer 1 as well.

Nevertheless, we should not underestimate the persistence of telecom technologies. Changes never take place suddenly, as we have seen during the past 30 years. Evolution and migration are always possible, whenever the installed base is not forced to perform a quick change, but a smooth one. This means that the integration and convergence of existing services must be guaranteed during migration. If Ethernet can do this, it is possible that in 10 years' time most of the metro and core networks will use the technology that LANs made universal.

Selected Bibliography

[1] Allan D., Bragg N., McGuire A., Reid A., Ethernet as Carrier Transport Infrastructure, *IEEE Communications Magazine*, 2006, pp. 134–140.
[2] Rosen E., Viswanathan A., Callon R., Multiprotocol Label Switching architecture, IETF Request For Comments RFC 3031, January 2001.
[3] Rosen E., Tappan D., Fedorkow G., Rekhter Y., Farinacci D., Li T., Conta A., MPLS Label Stack Encoding, IETF Request For Comments RFC 3032, January 2001.
[4] Andersson L., Doolan P., Feldman N., Fredette A., Thomas B., LDP Specification, IETF Request For Comments RFC 3036, January 2001.

[5] Awduche D., Berger L., Gan D., Li T., Srinivasan V., Swallow G., RSVP-TE: Extensions to RSVP for LSP Tunnels, IETF Request For Comments RFC 3209, December 2001.
[6] Jamoussi B., Andersson L., Callon R., Dantu R., Wu L., Doolan P., Worster T., Feldman N. Freddete A., Girish M., Gray E., Heinanen J., Kilty T., Malis A., Constraint-Based LSP Setup using LDP, IETF Request For Comments RFC 3212, January 2002.
[7] Bryant S., Pate P., Pseudo Wire Emulation Edge-to-Edge (PWE3) architecture, IETF Request For Comments RFC 3985, March 2005.
[8] Martini L., IANA Allocations for Pseudowire Edge to Edge Emulation (PWE3), IETF Request For Comments RFC 4446, April 2006.
[9] Martini L., Rosen E., El-Aawar N., Smith T., Heron G., Pseudowire Setup and Maintenance Using the Label Distribution Protocol (LDP), IETF Request For Comments RFC 4447, April 2006.
[10] Martini L., Rosen E., El-Aawar N., Heron G., Encapsulation Methods for Transport of Ethernet over MPLS Networks, IETF Request For Comments RFC 4448, April 2006.
[11] Lasserre M., Kompella V., Virtual Private LAN Services Using LDP, IETF Internet Draft Document draft-ietf-l2vpn-vpls-ldp, June 2006.
[12] Awduche D. *et al.*, Overview and Principles of Internet Traffic Engineering, IETF Request For Comments RFC 3272, May 2002.

Chapter 11: Next-generation SDH/SONET

11.1 Streaming Forces

The financial and technological cycle of the telecoms industry is forcing manufacturers, carriers, operators and standards organizations to move towards a new network that reduces costs while offering new services.

The new services, mostly relying on data packet technology, offer easy implementation and access to applications based on Internet, VOIP, IPTV, DVB, SAN, Ethernet or VPN. The architectures are increasingly requiring long-haul transport. However, packet technologies, like Ethernet, may not be appropriate for efficient optical transport and SDH/SONET has such a massive installed base that carriers cannot simply get rid of it, for many reasons:

- Most have their transport infrastructure entirely based on SDH/SONET.
- There is a lot of experience in managing SDH/SONET.
- No other technology has this maturity grade at the optical physical layer.

Triple Play: Building the Converged Network for IP, VoIP and IPTV Francisco J. Hens and José M. Caballero

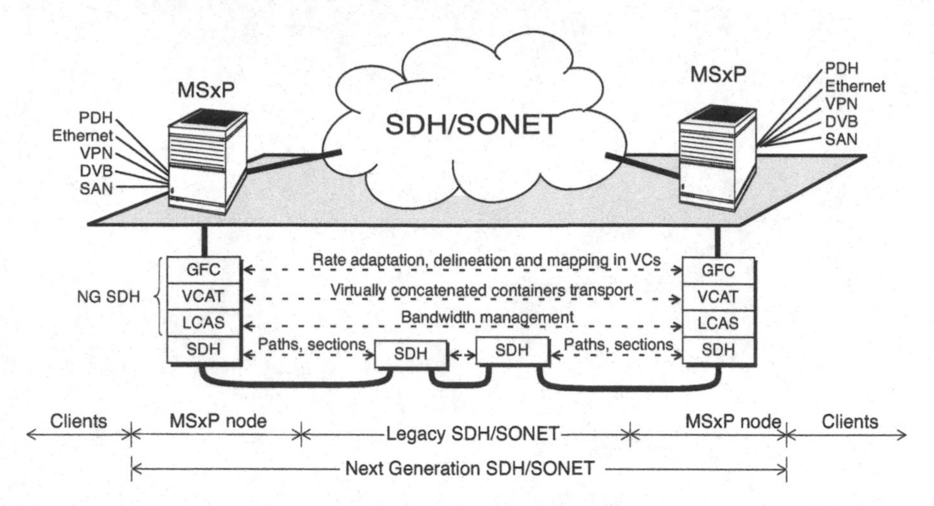

Figure 11-1. Next generation SDH/SONET enables operators to provide more data transport services while increasing the efficiency of the installed SDH/SONET base, by adding just the new MSxP edge nodes. This means that it will not be necessary to install an overlap network or migrate all the nodes or fibre optics. This reduces the cost per bit delivered, and will attract new customers while keeping legacy services.

Luckily, SDH/SONET has also evolved to more efficiently adapt statistical multiplexing traffic based on data packets (see Figure 11-1).

11.2 Legacy and Next-generation SDH

Telecom service providers are ready to include Ethernet/IP on their portfolio of solutions provided to the enterprise. This does not mean exactly that core network will soon be part of the past. No, it really means that a new generation of SDH/SONET nodes is offering a comprehensive combination of data-packet and TDM interfaces, optical physical layers (often based on DWDM), and a number of new functionalities to support efficiently any type of traffic (see Section 11.3.1). It also means that the Ethernet/IP tandem can obtain the most remarkable features of SDH/SONET, including resilience, reliability, scalability, built-in protection, management and rerouting.

11.2.1 Evolution of the Transmission Network

Most of the carriers and operators have been using SDH/SONET for several decades, mainly to transport voice and circuit-oriented data protocols. Since then, one of the challenges has been to efficiently transport statistically multiplexed, packet-oriented data services. Despite a number of architectures developed to do this, none of them have been widely accepted by the market. Sometimes this was because of the cost, other times because of the complexity and sometimes because of the poor efficiency.

Now, with the adoption of NG SDH/SONET, there emerges a new opportunity driven by two factors: first, its simple encapsulation method, capable of accommodating any data packet protocols, and secondly, its demonstrated bandwidth efficiency. This means that a

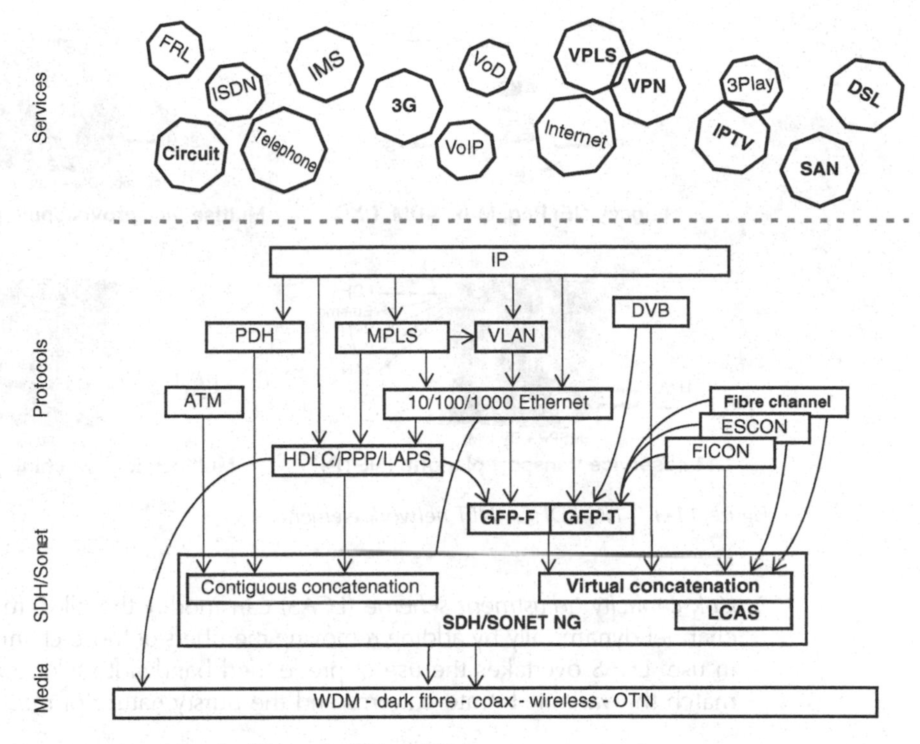

Figure 11-2. *Versatile, flexible and efficient SDH next generation.*

new adaptation protocol has been developed, as well as a new mapping mechanism for controlling bandwidth, while keeping the reliability of the legacy SDH/SONET transport and its centralized management (see Figure 11-2).

11.3 The Next-generation Challenge

Compared with the legacy SDH/SONET, the next generation offers a much more flexible architecture which has turned out to be compatible with data packet networks, and not only with circuit-oriented networks. Three new protocols have made this kind of evolution possible:

1. *Generic framing procedure* (GFP) is a robust and standardized encapsulation procedure for the transport of packetized data on SDH/SONET and OTN as well. In principal, GFP performs bitrate adaptation managing features, such as priorities, discard eligibility, transport channel selection and submultiplexing (see Figure 11-6). GFP replaces legacy mappings, many of them of a proprietary nature.
2. *Virtual concatenation* (VCAT) can create channels or pipes of granular bandwidth sizes rather than the exponential bandwidth provision of the contiguous concatenation. Therefore VCAT is more flexible and efficient.

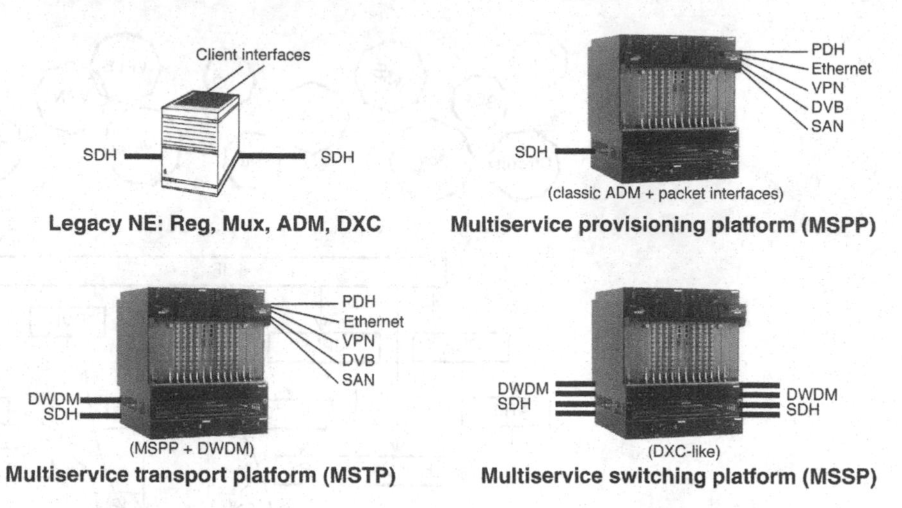

Figure 11-3. *NG SDH/SONET network elements.*

3. *Link capacity adjustment scheme* (LCAS) can modify the allocated VCAT bandwidth channel dynamically by adding/removing members of those channels that are already in use. LCAS overtakes the use of predefined bandwidth allocation, which does not match the variable bitrate patterns and the bursty nature of most data networks.

11.3.1 The New Network Elements

Under the generic name of *multiservice platforms* (MSxP) we find a new set of nodes that, combined with the legacy REG, ADM and DXC network elements, configure the topologies of the next generation SDH/SONET.

The MSxP nodes have a selection of data, circuit and optical interfaces, and they support the new GPF, VCAT and LCAS protocols. They can be classified as three types: MSPP, MSTP and MSSP (see Figure 11-3).

11.3.1.1 Multiservice Provisioning Platform

A *multiservice provisioning platform* (MSPP) is the result of the evolution of legacy ADM with TDM interfaces and optical interfaces, to a type of access node that includes a set of:

- legacy TDM, or circuit, interfaces;
- data interfaces, such as Ethernet, GigE, fibre channel or DVB;
- NG SDH/SONET functionalities such as GFP, VCAT and LCAS;
- optical interfaces from STM-0/STS-1 to STM-64/OC-192.

11.3.1.2 Multiservice Transport Platform

A *multiservice transport platform* (MSTP) can be defined as an MSPP with DWDM functions to drop selected wavelengths at a site that will provide higher aggregated capacity to transport client signals. MSTP allows to integrate circuit and data services, with efficient WDM transport and wavelength switching. Typically, MSTPs are installed in the metro and core network.

11.3.1.3 Multiservice Switching Platform

A *multiservice switching platform* (MSSP) is the next generation equivalent for cross-connect, performing efficient traffic grooming and switching at STM-N/OC-M levels but also at VC level. MSSPs should support more than just data service mapping, namely true data services multiplexing and switching. MSSP is still emerging as a next-generation network element, while MSPP and MSTP are quite mature.

11.4 Core Transport Services

Today's telecommunications services are based on a diverse combination of technologies such as Ethernet, leased lines, IP, ATM, etc. Many of these technologies have always been clients of SDH, and others can probably also be when they need to extend their service range to wider areas.

Channelized networks organized in $n \times 64$ kbit/s circuits like POTS, ISDN or GSM have been mapped efficiently into SDH/SONET containers in a natural way, because all of them are circuit-oriented, just like SDH/SONET. It has been more difficult to match the transport of data packets and best-effort technologies like Ethernet, IP or DVB, because the use of statistical multiplexing makes traffic variable and unpredictable. This is the opposite of SDH/SONET, which is constant and predictable, because it is based on *time-division multiplexing* (TDM).

11.4.1 Next-generation SDH

Among all the technologies that moved towards the NG SDH, Ethernet is the most remarkable. Ethernet, the standard technology for LANs, is cheap, easy to use, well-known and always evolving towards higher rates. Now, when it is also being considered for access and metro networks, carriers have started to look at SDH/SONET for transporting high volumes of Ethernet traffic to achieve long haul transport. To make that possible, SDH has needed to match the bursty and connectionless nature of Ethernet using a number of protocols:

- *Ethernet over LAPS*, defined in ITU-T X.86, is a mapping protocol of the HDLC family, which provides dynamic bit rate adaptation and frame delineation as well. It calls for contiguous concatenated bandwidth techniques (see Section 11.6) that do not match the bursty nature of Ethernet. However, what makes it inconvenient is the use of the 0x7E tag as frame delineation that forces byte stuffing on the payload: every 0x7E occurrence is swapped with 0x7D5E. The consequence is that the frame length is data-dependent and it cannot be mapped or demapped without reading the payload byte by byte.
- *Generic framing procedure (GFP):* defined in ITU-T Rec. G.7041, GFP is a protocol for mapping any type of data link services, including Ethernet, *digital video broadcasting*

(DVB) and *storage area networks* (SAN).[1] GFP provides bitrate adaptation and frame delineation. Compared with X.86, GFP has two important advantages. First, it uses an HEC-based delineation method independently of the data carried in the payload, based on a length indicator and a CRC-16 code (see Figure 11-5). Secondly, GFP uses the more efficient virtual concatenation, making GFP very popular compared with LAPS.

- *Virtual concatenation (VCAT):* defined in ITU-T Rec. G.707, VCAT creates the right sized pipes for the traffic, providing granularity and compatibility with legacy SDH (see Section 11.6).
- *Link capacity adjustment scheme (LCAS)*: defined in ITU-T Rec. G.7042, LCAS allocates or de-allocates bandwidth units to match data transport requirements, or to implement additional resiliency between two transport points. VCAT can be used without LCAS, but LCAS requires VCAT (see Section 11.7).

These functions are implemented on the new MSPP nodes which are located at the edges of the network. They interact with the client data packets that are aggregated over the SDH/SONET backplane, which continues unchanged. This means that the MSPPs represent the SDH next generation embedded in the legacy SDH network.

11.5 Generic Framing Procedure

The GFP is a very simple standard protocol to map layer 2 and layer 1 signals onto SDH/SONET containers. It is point-to-point oriented and provides dynamic bit rate adaptation, delineation and framing for client data signals. Another advantage of GFP is an improved bandwidth efficiency when compared with ATM. GFP is implemented only at the edge nodes where client data interfaces are (see Figure 11-4), while the rest of the SDH/SONET network remains unchanged.

GFP supports many types of protocols, including those used in LAN and SAN (see Figure 11-5). GFP uses an HEC-based delineation technique similar to ATM, and it therefore does not need byte stuffing like LAPS. Then, the frame size can be easily set up to a predictable length without random increments that would happen if byte stuffing was used. The GFP HEC-based delineation has the additional capability for correcting single errors and detecting multiple errors within the core header (see Figure 11-5).

Currently, two modes of client signal adaptation are defined for GFP:

- *Frame-mapped GFP* (GFP-F) is a layer-2 encapsulation PDU-oriented adaptation mode. It is optimized for data packet protocols (e.g. Ethernet, PPP, DVB) that are encapsulated onto variable size frames.
- *Transparent GFP* (GFP-T) is a layer-1 encapsulation or block-code oriented adaptation mode. It is optimized for protocols using 8B/10B physical layer (e.g. fibre channel, ESCON, 1000BASE-X, etc.), that is encapsulated onto constant size frames.

---

[1]SAN, as a generic acronym, can include ESCON, FICON and fibre channel as well.

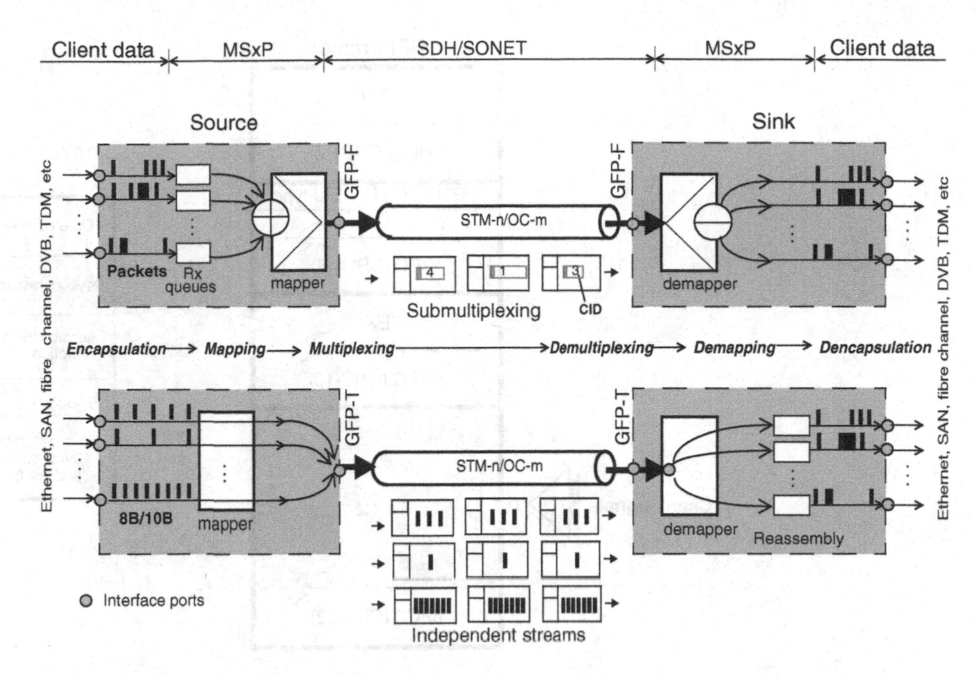

Figure 11-4. Data packet aggregation using GFP. In GFP-F packets are in queues waiting to be mapped onto an SDH channel. At the far end packets are dropped again to a queue and delivered. In GFP-T packets are encapsulated directly to the channel without waiting for the end of packet; at the far end the packet is reassembled and sent to the receiver.

11.5.1 Frame-mapped GFP

GFP-F drops the entire client packet into a GFP frame. The encapsulation process must receive the complete client packet, then, depending on the client protocol, specific signals are removed such as certain headers, idle codes and interframe gaps to minimize the transmission size (see Figure 11-6).

GFP-F has an optional GFP extension header (see Figure 11-5), and there are some additional fields that can be used here, such as source/destination address, port numbers, *class of service* (CoS), etc. The *extension header identifier* (EXI) linear type supports sub-multiplexing onto a single channel, by means of the *channel ID* (CID) (see Figure 11-4). This feature is intended for carriers that want to improve bandwidth usage along simple point-to-point paths by adding traffic from different low-rate sources in a single channel.

Carriers who not only need multiplexing but switching traffic as well will probably need a different approach. Traffic can be switched by means of MAC addresses or MPLS labels. These layer-2 switching capabilities are integrated in some MSPPs.

GFP and SDH/SONET are a good choice for transporting Ethernet and implementing the Ethernet Private Line (EPL) service defined by the Metro Ethernet Forum (MEF) in intercity or metropolitan networks.

GFP-F results in a more efficient transport; however, the encapsulation processes described above increase latency, making GFP-F inappropriate for time-sensitive protocols. This is the reason why GFP-F is used for Ethernet, PPP/IP and HDLC-like protocols where efficiency and flexibility are more important than delays.

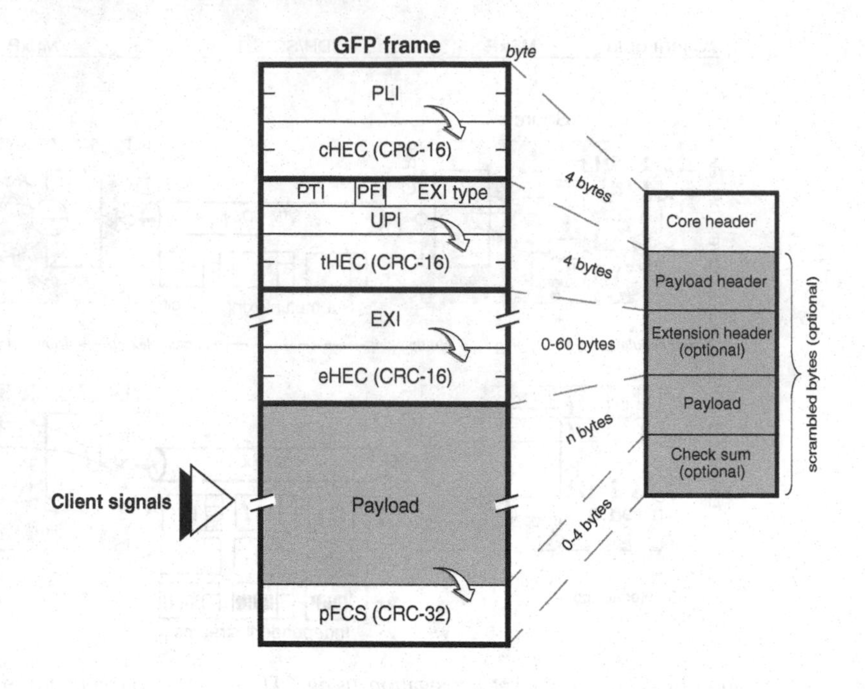

PLI: PDU length indicator
cHEC: core HEC protection

PTI: payload type identifier
000: client data
100: client management

PFI: Payload FCS indicator
1: presence of FCS
0: absence

EXI type: extension header identifier
0000: null
0001: linear
0010: ring

UPI: user payload identifier (PTI=0)
01x: ethernet (GFP-F)
02x: PPP (GFP-F)
03x: fibre channel (GFP-T)
04x: FICON (GFP-T)
05x: ESCON (GFP-T)
06x: gigabit Ethernet (GFP-T)
08x: MAPOS (GFP-F)
09x: DVB (GFP-T)
0Ax: RPR (GFP-F)
0Bx: fibre channel (GFP-F)
0Cx: async fibre channel (GFP-T)

tHEC: type HEC protection

EXI: extension header identifier

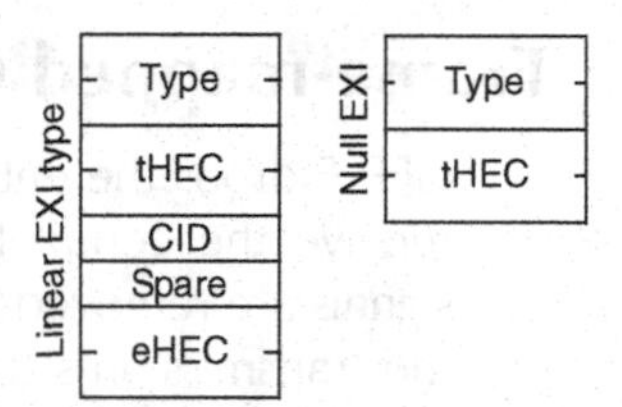

tHEC: type HEC protection
CID: channel ID for submultiplexing
eHEC: extension HEC protection

Payload: space for the framed PDU
pFCS: payload FCS

Figure 11-5. GFP frame formats and protocols. Frame delineation is done by the PLI indicating the length and cHEC, which is the CRC calculated over the two octets of the PLI in a similar way to ATM.

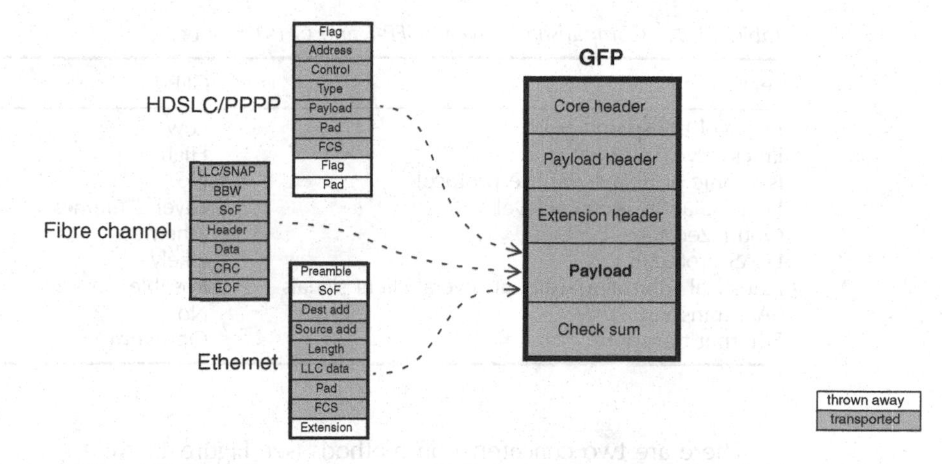

Figure 11-6. GFP mapping of client signals. Depending on the type of GFP in question, the mapping function can drop the whole signal (in the case of GFP-T), or can throw away certain fields (GFP-F).

11.5.2 Transparent GFP

GFP-T is a protocol-independent encapsulation method in which all client signals are mapped onto fixed-length GFP frames. Once the GFP frame is filled up, it is transmitted immediately without waiting for the entire client data packet to be received.

GFP-T encapsulates any protocol as long as it is based on 8B/10B line coding, which is why it is often called protocol-agnostic. 8B/10B symbols are decoded, coded again to 64B/65B, and finally dropped into fixed size GFP-T frames. Therefore, it is a layer-1 mapping mechanism, because all the client characters, without exception, are transported to the far end. GFP-T is completely blind to the meaning of the codes, and it does not distinguish between information, inter-frame gaps, headers, flow control characters, overhead and idle codes.

GFP-T is very good for isocronic protocols (time and delay sensitive), such as ESCON or FICON, and also for Gigabit Ethernet. This is because it is not necessary to process client frames or to wait for arrival of the complete frame. This advantage is counteracted by loss of efficiency, because the source MSxP node still generates traffic when no data is being received from the client (see Table 11-1).

Advantages of statistical submultiplexing or LCAS protection are limited to GFP-F transporting variable-rate client signals. GFP-T signals are not included in this group.

11.6 Concatenation

Concatenation is the process of summing the bandwidth of *X* containers (*C-i*) into a larger container. This provides a bandwidth *X* times bigger than *C-i*. It is well indicated for the transport of big payloads requiring a container greater than VC-4, but it is also possible to concatenate low-capacity containers, such as VC-11, VC-12 or VC-2.

Table 11-1. Comparison between GFP-F and GFP-T modes

Feature	GFP-F	GFP-T
Protocol transparency	Low	Highest
Efficiency	High	Low
Isocronic or delay sensitive protocols	No	Yes
Encapsulation protocol level	Layer 2 (frames)	Layer 1 (physical)
Optimized for	Ethernet	SAN, DVB
LCAS protection	Likely	Unlikely
Statistical submultiplexing of several client signals	Possible	Impossible
SAN transport	No	Yes
Ethernet transport	Optimum	Possible

There are two concatenation methods (see Figure 11-8):

1. *Contiguous concatenation*, which creates big containers that cannot split into smaller pieces during transmission. For this, each NE must have a concatenation functionality.
2. *Virtual concatenation*, which transports the individual VCs and aggregates them at the end point of the transmission path. For this, concatenation functionality is only needed at the path termination equipment.

11.6.1 Contiguous Concatenation of VC-4

A VC-4-*X*c provides a payload area of *X* containers of C-4 type. It uses the same HO-POH used in VC-4, and with identical functionality. This structure can be transported in an STM-*n* frame (where $n = X$). However, other combinations are also possible; for instance, VC-4-4c can be transported in STM-16 and STM-64 frames. Concatenation guarantees the integrity of a bit sequence, because the whole container is transported as a unit across the whole network (see Table 11-2).

Obviously, an AU-4-*X*c pointer, just like any other AU pointer, indicates the position of J1, which is the first byte of the VC-4-*X*c container. The pointer takes the same value as the AU-4 pointer, while the remaining bytes take fixed values equal to $Y = 1001SS11$ to indicate concatenation. Pointer justification is carried out in the same way for all the *X* concatenated AU-4s and $X \times 3$ stuffing bytes (see Figure 11-7).

Table 11-2. Contiguous concatenation of VC-4-Xc. X indicates the number of VC-n

SDH	SONET	X	Capacity	Justification unit	Transport
VC-4	STS3c-SPE	1	149 760 kbit/s	3 bytes	STM-1/OC-3
VC-4-4c	STS12c-SPE	4	599 040 kbit/s	12 bytes	STM-4/OC-12
VC-4-16c	STS48c-SPE	16	2 396 160 kbit/s	48 bytes	STM-16/OC-48
VC-4-64c	STS192c-SPE	64	9 584 640 kbit/s	192 bytes	STM-64/OC-192
VC-4-256c	STS768c-SPE	256	38 338 560 kbit/s	768 bytes	STM-256/OC-768

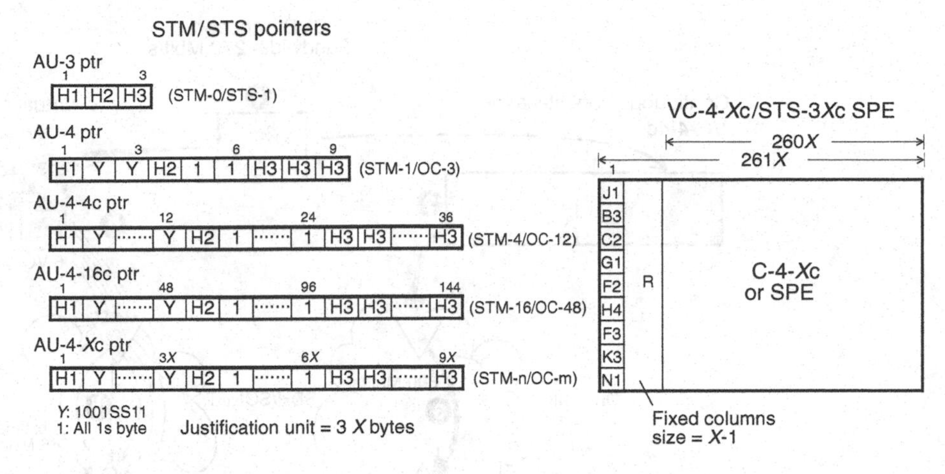

Figure 11-7. Contiguous concatenation: pointers and containers. A VC-4-Xc (X = 1, 4, 16, 64, 256) structure, where X represents the level. The increment/decrement unit (justification) is 3 X, as it depends on the level: AU-4 = 3 bytes, AU-4-256c = 768 bytes.

11.6.2 Virtual Concatenation

Packet-oriented, statistically multiplexed technologies, such as IP or Ethernet, do not match well the bandwidth granularity provided by contiguous concatenation. For example, to transport 1 Gbit/s, it would be necessary to allocate a VC4-16c containers, which has a 2.4 Gbit/s capacity. About 58% of the capacity would be wasted! (see Table 11-4).

VCAT is an inverse multiplexing technique that allows granular increments of bandwidth in single VC-*n* units. At the source node VCAT creates a continuous payload equivalent to *X* times the VC-*n* units (see Table 11-3). The set of *X* containers is known as a *virtual container group* (VCG), and each individual VC is a *member* of the VCG. All the VC members are sent to the destination node independently, using any available path if necessary. At the destination, each VC-*n* is organized according to the indications provided by the H4 or the V5 byte, and finally delivered as a single stream to the client (see Figure 11-9).

Differential delays between VCG members are likely, because they are transported individually and may have used different paths with different latencies. Therefore, the destination node must compensate for the different delays before reassembling the payload and delivering the service.

Table 11-3. Capacity of virtually concatenated SDH VC-n-Xv or SONET STS-3Xv SPE

SDH	SONET	Individual capacity	Number (X)	Virtual capacity
VC-11	VT.15 SPE	1 600 kbit/s	1 to 64	1 600–102 400 kbit/s
VC-12	VT2 SPE	1 276 kbit/s	1 to 64	2 176–139 264 kbit/s
VC-2	VT6 SPE	6 784 kbit/s	1 to 64	6 784–434 176 kbit/s
VC-3	STS-1 SPE	48 384 kbit/s	1 to 256	48 384–12 386 kbit/s
VC-4	STS-3c SPE	149 760 kbit/s	1 to 256	149 760–38 338 560 kbit/s

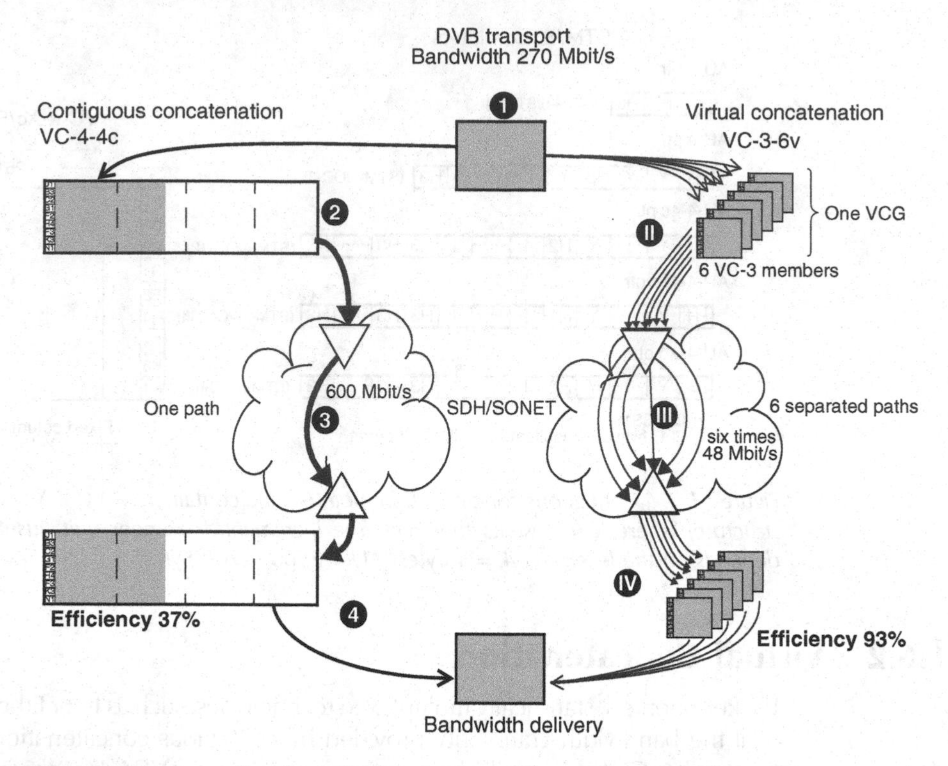

Figure 11-8. An example of contiguous concatenation and virtual concatenation. Contiguous concatenation requires support by all the nodes. Virtual concatenation allocates bandwidth more efficiently, and can be only needs support of edge nodes.

Virtual concatenation is required only at edge nodes, and it is compatible with legacy SDH networks, despite the fact that they do not support any concatenation. To get the full benefit of VCAT, individual containers should be transported by different routes across the network, so if a link or a node fails, the connection is only partially affected. This is also a way of providing a resilience service (see Figure 11-9).

Table 11-4. Comparison between contiguous and virtual concatenation efficiency

Service	Bitrate	Contiguous concatenation	Virtual concatenation
Ethernet	10 Mbit/s	VC-3 (20%)	VC-11-7v (89%)
Fast Ethernet	100 Mbit/s	VC-4 (67%)	VC-3-2v (99%)
Gigabit Ethernet	1000 Mbit/s	VC-4-16c (42%)	VC-4-7v (95%)
Fiber Channel	1700 Mbit/s	VC-4-16c (42%)	VC-4-12v (90%)
ATM	25 Mbit/s	VC-3 (50%)	VC-11-16v (98%)
DVB	270 Mbit/s	VC-4-4c (37%)	VC-3-6v (93%)
ESCON	160 Mbit/s	VC-4-4c (26%)	VC-3-4v (83%)

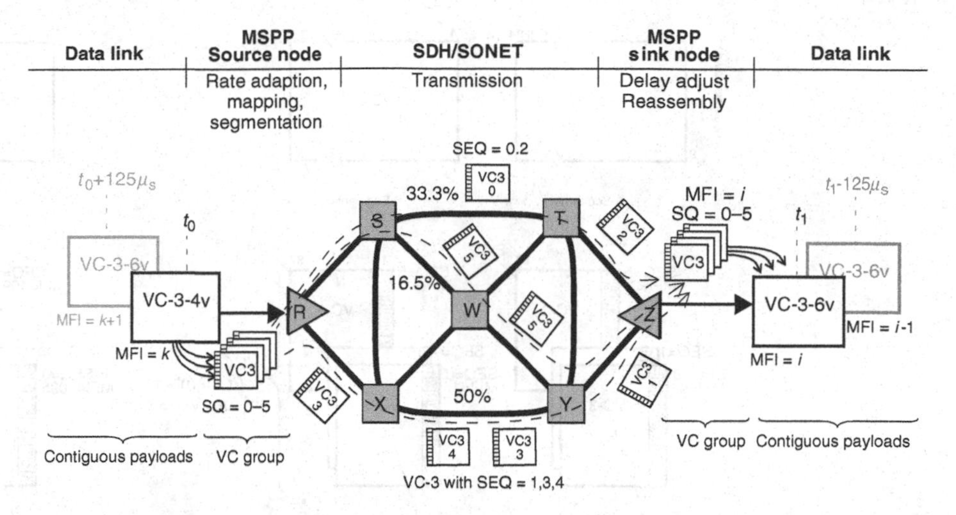

Figure 11-9. Virtual concatenation uses bandwidth more efficiently. In the sample, individual VC-3s are routed across different paths on the network. If a path fails, only part of the bandwidth is affected and the payload is rerouted through remaining paths.

11.6.2.1 Higher-order Virtual Concatenation

Higher-order virtual concatenation (HO-VCAT) uses X times VC3 or VC4 containers (VC3/4-*X*v, $X = 1$–256), providing a payload capacity of *X* times 48 384 or 149 760 kbit/s.

The virtual concatenated container VC-3/4-Xv is mapped in independent VC-3 or VC-4 envelopes that are transported individually through the network. Differenciated delays could occur between the individual VCs, and this obviously has to be compensated for when the original payload is reassembled (see Figure 11-10). A multiframe mechanism has been implemented in H4 to manage differential delays of up to 256 ms:

- Every individual VC has an H4 *multiframe indicator* (MFI) that denotes the virtual container they belong to.
- The VC also traces its position *X* in the VCG using the SQ number which is carried in H4.

The H4 POH byte is used for the virtual-concatenation-specific sequence and multiframe indication (see Figure 11-11).

11.6.2.2 Lower-order Virtual Concatenation

Lower-order virtual concatenation (LO-VCAT) uses *X* times VC11, VC12 or VC2 containers (VC11/12/2-*X*v, $X = 1$–64).

A VCG built with V11, VC12 or VC2 members provides a payload of *X* containers C11, C12 or C2; that is, a capacity of *X* times 1600, 2176 or 6784 kbit/s. VCG members are transported individually through the network; therefore differential delays could occur between the individual components of a VCG, that will be compensated for at the destination node before reassembling the original continuous payload (see Figure 11-9).

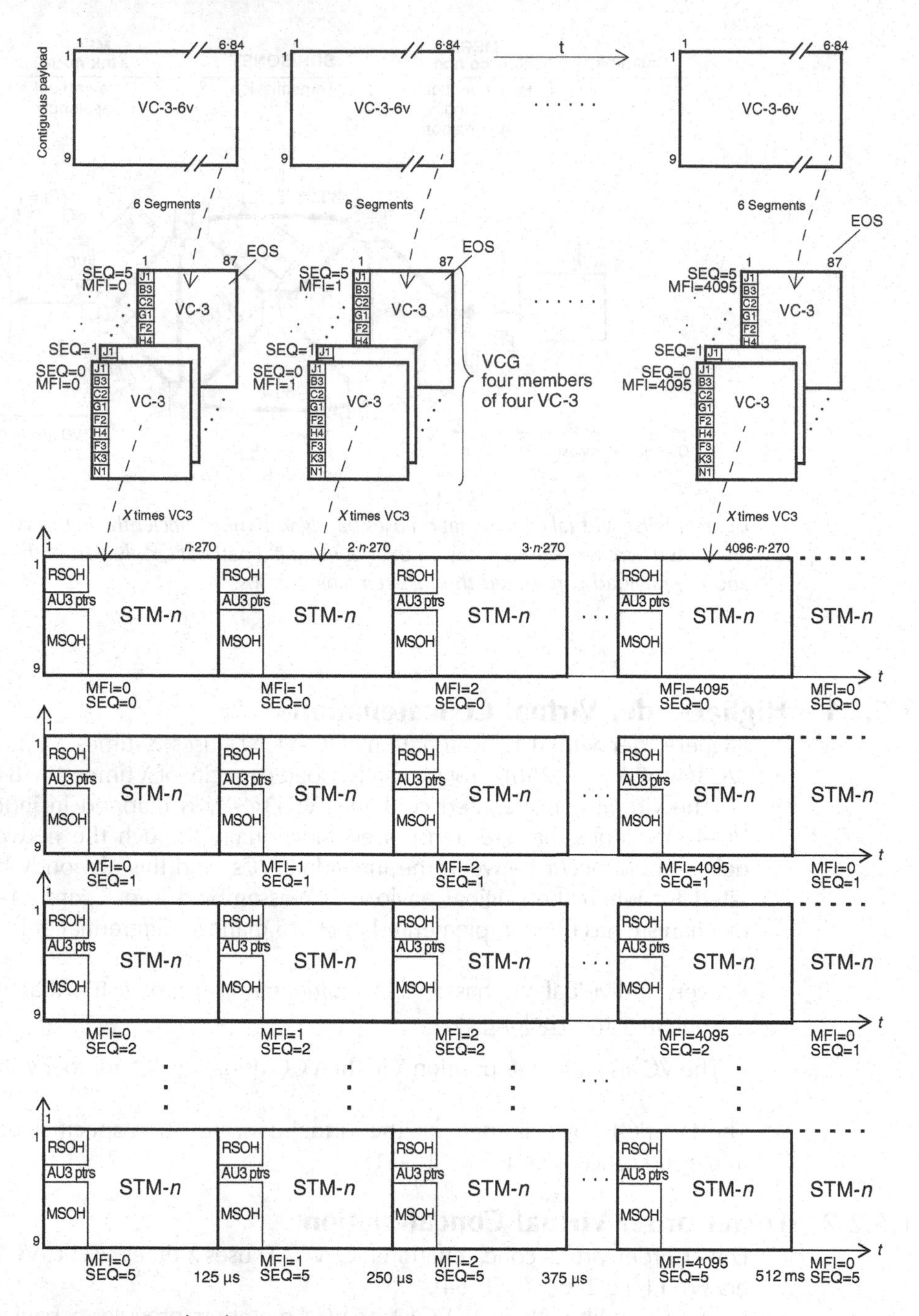

Figure 11-10. Graphical representation at the transmission (source) side of a virtual concatenation using VC-3-4v (X = 4). Includes sequence (SQ), multiframe indicator (MFI), envelopes and timings of the four VC3 containers.

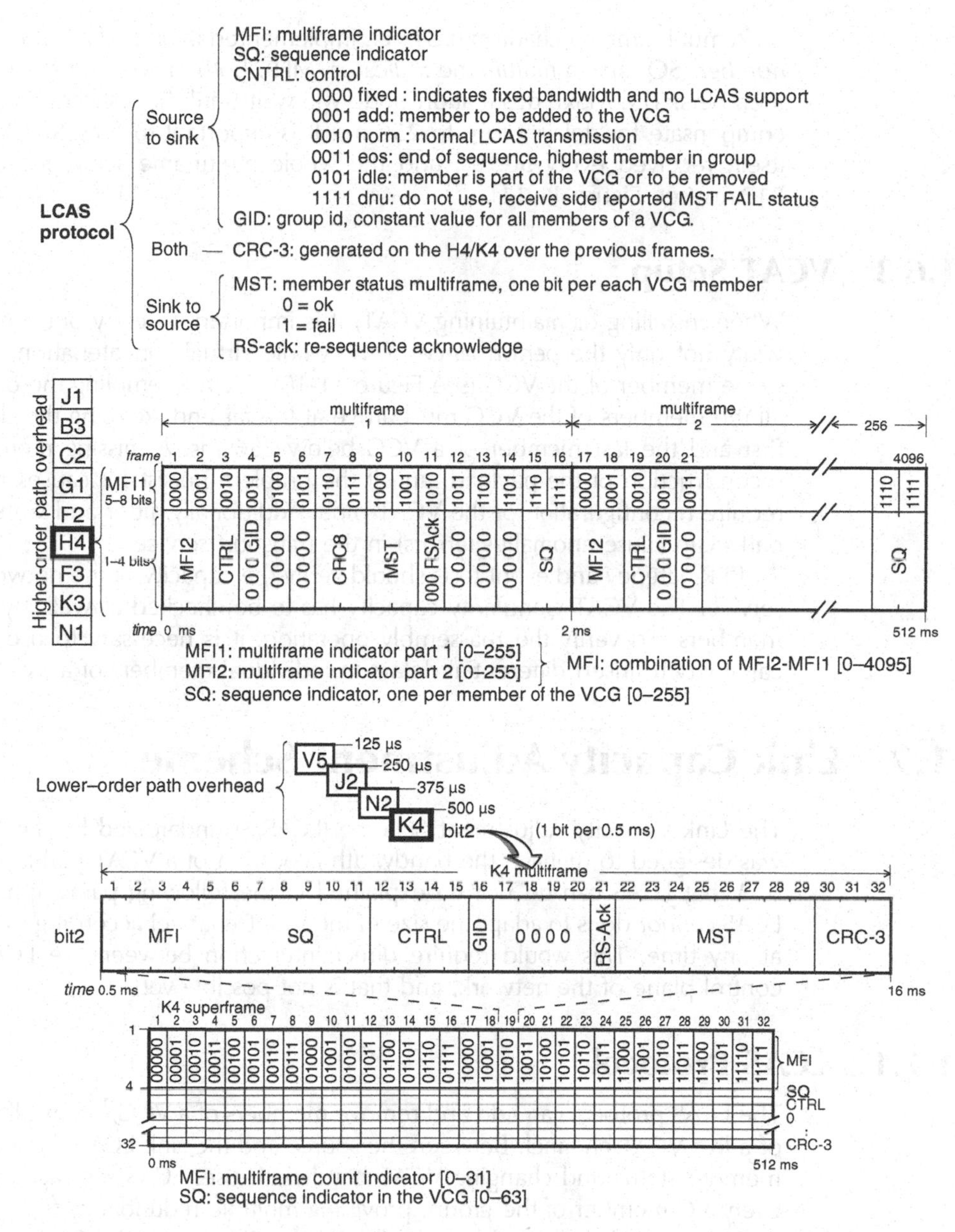

Figure 11-11. H4 and K4 codification of multiframes. SEQ identifies each member of the group, MFI allows the differential delay calculation. H4 is part of the HO-POH overhead, which is repeated every 125 ms, so the 16-byte multiframes takes 16 ms. A complete multiframe of 4096 bytes takes 512 ms to repeat (125 × 4096 = 512 ms). K4 is part of the LO-POH overhead and is repeated every 500 ms; 32 bits are sent in a complete multiframe which takes 16 ms to repeat (500 × 32 = 16 ms). The bit-2 superframe is made up of a sequence of 32 multiframes and takes 512 ms to repeat.

A multiframe mechanism has been implemented in bit 2 of K4. It includes a *sequence number* (SQ) and a *multiframe indicator* (MFI); both enable the reordering of the VCG members. The MSxP destination node will wait until the last member arrives and then compensate for delays of up to 256 ms. It is important to note that K4 is a multiframe itself, received every 500 μs, and the whole multiframe sequence is repeated every 512 ms (see Figure 11-11).

11.6.3 VCAT Setup

When installing or maintaining VCAT, it is important to carry out a number of tests to verify not only the performance of the whole virtual concatenation, but also of every single member of the VCG (see Figure 11-10). For reassembling the original client data, all the members of the VCG must arrive at the far end, keeping the delay between the first and the last member of a VCG below 256 ms. A missing member prevents the reconstruction of the payload, and if the problem persists, it causes a fault that would require reconfiguration of the VCAT pipe. Additionally, jitter and wander on individual paths can cause anomalies (errors) in the transport service.

BER, latency and event tests should verify the capacity of the network to provide the service. The VCAT granularity capacity has to be checked as well, by adding/removing members. To verify the reassembly operation, it is necessary to use a tester with the capability to insert differential delays in individual members of a VC.

11.7 Link Capacity Adjustment Scheme

The Link Capacity Adjustment Scheme (LCAS), standardized by the ITU-T as G.7042, was designed to manage the bandwidth allocation of a VCAT path.

What LCAS can do will be explained in the following paragraphs. However, what LCAS *cannot* do is to adapt the size of the VCAT channel according to the traffic pattern at any time. This would require direct interaction between the LCAS node and the control plane of the network, and that is not possible yet.

11.7.1 LCAS Protocol

The LCAS protocol can add and remove members of a VCG,[1] controlling the bandwidth of a live VCAT channel. Between the source and the sink LCAS is executed, monitoring member status and changing VCAT bandwidth use. LCAS messages are embedded in every VC member of the group, providing multiple redundancy.[2]

Using messages embedded in H4 or K4 (see Figure 11-11), LCAS establishes a protocol between the source node and the sink node (see Figures 11-14 and 11-12) to control the

---

[1] This means that LCAS needs VCAT to work. However, VCAT does not necessarily need LCAS.

[2] It is of key importance to realize that all LCAS messages from sink to source are replicated over each member path of the VCG. That is not true for the messages from source to sink, which are specific for each member. The sink-to-source redundancy enables the management of all members, even under a severe failure state.

LCAS states machine

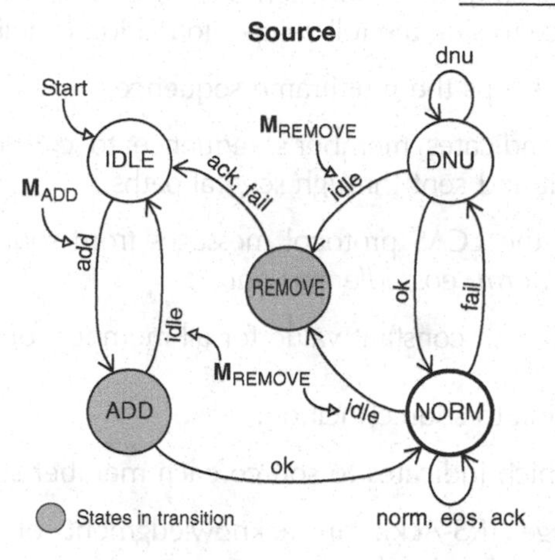

Source states

IDLE: Not provisioned member
ADD: In process of being added to the VCG
NORM: Active and provisioned member, good path
DNU (Do Not Use): Provisioned but its path has failed
REMOVE: In process of being removed of the VCG

Sink states

IDLE: Not provisioned member
OK: Provisioned and active member
FAIL: Provisioned member but its path has failed

LCAS protocol messages

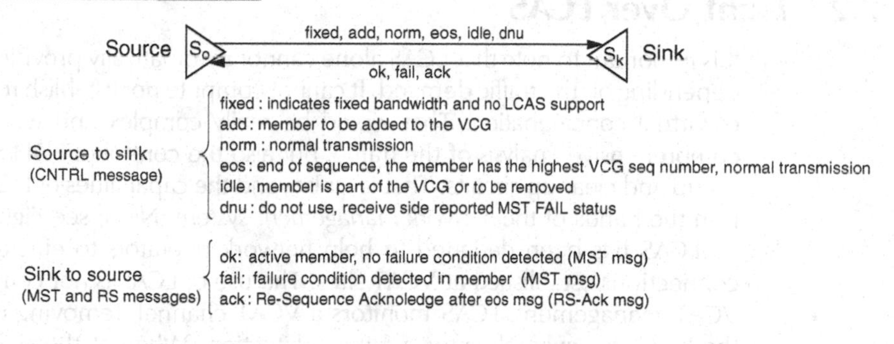

Management system command

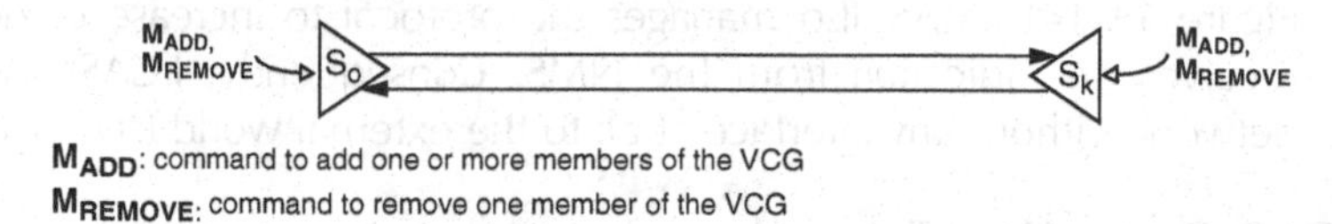

M_{ADD}: command to add one or more members of the VCG
M_{REMOVE}: command to remove one member of the VCG

Figure 11-12. Simplified LCAS source and sink state machine. ITU-T rec. G.7042 can be confusing because the same IDs can be a state, a command or a message. To minimize this problem, we have used lower case for protocol messages, uppercase for states and subscript for management commands (M_{xx}).

concatenated group. Control messages between source and sink are a point-to-point communication. From source to sink the following information is delivered:

- The *multiframe indicator* keeps the multiframe sequence.
- The *sequence indicator* indicates member's sequence to correctly reassemble the client signal that was split and sent through several paths.
- *Control* (CTRL) contains the LCAS protocol messages from source to sink. Possible messages are *fixed, add, norm, eos, idle* and *dnu*.
- *Group identification* (GID) is a constant value for all members of a VCG.

Control messages from sink to source include:

- *Member status* (MST), which indicates to source each member status: *fail* or *OK*.
- *Re-sequence acknowledge* (RS-Ack), an acknowledgment of renumbering of a sequence when a new member has been selected as *eos*.

With these control messages LCAS is also able to modify the pipe size of a VCAT channel, for example eliminating a member of the VCG that arrives with an unacceptable bit error rate, or does not arrive at all. When a capacity adjustment happens, the nodes restore channel capacity in a few milliseconds.

11.7.2 Light Over LCAS

It is important to note that LCAS alone cannot automatically provision dynamic bandwidth depending on the traffic demand. It cannot compute nor establish routes to take advantage of virtual concatenation. That would be really complex and would require not only a comprehensive analysis of the traffic, but also the control of SDH/SONET architecture to set up and clear up circuits. That is far beyond the capabilities of LCAS. Route provisioning is in the hands of the *network management system* (NMS; see Figure 11-12).

LCAS has been designed to help network operators to efficiently control NG SDH connections established at VCAT sites. The use of LCAS is not compulsory, but improves VCAT management. LCAS monitors a VCAT channel, removing or adding members to the VCG, for example after a failure detection. When at the sink side LCAS detects a failure, it sends a message to the source to remove affected member(s) of the group (see Figure 11-14). LCAS also manages the protocol to increase or decrease VCGs when it receives an indication from the NMS. Consequently, LCAS operates inside the core network without any interface at all to the external world (see Figure 11-13).

11.7.2.1 LCAS Operation Example

Imagine a Gigabit Ethernet service connecting two points, CP1 and CP2, that use a VC3/4v channel.[3] The VC group of four VC3 members has been set up from R to Z following diverse paths in our example (see Figure 11-13):

[3] The capacity of a VC-3-4v channel is 193.5 Mbit/s, that is 4 × 48.4 Mbit/s (see Table 11-3).

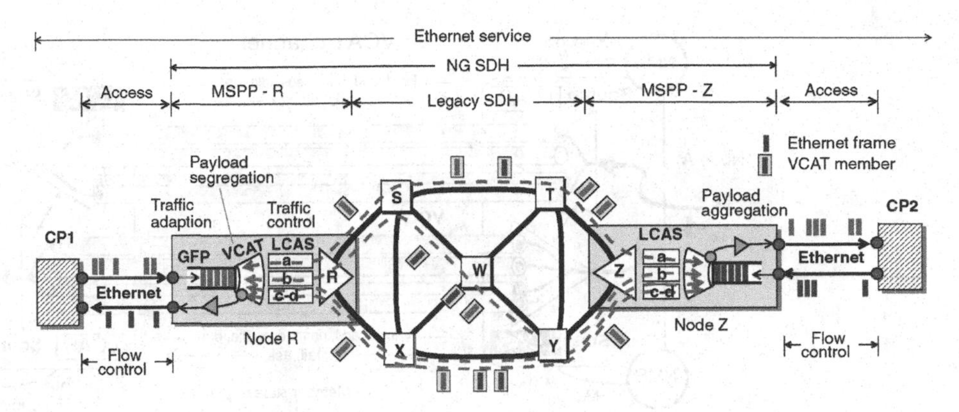

Figure 11-13. The link between node A and node Z transports Ethernet frames using a virtual concatenation group of three members. Three separate LCAS protocols constantly monitor each peer connection: LCAS-a of node R talks with LCAS-a of node Z, LCAS-b(R) with LCAS-b(Z), . . ., LCAS-n (R) with LCAS-n (Z).

- one VC3 follows the R–S–T–Z;
- one VC3 follows the R–S–W–Y–Z;
- two VC3s follow the R–X–Y–Z.

Once started, the LCAS source will send the message *add* to all members, to indicate their participation in the VCAT pipe (see Figure 11-14).

The sink has two key messages: *MST* and *RS-ACK*. The MST message can be *fail* or *OK*, and that is enough to manage failures, serious degradations, or changes to the pipe size. For example, a fault in the R–S–W–Y–Z route would cause a failure of the path b. The source will be notified of this fault with 'MST = fail (b)'. Immediately the VCAT removes the faulty path from the VCG and stops using it (see Figure 11-14). Despite this, the service will still be up, and after a short interruption (maximum 128 ms), all the client data will have been rerouted to the three paths still alive. The interface queues at GFP will probably be larger, because the capacity is now just 75% of the original (see Figure 11-17).

It is interesting to realize that most of the packet technologies, like Ethernet, are best effort. A burst is followed by periods of low activity that, on average, compensate for traffic peaks. Therefore, loss of capacity at the core network can only mean larger queues at the adaption layer, and often the only consequence is more delay but not any frame loss. Many data services are very tolerant of delays, so the situation of losing capacity can be managed perfectly. In the case of intense traffic, MSxP can manage the flow control between the CPs by using PAUSE protocol, in the case of Ethernet, or an equivalent for other protocols.

11.7.3 LCAS Applications

Most of the LCAS applications are related to the transport of data networks. This is of special interest in networks using GFP-F by making use of the statistical multiplexing gain.

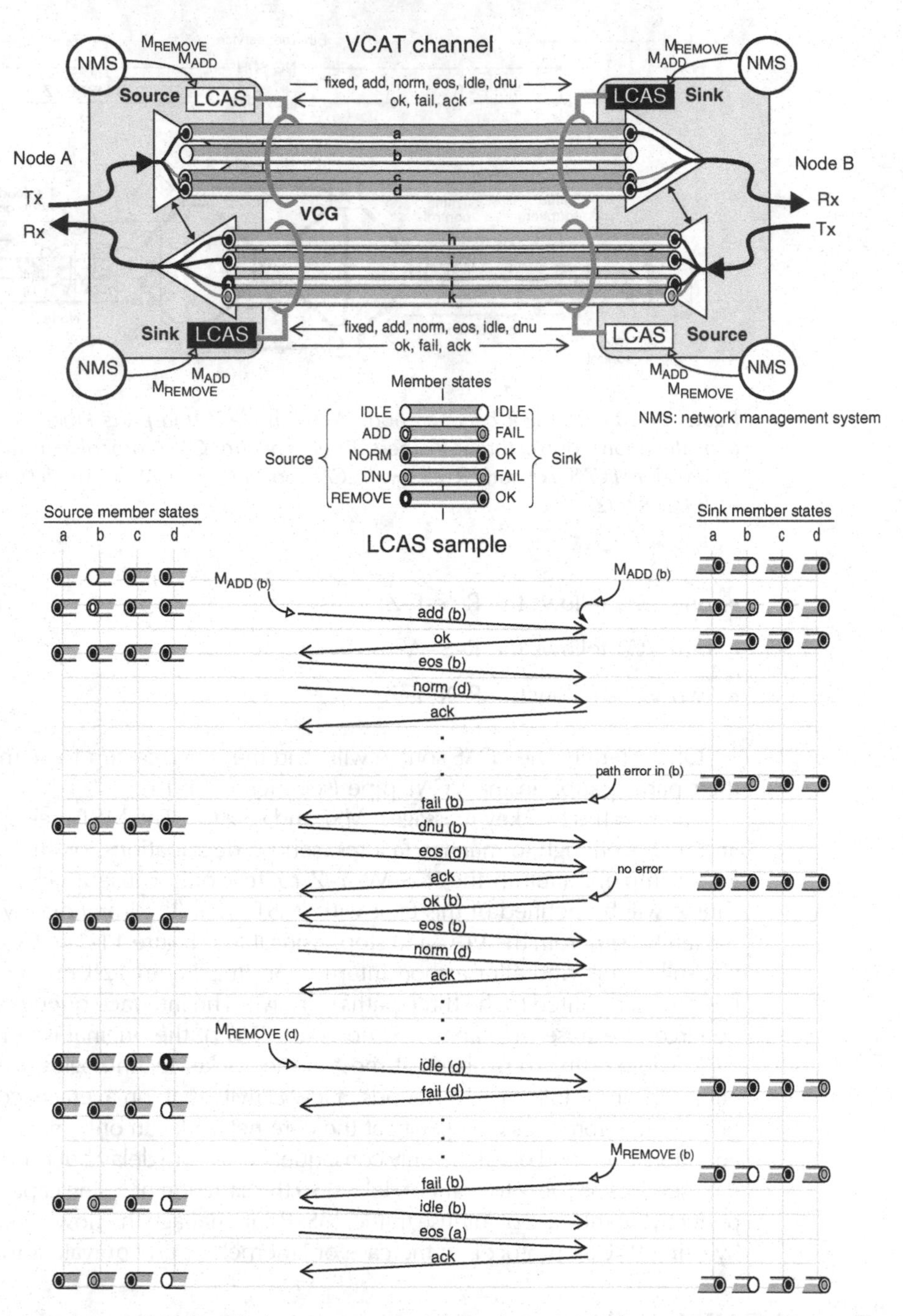

Figure 11-14. While virtual concatenation is a labelling of individual VC containers within a channel, LCAS is a two-way handshake protocol resident in H4 and K4 and executed permanently between source and sink as many times as there are VCAT members.

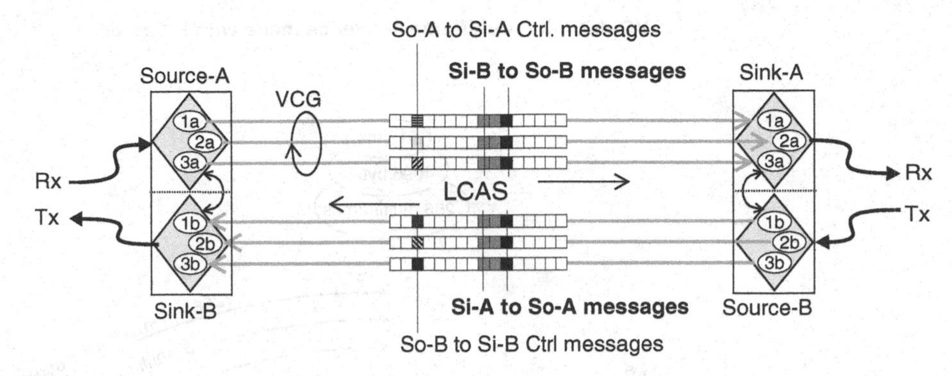

Figure 11-15. Notice that sink (Si) to source (So) messages (MST, RS-Ack) are redundant while source to sink are specific to each member. This means that sink messages are repeated as many times as members in the group. It also means that the origin of sink messages is irrelevant, because all the members are sending the same information in a multiframe. The result is a fault-tolerant protocol, which is necessary when one or more members fail. This allows active members to notify the source of the situation of the whole VCG.

11.7.3.1 VCAT Bandwidth Allocation

LCAS enables the resizing of the VCAT pipe in use when it receives an order from the NMS to increase or decrease the size. It can also automatically remove a specific VCG member, that is failing and add it again when it is recovered so that the VCAT connection is always kept alive.

11.7.3.2 Network Resilience

Resilience is probably the main LCAS application implementing a strategy known as diversification (see Figure 11.17). This strategy consists of sending traffic using several paths. In the case of a partial failure of one path, LCAS reconfigures the connection using the members still up and able to continue carrying traffic (see Figure 11-15).

Diversification is especially important for packet data networks using statistical multiplexing like Ethernet. Transported signals should not be particularly sensitive to delays because the reduction of available bandwidth can increase the queues at GFP-F. LCAS restoration time runs from 64 ms for VC-4 and other higher-order virtual concatenations, and 128 ms for VC-12 and other lower-order virtual concatenations.

In the case of an IP network, router topology would continue to be active, but less bandwidth would be available and consequently delay would be increased. However, constant complex configurations and reconfigurations between routers are avoided.

LCAS diversification can combine, and even replace, the existing protection architectures such as MSSPRING or MSDPRING which can also be used with NG SDH/SONET, but they are expensive, because they require spare resources which are never used except under a failure situation.

11.7.3.3 Asymmetric Configurations

It is important to note that LCAS is a unidirectional protocol that is executed independently at the two ends (see Figure 11-16). This feature allows the provision of asymmetric

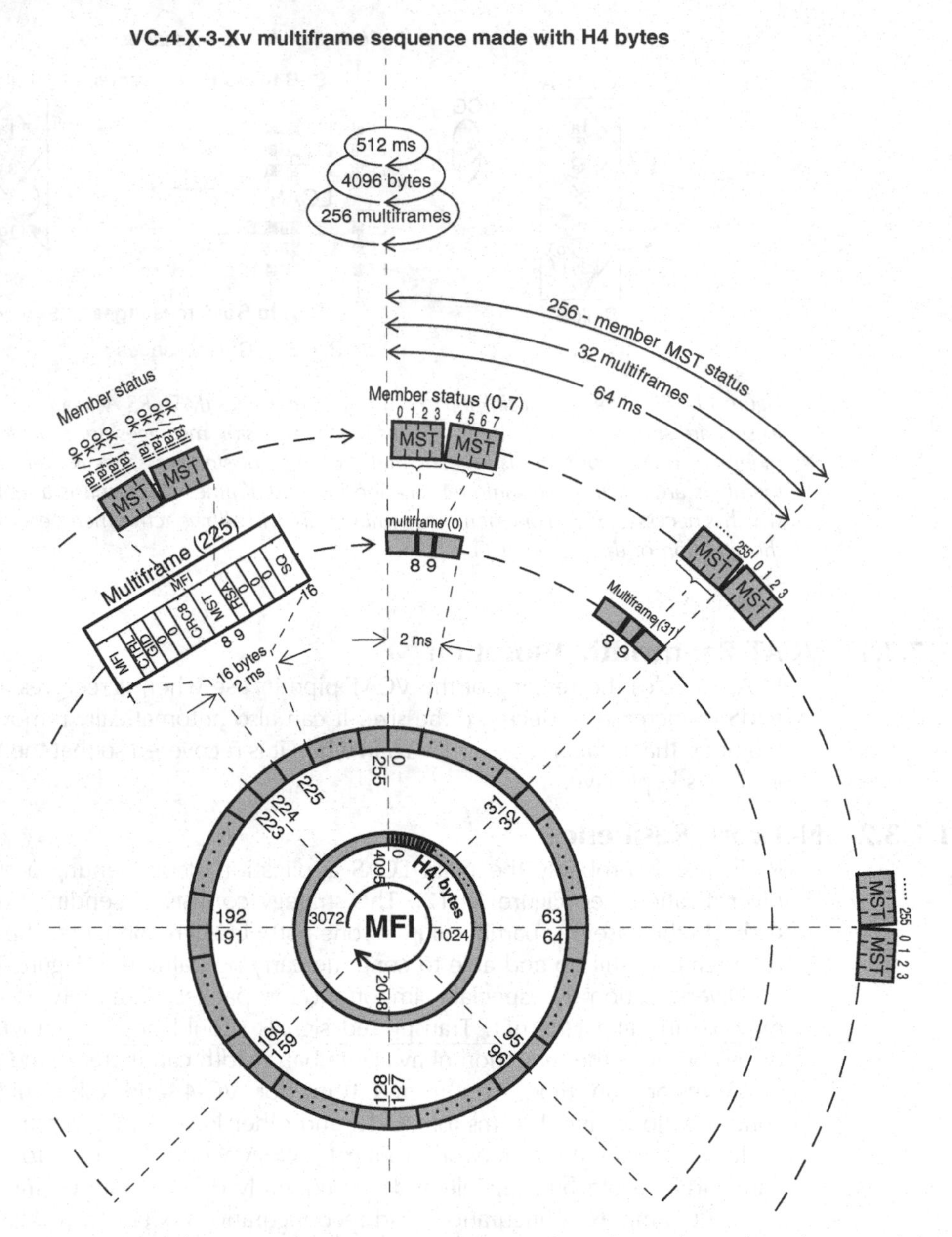

Figure 11-16. Sink (Si) to source (So) messages are a replicated in each member. The H4 multiframe has 8 status bits, therefore 32 multiframes (or 64 ms) are necessary to refresh the status of 256 members. If H4 is being used eight multiframes (or 128 ms) are needed to refresh up to 64 members.

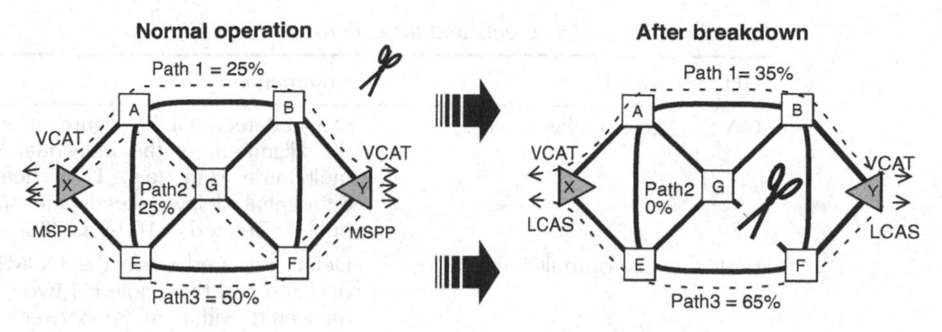

Figure 11-17. Diversification strategy between points X and Y using VCAT and LCAS.

bandwidth between two MSPP nodes to configure asymmetric links adapted to customer requirements. Asymmetric links are an interesting possibility for DSLAM connections to Internet service providers.

11.7.3.4 Cross-domain Operation

LCAS eliminates the slow and inefficient provisioning process of legacy SDH/SONET networks. In particular, if a service crosses several operators (for example international links), it is necessary to coordinate more than one configuration centre.

By using VCAT with LCAS, the configuration is easier because LCAS resides only at edge nodes. The applications mentioned in the previous paragraphs, like network resilience and asymmetric links, can also be implemented in cross-domain services. It is also possible to add and remove paths from a route automatically, in real time, and from both sides of the VCAT pipe.

11.7.4 NG SDH Event Tables

Tables summarize events and indications associated with next-generation SDH/SONET. Ethernet events have also been included (see Table 11-5).

11.8 Conclusions

The technologies that today challenge the supremacy of SDH are a combination of IP with Carrier Ethernet and MPLS. Apparently, they can replace SDH without being compatible with the installed base. There is a very good reason behind this strategy: if 95% of the traffic is generated and terminated in IP/Ethernet, why do we need intermediate protocols?

These challenger technologies are being installed and commissioned in new metropolitan environments that provide data services. Voice and access are also on the road map. IP and Ethernet completely dominate data communications networks, the Internet, VPNs LANs VoIP and IPTV. They have the following key features:

- low cost, easy installation, simple maintenance;
- direct bandwidth provisioning;
- high flexibility in topologies;
- scalability in terms of speed and distance.

Table 11-5. NG SDH events and indications

SDH	Means	Comments
LOA	Loss of alignment	Defect detected if the alignment process cannot perform the alignment of the individual VC-4s to a common multiframe start (e.g. LOA defect activated if the differential delay exceeds the size of the alignment buffer). Defined in ITU-T G.783
LOM	Loss of multiframe	Defect declared when the OOM1 or OOM2 status is declared and the whole H4 two-stage multiframe is not recovered within *m* (*m* between 40 and 80) VC-3/4 frames. Defined in ITU-T G.783
OOM1	Out of multiframe 1	Status declared when an error is found in the multiframe (16 frame) indication 1 (MFI1) sequence carried in bits 5–8 of H4 byte as defined in ITU-T G.783
OOM2	Out of multiframe 2	Status declared when an error is found in the multiframe (256 frame) indication 2 (MFI2) sequence carried in bits 1–4 of H4 byte as defined in ITU-T G.783.
LFD	Loss of frame delineation	Loss of GFP frame delineation
cH-U	Uncorrectable cHEC error	Uncorrectable error in the GFC core header
cH-C	Correctable cHEc error	Correctable error in the GFC core header
tH-U	Uncorrectable tHEC error	Uncorrectable error in the GFC type header
tH-C	Correctable tHEC error	Correctable error in the GFC type header
eH-U	Uncorrectable eHEC error	Uncorrectable error in the GFC extension header
eH-C	Correctable eHEC error	Correctable error found in the GFC extension header
pFCS	Payload frame checksum error	Error in the optional pFCS field of the GFC frame
Align	Frame alignment errors	Ethernet frame alignment error
Under	Under-sized frames	Frames with size smaller than 64 bytes with valid CRC as per 802.3 and RFC2819
Over	Over-sized frames	Frames with size greater than 1518/1522 bytes with valid CRC as per IEEE 802.3-2002 and RFC 2819
Fragm	Fragments	Frames with size smaller than 64 bytes with not valid CRC as per RFC 2819
FTL	Frames too long	Frames with size greater than the maximum permitted as per RFC1643 and RFC2665
FCS	Frame check sum errors	Frames with correct size but invalid CRC

Unfortunately, some weaknesses (poor management, lack of quality of service, jitter) still limit their application. However, they are moving in two directions:

1. Optical integration to improve performance and resilience. Several solutions have already been developed, with different levels of acceptance, including EoFibre, EoRPR, EoDWDM and EoS.

2. IP delayering process to provide quality voice and video as well.

Originally designed for telephony services, SDH has dominated the transmission networks of the world since the early 1990s, providing high-quality connections. Their key features and benefits are:

- comprehensive OAM functions;
- resilience mechanisms to configure fault-tolerant architectures;
- performance monitoring and hierarchical event control;
- synchronization, reducing jitter and wander below limits set in standards.

Today, SDH continues to improve its granularity and flexibility for data transport by means of such new standards as GFP, virtual concatenation and LCAS. In addition a new higher hierarchy, STM-256/OC-768, is ready to be installed.

The answer to the dilemma between the circuit-based and packet-based approaches is a new generation of multiservice nodes providing the best of both worlds by a process of convergence (see Figure 11-2). They integrate all the remote access, routing and switching capabilities into the same pipe. In addition, there are:

- network interfaces – Ethernet, STM-*n*/OC-*m* and DWDM;
- protocols – MPLS, ATM, IP and VLAN;
- client interfaces – TDM, VC, VT, Ethernet, ATM, IP and DVB.

They maintain resilience based on protection schemes, while adding a new resilience strategy based on LCAS which is far more interesting for data packet networks. Any topology is possible, be it linear, UPSR, BLSR or mesh.

Selected Bibliography

[1] ITU-T Rec. G.707/Y.1305, *Network Node Interface for the Synchronous Digital Hierarchy (SDH)*, December 2003.
[2] ITU-T Rec. G.7041/Y.1303, *Generic Framing Procedure (GPF)*, December 2003.
[3] ITU-T Rec. G.7042/Y.1305, *Link Capacity Adjustment Scheme (LCAS) for Virtual Concatenated Signals*, February 2004.
[4] Caballero J. et *al.*, *Installation and Maintenance of SDH/SONET, ATM, xDSL and Synchronization Networks*, Artech House, August 2003.
[5] IEEE 802.3-2002, *Carrier Sense Multiple Access with Collision Detection (CSMA/CD) Access Method and Physical Layer Specifications*, March 2002.
[6] Valencia, E. H., The Generic Framing Procedure (GFP), *IEEE Communications Magazine*, May 2002.
[7] ISO/IEC 3309 *Information Technology – Telecommunications and Information eXchange Between Systems – High-level Data Link Control (HDLC) Procedures – Frame Structure*, 1993.
[8] ITU-T Rec. X.85/Y.1321, *IP over SDH using LAPS*, October 2000.
[9] ITU-T Rec. G.803, *Architecture of Transport Networks based on the SDH*, March 2000.

[10] ITU-T Rec. G.841, *Types and Characteristics of SDH Network Protection Architectures*, October 1998.

[11] ITU-T Rec. G.957, Optical interfaces for equipment and systems relating to the SDH, June 1999.

[12] MEF, Metro Ethernet Networks – A Technical Overview – White paper, http://www.metroethernetforum. org/, December 2003.

[13] MEF, Metro Ethernet Services – A Technical Overview – White paper, http://www.metroethernetforum. org/, January 2004.

[14] Churuvolu G., Ge A., Elie-Dit-Cosaque D., Ali M., Rouyer J., Issues and Approaches on Extending Ethernet Beyond LANs, *IEEE Communications Magazine*, March 2004, pp. 80–86.

[15] Valencia E. H., Rosenfeld G., The Building Blocks of a Data-Aware Transport Network: Deploying Viable Ethernet and Virtual Wire Services via Multiservice ADMs, *IEEE Communications Magazine*, March 2004, pp. 104–111.

[16] Bonenfant P., Rodriguez-Moral A., Generic Framing Procedure (GFP): The Catalyst for Efficient Data over Transport, *IEEE Communications Magazine*, May 2002, pp. 72–79.

[17] Gorshe S., Wilson T., Transparent Generic Framing Procedure (GFP): A Protocol for Efficient Transport of Block-Coded Data through SONET/SDH Networks, *IEEE Communications Magazine*, May 2002, pp. 88–94.

Index

1000BASE-PX 230
10PASS-TS 235
1G 244
2BASE-TL 235
2G 244–245, 268
3G 3, 12, 29, 37, 51, 53, 68, 245, 256, 263, 268, 282, 286
3GPP 256, 269, 272, 278–280, 282, 285, 288, 293, 298, 300
64/65-octet 216, 236
64B/65B 359
64B/66B 25
8B/10B 230, 356, 359

AAA 279, 284, 289–290, 292
  Proxy 279
  Server 279
AAL-5 313
ABR 179–180
AbS 36
ABT 179
AC 250
Access Point (AP) 265
Accounting 48, 289
ACR 42
Add/Drop Multiplexer (ADM) 319
Address
  Broadcast 55
  Complete Message (ACM) 134
  Ethernet
    Multicast 144
  Group 145
  IP 103–104, 106, 142
    Multicast 88, 143
  MAC 143
  Resolution Protocol (ARP) 331
ADM 319
Admission control 90, 154, 166, 302
  token bucket 168
  See also CAC
ADPCM 35
ADSL 9, 14, 56, 207
  G.dmt 206
  G.lite 206–207
ADSL2 63, 209
ADSL2+ 14, 19–20, 63, 212
Adspec 188–189
Advanced Mobile Phone System (AMPS) 244
AF 294
AIPN 262
AKA 295
  IMS 295, 297
  UMTS 295
Alarms 229
  SDH 373
Algorithm
  Bellman-Ford 149
  Dijkstra 149
  KMB 149
  Kruskal 149
  Prim 149
All IP Network (AIPN) 262
Alliance for Telecommunications Industry Solutions (ATIS) 256
All-to-one bundling 309, 311
  See also Bundling
Alphabet 33
Alternative Best Effort (ABE) 198
AMPS 244
AMR 37
Anomaly 91

Triple Play: Building the Converged Network for IP, VoIP and IPTV Francisco J. Hens and José M. Caballero

ANSI
  T1.403 206
  T1.413 206
  T1.419 206
  T1.422 206
  T1.424 235
  T1.601 206
  T1.TR.28 206
  T1.TR.59 206
Answer Message (ANM) 134
Antenna 263, 272
  Smart, See MIMO
AP 265
APP 109
Application Function (AF) 294
Application-specific packet, See APP
ARIB 256, 265
ARP 331
ARPU 4, 17, 26
ARQ 261
AS 284–286
ASN 278
  Gateway (ASN-GW) 278
ASN-GW 278
Association of Radio Industries and Businesses (ARIB) 256
Assured Forwarding (AF) 198
Asymmetric DSL (ADSL) 207
Asynchronous Transfer Mode (ATM) 210
ATIS 256
ATM 29, 31, 48, 68, 179–181, 210, 212, 216, 254–255, 262, 275, 312, 317, 327
  Block Transfer (ABT) 179
  cell 178, 181
  Forum 178
  Inverse Multiplexing, See IMA
  QoS 177
  Resource Management 181
  service categories 179
  SLA 180
  VCI 314
  VPI 314, 327
Attenuation 20–21, 207, 274
Auditing 289
Authentication 108, 254, 265, 270, 279, 284, 287–289, 292, 295, 297
  and Key Agreement (AKA) 295
  Authorization, and Accounting (AAA) 279
  Center (AC) 250
  Mutual 295
Authorization 265, 270, 283–284, 289, 292, 294
Automatic Repeat Request (ARQ) 261
Availability 24
Available Bit Rate (ABR) 179
Average Revenue Per User (ARPU) 4

Band plan
  997 216
  998 216
  VDSL 216
  VDSL2 216
Bandwidth 18, 35, 104, 160, 190, 257, 263, 272, 274, 277, 282
  granularity 361
  profiles 178, 187
Base
  Station 244, 249, 257, 277
    Controller (BSC) 250, 261
    Subsystem GRPS application Protocol (BSSGP) 255
    Subsystem Management Application Part (BSSMAP) 251
  Transceiver Station (BTS) 250
BD 294
BE 277
Beamforming 263
BER 37, 154
Best-effort, See BE
BGCF 285
Billing 252, 279
  Domain (BD) 294
  Unified 16, 285
  See also Charging
Binary Phase Shift Keying (BPSK) 267
Binder 208
Bit
  Error Ratio (BER) 37, 154
  rate
    Available, See ABR
    Constant, See CBR
    Non-real-time Variable, See nrt-VBR
    Real-Time Variable, See rt-VBR
    Unspecified, See UBR
    Variable, See VBR
  Redundancy 249
Blacklist 288
Bluetooth 16
Bonding 236
BPON 221, 225

BPSK 267, 274
Breakout Gateway Control Function
(BGCF) 285
Bridged tap 20
Broadband 3, 18, 300
PON (BPON) 221
Broadcast 144, 312, 322
BSC 250, 261, 270
BSSGP 255
BSSMAP 251
BTS 250, 261
Buffer 98
Capacity 94
Overflow 78, 94
Underflow 78, 94
Bulk Transfer Capacity (BTC) 161
Bundling 308, 311
all-to-one 309, 311
BYE 109, 115

Cable
operator 12
CAC 181
Call
Control 46, 285, 298
International 15
Local 15
Multiparty 103
Phone 15–16
Routing 46, 181
Session Control Function (CSCF) 281
Setup 294
Short distance 15
Termination 294
to mobile 15
CallManager 46
CAM 209
CAMEL 285
Canonical end-point identifier, See CNAME
CAPEX 11, 19, 62
Capital Expenditure (CAPEX) 11
Carrier Sense Multiple Access
with Collision Avoidance
(CDMA/ CA) 268
with Collision Detection (CSMA/CD)
223, 268
Carrierless Amplitude Modulation (CAM) 209
CAT 85
CATT 257
CBR 78, 179–180
CBS 169
CC 288
CCK 266–268
CCSA 256
CDMA 256, 263, 274
Wideband, See WCDMA
cdma2000 246–247, 257, 265, 281
cdmaOne 245
CD-ROM 81
CDV
peak-to-peak 180
Cell
Delay Variation (CDV)
peak-to-peak 178
Loss Priority (CLP) 181
tagging 181
CELP 36
A 37
Conjugate Structure, See CS-CELP
Low-Delay, See LD-CELP
Q 37
CEP 319
Challenge 297
Channel
Bearer 295
Broadcast 258
Coding 253, 255
Control 258
Dedicated 258
Dedicated Physical, See DPCH
Dedicated Physical Control, See DPCCH
Dedicated Physical Data, See DPDCH
Digital 33
Feedback 261, 274
ID(CID) 357
Radio 247, 263, 266, 272, 274–275
Shared 258
Traffic, See TCH
Charging 265, 280, 285, 293–294
Event based 293
Interconnection 32
Offline 293
Online 294
Rule 293
Rules Function (CRF) 293
System
Offline, See OfCS
Online, See OCS
Time based 293
Volume based 293

cHEC 358
China
  Academy of Telecommunication Technology (CATT) 257
  Communications Standards Association (CCSA) 256
Chroma, See Chrominance
Chrominance 68–69, 74
Churn 4
  key 226
CID 276
CIR 169, 310
Circuit Emulation over Packet (CEP) 319
Circuit-switched Data (CSD) 252
Class of Service (CoS) 152, 357
Clean-up timer 187
CLEC 207
Clock 111
  accuracy 156
  drift 157
  frequency offset 157
  offset 156
  resolution 157
  skew 157
  System Timing, See STC
CLP 181
CLR 180
CM 252
CNG 37
Code
  Chipping, See Secuence Chipping
  Division Multiple Access (CDMA) 245
  Hoffman 75
  Space Time, See STC
  Trellis 211
  Variable Length, See VLC
Codebook 36
Codec 38, 40, 298
  Adaptive 36
  ADPCM 36
  Analysis by Synthesis, See AbS
  Hybrid 34, 36
  Source 34
  Voice 37, 42
  Waveform 34–35
Code-Excited Lincar Predictive, See CELP
Collision 268, 277
Committed Information Rate (CIR) 310
Competitive Local Exchange Carrier (CLEC) 207
Complementary Code Keying (CCK) 266
Concatenation 359
  contiguous 360
  virtual 356, 360, 362, 375
Conferencing
  Multiparty 282
Confidentiality 108, 292
  See also Encryption
Congestion 39, 102
  avoidance 154, 165
    admission control 154, 166–167
    resource management 154, 166, 170
  control 90, 108, 152–154, 162
Connection
  Admission Control (CAC) 181
  Identifier (CID) 276
  Management (CM) 252
Connectivity Service Network (CSN) 278
Constant Bit Rate (CBR) 179
Constraint-based Routed LDP (CR-LDP) 330
Contention 10, 93, 277
Contributing Source, See CSRC
    Control plane 324
Copper
  local loop 22
  voice grade 236
CoS 152
CPE 19, 21
CPVR 67
CRC 232
Credit-control 294
CRF 293
CR-LDP 330
Crosstalk 20–21, 40, 208
CS-CELP 36, 39
CSCF 281–283, 290
  Interrogationg, See I-CSCF
  Proxy, See P-CSCF
  Serving, See S-CSCF
CSD 252
CSMA/CA 268
CSMA/CD 223, 268
CSN 278–280
CSRC 104
  Count 105
  List 106
CTD 180
Customized Applications for Mobile networks Enhanced Logic (CAMEL) 285

DAMPS 244, 252
Data
Encryption Standard (DES)
Link Connection Identifier (DLCI) 312, 327
Over Cable Service Interface Specifications (DOCSIS) 12, 272, 281
DBR 179, 229
DCR 42
DCT 74–76, 78
DECT 246
Deficit Round Robin, See Scheduler
Delay 10, 18, 23, 33, 38–41, 43–44, 46, 62, 73, 91–92, 94, 109, 112, 148
Assembly 39
Buffering 95
Coding 34, 36, 39
Decoding 39, 95
Dejittering 40
End-to-end 33, 37–39, 42
Factor (DF) 98
Lookahead 39
one-way 38, 111–112, 155–158
Processing 39, 156
Propagation 40, 156
queuing 156
Reordering 79
Serialization 39, 156
STB 95
Switching 40, 156
Variation 38
Denial of Service (DoS) 290
De-registration, See Registration
DES 108
Deterministic Bit Rate (DBR) 179
DF 98
DHCP 239, 295
DHSS 266
Dial-up 290
Diameter 262, 284, 289–290, 292–295, 297
Agent 291
Proxy 291
Redirect 292
Relay 291
Translation 292
Client 291
Differential delays 361
Differentiated Services (DS) 183, 194
Diffserv 62
Digital
AMPS (DAMPS) 244
Cross-Connect, See DXC
Enhanced Cordless Telecommunications (DECT) 246
Rigths Management (DRM) 286
Signal Processor (DSP) 31
Subscriber Loop (DSL) 203
Video Broadcasting (DVB) 85, 355
Direct Sequence CDMA (DS-CDMA) 256
Discontinuous Transmission (DTX) 37
Discrete
Cosine Transform (DCT) 74
Multitone (DMT) 209
Distortion 10, 40, 44–45
Coding 93
Edge 89
Voice 89
DLCI 312, 327
DMOS 42
DMT 209
DNS 295
DOCSIS 12, 21, 272
Domain name 295–296
DoS 290
Downsampling 104
Downstream
on Demand 331
unsolicited 331
DPCCH 259
DPCH 260
DPCM 35
DPDCH 259
Drift 157
DRM 286
Drop Tail (DT) 173
DRR, See Scheduler
DS 183
codepoint 183, 195, 326
DS-CDMA 256
DSCH 260
DSL 2, 11–12, 15, 17–18, 20, 270, 272, 302
Access Multiplexer (DSLAM) 219
advantages 206
All-digital mode 210
Bonded 212
limits 207
DSLAM 20–21, 219, 240
DSP 31, 40
DSSS 266, 268

DTMF 34, 136
DTX 37
Dual Tone Multi-Frequency (DTMF) 34, 136
Duplex
  Half 275
Duplexing 257, 275
DVB 85, 96, 354–355
DVD 8, 67, 71, 81
  Player 67
DWDM 24
DXC 319
Dynamic
  Bandwidth Report (DBR) 229
  Host Configuration Protocol (DHCP) 239

E1 12, 251
Early Packet Discard (EPD) 174
Eavesdropping 288
EBS 169
Echo 10, 40, 44, 89
E-DCH 261
EDGE 246, 255, 261
EFM 212, 216, 230, 233–234, 236, 315
EFR 37
EGPRS, See EDGE
Egress LER 333
  See also LER
EIR 250
E-LAN 24, 309–310, 348
Electromagnetism 31
E-Line 24, 309–310
  QoS 309
E-mail 141, 270
EMM 85
E-model 43–44
eNB 264
Encapsulation
  ATM 240
  Ethernet 236, 240
  Martini 333
Encryption 46, 108, 270, 277
  Security Envelope (ESP) 270
Enhanced
  ADM, See ADM
  Data rate for Global Evolution (EDGE) 246
  Dedicated Channel (E-DCH) 261
  GPRS, See EDGE
  Node-B, See eNB
  UTRAN (E-UTRAN) 262
Entitlement Management Message (EMM) 85
EoS 316
EPL 310–311, 348, 357
  dedicated 311
  shared 311
EPLAN 310, 312
EPON 222, 230–231
  1000BASE-PX 230
  emulation 233
Equipment
  Approval 60
  Identity Register (EIR) 250
Error
  Bit 91
  Buffer 98
  CAT 97
  Continuity Count 97
  Control 85
  CRC 97
  protection 252, 255
  recovery 18, 255
  SDT 98
ErtPS 277
ESCON 359
ESP 270, 292
Ethernet 15, 18–19, 22–24, 46, 62, 80, 93, 182, 224, 235, 255, 275, 305, 309, 321, 351, 355, 361
  10 Gbit/s 315
  10 PASS-TS 235
  2BASE-TL 235
  64/65-octet 236
  access networks 222
  Active 21, 204
  bonding 236
  Broadcast 312, 322
  bundling 308, 311
  Carrier-class 24, 62, 305, 321
  Connectivity Services 310
  Copper 235
  CSMA/CD 223
  Dedicated bandwidth 311
  DSLAM 219
  E-LAN 309
  E-Line 309
  end-to-end 307, 312
  EPON 231
  frame formats 323
  GbE 373
  Gigabit 302
  in the First Mile (EFM) 212, 233, 315
  LAN 309, 312
  Line, See E-Line

LLID 231
MAC address 308
multicast 143, 312
multiplexing 308, 311
NG-SDH 314
OAM 182, 315
Optical 315
over ATM 312, 349
over dark fibre 316
over FR 349
over LAPS 355
over NG-SDH 319
over SDH 312, 316, 320
over UTP 204
over WDM 316
point-to-point 311
PON (EPON) 222
preamble 232
Private LAN, See EPLAN
Private Line (EPL) 310–311, 348
QoS 315, 324
scalability 322
shared bandwidth 311
Virtual Connection (EVC) 308
Virtual Private LAN (EVPLAN) 310
Virtual Private Line (EVPL) 310
ETSI 43, 245, 252, 256, 265, 273, 286, 300
ETS 300 726 107
ETS 300 961 107
TR 101 290 96
TS 101 135 206
TS 101 524 206
TS 23.234 279
TS 29.229 292
TS 32.240 294
TS 33.234 279
European
Commision 22
Telecommunications Standards Institute (ETSI) 245
E-UTRAN 262, 264
EVC 307–308
point-to-point 309
VLAN 309
EVPL 310–311
EVPLAN 310, 312, 343
Excess Information Rate (EIR) 170
Expedited Forwarding (EF) 198
Expenditure
Capital, See CAPEX 280
Operating, See OPEX 280
Explicit LSP, See LSP
Extended Real-time Polling Service (ErtPS) 277
Extension Header Identifier (EXI) 357
External NNI (E-NNI), See also NNI 308

Fading 263, 267, 274
Failure 91
Far End Crosstalk (FEXT) 208
Fast Fourier Transform (FFT) 44
FBC 293
FCC 22
FDD 208, 216, 257, 275
FDM 6, 31, 206
FDMA 245, 247, 256, 274
Orthogonal, See OFDMA
FEC 103, 230, 261, 329, 331
FEXT 208
FGNGN 300
FHSS 266
Fiber
to the Home (FTTH) 204
Fibre
Channel 356
Dark 24, 320
to the Building (FTTB) 21, 204
to the Cabinet (FTTCab) 218
to the Curb (FTTC) 20
to the Node (FTTN) 20
FICON 359
FIFO 163
Filter
coeficient 36
Dejittering 33
Linear 34
Firewall 10, 47, 127
First In, First Out (FIFO) 163
First mile 6, 59, 63, 272
Fixed Mobile
Convergence (FMC) 244
Substitution (FMS) 243
Flow 277, 293, 295
Based Control (FBC) 293
control 103, 108
Closed-loop 165
Open-loop 165
label 188
See also Traffic
FMC 244, 268, 282
FMS 243
Forward Error Correction (FEC) 103, 230, 261

Forwarding Equivalence Class (FEC) 329
Forwarding plane 324
FQDN 118–119
FR 12, 31, 255, 312, 327
Frame
  Freezing 89
  Relay (FR) 255, 305–306
Frame-mapped GFP (GFP-F) 317, 356–357
Frequency 35, 75, 111, 275
  Carrier 257, 266
  Division Duplexing (FDD) 208, 257
  Division Multiple Access (FDMA) 245
  Division Multiplexing (FDM) 31, 206
  Licensed 273
  Sampling 34
  Unlicensed 266, 273
  Warping 44
FSAN 214, 221, 225
FTP 277, 293
FTTB 21, 204, 218–219, 235
FTTC 20
FTTCab 204, 218, 235
FTTH 63, 204
FTTN 20, 22, 63
FTTP, See FTTH
Full-Service Access Network (FSAN) 214, 221

GA-CSR 272
GAN 256, 269–270, 278
  Controller (GANC) 269
GANC 269–270
GA-PSR 272
GA-RC 271
Gateway
  Decomposition 134
  GPRS Support Node (GGSN) 253
  Media (MGW) 134
    Control Protocol (MGCP) 101, 136
    Controller (MGC) 135
  Residential 48
Gaussian Minimum Shift Keying (GSMK) 249
GEM 222, 228
General Packet Radio Service (GPRS) 252
Generalized MPLS (GMPLS) 327
Generic
  Access
    Circuit Switched Resources (GA-CSR) 271
    Network (GAN) 256
    Packet Switched Resources (GA-PSR) 271
    Resource Control (GA-RC) 271
  Framing Procedure (GFP) 222, 316, 353, 355–356
Routing Encapsulation (GRE) 336
GFP 23, 222, 316
  GFP-F 317–318, 327
  GFP-T 316
GGSN 253, 261, 293
Gigabit PON (GPON) 221
Global
  Positioning System (GPS) 157
System for Mobile Communications (GSM) 246–247
Globally Routable User agent URI (GRUU) 289
GMPLS 327
GOP 79
GPON 221, 227
Encapsulation Mode (GEM) 222, 228
GPRS 252, 261, 270, 281, 295
  Backbone 253
  Support Node (GSN) 253
  Tunneling Protocol (GTP) 253
GPS, See Scheduler
GRE 336
Group of Pictures (GOP) 79
Groupe Spécial Mobile (GSM) 245
GRUU 289
  Public 289
  Temporary 289
GSM 9, 11, 17, 37, 247, 256–257, 261, 270, 281, 285–287, 295
  Evolution 255
GSMK 249, 255
GSN 253
GTP 253–254
Guaranteed service 184, 189

Handoff 48, 244, 250, 258, 263–264, 269, 272, 277, 286, 300
  ASN anchored 279
  CSN anchored 279
  Hard 257
  Inter-frequency 258
  Intra-frequency 257
  Soft 257
  Softer 257

Hard QoS, See QoS
HARQ 261
HD, See HDTV
HDSL 207
HDTV 8–9, 56, 61, 73, 203
Head-end, See Network Contribution
Header Error Control (HEC) 229
HEC-based delineation 356
HFC 20–21
High Speed
  CSD (HSCSD) 252
  Downlink Packet Access (HSDPA) 256
  Downlink Shared Channel (HS-DSCH) 260
  OFDM Packet Access (HSOPA) 263
  Packet Access Evolution, See HSPA+
  Uplink Packet Access (HSUPA) 256
High-bit-rate DSL (HDSL) 207
High-Definition TV (HDTV) 203
Higher-Order Virtual Concatenation (HO-VCAT) 363
HiperLAN 265
HiperMAN 273
HiSWAN 265
HLR 250, 261, 279, 284, 287–288
Home
  Location Register (HLR) 250
  Subscriber Server (HSS) 281
Hop-by-hop LSP, See LSP
Hotspot 265, 279
HO-VCAT 363
HSCSD 252
HSDPA 256, 260–261
HS-DSCH 260
HSOPA 263–264
HSPA 262
HSPA+ 256
HSS 281, 284, 287–288, 290, 295–298
HSUPA 256, 260–261
HTTP 118, 141, 162
Hub-and-spoke architecture 309
HVPLS 343
Hybrid
  Automatic Repeat Request (HARQ) 261
  circuit 40
  Fibre Coaxial (HFC) 20
Hypertext Transfer Protocol (HTTP) 141

ICMP 156
I-CSCF 295, 297–298
  See also CSCF 295
IEEE 240, 263–264, 272, 278, 306
  802.11, See Wi-Fi
  802.11–1997 266
  802.11a 266–267, 273
  802.11b 266
  802.11g 267–268
  802.1 1n 268
  802.16 278
  802.16, See WiMAX
  802.16–2001 272
  802.16a-2003 272
  802.16d 272
  802.16e 2005 275
  802.16e-2005 272
  802.17 322
  802. 1ad 322, 343
  802.1ag 234
  802. 1D 230, 240, 322, 324
  802. 1p 162
  802.1Q 309, 322, 324
  802. 1q 311
  802.1Q/p 324
  802.3 25, 204, 230, 326
  802.3, See Ethernet
  802.3ah 212, 222, 233, 236
IETF 101, 115, 145, 270, 278, 280
  IPPM 154
  Megaco/H.248 137
IGMP 63, 88, 92, 95, 143–144
  Message 145
  Snooping 144
IKE 292
ILEC 207
IMA 213
IMEI 288
IMP 117, 126, 281, 286
IMS 244, 256, 262, 269–270, 278, 280–282, 284–288, 290, 292, 294–295, 298, 300, 302
  Procedure 294
  Service Control (ISC) 285
IMSI 287
IMT-2000 257
IN 285
Incumbent Local Exchange Carrier (ILEC) 207
Industrial, Scientific, and Medical (ISM) 273
Ingress LER 332
  See also LER

INI 181
Initial Address Message (IAM) 134
Instant Messaging and Presence (IMP) 126, 281
Institute of Electrical and Electronics Engineers (IEEE) 306
Integrated Services (IS) 183
  Digital Network (ISDN) 134, 210
Integrity 270, 292
Intellectual Property Rights (IPR) 57
Interactive Connectivity Establishment (ICE) 133
Interface
  A 250–251, 269–270
  Abis 250–251, 262
  Air 247, 276, 278
  Cx 284, 290, 295
  Dx 284
  Gb 269–270
  Gm 283
  Gy 294
  Gz 294
  Iub 262
  Iur 262
  Mn 285
  Mp 285
  Radio, See Interface Air
  S1 264
  Up 270
  Uu 262
  Wn 280
  Wp 280
  X2 264
Interference 253, 266
  Inter-Symbol, See ISI
  Narrowband 257, 266–267
Interim Standard-54 (IS-54) 244
Intermediate System - Intermediate System (IS-IS) 324
Internal NNI (I-NNI), See also NNI 308
International
  Mobile
    Communications 2000 (IMT-2000) 246
    Equipment Identity (IMEI) 288
    Subscriber Identity (IMSI) 287
  Telecommunications Union, See ITU-T
Internet 9, 11–12, 16–17, 21–22, 26, 32, 45, 51–52, 60–61, 66, 71, 87, 90, 142, 182, 265, 270, 277, 279–280
  Assigned Numbers Authority (IANA) 143
  Broadband 12
  Engineering Task Force (IETF) 306
  Group Management Protocol (IGMP) 63, 88, 137, 142
  Key Exchange (IKE) 292
  Protocol (IP) 152
    Performance Metrics (IPPM) 154
  Service Provider (ISP) 12
Internetwork Interface, See INI
Interworking 47, 134, 278–279, 285, 300
INVITE 118, 293
IP 22–23, 103, 156, 166, 182, 262, 264, 270, 275, 280, 324, 327
  Address
    Class D 143
  DS field 196
  flow label 188
  IPv6 191
  Multicast, See Multicast
  Multimedia Subsystem (IMS) 48, 244
  QoS 152, 182
  security protocol (IPsec) 108
  Television, See IPTV
IPPM 156–157
IPR 57–58
IPsec 270, 292
IPTV 7–8, 11, 15, 17, 52–53, 55, 57, 59–63, 65, 67, 71, 73, 79–81, 87–90, 92–93, 98, 141, 240, 282
  Business model 53
  Early implementations 71
  Quality 89
  Zapping 63, 65
IPv4 240, 302
IPv6 281
IS 172, 183, 186
IS-136 245, 255
IS-54 244
IS-95, See cdmaOne
ISC 285
ISDN 9–10, 29, 46, 68, 134, 210, 215, 250, 285, 288, 300
  Country Code, See CC
  National Destination Code, See NDC
  Subscriber Number, See SN
  User Part, See ISUP
IS-IS 324
ISM 266
ISO 71
  10918–1 107
  10918–2 107
  11172–3 107

Isochronous 270
ISP 11–12
ISUP 115–116
ITU-R 246
BT.601 69
ITU-T 137, 300, 306
E.164 288
G.107, See E-model
G.114 40
G.703 206
G.7041 316, 355
G.7042 316, 356, 366–367
G.707 316, 356
G.711 107
G.711, See PCM
G.721, See Codec ADPCM
G.722 107
G.723.1 37–38, 107
G.726 107
G.726, See ADPCM
G.728 107
G.728, See LD-CELP
G.729 annex D 107
G.729 annex E 107
G.729, See CS-CELP
G.783 374
G.961 206
G.983.1 221, 226
G.983.3 226
G.984.3 228
G.991.1 206
G.991.2 206, 235
G.992.1 206, 212
G.992.2 206–207
G.992.3 209–210, 212
G.992.4 209
G.992.5 212
G.993.1 204, 214
G.993.2 204, 215, 218
G.994.x 221
G.998.1 212–213
H.245 101
H.248, See Megaco
H.261 107
H.262 71
H.263 107
H.264, See MPEG-4 part 10
H.323 46, 101, 116, 126, 134
I.371 179
P.861, See PSQM
P.862, See PESQ
Q.931 115–116
SG13 300
X.25 254
X.86 312, 355
Y.1541 91
Y.2001 300
Y.2011 300
I-WLAN 256, 270, 279–280

Jitter 18, 28, 33, 46, 91–92, 98, 109, 112, 157, 199
Interarrival 94, 111
Packet 78
Statistics 111

K4 codification 365

L2TP 336
Label 326
distribution control
Independent 331
Ordered 331
Distribution Protocol (LDP) 172, 330
Edge Router (LER) 328
forwarding 324
stacking 326, 333
switching 329
Label-Switched
Path (LSP) 172, 319, 324
Router (LSR) 324
LAN 24, 46, 182, 265, 306
LAPD 251
modified, See LAPDm
LAPDm 251, 255
LAPS 312–313
Latency 89, 156, 190, 260
IGMP 92
Join 95
Leave 95
Network 95
Layer-2 Tunneling Protocol (L2TP) 336
LCAS 316
LD-CELP 36
LDP 330
Message
Advertisement 331
Discovery 331
Label Mapping 331
Label Request 331

LDP (*Continued*)
    Notification 331
    Session 331
LER 328
  egress 333
  ingress 332
Less than Best Effort (LBE) 198
Linear
  Prediction 36
  Predictive Coding (LPC) 34
Link
  Access Procedure - SDH (LAPS) 312
  Access Procedure channel D, See LAPD
  Bottleneck 161
  Capacity Adjustment Scheme (LCAS) 316, 354, 356, 366
LIR 298
LLC 255
LLID 230–231
Local Area Network (LAN) 355, 373
Location Info Request (LIR) 298
Logical Link
  Control (LLC) 255
  Identifier (LLID) 230
Logical Link Identifier (LLID) 231
Long Term Evolution (LTE) 262
Loss 159
  Frame 79
  Of Alignment (LOA) 374
  Packet 91–92
Lower-Order Virtual Concatenation (LO-VCAT) 363
LPC 34, 36
LSP 319, 324
  Explicit 329
    Loose 330
    Strict 330
  Hop-by-hop 329
LSR 324, 329, 331–332
  label distribution 330
LTE 262, 264
Luma, See Luminance
Luminance 68–69, 74

MAA 297
MAC 231, 255, 268, 272, 276, 308, 314, 318
  address explosion 348
  Address Stacking (MAS) 322
Macrocell 258
MAN 24, 306, 312
Management
  IP address 278
  Key 108, 277, 292
  Profile 280
Map
  Downlink 277
  Uplink 277
Mapping 144
MAR 297
Marking 102, 161, 167
MAS 322
Maximum
  Burst Size (MBS) 178
  Cell Transfer Delay (max CTD) 178
  Transfer Unit (MTU) 190
MBone 146
MCC 288
MCR 180
MDI 98–99
Mean Opinion Score (MOS) 38
Media
  Access Contol (MAC) 255
  Delivery Index (MDI) 98
  Gateway (MGW) 285
    Controller Function (MGCF) 285
  Loss Rate (MLR) 98
Megaco 101, 137, 285
MEN 306, 308
  NNI 308
  UNI 308
Message
  Integrity 108
  Transfer Part (MTP) 251
Messaging, See IMP
Metro
  Ethernet Forum (MEF) 306, 357
  Ethernet Network (MEN), See also Ethernet 306
MGC 135, 137
MGCF 285–286
MGCP 101, 136
MGW 134, 137, 285
  Access 137
  Trunking 137
Microcell 258
Middleware 56
MIME 121, 123
MIMO 256, 263, 268
Minimum Cell Rate (MCR) 178
Mixer 104, 113

MLR 98
MM 251
MME 264
MMS 286
MMUSIC WG 117
MNC 288
Mobile
  Country Code (MCC) 288
  Network Code (MNC) 288
  operator 12
  Station 249, 257
    Integrated Services Digital Network (MSISDN) 288
    Roaming Number (MSRN) 250
  Subscriber Identity Number (MSIN) 288
  Switching Centre (MSC) 250
  TV 243
Mobility 12, 243
  Generalized 300
  Management (MM) 251
    Entity (MME) 264
Modulation 255, 266
  DMT 209
  Multi-carrier 274
  Phase, See PSK
  Pulse Code, See PCM
  Single carrier 273
Monitoring 43, 48, 62
MOS 38, 42–45
  Conversational 42–44
  Degradation, See DMOS
  Listening 42–44
  Objective, See OMOS
  Video 99
Motion vector 77
Moving Pictures Expert Group (MPEG) 8, 71
MP3 71
MPCP 230
MPEG 8–9, 52, 59, 61–62, 71, 73, 76, 78, 86, 277
  Coding
    Interframe 74
    Intraframe 74
  Macroblock 74
  Processing
    Spatial 74
    Temporal 74, 76
  Stream 79
    Elementary, See ES
  Timing 87
MPEG-1 61, 71
MPEG-2 9, 61, 71, 81, 85, 96, 98
  B-frame 73, 76, 78–79, 93
  Frame
    Bidirectional-prediction, See MPEG2 B-frame
    Intra, See MPEG-2 I-frame
    Prediction, See MPEG-2 P-frame
  I-frame 73, 78–79, 93
  Level 71–72
    High 1440 72
    High 1920 72
    Low 72
    Main 72
  P-frame 73, 76, 78–79, 93
  Profile 71, 73
    High 73
    Main 73
    Simple 73
    SNR 73
    Spatial 73
MPEG-4 9, 61, 65, 71, 81, 85, 96
  Part 10 71
MPFP 285
MPLS 15, 18, 62, 240, 262, 324, 329, 331
  advantages 345
  label 326, 328
  pseudowire 335
  See also GMPLS
  with SDH 327
MP-MLQ 37
MRFC 285
MSC 250, 261–262, 270, 286
MSIN 288
MSISDN 288
MSPP 348, 354
MSRN 250
MSSP 355
MSTP 354
MTP 251
MTU 343
Multicast 88, 141–142, 187, 312
  Agent 55, 143–144
  Backbone, See MBone
  Ethernet 143–144
  Group 63, 88, 142
    Dynamic 148
    Management 144
    Private 145
    Public 145

Multicast (*Continued*)
IGMP 144
RSVP 193
Tree 147
Multipoint-to-multipoint 147, 150
Point-to-multipoint 147
Steiner 148–149
Multiframe Indicator (MFI) 363–364, 366, 368
Multimedia 297, 300
Authorization Answer (MAA) 297
Authorization Request (MAR) 297
Conferencing 243
Messaging Service (MMS) 286
Resource Function Controller (MRFC) 285
Resource Function Processor (MRFP) 285
Service 280
Multiple
access 244, 247, 253, 255, 268, 274–275
Input and Multiple Output (MIMO) 256
Multiplexing 109, 308, 311
Hierarchy 35
Statistical 22–23, 92
Stream 106
WDM 222
Multi-Point Control Protocol (MPCP) 230
Multiprotocol Label Switching (MPLS) 240, 373
Multipurpose Internet Mail Extension (MIME) 121, 161
Multiservice
Platform (MSxP) 354
Provisioning Platform (MSPP) 354
Switching Platform (MSSP) 352, 355
Transport Platform (MSTP) 348, 354
Multitenant Unit (MTU) 343

Narrow link 160
NAS 290, 292
NASS 302
NAT 127, 240
Problems 129
Traversal 302
Using Relay (TURN) 132
Types of 128
NDC 288
Near End Crosstalk (NEXT) 208
Network
Access 4, 22, 59–60, 63, 88, 203, 222, 252, 269, 280, 282, 286, 295, 302
Broadband 34
Server (NAS) 290
Address Translation (NAT) 127, 240
Ad-hoc 265
Aggregation 59
Attachment Sub-System (NASS) 302
availability 23
Best-effort 32, 53, 90–91
bridged 320
Cellular 17, 34, 281, 290
Circuit-switched 151, 170
QoS 151
Commissioning 60
congestion, See Congestion
Contribution 62, 92–93
Converged 1, 4–5, 11–12, 15, 18, 37, 53, 60
Core 279
Corporate, See Network Enterprise
Data 252
Discovery 278
Distribution 59, 62, 66
Enterprise 46, 267
Fixed 28, 48, 285
Home 283, 295, 297
Impairment 92–93
Intelligent, See IN
Inter-Networking NNI (NI-NNI), See also NNI 308
IP 4, 8, 19, 23, 32–33, 40, 45, 48, 62, 80, 93–94, 102, 142, 254, 285, 290, 293
Legacy 27
Management System (NMS) 329, 368
Mesh 265, 268
Metro 24
Mobile 28, 37, 48, 285, 301
Monitoring 93
Multicast 88
Multiservice 27
Next Generation, See NGN
Packet-switched
Parameter Control (NPC) 181
Protection 91
Resource 293, 295
Roll-out 93
Scalability 148
Service
Access, See ASN
Time Protocol (NTP) 111, 157
Transport 47
Unified, See Network Converged
Visited 250, 283, 295

Wireless 281, 300
Wireline 281, 300–301
Network-to-Network Interface (NNI) 137
NEXT 208
Next-Generation SDH (NG-SDH) 306
NG SDH
Events 373
NGN 244, 298, 300
Focus Group on, See FGNGN
NG-SDH 24, 306, 314, 316, 319, 344, 348, 351
MPLS 327
NIT 86
NMS 329
NMT 244
NNI 137, 308
E-NNI 308
I-NNI 308
NI-NNI 308
SI-NNI 308
Node-B 261–262
Noise 10, 21, 31, 44, 253
Background 38, 44
Chroma 90
Comfort
Generation, See CNG
Electrical 44
Quantizing 35, 44
Visual 89
Non-real-time
Polling Service (nrtPS) 277
Variable Bit Rate (nrt-VBR) 179
Nordic Mobile Telephone (NMT) 244
NPC 181
NPVR 67
nrtPS 277
nrt-VBR 179–180
NTP 111, 157
NTT 43
NVPR 67
Nyquist theorem 34

OAM 234, 305, 307
discovery 235
Ethernet 182
link monitoring 235
remote failure indication 235
remote loopback 235
OCS 294
OFCOM 22
OfCS 293
OFDM 246, 263, 267–268, 273–275
OFDMA 274
Offset 156
OLT 222, 224, 229, 232
OMA 286
One Pass with Advertising (OPWA) 191
One-Way Active Measurement Protocol (OWAMP) 155
One-way delay
See also Delay 155
ONT 222
ONU 222, 224, 232
Bridging 231
Open
Mobile Alliance (OMA) 286
Services Architecture (OSA) 285
Systems Interconnection (OSI) 251
Open Shortest Path First (OSPF) 324
Operation, Administration and Maintenance (OAM) 234
Operational Expenditure (OPEX) 11
Operator
Cable 21, 27–28
Dominant 22
Fixed line 28
Incunbent 32
OPEX 11, 19, 62, 89
OPINE 43
Opinion model 43
Optical
access 21
Ethernet, See Ethernet
Line Termination (OLT) 222
loss 230
Network Termination (ONT) 222
Network Unit (ONU) 222
splitter 222
Transport Network (OTN) 308
Orthogonal Frequency Division
Multiple Access (OFDMA) 274
Multiplexing (OFDM) 246
OSA 285–286
OSI 251
OSPF 324
Out of Multiframe (OOM) 374

P2MP 237
P2P 237, 280
Packet
ADM, See ADM
Conforming 154
count 109

Packet (*Continued*)
Data Gateway (PDG) 280
Data Protocol (PDP) 253
Dropped 97
Dropping Policy (PDP) 172
Duplicated 97
Identifier (PID) 85
IP 39, 43
jitter 157
loss 18, 38, 41, 44, 97–98, 109, 112, 159
defined 159
distance 159
period 159
ratio 38
statistics 112
Loss Concealment (PLC) 38
marking 156, 161
non-conforming 154, 167, 185
Null 86
Out of order 97
Out-of-order 98
periodicity 158
sequencing 159
switching 31
Temporary Subscriber Identity (P-TMSI) 254
timeout 159
timestamping 156–157, 159
Transfer Mode (PTM) 210, 235
See also Traffic
Packet-layer model 43
Packet-switched networks 170
PAMS 44–45
Parlay Group 286
Partial Packet Discard (PPD) 173
Pass-band 266
Passive Optical Network (PON) 19, 204
PAT 85, 87, 95–96
Error 96
PATH message, See RSVP
Payload
Control Block (PCBd) 229
Length Indicator (PLI) 229
Pay-per-view 11
PBX 46
PC 11
PCBd 229
PCC 293–294
PCEF 293
PCM 31, 33–36, 38, 42–43, 277
Adaptive Differential, See ADPCM
Differential, See DPCM
PCR 85, 94, 97, 180
PCRF 293–294
P-CSCF 282–283, 294–295, 297–298
PDA 63, 281, 289
PDF 293
PDG 280
PDH 254
PDP 253
Context Activation 254, 295
Peak Cell Rate (PCR) 178
Peak-to-peak Cell Delay Variation (CDV) 178
Peer discovery 290
Peer-to-Peer, See P2P
PEH 80
Perceptual
Analysis Measurement System (PAMS) 44
Evaluation of Speech Quality (PESQ) 44
Speech Quality Measure (PSQM) 44
Per-Hop Behaviour (PHB) 196
Permanent Virtual Circuit (PVC) 238, 309
Personal Digital Assistant (PDA) 281
PES 80
PESQ 44
Phase Shift Keying (PSK) 255
PHB
ABE 198
AF 198–199
default 198
EF 198, 200
LBE 198
Physical
Layer
Operation, Administration and Maintenance (PLOAM) 226
Overhead (PLO) 229
Link Layer (PLL) 255
Media Specific / Transmission Convergence (PMS-TC) 210
Physical-Media Dependent (PMD) 210
PID 85–86, 98
Ping 156, 159
PINT 117
Pixel 69
Pixelation 89
Plain Old Telephone Service (POTS) 210

Plane
  Control 278
  User 278, 293
PLC 38
Plesiochronous Digital Hierarchy (PDH) 254
PLI 229, 358
PLL 255
PLO 229
PLOAM 226, 229
PLS 229
PMD 210
PMT 85, 95, 97
  Error 96
PoC 281, 286
Point of Presence (PoP) 238
Point-to-MultiPoint, See P2MP
Point-to-Point
  Protocol (PPP) 238
  See also P2P 237
Policing 167
Policy 293
  and Charging
    Control (PCC) 293
    Enforcement Function (PCEF) 293
    Rules Function (PCRF) 293
  Control 293–294, 302
  Decision Function (PDF) 293
Polling 277
  Broadcast 277
  Multicast 277
  Unicast 277
PON 19, 21, 204, 219, 272, 281, 302
  advantages 224
  bandwidth sharing 224
  CSMA/CD 223
  definition 221
  range 222
  rates 222
  TDM 223
  WDM 223
Port ID 229
POTS 9–11, 46, 215
Power
  Control 260
  Leveling Sequence (PLS) 229
  Transmitted 263
PPP 238, 254, 290, 356
PRACK 298
Prediction
  Backwards 76
  Forward 76
  Interframe 76
  Motion compensated 76, 79
Preemption 216
Presence, See IMP
PRI 134
Primary Rate Interface (PRI) 134
Priority scheduler, See Scheduler
Private Branch Exchange (PBX) 45
Processing delay, See Delay
Program
  Clock Reference (PCR) 85, 87
  Specific Information (PSI) 85
Propagation
  delay, See Delay
  Multipath 257, 267, 274
Provisional Acknowledge, See PRACK
PS 81
Pseudowire 335–336
  topologies 341
PSI 85
  Table 85
PSK 255, 266
PSQM 44–45
PSQM+ 45
PS-TC 209
PSTN 9, 32, 34, 45–46, 48, 133, 250, 280, 285, 287, 300
  substitution 10
PTM 210, 235
P-TMSI 254
PTM-TC 216
Public Switched Telephone Network (PSTN) 9, 280
Push to talk over Cellular (PoC) 281
PVC 238, 240, 309
PVR 66–67
PVS 55, 59
PW, See pseudowire
PWE3 344

QAM 209, 214–215, 267, 274
Q-in-Q 322
QoE 89, 99
QoS 5, 15, 18, 23–24, 29, 32–33, 38, 40–41, 45–47, 53, 66, 90–91, 101–102, 106, 151–152, 240, 254–255, 262–264, 270, 275–276, 279–280, 283, 293, 298, 300

QoS (*Continued*)
  architectures 177
    DS 194
    IS 183
  ATM, See ATM
  Authorization 298
  congestion management 153
  delay 155
    See also Delay 155
  Differentiated 59, 153
  E-LAN 310
  Guaranteed 8, 153, 285
  hard 153, 184
  IP 177, 182
  Monitoring 109
  negotiated parameters 178
  PCR 180
  soft 153, 184
  traffic differentiation 152
  See also Congestion
QPSK 259–261, 267, 274
Quadrature Amplitude Modulation (QAM) 209, 267
Quadruple Play 16, 48, 243, 282
Quality 151
  of Experience (QoE) 89
  of Service (QoS) 151
Quantum 164
Quaternary Phase Shift Keying (QPSK) 259, 267

RACS 302
Radio
  Band 273
  FM 33
  Frequency Layer (RFL) 255
  IP 33
  Link Control (RLC) 255
  Network Controller (RNC) 261
  Resource (RR) 251
RADIUS 290, 292
Rake receiver 257
Random Early Detection (RED) 174–175
Ranging 227
Rapid Spanning Tree Protocol (RSTP) 322
RAS 238
Rating
  Absolute Category, See ACR
  Algorithmic 42
  Degradation Category, See DCR
  Objective 42
  Opinion 40–41
Raw mode 339
RC 109
Real-time
  Control Protocol (RTCP) 108
  Polling Service (rtPS) 277
  Streaming Protocol (RTSP) 101
  Transport Protocol (RTP) 101
  Variable Bit Rate (rt-VBR) 179
Receive diversity 263
Reception Report Count, See RC
Reference point, See Interface
Registration 279, 283–284, 288, 294–295, 297
Remote Access Server (RAS) 238
Remote Authentication Dial-In Service (RADIUS) 290
Request for Comments (RFC) 155
Re-registration, See Registration
Reservation Protocol (RSVP) 186
Resilient Packet Ring (RPR) 319
Resource
  and Admission Control Sub-system (RACS) 302
  Management 46
  Reservation 298, 302
Resource Management (RM) 154, 181
RESV messages 187
Re-timing 102–103
R-factor 43
RFC 155, 280
  2198 107
  2205 186
  2327 121
  2406, See ESP
  2409, See IKE
  2427 349
  2486 288
  2543 117
  2640 159
  2679 156
  2684 235, 312, 349
  2705 136
  2806 116
  3036 330
  3046 239
  3118 161
  3209 330
  3212 330
  3261 117
  3329 295
  3389 107

3393 157
3551 107
3588, See Diameter
4005 292
4006 294
4445 98
4447 338
4740 292
822 114
RFL 255
RLC 255
RM 181
RNC 261–262
Roaming 283, 285, 295, 298
Round
Robin, See Scheduler
Trip Delay (RTD) 231
Round Trip Delay (RTD) 111
Router 17, 45, 88, 143, 183
Broadband 11
DSL 20
PE 339
RSVP PATH 187
VoIP enabled 46
Routing 5, 147, 284, 288, 298
Multicast 146–147
Algorithms 147–148
Protocol 143
RSVP 186
RPR 319, 322
RR 109, 111–112, 251
RR, See Scheduler
RS encoding 211
RSTP 322
RSVP 172, 186–188, 190–191, 193–194
Adspec 191
filter specification (Filterspec) 191
flow specification (Flowspec) 191
message format 186
multicasting 193
object 186
PATH 187–189
reservation styles 191, 194
fixed filter 194
shared explicit 194
wildcard filter 194
RESV 187–188, 191
routing 186
Sender template 188
slack term 192
traffic specification (Tspec) 188
RSVP-TE 330
RTCP 23, 83, 108–109, 113
BYE 109
Packet 109
Receiver Report (RR) 109
SDES 113
Sender Report (SR) 109
Source Description (SDES) 109
RTD 111, 156, 231
RTP 23, 37, 62, 80, 83, 88, 92, 101–104, 106, 108–109, 111–112, 270
header 104–105
Mixer 104, 106
Payload
Identification 102
Timestamping 102
Type 105
Profile 105
Translator 104
rtPS 277
RTSP 62–63, 66, 92, 95, 101
rt-VBR 179–180

S/MIME 125
SA 295
SAA 297
SAP 121
SAR 297
SBLP 293
SBR 179
SC 113
Scalable OFDMA (S-OFDMA) 274
SCB 233
SCCP 251
SC-FDMA 263
Scheduler 102, 163, 183, 277
Deficit Round Robin 164
FIFO 163
GPS 165
priority 165
Round Robin 164
WFQ 165
S-CSCF 283–285, 296, 298
SCTP 290
SDES 109, 113
SDH 7, 18, 22–23, 254, 307, 314, 319
events 373
Next Generation, See NG-SDH
TDM 310
with MPLS 327

SDP 101, 121, 262, 280, 293, 298
Security 24, 108
  Association (SA) 295
  Certificate 125
  Encryption 7, 85
  End-to-end 123, 292
  Gateway (SEGW) 270
  Hop-by-hop 123, 292
  Key
    Private 125
    Public 125
SEGW 270
Sequence
  Barker 266
  Chipping 257–258, 266
  Complementary 266
  Complex 266
  Identifier (SID) 213
  Indicator 368
  Number 105, 366
  Orthogonal 245, 258
  Power Leveling (PLS) 229
  Pseudo-random 266
  Scrambling 258
  Spreading 258
Sequencing 102
Serialization delay, See Delay
Server
  Application 284
  Assignment Answer (SAA) 297
  Assignment Request (SAR) 297
  DNS 295
  Overload 93
  SIP 286
Service
  Assurance 60
  Asymmetric 252
  ATM 179
    ABR, See ABR
    CBR, See CBR
    nrt-VBR, See nrt-VBR
    rt-VBR, See rt-VBR
    UBR, See UBR
  Best-effort 66
  bundling 15, 28
  Capability Server (SCS) 286
  Controlled-load 184, 189
  Data 252
  Definition 59
  Fixed 17
  Group-shared 147
  Guaranteed 191
  IP 12
  Mobile 17
  Multimedia 11, 15
  multiplexing 308, 311
  Packaging 4
  Provision 60
  Qualification 41
  quality, See QoS
  Source-specific 147
  Switching Function (SSF) 285
  Value added 285
  Voice 28, 286
Service-Level Agreement (SLA) 154, 180, 310
Services Inter-Networking NNI (SI-NNI), See also NNI 308
Serving GPRS Support Node (SGSN) 253
Session
  Announcement Protocol (SAP) 121
  Control 108
  Description Protocol (SDP) 101
  Handling, See Call control
  Initiation Protocol (SIP) 101, 241
  management 115
  Media 298
  Progress 298
Session-Based Local Policy (SBLP) 293
Set Top Box (STB) 63
SGSN 253, 261–262
Shaping 102, 167
SHDSL 235
Short Message Service (SMS) 247
SID 213
Signal
  Analog 31, 33
  Coding 33, 44, 255
  Compression 31, 38
  Decoding 40
  Digital 31, 33
  Encoding 40, 42
  Encryption 31
  Modulation 255
  Quantizing 35
  Reference 42, 44–15
  RGB 68
  Sampling 33
    Rate 33
    Voice 33

to Noise Ratio (SNR) 37
Transcoding 34, 40
Distortion 34
Voice 33, 35, 44
YUB 68
Analog 69
Signaling 101
Connection Control Part (SCCP) 251
Protocols for Advanced Networks (SPAN) 300
System 7 (SS7) 134, 251
Signal-to-Noise Ratio (SNR) 211
SIMPLE 117
Simple
Mail Transfer Protocol (SMTP) 141
Traversal of UDP through NAT (STUN) 132
Single
Carrier FDMA (SC-FDMA) 263
Copy Broadcast (SCB) 233
Sign on 285
Single-rate Three-Color Marker (srTCM) 169
SIP 23, 46–7, 101, 115, 118, 241, 262, 280–282, 285, 288, 294–295
ACK 298
Architectural Entities 117
Authentication 124
Call
Hold 127
Transfer 127
Extensions 125
IMP 126
INVITE 118, 298
MESSAGE 126
MGC 137
NOTIFY 127
PRACK 298
Procedure 295
Proxy 46, 282
REFER 127
REGISTER 125, 295, 297
Registrar 118, 282
Security 123, 125
Server
Proxy 117
Redirect 117
Standardization 117
SUBSCRIBE 127
Supplementary Services 127
UPDATE 298
User Agent 117
Skew 157
Skype 12
SLA 154, 167, 180, 310
Slack term, See also RS VP 193
SLD 232
Small Office/Home Office (SOHO) 316
Smart card 287–288
SMPTE 71, 88
VC-1 71
SMS 247, 272
SN 288
SNDCP 254, 270
SNR 37, 44, 211
Society of Motion Picture and Television Engineers (SMPTE) 71
S-OFDMA 274–275
Soft QoS, See QoS
SOHO 316
SONET, See SDH
Source Count (SC) 113
Space Time Code (STC) 263
SPAN 300
Spanning Tree Protocol (STP) 311, 322
Spatial multiplexing 263
Spectral compatibility 209–210, 216, 221, 236
Spectrum 57, 273–274
Efficiency 249, 261, 274
Licensed 266
Spread 256, 266
Direct Sequence, See DSSS
Frequency Hopping, See FHSS
Speech 34, 43–44
Speech-layer model 43–14
Split-horizon rule 341
Splitter 221
See also Optical splitter
Spreading factor 258
SR 109, 111–112
SS7 134, 251, 285–286
SSRC 103–104, 106, 109
Statistical Bit Rate (SBR) 179
STB 19, 63, 79, 87–88, 95, 98
STC 87, 263
Stereophony 33
STM 210
Storage area networks (SAN) 359
STP 322

Stream
  Control Transmission Protocol (SCTP) 290
  Elementary
    Packetized, See PES
  Program, See PS
  Transport
    Description Table, See TSDT
    Packet, See TSP
  Transport, See TS
STUN 47
Sub-carrier 267, 274–275
Sub-channel 274
Subnetwork Dependent Convergence Protocol (SNDCP) 254
Subscriber Profile 284, 287
Subscription Locator Function (SLF) 284
Sustainable Cell Rate (SCR) 178
Synchronization 173
  Source, See SSRC
Synchronous
  Digital Hierarchy (SDH) 254
  Transfer Mode (STM) 210

T1 12
Table
  Conditional Access, See CAT
  Network Information, See NIT
  Program Association, See PAT
  Program Map, See PMT
TACS 244
Tagged mode 339
Targeted advertising 54
TCH 249
TCP 166, 290
TCP/IP, See IP
TD-CDMA 246
TDD, See TD-CDMA
TDM 6, 31, 33, 35, 134, 247, 258, 276, 309–310
  Deterministic 31, 33
  Statistical 31
TDMA 244, 255–256, 274–276
TD-SCDMA 246, 256–257
Telecommunications and Internet Protocol Harmonization Over Networks (TIPHON) 300
  Industry Association (TIA) 43, 245
  Technology Association (TTA) 256
  Technology Committee (TTC) 256
Telecoms & Internet converged Services & Protocols for Advanced Network (TISPAN) 286
Telephone number 287
Telephony 18, 31, 33–34, 38, 40, 60
  Cellular 2
  Converged 67
  Digital 33
  Fixed 1–2, 12, 32
  IP, See VoIP
  Mobile 32, 53
  Personal 12
Television
  Digital 71
  IP(IPTV) 1
  Service 6
Television, See TV
Temporary Mobile Subscriber Identity (TMSI) 288
TIA 43, 245
Tight link 161
Tiling 89
Time 355
  Arrival 111
  Division
    CDMA (TD-CDMA) 246
    Multiple Access (TDMA) 244
    Multiplexing (TDM) 31, 247, 310
    Synchronous CDMA (TD-SCDMA) 246
  Division Duplexing (TDD) 257
  Host 157
  source, See Clock
  stamp 87, 104–105, 109, 111
    NTP 111
    RTP 111
  To Live (TTL) 123, 326
  Wire 157
Timestamping, See Packet
TIPHON 300
TISPAN 286, 300, 302
TLS 123, 292
TMSI 288
Token bucket 167, 184, 188
  defined 168
  depth 189
  EF traffic 200
  maximum packet size 189
  minimum policed unit 189
  peak rate 189
  rate 189, 192

ToS 326
Total Access Communication System (TACS) 244
Traffic
Best-effort 61
Broadcast 87
Bursty 275
class 46, 162
controlled load 185
differentiation 152
filtering
marking 167
policing 167
shaping 167
flow 162
Green 169
marking 161
Multicast 61–62, 87
non-conforming 167
out-of-profile 179
Pattern 62
policing 167, 183, 185
srTCM 169
trTCM 170
priorization 61, 90
profile 28
Real-time 103
re-classifying 185
Red 169
re-shaping 185
See also Packet
shaping 167, 183, 185, 188, 193
Unicast 61–62, 87
Yellow 169
Transcoding 285
Transfer Control Protocol (TCP) 161
Transmission Protocol Specific / Transmission Convergence (TPS-TC) 209
Transparent GFP (GFP-T) 316, 356, 359
Transport Layer Security (TLS) 123, 292
Triple Play 1, 3–5, 8, 11–12, 14–17, 22–23, 27–28, 56, 59, 66, 68, 305–306, 316
Challenges 101
market 26
TS 80–81, 85, 88, 92, 96, 98
Sync loss 96
TSDT 86
TSP 80, 85, 93
TTA 256
TTC 256
TTL 123, 326
Tunnel 253
IPsec 270
label 333
TURN 47
TV 6, 12, 17–18, 21, 26, 28, 51, 62, 72, 89, 95, 103
Analog 68
Bidirectional 7, 65
Broadcast 11, 27, 53, 57
Broadcaster 7
Cable 11, 27, 56
Color 68
Digital 11, 27, 52, 69, 96
Terrestrial 14, 73
High Definition, See HDTV
Interactive 53
Last week 67
Model
Bidirectional 52
Unidirectional 52
Multicamera program 80
Operator 51
Pay-to-view 14
Private 67
Receiver 8
Satellite 56
Service
Linear 57
Non-linear 58
Standard Definition 8
Terrestrial 27, 56
Time shift 67
Unidirectional 7
Two-rate Three-Color Marker (trTCM) 170
Two-Way Active Measurement Protocol (TWAMP) 155
Type of Service (ToS) 196, 326

UAA 296–297
UAR 295, 297
UBR 179–180
UDP 80–81, 83, 102, 106, 109, 270, 290
UGS 277
UMA, See GAN
UMTS 246–247, 256–257, 261, 270, 285–287, 295
Terrestrial Radio Access Network (UTRAN) 261

UNI 181, 307–309
  EVC 308
  type 1 307
  type 2 307
  type 3 307
Uniform Resource Identifier (URI) 115
U-NII 267, 273
Universal
  Mobile Terrestrial System (UMTS) 246
  Plug and Play (UPnP) 132
  Time Coordinated (UTC) 111
Unlicensed Mobile Access (UMA) 269
Unshielded Twisted Pair (UTP) 204
Unsolicited
  Grant 277
    Service (UGS) 277
Unspecified Bit Rate (UBR) 179
UPC 181
URI 115
  SIP 289
  SIPS 116
  tel 116, 289
URL 161
Usage Parameter Control (UPC) 181
User
  agent 282
  Authorization Answer (UAA) 296
  Authorization Request (UAR) 295
  Credential 290
  Datagram Protocol (UDP) 102
  Equipment 289
  Identity
    Private 288–289, 295
    Public 288–289, 295
  Subscription 290
User-to-Network Interface (UNI) 181, 307
UTC 111
UTP 204
UTRAN 256, 261–262, 272

VAD 37
Variable Packet Size (VPS) 161
VBR 78
VC 170, 308
  label 334
VC-1 65, 71, 88
  Level 88
  Profile 88
    Advanced 88
    Main 88
    Simple 88
VCAT 316, 366
VCC 256, 269–270, 278, 286
VCG 361, 368
  Member 361
VCI 312, 327
VCR 65–67
VDSL 22, 204, 207, 214, 235
  band plan 216
VDSL2 14, 19–20, 63, 215, 218
  band plan 216
  Ethernet 216, 219
  FDD 216
  low frequency band 218
  PON 219
  PTM-TC 216
  remote DSLAM 220
Vector
  Motion 76
Very-high-bit-rate DSL (VDSL) 204, 207
VHS 71
Video 3, 11–12, 27
  Blurring 89
  Broadcast 6
  Coding 6, 93
    Interpolative 93
    Predictive 93
    Spatial 93
    Statistical 93
  Compression 73, 93
    Lossy 93
  Conference 15, 18, 67
  Digital 7
  Interlaced 88
  Multicast 6, 8
  non-real-time 199
  on Demand (VoD) 1, 8, 61
  Pay-per-view 7
  Personal Recording (PVR) 66
    Client (CPVR) 67
    Networked (NPVR) 67
  Progressive 88
  Resolution 6
  Services
    Personal, See PVS
  Streaming 268
  Synchronization 94
  Timing 97
Videotelephony 32

Virtual
Circuit (VC) 308
Identifier (VCI) 312, 327
Concatenation (VCAT) 353, 361
Container Group (VCG) 361, 366
LAN (VLAN) 309
Path Identifier (VPI) 312
Private LAN Service (VPLS) 340
Private Network (VPN) 46, 305, 340, 373
Private Wire Services (VPWS) 336
Visiting Location Register (VLR) 250
VLAN 309, 322
ID 309, 343
Q-in-Q 322
tagging 315, 322, 324, 343
VLC 75, 77
VLR 250, 270
Vocoder, See Codec Source
VoD 8, 11, 14, 17, 21, 63, 65–67, 87, 92
Downloading 8, 66
Streaming 8, 66
Voice
Activity Detection (VAD) 37
Business 1
Call Continuity (VCC) 256
Compression 249
Distortion 40, 42
over IP (VoIP) 31
Quality Rating 41
Voice, video and data, See Triple Play
VoIP 9–10, 12, 15, 17, 34, 37–38, 40, 43, 45–17, 103, 141, 162, 268, 270, 300
Enterprise 46
Gateway 45
Phone 281
policing 167
Session 118
VPI 312, 327
VPLS 15, 18, 23, 62, 340
Hierarchical, See HVPLS
Pseudowire topologies 341
VPN 10, 46, 305, 340, 351
VPWS 336

WAG 280
WAN 24, 46, 306, 312
Interface Subsystem (WIS) 315
WAP 51
Waveform 34
Wavelength Division Multiplexing (WDM) 222, 316
WCDMA 246, 256, 258, 260–263, 265, 270, 281
WDM 18, 222, 316
Web 270
Weight function 147
Weighted
Fair Queuing (WFQ) 165
Random Early Discard (WRED) 199
RED (WRED) 175
Round Robin (WRR) 164, 199
WFQ 184, 199–200
WiBro 273, 275
Wideband CDMA (WCDMA) 246
Wi-Fi 16–18, 264–268, 270, 272–273, 278, 281–282
WiMAX 18, 63, 246, 263–264, 270, 272–275, 277–281
Forum 272, 275, 278–279
MAC layer 275
Mobile 272, 275
Windows Media 9, See WM-9
Wireless 16, 264
Communications Service (WCS) 273
LAN (WLAN) 243
Local Area Networks (WLAN) 265
Metropolitan Area Networks (WMAN) 272
Wireline 264, 270, 282
WIS 315
WLAN 243, 265, 267–268, 270, 278–279
Access Gateway (WAG) 280
Interworking, See I-WLAN
WM-9 61, 71, 81, 85, 96
WM9V, See SMPTE VC-1
WMAN 272, 278
World Radio Conference (WRC) 273
World Wide Web (WWW) 141
WRC 273

Zapping 88–89, 92
Delay 94